Theories of Meaningfulness

SCIENTIFIC PSYCHOLOGY SERIES

Stephen W.Link and James T.Townsend, Series Editors

MONOGRAPHS

Louis Narens • *Theories of Meaningfulness*

R.Duncan Luce • *Utility of Gains and Losses: Measurement-Theoretical and Experimental Approaches*

William R.Uttal • *The War Between Mentalism and Behaviorism: On the Accessibility of Mental Processes*

William R.Uttal • *Toward a New Behaviorism: The Case Against Perceptual Reductionism*

Gordon M.Redding and Benjamin Wallace • *Adaptive Spatial Alignment*

John C.Baird • *Sensation and Judgment: Complementarity Theory of Psychophysics*

John A.Swets • *Signal Detection Theory and ROC Analysis in Psychology and Diagnostics: Collected Papers*

William R.Uttal • *The Swimmer: An Integrated Computational Model of a Perceptual-Motor System*

Stephen W.Link • *The Wave Theory of Difference and Similarity*

EDITED VOLUMES

Michael Wenger and James Townsend • *Computational, Geometric, and Process Perspectives on Facial Cognition: Contests and Challenges*

Jonathan Grainger and Arthur M.Jacobs • *Localist Connectionist Approaches to Human Cognition*

Cornilia E.Dowling, Fred S.Roberts, and Peter Theuns • *Recent Progress in mathematical Psychology*

F.Gregory Ashby • *Multidimensional Models of Perception and Cognition*

Hans-Georg Geissler, Stephen W.Link, and James T.Townsend • *Cognition, Information Processing, and Psychophysics: Basic Issues*

TEXTBOOKS

Norman H.Anderson • *Empirical Direction in Design and Analysis*

Theories of Meaningfulness

Louis Narens

University of California, Irvine

Psychology Press
Taylor & Francis Group
New York London

Camera ready copy for this book was provided by the author.

First published by
Lawrence Erlbaum Associates, Inc., Publishers
10 Industrial Avenue
Mahwah, NJ 07430

This edition published 2012 by Psychology Press

Psychology Press
Taylor & Francis Group
711 Third Avenue
New york, NY 10017

Psychology Press
Taylor & Francis Group
27 Church Road, Hove
East Sussex BN3 2FA, UK

First issued in paperback 2014

Psychology Press is an imprint of the Taylor and Francis Group, an informa business

Cover design by Kathryn Houghtaling Lacey

Library of Congress Cataloging-in-Publication Data
Narens, Louis.
Theories of meaningfulness/Louis Narens.
p. cm.
Includes bibliographical references and index.
ISBN 0-8058-4045-1 (cloth: alk. paper)
1. Mensuration. 2. Philosophy of Science. 3. Psychology—Mathematical models. I. Title.
BF39 .N37 2001
300'.151—dc21 2001033183
CIP

ISBN 13: 978-0-8058-4045-2 (hbk)
ISBN 13: 978-0-415-65456-2 (pbk)

For a Rainbow Chaser and an Echo

Contents

Foreword

by R.Duncan Luce

During the second half of the 20th century, a number of us in the behavioral and social sciences have focussed a good deal of attention on the conditions giving rise to scientific measurement. Among the issues addressed were:

- Constructing measurement structures far more general, and yet sufficiently restrictive to be of scientific interest, than the extensive measurement axiomatizations of Helmholtz in the late 19th century and Hölder at the turn of the century. Concatenation operations with additive representations simply have not been the bread and butter of behavioral measurement as they were for classical physics. (See Chapter 5.)
- Developing an increased understanding not only of the measurement of single attributes, but also how two or more such attributes relate to one another. A good example is the linking of extensive structures, such as underlie the measurement of mass, length, volume based on a binary operation of concatenation, and additive conjoint structures, which describe how attributes are affected by two or more independent variables. One such link is the property of distribution that leads to products of powers and that underlie the classical model used in dimensional analysis (see Section 5.10). A different sort of link has arisen in my work on utility theory. An adequate qualitative link that still eludes us is among length, time, and velocity in the relativistic case.
- Gaining a deeper understanding of the scale-type classification made familiar by the psychologist S.S.Stevens (Sections 2.2–2.4). This classification is less transparent than it first seems. For one thing, Narens reformulated it in terms of properties of rich groups of automorphisms (symmetries). He and T.Alper established that nothing lies between interval and ordinal scales and that a large number of structures—so far unused in measurement—fall between ratio and interval. For another, in linking structures we find, for example, two representations of velocity concatenation in relativity called velocity and rapidity, and at the numerical level both are ratio scales. Yet they are non-linearly related. Much the same thing holds in utility theory. Which representation is relevant in, for example, Stevens' arguments about permissible

statistics? For a third, many important structures have isolated singular points—zeros or infinities—that are fixed points of all au-tomorphisms, but otherwise the structures are homogeneous. Examples are bounded extensive structures and utility structures with an inherent zero (no change from the status quo).

- Attempting to understand better the role of invariance arguments, such as those proposed by Klein in the definition of geometric objects (Erlanger Program) and are the key aspect (dimensionally invariant functions) of dimensional analysis. We came to view these invariant entities as manifestations of a concept of meaningfulness in the sense that the things we talk about should derive in some sense from the primitives of the structure we are using to describe an aspect of the empirical world. This typically was shown to amount to invariance under automorphisms of the underlying structure. (Chapters 2 and 5)
- Applying these ideas to various scientific problems, mostly in the behavioral sciences. Examples of these are magnitude estimation and production, threshold structures and Weber-like laws, and the role of dimensionless quantities. (Chapters 2 and 5)

Louis Narens has, in earlier publications, contributed to all of these areas, as have a number of us. Here he undertakes far, far more—more than any of the rest of us have achieved—in seeking a deeper understanding of these and other difficult intellectual issues at the foundations of science and mathematics. He brings a degree of unity and breadth to the enterprise that few of us anticipated could be done.

His work was initially stimulated, I believe, by the following observation. When the structure in question has a rich group of automorphisms, invariance arguments seem to characterize those entities that can be defined in terms of the primitives of the structure being studied. The power of dimensional analysis attests to that. But, as was first recognized in geometry, there are numerous interesting structures for which the automorphism group is trivial, and so any proposition formulated within the system is invariant under the trivial automorphism, the identity. That helps not at all in answering the real question of how to characterize those statements that can in some sense be defined in terms of the primitives of the structure under consideration. Making sense of the phrase "can in some sense be defined" was the challenge that Narens addressed. The book you hold provides his answer and its remarkable implications.

Narens' approach goes back to debates on the foundations of mathematics that took place in the late 19th and early 20th centuries. He takes as the basic language of science the Zermelo-Fraenkel axiomatization of set theory, but modifies it in two novel ways. First, he partitions sets into two types, intuitively those that are purely mathematical (pure sets) and those that are empirical. This permits an explicit distinction between pure and applied mathematics. Second, he adds as a primitive a partition of entities and propositions into those that are meaningful and those that are not. Meaningfulness and truth are orthogonal concepts. Both meaningful and meaningless statements may be true or false. The basic task is to explore axiomatically various intuitively plausible properties characterizing aspects of meaningfulness. He studies how these conditions relate and, in particular, he establishes which combination of conditions correspond to the usual invariance concept of meaningfulness. Chapters 3 and 4 present that development after first motivating it by an historical chapter on meaningfulness as treated in the published literature. The four remaining chapters are in some sense applications, both to measurement problems but also to deep philosophical ones including shedding light on the important and elusive Axiom of Choice of set theory. (Chapter 8)

In my opinion this is one of the most important volumes on the foundations of science to appear during my career, and I venture to guess that it will stimulate much critical discussion both at philosophical and scientific levels. I am grateful to have worked with Louis, to have been exposed at close range to his subtle and penetrating mind, and to have witnessed for over a decade the creation of this great accomplishment.

Irvine, California
March 8, 2001

Acknowledgments

This volume was improved by the suggestions and corrections of many people, especially Eric Cosyn, Rolf Johnson, Tony Marley, Michael Regenwetter, Daniel Wollschlaeger, and Ryozo Yoshino. I am particularly indebted to my coworkers in the theories of meaningfulness and measurement—William Batchelder, Michael Cohen, Jean-Claude Falmagne, Rainer Mausfeld, and Duncan Luce—who contributed to many of ideas presented here, and to many others whose research influenced developments described in the volume, especially Ted Alper and Reinhard Niederée.

I want to express my special thanks to Duncan Luce, who introduced me to measurement theory and encouraged me to embark on the research in this volume. Duncan, with David Krantz, Patrick Suppes, and Amos Tversky, demonstrated through their monumental three volume treatise, *Foundations of Measurement,* that measurement theory was a coherent, intellectually rich enterprise for the philosophy and foundations of science. Some of Duncan's other work pointed to interconnections between the meaningfulness concept employed by measurement theorists, invariance concepts employed by mathematicians, the form of laws in physics, and relationships between physical stimuli and their subjective intensities. Motivated by these results, I sought to find a formal theory that would encompass the intuitions behind the interconnections and illuminate the role of mathematics in science. Duncan, through many conversations over a number of years, greatly helped this endeavor with incisive insights and suggestions.

Finally, I want to thank the Center for the Advanced Study in the Be-havioral Sciences and the National Science Foundation for supporting the research leading to this volume.

1. Introduction and Historical Background

1.1 Introduction

Since ancient times, the usefulness, power, and certainty of mathematics have aroused wonder. Many of the best minds in western civilization have speculated about the distinctive character of mathematics and its relationship to science. But today many issues remain unresolved. Throughout this book, I explore theories about the relationship of a qualitative situation to mathematical models of the situation, with special emphasis on the qualitative significance of quantitative concepts and statements about the mathematical models. Such theories are called "theories of meaningfulness," and as is shown, related theories under various guises have appeared since the beginnings of mathematics and science.

The psychologist S.S.Stevens in the mid-twentieth century was probably the first person to systematically employ the term "meaningfulness" in the scientific literature. He was concerned with issues involving the matching of statistical procedures used to analyze empirical phenomena with the procedures of measurement used to quantify them. He called statistics that matched appropriately with the underlying scaling processes "meaningful" and those that did not "meaningless." His concept was later generalized in several directions by other measurement theorists. These generalizations are discussed in detail in Chapters 2 and 5.

The meaningfulness theories developed in this book are related to the ones advanced by the measurement theorists. They are, however, motivated by a much broader set of concerns, and correspondingly are applicable to much wider sets of scientific and philosophical issues. They focus on a fundamental problem that is pervasive throughout mathematics and science: *If certain concepts are "meaningful" (i.e., are "real," "intrinsic," "qualitative," or "empirical," etc.), then what other concepts are "meaningful" ("real," etc.)?* Instances of this general schema of problems have appeared repeatedly throughout the development of mathematics and science, and the resolutions of some important ones have been the impetus for major intellectual revolutions. The following are examples of issues that gave rise to meaningfulness problems that can be cast in terms of the general schema: Pythagorism, a philosophical tradition with ancient roots that maintains that reality is reducible to number; the longstanding problem of deciding what curves are geometrical; the use and effectiveness of allegedly "fictitious" mathematical quantities such as -1, $\sqrt{-1}$, and infinitesimals, and debates about their nature and relationship to (mathematical) reality; the role of invariance in science and the nature of scientific laws; the use of absolute infinity and infinitistic methods in mathematics and science; controversies surrounding the existence of sets that are not Lebesgue measurable and other counter intuitive consequences of the axiom of Choice; and methods for the proper scientific assignment of numbers and other mathematical entities to empirical and qualitative objects and concepts.

The aforementioned examples are discussed in the book, some to a much greater extent than others. One of the central themes of the book is the development of general, formal theories of meaningfulness that can be productively applied to examples like those previously mentioned and other issues in foundations of science.

To formalize meaningfulness concepts adequately, a position has to first be taken on the nature of mathematical entities and the permissible methods for constructing mathematical models of qualitative phenomena. The one adopted in later chapters of this book is common and widely held: axiomatic set theory is sufficiently powerful to encompass all of ordinary mathematics and mathematical science. For technical reasons, the particular version of set theory chosen is slightly different from those generally encountered in philosophy and mathematics—one that has in addition to the usual sets a certain set of nonsets called the "set of atoms." The reason for this choice of set theory is that it helps to provide an easy means for distinguishing qualitative from (purely) mathematical concepts by allowing the qualitative ones to be linked directly to the set of atoms.

The next step in the formalization process is to axiomatize the concept "x is meaningful." Several axiomatizations of this are given throughout the book. They all revolve around two basic ideas. The first is that meaningfulness can be captured by a definability concept. According to this view, certain qualitative concepts are initially given as meaningful. They are called the "primitives," and presumably are sufficient in number and richness to capture the qualitative situation under investigation. Meaningfulness is then identified with definability from the primitives through specific kinds of formulae. The second idea is that meaningfulness is an invariance concept. According to this approach, certain transformations that leave the primitives invariant are selected, and meaningfulness is defined to be invariance under these transformations. Both kinds of meaningfulness have been widely used in mathematics and science, and, as is shown, they are tightly related but are often not identical.

The final step is to show that axiomatizations of meaningfulness are useful for the analysis of meaningfulness issues that arise in the philosophies of mathematics and science. Examples of such applications are given throughout, and various formal theories of meaningfulness are applied to such seemingly diverse topics as the axiom of Choice, the nature of classical physical laws and dimensional analysis, and the foundations of measurement.

The book is essentially self-contained, though it requires some mathematical sophistication and a familiarity with elementary set theory.

Sections 1.2 to 1.8 of this chapter are concerned with presenting historical background material relating to meaningfulness. The presentation is not intended to be either complete or systematic. Rather it is designed to illustrate ideas that are useful for understanding various meaningfulness issues and the types of philosophical positions they generate, and to demonstrate that the topic of the book is about a very ancient and pervasive problem in the foundations of mathematics and science.

1.1.1 Plan and Objectives of This Book

The discussions in Sections 1.2 to 1.8 show that the concept of meaningfulness is multifarious and profound. This conclusion is confirmed even more strongly by the many mathematical applications of the meaningfulness concept given in Chapter 2. However, a *theory* of meaningfulness must consist of much more than a listing of examples, no matter how important and interesting they are individually; for to have a theory, one must have reasonably interrelated phenomena to theorize about. It is not immediately obvious that the diverse subject matters of this and the next chapter are so interrelated. It will be shown that they are.

The central idea that produces the common linking thread is that a framework can be given in which mathematics, qualitative matters, and relationships between them are treated simultaneously. In such a setting, meaningfulness can be introduced as an undefined term, and some of its basic properties axiomatized. Different uses of meaningfulness will have different (but interrelated) axiomatizations.

In order to carry out such a program, a powerful framework is needed, for it must be able to *(i)* describe all of mathematics, *(ii)* provide for qualitative objects and relations, *(iii)* be able to talk about relationships between qualitative objects and relations and pure mathematical ones, and *(iv)* be adequate for formulating meaningfulness concepts.

The groundwork for such frameworks is provided in Chapter 3, and an example of a powerful framework describing the famous Erlanger Program for geometry and some of its natural generalizations is presented in Chapter 4. Roughly, the setting up of these frameworks goes as follows: A set A is specified at the beginning. The elements of A are the qualitative objects to be investigated. Then a structure is formed by creating a transfinite sequence of sets, $A_0, A_1, \ldots, A_\alpha, \ldots$, where $A_0=A$ and $A_{\alpha+1}=A_\alpha \cup B_\alpha$ where B_α is the power-set of A_α. The framework structure is then the union of this sequence of sets. It clearly contains the qualitative objects since the set A belongs to it. By using the standard methods of constructing relations in set theory (which are described in Chapter 3), it easily follows that it contains all qualitative relations between elements of A, all relations of such relations, etc. Thus *(ii)* is satisfied. *(i)* follows for a somewhat analogous reason: A_1 contains the empty set Ø. Let C_0, $C_1,\ldots$, $C_\alpha,\ldots$ be the transfinite sequence of sets where $C_0=\emptyset$ and $C_{\alpha+1}=C_\alpha \cup D_\alpha$, where D_α is the power-set of C_α. The union of this latter sequence of sets is denoted by P. Elements of P are called "pure sets". They are sets that are constructible out of the empty set, and it is easy to show that they belong to the framework structure. They really do not depend, however, on the set of qualitative objects, A. For if a different set of qualitative objects were selected initially, and the above procedures were repeated, then the exact same collection of pure sets P would result. The pure sets can be identified with (pure) mathematics through well-known constructions of set theory (which are outlined in Subsection 4.1.3). Thus *(i)* is satisfied. Important relationships between qualitative objects and relations and pure mathematical ones can easily be described in this framework. For example, by *(i)* the individual real numbers can be identified with particular pure sets. Measurements of elements of A are usually thought of as functions from A into the real numbers, and such functions abundantly exist in this framework. Thus *(iii)* also satisfied. *(iv)* is accomplished by introducing a new undefined concept, M (where "$M(x)$" is read "x is meaningful"), and giving relevant axioms about it; the details of this are discussed in Chapter 4.

This Chapter and Chapter 2 consider a wide range of meaningfulness-like issues, and to deal with these appropriately, the meaningfulness concept often needs to be further specialized. This is done formally in Chapters 6 and 7, where two additional formal concepts are introduced, *intrinsicness* and *qualitativeness*. Intrinsicness deals with meaningful concepts and relations that are lawful, and qualitativeness is concerned with meaningful concepts and relations that are qualitative. In Chapter 7, the additional concept of *empiricalness* is introduced, primarily to distinguish it from meaningfulness and qualitativeness. The failure to distinguish empiricalness from meaningfulness and qualitativeness has produced much confusion in the foundations of science literature and has generated many pseudo controversies. Empiricalness can be intersected with the other concepts to produce "meaningful and empirical relations," "empirical laws," and "qualitative and empirical concepts."

The formal theories of meaningfulness and intrinsicness developed in this book are based on rich frameworks and have inherent in them a logicist perspective regarding mathematics and a platonic perspective regarding the role of mathematics in science. In the theory of qualitativeness of Chapter 7, "numbers" are much less platonic and arise out of qualitative considerations.

There are many traditional tests of the goodness of theories. I believe that the formal theories of meaningfulness, intrinsicness, and qualitativeness presented in this book satisfy many of them. In particular, I will show that these theories are not only descriptive but also are potent. By this is I mean that they do more than codify already existing concepts; in particular, (1) they work out logical relationships between meaningfulness concepts that were previously unrecognized; (2) they clarify certain well-known and important debates by providing rich languages with new concepts and technical results (theorems) that yield insights into the debated issues and positions taken on them; and (3) they provide new techniques and results in substantive scientific areas of inquiry.

There are three areas in which meaningfulness concepts have been extensively applied and theorized about—geometry, physics, and the behavioral sciences. Primarily for the purposes of exposition, this book concentrates on examples and theorizing from the behavioral sciences. This is because the relevant behavioral situations are often mathematically simpler than those in geometry and physics, and because most of the foundational research in the theory of measurement, which provides the critical theoretical link between qualitative and quantitative concepts, has been done by behavioral scientists.

1.2 Philosophies of Mathematics

The resolutions of meaningfulness issues are necessarily linked to metaphysical views about the nature of the entities involved. Because most situations in science involve some mathematics, such resolutions, if they are well thoughtout, should also take into account the nature of the underlying mathematical entities. There are various current views about the nature of mathematics. Four are particularly relevant for the issues covered in the book: *intuitionism, formalism, logicism,* and the *"received view."* The first three are full-fledged philosophies of mathematics with precisely articulated positions. The "received view" is a term I use to describe mathematics as conceived by most good, reflective mathematicians today. Because such a view is not written down anywhere, the reader who objects to its accuracy may consider it as just my view of the beliefs of most modern mathematicians.

The basic tenet of *intuitionism* is that mathematical objects are constructions of the human mind and have no existence outside of human mental experience. Its principal proponent in the twentieth century was the Dutch mathematician L.J.Brouwer (1882–1966). Its intellectual orientation is reminiscent of the conceptualist view of general terms of medieval scholasticism, as well as the philosopher Kant's approach to geometry. The serious adoption of its basic tenet leads to forms of mathematics and logic that are very alien to the ones used by conventional mathematicians and scientists. Thus although exciting philosophically, intuitionism has had almost no impact on mathematical science.

Formalism, broadly stated, holds that mathematics can be conceived of as a type of game, the object of which is to produce according to specific rules finite sequences of strings of predetermined symbols. In ordinary mathematics, the symbol strings would correspond to propositions, the rules of the game to logic, and the finite sequence of strings to proofs. The last string of such a sequence corresponds to a theorem, the sequence being its proof. Because mathematical logic too can be set up in this way, there is actually no difference in the totality of theorems of formalist mathematics (using the "rules" of mathematical logic) and ordinary mathematics. However, there may be a vast difference in how the theorems are to be interpreted. For example, ordinary mathematicians generally assume that the theorems state propositions about mathematical entities that have a timeless existence outside of human experience; the formalists do not necessarily assume this. A branch of formalism called *nominalism* holds in its extreme form that a theorem of mathematics is nothing more than a string of symbols, and has no proper mathematical significance beyond this.

A more liberal kind of formalism—and one that is particularly attractive for the kind of issues considered in this book—considers the strings of symbols to be *initially* uninterpreted, so that the "correctness" of a mathematical theorem depends not on what interpretation is given to the symbols, but on its method of derivation. Interpretations—if they are needed or desired—can be given at later stages, and these may vary with intended applications, from cases of no interpretation *(extreme nominalism)* to cases that interpret the symbols in the same manner as ordinary mathematics. As an example of a case between these extremes, consider a formal version of axiomatic set theory. Suppose that there is reason for considering the positive integers and certain of their subsets as "real," in the sense that they exist independently of human experience, and infinite sets of infinite sets of positive integers as "fictitious," in the sense that they have no interpretation in reality. The fictions, although not real, may nevertheless be useful—and even indispensable—for the derivation of certain statements about real entities. Historically, various mathematical concepts were handled this way, the imaginary numbers being a particularly well-known instance. This kind of approach probably has even more bite in certain scientific contexts where empirical reality can be assigned to some of the mathematical concepts and no (empirical) reality to others.

Formalism has also given rise to some interesting mathematics in its own right. Because various parts of mathematics (and for that matter mathematical science) can be looked at as formal manipulations of strings of symbols by highly specified rules, such systems of symbol manipulations themselves can be considered as mathematical structures, and thus have their mathematical properties revealed by ordinary mathematical means. This sort of mathematics of formal mathematical systems, called *metamathematics* by the mathematician D.Hilbert (1862–1943), has proven in recent times to be a fruitful source of ideas for the development of new techniques for proving propositions of ordinary mathematics.

Logicism takes a rather different position. Its basic tenet is that mathematical concepts are reducible to logical ones. The first person who developed a program for such a reduction was the logician G.Frege (1848–1925). Drawing on Cantor's set theory and Dedekind's and Peano's elegant con- structions of mathematical entities from very elementary arithmetic considerations, Frege was able to show, in the system of logic that he developed, the existence of a representation that was isomorphic to ordinary mathematics. He further argued that the "natural numbers" of his system were true ones, and thus that the mathematics based upon them (which by Peano's program was all of ordinary mathematics) was part of his logic. The same claim was made more forcefully in Whitehead and Russell's highly influential *Principia Mathematica* (1910–1913). A spin-off of this program is a rationale for why the truth of mathematical propositions are so convincing psychologically: because logical truths are necessary and mathematics is a part of logic, it follows that mathematical truths are necessary, and thus when properly understood by a rational being, they must be perceived as true.

Although logicism has been the impetus for much modern philosophical thought, it too has had very little impact on mathematics. The reasons for this undoubtedly have to do with claims that its definitions of mathematical entities correspond to the real nature of such entities—claims that run strongly counter to the intuitions and practices of mathematicians. Also, one of its principal contributions, providing a universal and rigorous foundation for mathematics, can just as easily be achieved through axiomatic set theory, and most mathematicians find the latter approach preferable.

The set-theoretical framework for the formulation of mathematics was a natural outgrowth of a nineteenth century movement to "arithmetize analysis." During the eighteenth century, real numbers were identified with geometrical magnitudes, and functions of real numbers were analyzed in terms of their graphs. By modern standards there was nothing approaching an acceptable definition of "function." This lack became very apparent during the early part of the nineteenth century when the important work of J.B.Fourier (1768–1835), which specified functions by infinite trigonometric series, started to have a strong impact on mathematics and science. This new and powerful method of definition immediately led to conceptual problems concerning the nature of functions. To resolve these and put Fourier methods on a rigorous basis, the mathematician J.Dirichlet (1805–1859) in 1837 formulated what has become the familiar and standard modern definition of a function as a rule that assigns to each object x in a specified domain, a unique object y. One consequence of this definition—as Dirichlet clearly noted—was that it allowed for functions of real numbers that had no graph, for example, functions like $f(x)=0$ if x is rational, and $f(x)=1$ if x is irrational. Such functions were considered to have no geometrical significance, and as they became increasingly prevalent in analysis, it was quickly realized that a nongeometric concept of real number was needed if analysis were ever to be put on a rigorous foundation.

Such a new concept was provided by R.Dedekind (1831–1916) in 1888. Dedekind started with the natural numbers as given. From these he constructed integers as sets of ordered pairs of natural numbers. He then con- structed the rational numbers as certain sets of ordered pairs of integers. Finally, the real numbers were constructed as ordered pairs of certain nonempty sets of rational numbers. Dedekind found natural ways of extending the ordering relation and the operations of addition and multiplication of the natural numbers to the integers, the rationals, and the reals, and showed that the reals endowed with these operations and ordering relation became an ordered field with substructures isomorphic to the number systems of the rationals, integers, and natural numbers. By use of Dedekind's system of real numbers, a systematic and rigorous theory of limits was developed by other mathematicians, and analysis was finally put on a solid foundation. Inevitably, however, there was still the lingering question of whether the real numbers of Dedekind were the "real" real numbers. For many mathematicians this became a problem of philosophy that had little intrinsic mathematical interest or value, because for them the issue had been settled by Dedekind when he showed that his system of real numbers had all the formal mathematical properties that one would want the "real" real numbers to have. This pleasant state of affairs is an instance of the modern concept of *categoricity,* which is discussed next.

A *model* of a set of axioms is any structure in which the axioms are true. A set of axioms is said to be *categorical* if and only if it has a model and every two models of it are isomorphic. Reasonable and intuitive axioms can be given for which there is virtually universal agreement among mathematicians that they are true and characteristic of the "real" real number system, and it can be shown that these axioms are *(i)* also true of Dedekind's construction and *(ii)* categorical. Thus for all *mathematical* purposes, Dedekind's model of the categorical axioms is just as good as any other model of these axioms, including the model of the "real" real number system.

The next steps of reductionism and axiomatization for the real number system are obvious—at least from today's perspective: give a categorical description of the natural numbers, axiomatize the theory of sets, and within this axiomatized theory of sets construct a model that satisfies the categorical axioms for the natural numbers. Such a program provides a framework in which all of mathematics that is based on the natural or real numbers can be developed. It was successfully carried out in the early part of the twentieth century.

It should be noted that in establishing the categoricity of axioms with infinite models, some fragment of set theory or an equivalent is necessarily needed in the axiomatization. For example, in establishing the categoricity of axiom systems for the natural numbers, the following axiom is often used: Every nonempty subset of natural numbers has a least element. This general feature of most axiomatizations—that to achieve categoricity a portion of set theory is required—is a consequence of some rather profound theorems of mathematical logic that are due to the logicians Löwenheim, Skolem, and Tarski. Their theorems also yield the rather startling result that axiom systems for set theory are themselves never categorical. Thus categoricity necessarily becomes a relative concept: once a particular model of set theory $\mathfrak{S}$ is selected, then the categoricity of an axiom system may be established *within* $\mathfrak{S}$ by showing that every two models of the axiom system that exist *within* $\mathfrak{S}$ are isomorphic. Categoricity does not—and cannot—demand that models of axioms constructed *in different models of set theory* must be isomorphic.

To summarize, it is believed by most mathematicians today that the axiomatic version of set theory described during the first part of this century by E.Zermelo (1871–1956) and A.Fraenkel provides a rigorous framework in which isomorphic copies of all the objects and concepts of ordinary mathematics can be constructed. This view of mathematics hereafter is called the *received view*. It is a weaker position than logicism, since it neither identifies the theorems of set theory as being logically necessary nor claims that mathematical entities such as the natural numbers are exactly the same as their set-theoretic counterparts—even though they and their counterparts share mathematically the same f ormal characteristics.

1.3 Pythagorism

Of considerable importance to the developments of mathematics and science is an ancient form of mystical thought called *Pythagorism*. It has had a tremendous impact on our thinking about mathematics and nature, and generated some highly controversial and long lasting meaningfulness issues.

It began in the sixth century B.C. with a movement started by the philosopher Pythagoras. He was an Ionian who traveled and studied widely, finally settling in what is now called Sicily, where he established a religious-political movement that became known as the Pythagorean Brotherhood. Today the Brotherhood would be looked upon as a cult. For a time it controlled a considerable portion of southern Italy. It was autocratically ruled by Pythagoras, who was believed by his followers to be semidivine. This gave rise to the well-known saying, "Among rational creatures there are gods and men and beings like Pythagoras." Pythagoras demanded strict codes of conduct among his followers, which among other things included vegetarianism. He also required them to attend lectures on "philosophy" and "mathematics"— terms he coined to describe the type of intellectual activities he engaged in to achieve knowledge—so that they too could obtain the correct insights and understandings that lead to a moral life.

Pythagoras was a visionary par excellence, and at the core of his philosophy were two mystical ideas that were to have a profound impact on the development of mathematics and science. First, that all domains of knowledge, be they religious, medical, scientific, musical, cosmological, or so forth, could be understood through mathematics; and second, when processes functioned properly, they exhibited a form of "harmony." The first idea still pervades much of science and continues to be advanced in various ways in scholarly publications. The second one, although not usually directly invoked in scientific publications anymore, is still widely used by scientists as a heuristic for discovering new phenomena and laws, and appears frequently in scientists' popular accounts of science.

Today, when some mathematically based description of a mental or worldly phenomenon is given, it is considered a "degrading" (or "reduction") of that phenomenon. For example, the mathematical description specifying which sequences of combinations of pure tones are heard as concordant and which as discordant would today be considered as a downgrading or reduction of the mental phenomenon of hearing "musical harmony." The Pythagoreans believed the exact opposite: For them, numbers were the purest of ideas, and thus the understanding of the arithmetical relations corresponding to the perceived musical sounds provided the mind with much greater insight into the inherent harmony/disharmony exhibited by concordant/discordant musical sounds than the sensory driven methods of hearing. Thus for the Pythagoreans, to an "attuned" mind the usual way of listening was a downgrading of a potentially far richer mathematical experience.

The Pythagoreans discovered that the pitch of a note depended on the length of the string that produced it, and that the concordant intervals of the scale were produced by simple numerical ratios, for example, the ratio of 2 to 1 for the octave, 3 to 2 for the fifth, and so forth. In their mysticism, they extended this empirically validated idea to astronomy: They believed that the sun, moon, and the 11 planets, including the earth, move in circular orbits about a "central fire," and that outside of all of this is the fixed sphere of stars, which also had the central fire as its center. Because the heavenly bodies should be—or perhaps had to be—in "harmony," the ratios of their orbit lengths should be in musical harmony: if a musical instrument were made with strings representing the orbits of the astronomical bodies (that is, with the strings in the same order as their assigned body's distance from the central fire, and their ratios of lengths being the same as the ratios of their assigned bodies' orbits), then this instrument when played would sound with the "Harmony of the Spheres." According to tradition, Pythagoras was able to hear this "Harmony" directly from the heavenly motions.

Individual whole numbers were very important to the Pythagoreans. Each had its peculiar attributes. The most holy number was ten, which to the Pythagoreans represented the number of the universe. It was the sum of the number of points needed to generate all possible geometrical dimensions, that is, of a single point, which is the generator of dimensions; of two points, which determines a line of dimension one; of three noncollinear points, which determine a triangle with area of dimension two; and of four nonplanar points, which determine a tetrahedron with volume of dimension three. Philolaus, a later Pythagorean (died ca. 390 B.C.), wrote that the number 10 was "great, all-powerful and all-producing, the beginning and the guide of the divine as of the terrestrial life." Philolaus used the number 10 as the inspiration for his astronomical system, which was the first non-geocentric one in history. In this system, eight bodies moved around the central fire: earth, moon, sun, and the five (visible) planets. This brought the number of heavenly bodies to nine, excluding the fixed sphere of stars. The earth moved around the central fire always with the same (uninhabited) hemisphere facing it. This is why the central fire has not been seen. A tenth planet called "antichthon" was assumed to exist. It revolved around the central fire in the exact same period as the earth (one day), and was always positioned collinear with the earth and the central fire. Thus it too was never seen from earth. But because it was in harmony with the other moving astronomical bodies, much could be deduced about its orbit. This line of reasoning may seem strange to empirically minded scientists, but it should be noted that analogous forms of it are in wide use today, for example, in quantum mechanics.

The entire Pythagorean system of metaphysics was based on whole numbers: In the words of Philolaus. "All things which can be known have number; for it is not possible that without number anything can be either conceived or known." Rational numbers were considered as ratios of whole numbers. Near the end of Pythagoras's life a startling discovery was made by the Pythagoreans: the length of the diagonal of a unit square cannot be expressed as a ratio of whole numbers; that is, $\sqrt{2}$ is irrational. The significance of this discovery to the Pythagoreans was that certain geometrical objects could not be expressed numerically, that is, that some things were not number. They tried to suppress the discovery. According to one story, the disciple who let the secret out, an unfortunate called Hippasos, was put to death for his transgression. The ancient geometer Proclos gives another version: "It is told that those who first brought out the irrationals from concealment into the open perished in shipwreck, to a man. For the unutterable and the formless must needs be concealed. And those who uncovered and touched this image of life were instantly destroyed and shall remain forever exposed to the play of the eternal waves."

The Pythagorean doctrine that "all is number" also came under strenuous attack by the Eleatic school of philosophy. For the Pythagoreans, numbers were discrete entities from which everything was constituted. The Eleatics, whose best known members were Parmenides and Zeno, challenged this by showing the contradictions inherent in such a view. (The famous paradoxes of Zeno were typical of the type of argumentation used by the Eleatics on this point.)

Ancient Greek mathematics found a way out from the challenges of irrational numbers and of the Eleatics, one that was to have a profound impact on subsequent mathematics. There are two ways of dealing with the problem of irrationals: *(i)* allow the existence of lengths to which no number could be assigned, or *(ii)* enlarge the number system to include the irrationals. The ancient Greek mathematicians decided in favor of the second, and the re- sult was Eudoxus' beautiful theory of proportions, which identified numbers with geometric constructs. By the time of Euclid's *Elements,* numbers were so geometrized that integers were represented as line segments. The switch from the discrete integers to the continuous geometric magnitudes as the basis of things mathematical also took much of the steam out of arguments of the Eleatics, because they generally used some discrete concept as a basis for contradictions. The idea of identifying (real) numbers with geometric magnitudes continued well into the nineteenth century.

Pythagorean philosophy had a great influence on the philosopher Plato (427–347 B.C.), who in turn had a strong influence on mathematicians of his era. He continually emphasized that mathematical theorems and the methods by which they were reasoned refer not to visible or worldly figures, but to pure Ideas of which the visible or worldly figures are at best a highly degraded representation. This view about the nature of mathematical entities—which today is called *Platonism*—continues to have sway over some mathematicians. This can be seen for example in the writings of the British mathematician G. H.Hardy, who in 1940 said that "mathematical reality lies outside us, that our function is to discover or *observe* it, and that the theorems which we prove, and which we describe grandiloquently as our 'creations,' are simply the notes of our observations." The logician Kurt Gödel (1906–1978) made in 1963 a similar point about mathematical intuition:

> It should be noted that mathematical intuition need not be conceived of as a faculty giving an *immediate* knowledge of the objects concerned. Rather it seems that, as in the case of physical experience, we *form* our ideas also of those objects on the basis of something else which *is* immediately given. Only this something else here is *not*, or not primarily, the sensations. That something besides the sensations actually is immediately given follows (independently of mathematics) from the fact that even our ideas referring to physical objects contain constituents qualitatively different from sensations or mere combinations of sensations, e.g., the idea of object itself, whereas, on the other hand, by our thinking we cannot create any qualitatively new elements, but only reproduce and combine those that are given. Evidently the "given" underlying mathematics is closely related to the abstract elements contained in our empirical ideas. It by no means follows, however, that the data of this second kind, because they cannot be associated with actions of certain things upon our sense organs, are something purely subjective, as Kant asserted. Rather they, too, may represent an aspect of objective reality, but, as opposed to the sensations, their presence in us may be due to another kind of relationship between ourselves and reality. *(Gödel, 1983, pg. 484)*

Like the Pythagoreans, Plato attached mystical significance to certain mathematical objects. The regular polyhedra were one of his favorites, so much so that throughout a good deal of history they were called *Platonic Solids*. As we will see shortly, his ideas about them had a profound impact on the astronomer Kepler.

In 427 B.C. a plague struck Athens, killing a quarter of its population (including the statesman Pericles). According to tradition, a delegation of Athenians was sent to the oracle of Apollo at Delos to ask how to stop it. The oracle replied that the cubical altar to Apollo must be doubled. The Athenians dutifully doubled the dimensions of the altar, but the plague did not cease. Of course, by one interpretation of the oracular pronouncement, they made an error and increased the *volume* of the altar eightfold rather than the twofold required. The problem of finding a method of exactly doubling the volume of a cube proved to be a very difficult one and became known as the *Delian problem*. A number of solutions for it were suggested. They all used either special instruments (other than compass and straightedge) or special geometric curves—conditions that Plato found unacceptable, because he believed that all geometrical constructions should be restricted to the use of compass and straightedge, because their applications resulted in circles and lines, the purest of ideas. Plato's view about geometrical constructions generally held, although there were some important exceptions. Today we know that the Delian problem cannot be solved by use of only the compass and straightedge.

The Delian problem brings into focus a far deeper problem of geometry: What is a proper geometric construction? Plato clearly stated his views on the subject, which were generally accepted by mathematicians for over two thousand years. This conceptual problem can be looked at as a particular instance of the general meaningfulness problem, which in this case specializes to: Given that certain objects are proper geometric constructions, then what other objects are? If one were to allow in addition to the constructions of lines through a given point and circles of specified radii and centers, the constructions of planes through three given points, spheres of specified radii and centers, and right circular cones through specified circles and with specified vertices, then the Delian problem can be solved. Such a solution was in fact achieved by the mathematician Menaechmus, a teacher of Alexander the Great, who initiated the study of conic sections for this purpose. Because Plato accepted planes, spheres, and cones as proper geometrical objects, his highly restricted view on geometrical constructions seemed inappropriate: there were other, less severe ways to avoid the use of mechanical instruments and special curves in geometry, a practice which—according to Plutarch— Plato found extremely distasteful and considered to be "the mere corruption and annihilation of the one good of geometry, which was thus shamefully turning its back upon the embodied objects of pure intelligence." However, Plato's approach did have the advantages of being clearly articulated and being certain that only geometrical objects would be constructed. This together with the lack of clearly stated alternatives gave his view the upper hand. How- ever, the difficult problem was not in knowing that points, lines, and circles were geometric, but knowing where to stop including types of objects as being geometric. Obviously not every curve was to be considered geometrical. This posed serious problems when new types of curves were introduced into geometric constructions. For example, the mathematician Archimedes (287–212 B.C.) used a spiral that now bears his name, in polar coordinates the graph of the function $\rho=\theta$, to solve the other two famous construction problems of antiquity, the squaring of a circle and the trisection of an angle. Archimedes showed that if he had such a spiral, then with a compass and straightedge he could construct for any circle a square that had the same area as that circle and he could trisect any angle. The problem with this approach is whether or not Archimedes' spiral is to be considered as a geometrical object on a par with the circle and line. It can be shown that it cannot be constructed with circles, lines, planes, spheres, and cones. In many ways it looks very different from the sort of object usually encountered in geometry. However, it is not an easy matter to describe where this difference might reside. Descartes thought he had a way to do it. He writes in his 1637 *La Géométrie:*

> It is true that the conic sections were never freely received into ancient geometry, and I do not care to undertake to change names confirmed by usage; nevertheless, it seems very clear to me that if we make the usual assumption that geometry is precise and exact, while mechanics is not; and if we think of geometry as the science which furnishes a general knowledge of the measurement of all bodies, then we have no more right to exclude the more complex curves than the simpler ones, provided they can be conceived of as described by a continuous motion or by several successive motions, each motion being completely determined by those which precede; for in this way an exact knowledge of the magnitude of each is always obtainable.
>
> Probably the real explanation of the refusal of ancient geometers to accept curves more complex than the conic sections lies in the fact that the first curves to which their attention was attracted happened to be the spiral, the quadratrix, and similar curves, which really do belong only to mechanics, and are not among those curves that I think should be included here, since they must be conceived of as described by two separate movements whose relation does not admit of exact determination. *(Descartes, 1952, pp. 43–44)*

Thus, in modern terms, Descartes in the previous passage is arguing that algebraic curves are geometrical whereas transcendental ones, which he calls "mechanical," are not.

These and other meaningfulness issues in geometry are considered in greater detail in Section 1.4. For now let us look at some of their impact on early astronomical thinking, a subject Plato again had a profound and long lasting influence.

Through purely metaphysical reasoning, Plato concluded that the shape of the universe had to be a perfect sphere and that all heavenly motions had to move in perfect circles at uniform speed. Aristotle arrived at similar conclusions, although by different lines of metaphysical reasoning. He and Plato, however, differed substantially on the arrangement of the universe. For example, Plato thought God (or the Pythagorean central fire) was its center, and Aristotle thought it was the earth, God (or the Prime Mover) being for him the outermost sphere. These two metaphysical positions about what was at the center of the universe greatly shaped subsequent astronomical theory.

In ancient times there were two mathematical models for the observed heavenly motions, one by Eudoxus and one by Ptolemy. Each used uniform circular motion as its fundamental concept. Ptolemy's model ultimately came to dominate astronomical thinking. In it, the earth was the center of the universe and heavenly motions were the result of compoundings of uniform circular motions. These compoundings occurred in the following way: in the vicinity of the earth (but centered outside of it), a circle called the "excentric" moved with uniform motion. Moving around it in uniform motion was the center of a very large circle, called the "deferent." For each heavenly body, there was a series of additional uniformly moving circles, the first being centered on the deferent, the second being centered on the first, and so forth, with the heavenly body's final motion being the compoundings of these circular ones.

The Ptolemaic system remained unchallenged for fourteen centuries, until Copernicus' *On the Revolutions of the Heavenly Orbs.* In Copernicus' system, the sun was fixed, and the earth revolved in a circular orbit about a point near the sun, and the planets moved—as in Ptolemy's system—on circles of circles. The planetary orbits, however, were centered on the center of the earth's orbit. Of course, as in Ptolemy's system, each circle moved with uniform speed. Thus Copernicus also had a rather complicated system. He claimed his was simpler than Ptolemy's: In a preliminary announcement of the *Revolutions,* he said that "altogether, therefore, thirty-four circles suffice to explain the entire structure of the universe and the entire ballet of the planets." However, by time the *Revolutions* reached publication, Copernicus actually used forty-eight circles. This was fewer than the number used by Ptolemy himself, eighty, but more than the thirty-nine used by his Ptolemaic contemporary, Peurbach. Thus contrary to the popular modern view, Copernicus had a complicated system of "circles on circles on circles" to explain the heavenly motions; also in his system the center of the motions was not the sun, but a point near the sun. Furthermore his system, though less complicated than Ptolemy's original one, was not simpler than the best Ptolemaic-like system of his day. Thus the reason for the ascension of the Copernican system over the Ptolemaic ones was not simplicity. It appears to have been due to two factors: *(i)* the perception that the Ptolemaic ones were largely arbitrary—in the words of the Arab philosopher Averroës, "The Ptolemaic astronomy is nothing so far as existence is concerned; but it is convenient for computing the non-existent," and *(ii)* the fact that certain astronomical phenomena, such as the apparent retrograde motions of the planets, were easily explained in the Copernican system, but lacked reasonable explanations in the Ptolemaic systems.

The Ptolemaic and Copernican systems both yielded accurate descriptions of apparent heavenly motions, and both were based on a few simple principles involving uniform circular motion. They differed primarily with respect to the body that was near the center of the motions. Neither of them did (or could) give reasons for the number of planets or the arrangement of their orbits. It was precisely these kinds of Pythagorean concerns that led the astrologer, astronomer, and mathematician J.Kepler (1571–1630) into his investigations that were to become the beginnings of modern science. His first work in astronomy was completely Pythagorean in nature, and is considered somewhat nonsensical today. Written when he was 25, it was called *Mysterium Cosmographicum.* Its goal was to explain why there were exactly six planets in the universe, and why their orbits were positioned in the manner they were. His answer was straightforward and in the classical Pythagorean mode: there were exactly *six* planets because there were exactly *five* Platonic solids.

It was well known in antiquity that there were exactly five regular solids, the tetrahedron, the cube, the octahedron, the dodecahedron, and the icosahedron, and that each such solid could be inscribed in a sphere and circumscribed about a sphere. Kepler conceived of each planet's orbit as being determined by a uniformly moving sphere centered at the sun. This gave him six spheres and five spaces between them. He computed the radii of these spheres from existing astronomical data. He then found a way of placing into each gap between spheres a different Platonic solid that circumscribed its inner sphere and was circumscribed by its outer one. The final result was justified by the following metaphysical argument: Because God could only create a perfect world, and since only five perfect solids existed, they were meant to be placed between the six planetary orbits—which in fact they are, as his system clearly shows.

Kepler held this cosmological belief in the five perfect solids throughout the rest of his life. Arthur Koestler makes the following observation about this:

> For Kepler's misguided belief in the five perfect bodies was not a passing fancy, but remained with him, in a modified version, to the end of his life, showing all the symptoms of a paranoid delusion; and yet it functioned as the *vigor motrix,* the spur of his immortal achievements. He wrote the *Mysterium Cosmographicum* when he was twenty-five, but published a second edition of it a quarter-century later, towards the end, when he had done his life work, discovered his three Laws, destroyed the Ptolemaic

universe, and laid the foundations of modern cosmology. The dedication to this second edition, written at the age of fifty, betrays the persistence of the *idée fixe:*

> Nearly twenty-five years have passed since I published the present little book… Although I was then still quite young and this publication my first work on astronomy, nevertheless its success in the following years proclaims with a loud voice that never before has anybody published a more significant, happier, and in view of its subject, worthier first book. It would be mistaken to regard it as a pure invention of my mind (far be any presumption from my intent, and any exaggerated admiration from the reader's, when we touch the seven-stringed harp of the Creator's wisdom). For as if a heavenly oracle had dictated it to me, the published booklet was in all its parts immediately recognized as excellent and true throughout (as is the rule with obvious acts of God).

(Koestler, 1968, pg. 254)

Kepler was of course interested in the Pythagorean harmony of the spheres, and he devotes a chapter of the *Mysterium* to it. This issue was considered more thoroughly in his later work, *Harmonice Mundi,* which was written well after he achieved the scientific successes for which today he is renowned. In this latter work, Kepler sees certain geometrical proportions as corresponding to the pure harmonies that guided God in the work of Creation. For Kepler the pentagon, when interpreted musically, produced ratios 1/5 and 4/5, which correspond to consonant sounding cords. However, the heptagon, for example, produced ratios of 1/7 and 6/7, which correspond to dissonant chords. Kepler explained this as follows: The pentagon could be constructed by use of compass and straightedge, but the heptagon could not. Because only through geometry could man understand the workings of the divine mind, those figures that could not be constructed with compass and straightedge—such as the heptagon, the 11-gon, 13-gon—were in a sense unspeakable and nonexistent. "Therein lies the reason," Kepler wrote, "why God did not employ the heptagon and the other figures of this species to embellish the world." It was also in this latter book that Kepler formulated what was to become known as his Third Law, a law that was very instrumental in helping Newton discover and formulate the universal law of gravitation. (The Third Law states that the ratios of the squares of the planets' periods of revolution to the cubes of their mean distances to the sun is the same constant number for each planet.)

The astronomer Tycho Brahe (1546–1601) collected much more precise data about the planetary motions than had previously existed. These data produced discrepancies with predictions derived from the accepted cosmological models of the time. Kepler, who had access to Tycho's data, played with various models to account for the discrepancies. One idea he hit upon was putting planets in circular orbits with centers different from the sun but with the sun always inside their orbits. Kepler realized that such a model still could not fit Tycho's data if the planets were to move with uniform speed about their centers. He also computed that in such a model the planets would not be moving with perceived uniform speed if viewed from the perspective of the sun. He finally decided to abandon the requirement of uniform motion, and calculated what would happen in his model if the planets moved with variable motions. To his own satisfaction, he verified that the areas of the sectors (with vertices at the sun) that planets sweep out in an interval of time was a constant number that depended only on the time interval. This became known as his "Second Law." (However it was discovered prior to his "First" and "Third.") Thus by keeping circular orbits but weakening the concept of uniform motion he was able to produce a cosmological model that closely fit the more precise data of Tycho while at the same time being immensely simpler than the ones of Ptolemy and Copernicus, since in his model there were no circles moving on circles. However, there still remained a small discrepancy in Mars' orbit that could not be explained away by observational error. In order to deal with this error, Kepler realized that he would have to abandon the concept of circular orbit. He finally settled on the concept of elliptical orbit, and then everything fell perfectly into place: for each planet, he could put the sun at one of the foci of the planet's elliptical orbit (his "First" Law), and all planets moved with variable motions in such a way that the areas of the sectors (with vertices at the sun) they swept out are the same for same periods of time (his "Second" Law). This model fit Tycho's data perfectly.

Kepler thought a great deal about the nature of his science, and systematically formulated many of the metaphysical principles upon which it was based. One of these was a meaningfulness principle that justified the intro duction of mathematics for the understanding of worldly phenomena. Kepler, like other philosophers before him, made a distinction between primary and secondary qualities. The following quote from E.A.Burtt's *The Metaphysical Foundations of Modern Science* describes the fundamental role that these concepts played in his thinking:

> …Knowledge as it is immediately offered the mind through the senses is obscure, confused, contradictory, and hence untrustworthy; only those features of the world in terms of which we get certain and consistent knowledge open before us what is indubitably and permanently real. Other qualities are not real qualities of things, but only signs of them. For Kepler, of course, the real qualities are those caught up in this mathematical harmony underlying the world of the senses, and which, therefore, have a causal relation to the latter. *The real world is a world*

> *of quantitative characteristics only; its differences are differences of number alone.* In his mathematical remains there is a brief criticism of Aristotle's treatment of the sciences, in which he declares that the fundamental difference between the Greek philosopher and himself was that the former traced things ultimately to qualitative, and hence irreducible distinctions, and was, therefore, led to give mathematics an intermediate place in dignity and reality between sensible things and the supreme theological or metaphysical ideas; whereas he had found means for discovering quantitative proportions between all things, and therefore gave mathematics the pre-eminence. "Wherever there are qualities, there are likewise quantities, but not always *vice versa*"
>
> Again, Kepler's position led to an important doctrine of knowledge. Not only is it true that we can discover mathematical relations in all objects presented to the senses; *all certain knowledge must be knowledge of their quantitative characteristics, perfect knowledge is always mathematical.* "There are, in fact, as I began to say above, not a few principles which are the special property of mathematics, such principles as are discovered by the common light of nature, require no demonstration, and which concern quantities primarily; then they are applied to other things, so far as the latter have something in common with quantities. Now there are more of these principles in mathematics than in the other theoretical sciences because of that very characteristic of human understanding which seems to be such from the law of creation, that nothing can be known completely except quantities or by quantities. And so it happens that the conclusions of mathematics are most certain and indubitable." He notes certain practical illustrations of this fact in optics, music, and mechanics, which of course, best afforded him the confirmation he sought. "Just as the eye was made to see colours, and the ear to hear sounds, so the human mind was made to understand, not whatever you please, but quantity." Therefore, quantity is the fundamental feature of things, the *"primarium accidens substantiae,"* "prior to the other categories." Quantitative features are the sole features of things as far as the world of our knowledge is concerned.
> *(Burtt, 1954, pp. 67–68)*

To summarize, Kepler integrated modern scientific and ancient mystical perspectives by demanding that his mathematical models of worldly phenomena accurately fit empirically obtained data while at the same time exhibiting Pythagorean-like harmonies. He did all of this against a theological backdrop in which he believed God was a geometer who arranged creation according to precise, geometrical, harmonious principles. He clearly recognized the generality and flexibility of the mathematical modeling process, and restricted himself to using only certain mathematical concepts—namely those that exhibited harmony—for describing the true nature of things, that is, for doing science.

Thus for Kepler the mathematical concepts corresponding to the true nature of things correspond to what is here called meaningful concepts. These share a common structural characteristic: they exhibit harmony. Although Kepler used some principles to determine if the mathematical form of a given concept displayed harmony—for example, symmetry, regularity, constructibility by compass and straightedge—he had no precise, general definition of harmony. In particular, there were no principles for generating harmonious concepts from other ones: it is unlikely, for example, that an arbitrary collection of harmonious concepts is itself harmonious (they might be sounding different "tunes") or that an arbitrary object constructed by compass and straightedge is harmonious (it might be irregular, nonsymmetric, and ugly). Also it is not at all obvious that Keplerian harmony is itself "harmonious." Such reflexivity, while perhaps a debatable principle for the general concept of meaningfulness, is certainly a desirable one given Kepler's metaphysical perspective. At the qualitative level, Kepler makes the distinction between primary and secondary properties. For him, the primary ones were at a more fundamental level of reality, and knowledge of this kind of reality was sometimes obtainable by quantitatively representing concepts involving primary properties and noting the (mathematical) harmony within the representations. Thus in Kepler's system the level of reality of a qualitative concept could often be determined by the form of its quantitative representation.

Related views have been put forward by a number of other thinkers. One by an earlier Renaissance philosopher, Nicholas of Cusa (1401–1464), deserves special note. Cusa, like modern day intuitionists, believed that mathematical entities were human mental creations that did not have "another, still more real being over and above the mind." But instead of stopping with such observations, he delved further into the reasons why this should be. Following the biblical lead that man was created in God's image, Cusa reasoned that man's mind was created in the image of the divine mind, and thus certain human mental activities reflected analogous divine ones. Cusa concluded that such human mental activities coincided with mathematical cogitations. And although they were qualitatively different from divine cogitations, they nevertheless could be employed as a very good metaphor for them: The human mind brings forth mathematical entities with complete precision and understands them with perfect clarity; the divine mind does similarly with its cogitations. Of course because in Cusa's view the world was the end result of divine cogitations, it was of little surprise to him that the part of human mental activity which was the closest image of divine cogitation, namely mathematics, should also be the most useful in understanding the results of divine cogitations, that is, that mathematics yielded "an enigmatic image for the hunting out of God's work." A similar line of argument produced for him the reason for the perceived certainty of mathematical propositions.

Although from today's perspective Cusa seems highly mystical and therefore hopelessly out of step with modern science, the kinds of idealist views he espoused have counterparts in modern thought, particularly in speculations about the nature of quantum mechanics. For example, the astronomer Sir James Jeans (1877–1946) writes in his widely read *The Mysterious Universe:*

> To-day there is a wide measure of agreement, which on the physical side of science approaches almost to unanimity, that the stream of knowledge is heading towards a non-mechanical reality; the universe begins to look more like a great thought than like a great machine. Mind no longer appears as an accidental intruder into the realm of matter; we are beginning to suspect that we ought rather to hail it as the creator and governor of the realm of matter—not of course our individual minds, but the mind in which the atoms out of which our individual minds have grown exist as thoughts. *(Jeans, 1948, pg. 137)*

To summarize, the Pythagoreans thought that reality could be understood in terms of whole numbers, and made some progress towards this end. However, this position gave rise to a serious meaningfulness problem: $\sqrt{2}$, unlike individual rational numbers, could not be meaningfully represented in terms of whole numbers. It arose in geometry in a meaningful way as the length of a diagonal of a square that had sides of unit length. Thus if a square with unit sides could be defined purely out of whole numbers, which according to Pythagorean metaphysics it should be, then its diagonals must also be so definable, which Pythagorean mathematics showed it not to be. The conclusion is that the number corresponding to a diagonal of a unit square is not constructible (definable) out of whole numbers, that is, geometry, and therefore physical space is not part of arithmetic. To deal with this state of affairs, ancient mathematicians made geometry primary, by identifying numbers with lengths of line segments; and some considered arithmetic to be a part of geometry.

Other meaningfulness issues involving the nature of geometrical entities also arose in ancient mathematics. These were basically handled as definability problems, with various views as what the definability conditions should be. Plato's view of constructibility by compass and straightedge ultimately became the dominant one, and it lasted until at least the seventeenth century.

Astronomy provided early science with a different set of meaningfulness concerns. The Ptolemaic and Copernican models of the heavenly motions were founded on the Platonic principle that all such motions were uniform and circular. This necessarily made cumbersome any model that could accurately fit the apparent data. These types of models also generated meaningfulness/meaninglessness questions for which there were no satisfactory answers, for example, why the center of the entire class of motions did not appear as an astronomically significant point. Many such problems were resolved by Keplerian system, which discarded entirely the Platonic principle of circular motion.

Kepler wanted to produce a Copernican-like model in which the sun was the center of the heavenly motions. He realized that if the sun was the center, then planetary orbits looked oval-like, and the planets did not move with uniform speed. His basic problem was to give geometrical (i.e., meaningful) interpretations to the oval-like orbits and variable speeds, which he succeeded in doing in his First and Second Laws.

Kepler shared the Pythagorean belief that the "harmonies" in the world became much more apparent when they were represented mathematically. For him only certain mathematical objects displayed harmony, and in his later works he tried to develop a theory of what this essentially consisted of. He ultimately settled on the Platonic idea of construction by straightedge and compass. Using modern terminology, Kepler's theory is basically as follows: There is a proper mapping of reality into mathematics. Under this mapping, the "harmonies" or laws that determine the real nature of things are mapped into pure geometrical objects, that is, the platonic solids and objects that can be constructed by compass and straightedge. If we consider the meaningful mathematical objects as the purely geometrical ones, then this theory says that the laws that govern nature, when interpreted mathematically must be meaningful, where "meaningfulness" is a precisely given definitional concept built out of the primitive concepts of line and circle. This idea is very close to ones in modern dimensional analysis, which is discussed briefly in Sections 1.5 and 1.6 in more detail in Section 5.10 of Chapter 5.

Descartes also considered meaningfulness concepts in geometry, and in his *La Géometrié* concluded that those planar figures that could be described by algebraic equations were meaningful and those that could not were meaningless. Thus Descartes' concept of geometrical meaningfulness was also a definitional one, but rather than being defined directly in terms of geometrical entities, it is defined in terms of algebraic concepts that through a specialized correspondence refers to geometrical entities.

Inherent in the idealist positions of Nicholas of Cusa and Jeans are other meaningfulness concepts, although it is not exactly clear to me what they are. But because "thoughts" are usually linked somehow to "language," it seems likely that meaningfulness, if clearly specified, would for these two thinkers take the form of a definability concept.

Invariance concepts of meaningfulness have also played a prominent role in geometry and science, and some of these are investigated next.

1.4 Invariance in Geometry

Although many different kinds of geometries are studied in mathematics today, only one was studied throughout most of history: the geometry of physical space:

> Our current conception of mathematics as an ideal science, of geometry in particular as dealing with an ideal space, rather than the actual space in which the universe is set, was a notion quite unformulated before Hobbes, and not taken seriously till the middle of the eighteenth century, though it was dimly felt by a few Aristotelian opponents of Copernicus. The space of geometry appears to have been the space of the real universe to all ancient and medieval thinkers who give any clear clue to their notion of the matter. In the case of the Pythagoreans and Platonists the identity of the two was an important metaphysical doctrine; in the case of other schools the same assumption seems to have been made, only its bearings were not thought out along cosmological lines. Euclid takes it for granted that physical space (ψωριον) is the realm of geometry; later mathematicians use his terminology, and there is no clear indication anywhere in the available works that anybody thought differently. When some, like Aristotle, defined space in a quite different manner, it is noticeable that the definition is still such that the needs of geometers are fully met. *(Burtt, pp. 44–45)*

Recall that until the later part of the nineteenth century, all of continuous mathematics was ultimately based on geometrical concepts, because this type of mathematics was founded on real numbers, which at that time were conceived of as length of line segments existing in physical, Euclidean space. This allowed mathematics to be identified with reality and its methods justified metaphysically. However, the introduction of non-Euclidean geometries into mathematics in the early part of the nineteenth century greatly weakened this conceptual foundation for continuous mathematics, because it allowed the Euclidean nature of physical space to be seriously questioned. Prior to the introduction of non-Euclidean geometries, it was generally believed that Euclidean geometry was not only a valid description of physical space, but it was the only "thinkable" one, that is, it was unthinkable that space was not Euclidean. The non-Euclidean geometries of Lobachevsky (1827) and Bolyani (1831) quickly changed that point of view among mathematicians. Not only were they "thinkable," but they produced mathematical results of the same high degree of richness, elegance, and beauty as Euclidean geometry. Furthermore, as Beltrami (1868) showed, they were interpretable within Euclidean geometry, and thus their axiomatizations could not produce contradictions that were not already inherent in Euclid's axioms. These geometries were so easy to conceptualize and work with that mathematicians began to speculate about their appropriateness as models of physical space, and Gauss proposed that experiments be undertaken to determine which geometry best described physical space. Thus by the middle of the nineteenth century Euclidean geometry was no longer universally accepted by mathematicians as being necessarily descriptive of physical space.

Uncertainties about the Euclidean nature of physical space produced difficulties for the foundations of mathematics, because most of mathematics outside of arithmetic was founded on Euclidean concepts. The "arithmeti- zation of analysis," which was completed in the late part of the nineteenth century, resolved such difficulties by founding mathematics on arithmetic—a subject that had unquestioned mathematical significance and no philosophical doubts about its reality. After this "arithmetization" was achieved, the traditional roles between analysis and geometry became inverted: analysis now provided the basis for models of axiomatic geometry both Euclidean and non-Euclidean. Such a change, however, produced serious meaningfulness problems for geometry: In the geometries based on analysis—henceforth called "analytic geometries"—it was difficult to distinguish geometric concepts from nongeometric ones. Before, when geometry was based on physical space, metaphysical principles about physical reality could be invoked to define "geometrical." But in analytical geometry no analogous program could be carried out, because arithmetic and its metaphysics provided no insight into the geometrical nature of things. Many new geometries were discovered, and it became very clear that since there are so many different and varied types of geometries and only one physical space, only a very few geometries— perhaps not even including Euclidean geometry—could be naturally and directly interpreted in physical space. The upshot of all of this was that if the concept of "geometrical" were to be developed for the wide class of known geometries, then it had to be based on principles other than metaphysical ones about physical or mathematical reality. The mathematician Felix Klein (1849–1925) conceived of a program to accomplish exactly this.

In a famous address given at Erlangen in 1872, Klein identified geometries with groups of transformations and the concept "geometrical" with invariance under transformational groups. However, a satisfactory justification for this position, in my opinion, was not given either in the Erlangen Address or in Klein's subsequent publications on the subject. In fact, there is noticeable lack of any serious effort by Klein and his followers to justify it philosophically. Nevertheless the "Erlanger Program"—as it was subsequently known—had a powerful and positive impact on the field of geometry, and today is seen as one of the major advances in mathematics. This is primarily because the identification of geometries with transformation groups proved to be an enormously fruitful idea. Among other things, it gave ideas about the possible range of geometries and about how these were related to one another, and it provided techniques that transformed subtle geometric questions into straightforward, easy-to-solve group-theoretical ones. It also provided very interesting insights into meaningfulness issues.

Let Γ be Euclid's axioms for the plane. It is often said, "Γ axiomatizes the Euclidean plane," but what does this mean? Everyone will agree that it at least implies the following: (1) The Euclidean plane—whatever it is—is a model of Γ But if Γ is to axiomatize essentially one thing rather than a number of different things, then the following should also be true: (2) Every model of

Γ is isomorphic to the Euclidean plane. Together these two conditions imply that the axiom system Γ is categorical. Euclid believed (1) to be true, a portion of physical space with the undefined terms appropriately interpreted providing the Euclidean plane; (2) was not even considered by ancient mathematicians because they did not have the concept of "model."

The consequence of categoricity—that there is exactly one model of the axioms up to isomorphisms—is a reasonable requirement of an axiomatization of a geometry. Euclid's axioms were not categorical; but a slightly modified version is, one that takes into account some notions he used implicitly. This latter result can be established without having any idea what the "Euclidean plane" is. Its method of proof goes roughly as follows: Let Γ' be the modified axiom system. A model M of Γ' is constructed as follows: Let E be the set of ordered pairs of real numbers. Elements of E are called "points." The set of lines, L, and the set of circles, C, are defined as sets of points (x,y) that satisfy the respective equations $ax+by=c$ and $(x-d)^2+(y-e)^2=f^2$, where a,b,c,d,e and f are real numbers such that f is positive and not both a and b are zero. Then it is easy to verify that E, L, and C with these interpretations are a model of Γ' To show that all models of Γ' are isomorphic is more difficult, and we do not go into it here, but merely note that it consists in taking an *arbitrary* model of Γ' and showing it isomorphic to the aforementioned model M. Note that the construction of the model M depends only on assumptions about the real numbers. So if the real numbers are taken to be entities constructed out of the positive integers, then the construction of M does not depend on features of physical space.

The next example shows how naturally meaningfulness issues arise in geometry. Suppose we have two sets of axioms. When do they determine the same geometry? If the two axiom systems have the same number and type of undefined terms, then the answer is simple: Only when they are categorical, and each model of one axiom set is isomorphic to a model of the other. If the two axiom sets have different numbers or sorts of undefined terms, then the question becomes more difficult. There are well-known instances of this where they clearly do determine the same geometry, for example, Euclidean plane geometry can be axiomatized either by using point, line, and circle as undefined terms or by using point, line, and a binary relation of perpendicularity as undefined terms. The problem, however, quickly becomes more complicated if nonaxiomatic procedures are used for characterizing geometries, for example, describing them through the use of metrics. The Erlanger Program is able to handle all of these situations. But before considering how, let's look at a definability approach to this sort of problem.

Suppose $\mathfrak{X} = \langle X, P_1, \ldots, P_i, \ldots, P_m \rangle$ and $\mathfrak{Y} = \langle Y, Q_1, \ldots, Q_j, \ldots, Q_n \rangle$ are geometries. They may have arisen through categorical axiomatizations or other methods of specification. Because in general $\mathfrak{X}$ and $\mathfrak{Y}$ will be different and the P- and Q-relations may be quite different structurally, it may be a nontrivial matter to determine whether or not they capture the "same geometry." Of course describing "capturing the same geometry" is itself a deep problem. The following criterion seems to me to be a sensible and useful one for saying, "$\mathfrak{X}$ and $\mathfrak{Y}$ capture the same geometry": If relations $P'_1, \ldots, P'_j, \ldots, P'_n$ can be defined in terms of X, $P_1, \ldots, P_i, \ldots, P_m$ and relations $Q'_1, \ldots, Q'_i, \ldots, Q'_m$ in terms of Y, $Q_1, \ldots, Q_j, \ldots, Q_n$ so that the structures

$$\mathfrak{X}' = \langle X, P_1, \ldots, P_i, \ldots, P_m, P'_1, \ldots, P'_j, \ldots, P'_n \rangle$$

and

$$\mathfrak{Y}' = \langle Y, Q'_1, \ldots, Q'_i, \ldots, Q'_m, Q_1, \ldots, Q_j, \ldots, Q_n \rangle$$

are isomorphic, then we may say that $\mathfrak{X}$ and $\mathfrak{Y}$ "capture the same geometry." The intuitive justification for this is as follows: The primitive relations $P_1, \ldots, P_i, \ldots, P_m$ are meaningful in the geometry captured by the structure $\mathfrak{X}$ and because the relations $P'_1, \ldots, P'_j, \ldots, P'_n$ are definable from the primitives X, $P_1, \ldots, P_i, \ldots, P_m$, they too are meaningful, and I take it as a reasonable principle about geometry that for such instances,

$$\langle X, P_1, \ldots, P_i, \ldots, P_n \rangle$$

and

$$\langle X, P_1, \ldots, P_i, \ldots, P_m, P'_1, \ldots, P'_j, \ldots, P'_m \rangle$$

capture the same geometry. By a similar line of reasoning, $\mathfrak{Y}$ and $\mathfrak{Y}'$ also capture the same geometry. Because $\mathfrak{X}'$ and $\mathfrak{Y}'$ are isomorphic, $\mathfrak{X}'$ and $\mathfrak{Y}'$ also capture the same geometry, because geometrical concepts are preserved through isomorphisms. Thus it follows that $\mathfrak{X}$ and $\mathfrak{Y}$ capture the same geometry.

The difficulty with the above approach is that "definable" has not been specified. Of course, in a particular application one does not need a general definability concept but only a definability concept that is pertinent to the application, and in many situations by using special properties of the pertinent definability concepts it might be an easy matter to justify (or claim) that such definability concepts yield "geometrical" results. In particular it seems reasonable to me that if $T_1, \ldots, T_k, \ldots, T_p$ are meaningful concepts within a geometry and T is definable from $T_1, \ldots, T_k, \ldots, T_p$ through first-order logic, then T is also a meaningful concept within that geometry. Thus "first-order definability" appears to be a necessary condition for the general kind of definability concept needed. Examples can readily be produced to show that it is not a sufficient condition. Therefore, although we might be confident that certain methods of defining produce proper geometrical concepts, we are still faced with the problem of giving an exact description of what "geometrical definability" should be. This is similar to the problem that the ancient geometers faced in trying to define "proper geometric construction."

Klein's approach to meaningfulness issues is rather different. It is based on the concept of transformation group, which is defined next. Let Z be a nonempty set. A "transformation" on Z is a one-to-one function from Z onto itself. Transformations on Z have group properties, i.e., there is always an identity transformation ι on Z defined by $\iota(x)=x$ for all x in Z, and each transformation on Z has an inverse transformation on Z, and transformations on Z can be combined by function composition to produce other transformations on Z. If S is an n-ary relation on z and α is a transformation of Z, then S said to be "invariant under α" if and only if for each $x_1,\ldots, x_i,\ldots, x_n$ in Z,

$$S[x_1,\ldots,x_i,\ldots,x_n] \text{ iff } S[\alpha_1(x_1),\ldots,\alpha(x_i),\ldots,\alpha(x_n)]\,.$$

Suppose $\mathfrak{Z} = \langle Z, S_1, \ldots, S_i, \ldots, S_n\rangle$ is a structure, where Z is a nonempty set and $S_1,\ldots, S_i,\ldots, S_n$ are relations on Z. Then the "transformation group of $\mathfrak{Z}$ is the set of transformations on Z under which each of the relations, $S_1,\ldots, S_i,\ldots, S_n$, is invariant. Klein identified "geometrical" relations of $\mathfrak{Z}$ with those that were invariant under each element of its transformation group. Invariance under transformations can also be defined for more complicated relations, for example, relations on relations on Z, and the concept of transformation group easily extends to structures based on these sorts of "higher order" relations. The details are described in Section 3.6 of Chapter 3 and Section 4.2 of Chapter 4. For structures based on higher order relations, "geometrical" is defined in the same manner.

Let us now look at Klein's Erlanger Program applied to the situation described above, where there are structures $\mathfrak{X}$ and $\mathfrak{Y}$ and we want to know if they capture the same geometry. Klein's answer for this situation is simple and direct: they capture the same geometry if and only if there is a one-to-one function from X onto Y such that it defines (in the obvious way) an isomorphism of the transformation group of $\mathfrak{X}$ onto the transformation group of $\mathfrak{Y}$. Thus for Klein the essence of geometries is captured by their transformation groups. However, this approach leads to several philosophical difficulties.

The first is that it is too general for geometry. It was true that the important geometries of Klein's time could be characterized by their transformation groups, and as new geometries were discovered they too could so be classified. However, all of these geometries were based upon certain ideas that had "geometrical content." It is not at all clear that for an arbitrary transformation group G invariant properties that have conceptual geometrical content can be found so that the transformation group they determine is identical to G. This point is not a criticism of using the Erlanger Program in situations that are known to be geometrical, but rather a criticism about stretching the concept of "geometry" too far. If rather than "geometrical" we call those concepts that are invariant under transformation groups "meaningful," and save the term "geometrical" for those situations in which the transformation groups arose out of geometrical considerations, then this difficulty disappears.

The second, and a most serious difficulty, is that Klein nowhere justifies the identification of "geometrical" with invariance under transformation groups—even for the cases of well-known geometries. It might well be argued that such a justification is not needed, that the rich set of ideas and powerful methods produced by the Erlanger Program is justification enough. However, it seems to me that the connections between invariance and "meaningful" (or "geometrical") are sufficiently mysterious to deserve comment and, if possible, explanation. It is eminently reasonable that meaningful concepts should be invariant. However, why every invariant one should be meaningful is by no means obvious. In Chapter 4 it is shown that the Erlanger Program is essentially equivalent to a concept of meaningfulness based on definability, and this result may prove useful in philosophically justifying invariance as a form of meaningfulness.

The third difficulty is related to the second. There are geometries that have the identity as their only transformation. By the Erlanger Program any two such geometries that have their points in one-to-one correspondence are essentially the same, and for all such geometries, all relations are invariant and therefore are "geometrical." Examples of such geometries were constructed in the middle of the nineteenth century through differential geometric techniques developed by B.Riemann (1826–1866). Such geometries had a different character, because of this, perhaps should be separated from those with richer transformation groups and not be treated as "true geometries." However, with the introduction of the General Theory of Relativity by Einstein in 1916 this could no longer be the case, for it was clear that whatever a "geometry" was, Einstein's model—which at the time was the best description of physical space—was one of those Riemannian geometries that had the identity as its only transformation. Thus by the Erlanger Program every relation based on the points of physical space was meaningful, a proposition that was unacceptable to geometers and physicists of the time. I can find no clear statement in the literature as to why it was unacceptable. It appears to me that from some metaphysical perspectives it is legitimate to require all relations based on physical points for general relativity to be meaningful. But in any case this example was damning for the Erlanger Program, and except for a few feeble attempts to revive it, it quickly lost much of its influence.

One of the most successful uses of invariance in geometry was the characterization of spaces of constant curvature, which is discussed next.

In a famous lecture given in 1854 at Göttingen, Riemann (1826–1866) laid out the foundations for the modern study of differential geometry. His approach was completely quantitative, and basically consisted of defining geometries through coordinate systems and a very general concept of distance. In the 1860's, H.von Helmholtz (1821–1894) realized that Riemann's approach could be greatly simplified for the special case of physical space by using a qualitative condition that he considered to be "a fact of experience," namely, that there existed rigid bodies, that is, that there existed physical bodies that can move freely in space "without change of form or size." Helmholtz considered this "fact" to have nongeometrical aspects: "Geometric axioms do not speak about spatial relations only, but also at the same time about the mechanical behavior of our most rigid bodies in motion." By using the existence of rigid bodies, Helmholtz was able to demonstrate that physical space was of constant curvature, and from this it followed that it was either one of the three types of geometries that resulted from Euclid's axioms by appropriately changing (or leaving alone) the parallel postulate (Euclidean, hyperbolic, elliptic) or was a three dimensional version of spherical geometry.

At first sight, Helmholtz's assumption of the existence of rigid bodies appears to be a different kind of invariance principle than employed by the Erlanger Program. But Helmholtz was able to demonstrate that the existence of rigid bodies was equivalent to the existence of a particular transformation group, Thus Helmholtz's approach could be captured by the Erlanger Program. This fact was used later by S.Lie (1842–1899) to give an elegant characterization of geometries of constant curvature. Lie's approach was to give assumptions directly in terms of transformation groups rather than in terms of more intuitively plausible (and perhaps philosophically justifiable) principles such as the existence of rigid bodies. Later, a different and more elegant description was given by Hilbert (1921) for the cases of planar hyperbolic and Euclidean geometries. Hilbert's characterization also relied on transformation groups, but did not directly use a transformational analog of the existence of rigid bodies. An equally elegant description, which is much more a descendant of the ideas of Helmholtz and Lie, was given in 1952 by Tits and later improved upon by Freudenthal (1965). The improved version can be succinctly summarized as follows: Suppose $\mathfrak{S}$ is a locally compact, connected metric space, and suppose for all sufficiently small congruent triangles *u* and *v* of $\mathfrak{S}$ there exists an isometry of $\mathfrak{S}$ (that is, an isomorphism of the metric space $\mathfrak{S}$ onto itself) that maps *u* onto *v*. Then $\mathfrak{S}$ as a geometry is either Euclidean, hyperbolic, elliptic, or spherical.

A different use of invariance in geometry occurred in Einstein's (1879– 1955) 1905 papers about the special theory of relativity. This physical-geometrical theory was based on two invariance principles: *(i)* that the state of rest and a state of motion at constant speed cannot be distinguished by any experiment performed by observers in either system (the *principle of relativity*), and *(ii)* that the speed of light as measured by an observer is the same for all light emitting sources moving at constant speeds (the *principle of constancy*). The principle of relativity is clearly an invariance principle. By relying on it and the principle of constancy, Einstein was able to derive the Lorentz transformations, which describe how the measurements of objects change with speed. Later in 1908 the mathematician Hermann Minkowski (1864–1909) gave a purely geometrical formulation of the kind of space-time that resulted in the special theory of relativity. Basically, he noted that the Lorentz transformations left invariant the quantity

$$(x_1 - x_2)^2 + (y_1 - y_2)^2 + (z_1 - z_2)^2 - c^2(t_1 - t_2)^2 , \qquad (1.1)$$

where (x_1, y_1, z_1, t_1) and (x_2, y_2, z_2, t_2) are space-time points with spatial coordinates x_1, y_1, z_1 and x_2, y_2, z_2 respectively and temporal coordinates t_1 and t_2, and c is the velocity of light. The quantity in Equation 1.1 resembles the square of the distance between two points in four dimensional Euclidean space except for the negative sign in front of the term with the time coordinates. Minkowski noted this similarity and showed that Equation 1.1 could be used to define a "distance" function that looked in many ways like a metric, and that in the four dimensional geometry that resulted from its use the Lorentz transformations corresponded to rotations through imaginary angles. Because of such features, the geometry of special relativity could be adequately treated by the techniques of the Erlanger Program.

1.5 Dimensional Analysis

Another important use of invariance is to be found in the techniques of "dimensional analysis" of physics. It has been noted for some time that the kinds of physical quantities that appear in physical laws are expressible as products of powers of a relatively small number of basic physical quantities such as length, time, mass, charge, and so forth. The reasons for this are explained in Section 5.10 of Chapter 5. The basic physical quantities are often measured in different ways, for example, length is measured in centimeters, meters, inches, yards, and so forth. The following two principles, which take into account that physical quantities can properly be measured in different ways, were recognized early in the development of physics:

> *Principle 1:* proper physical measurements of a basic physical quantity are related by multiplications by positive reals.
>
> *Principle 2: (dimensional invariance)* a numerical relation that expresses a valid physical relationship between physical variables has the same mathematical form no matter what proper measurements are used to measure the physical variables.

If all physical quantities are products of powers of basic physical quantities and all valid physical relationships are really laws about physical quantities, then Principle 1 implies Principle 2, and in fact implies that different measurements of the basic physical quantities produce relatively minor changes in the physical relationship—ones that result in multiplications by positive reals. This latter fact forms the center piece of an important area of physics called *dimensional analysis.* It follows from results of Narens (1981a; discussed in Section 5.10 of Chapter 5) and Falmagne and Narens (1983; discussed in Section 2.6 of Chapter 2) that the product of powers representation is essentially a consequence of Principles 1 and 2. Thus for the purposes of this section we may consider dimensional analysis as being a set of techniques that systematically exploit Principles 1 and 2.

The following elementary example nicely illustrates the use of dimensional analysis.

Consider a simple pendulum consisting of a ball suspended by a string. The ball has mass $\boldsymbol{m}$, the string has length $\boldsymbol{d}$, and $\boldsymbol{g}$ is the gravitational constant that describes acceleration of small masses towards the center of the earth. The ball is displaced so that the pendulum makes an angle of $\boldsymbol{\theta}$ with the vertical and is then released. We want to find its period $\boldsymbol{t}$. The variables $\boldsymbol{m}$, $\boldsymbol{g}$, $\boldsymbol{d}$, and $\boldsymbol{t}$ can be measured in various ways—that is, can be properly assigned numerical values in various ways. Assume that they are measured in terms of a coherent set of units, i.e., if $\boldsymbol{d}$ is measured in terms of the unit u and $\boldsymbol{t}$ in terms of the unit v, then $\boldsymbol{g}$ is measured in terms of the unit u/v^2. In this application $\boldsymbol{\theta}$ is measured by the ratio of a measured length of arc divided by a measured radius, and is thus a real number—a "dimensionless" quantity—that does not depend on the unit in which length is measured. It should be emphasized that at this point that $\boldsymbol{t},\boldsymbol{m},\boldsymbol{g}$, and $\boldsymbol{\theta}$ are to be thought of as *qualitative* entities; they have not yet been assigned numbers. Now by using "physical intuition" we assume that we have enough information to determine $\boldsymbol{t}$—that is, we assume that $\boldsymbol{t}$ is physically determined by $\boldsymbol{m}$, $\boldsymbol{g}$, $\boldsymbol{d}$, and $\boldsymbol{\theta}$. Using functional notation, we write,

$$\boldsymbol{t} = \boldsymbol{F}(\boldsymbol{m}, \boldsymbol{g}, \boldsymbol{d}, \boldsymbol{\theta}) \,, \tag{1.2}$$

where $\boldsymbol{F}$ is a qualitative function. A proper numerical representation for Equation 1.2 consists in giving proper numerical measurements, t, m, g, d, and θ, to $\boldsymbol{t}$, $\boldsymbol{m}$, $\boldsymbol{g}$, $\boldsymbol{d}$, and $\boldsymbol{\theta}$ respectively through some coherent system of units and finding a numerical function $\boldsymbol{F}$ such that

$$t = F(m, g, d, \theta) \,. \tag{1.3}$$

Principle 2, dimensional invariance, requires any other proper representation for Equation 1.2 to be of the form:

$$t' = F(m', g', d', \theta) \,, \tag{1.4}$$

where t', m', g', and d' are proper numerical measurements in some other coherent system of units. Note that in Equations 1.3 and 1.4 the same numerical function F is used to represent the qualitative function $\boldsymbol{F}$. θ also appears in Equations 1.3 and 1.4. By the way it was defined, θ is a real number that does not depend on which unit is used to measure length, and so it has the same numerical value for each coherent system of units. By Principle 1, t', m', g', and d' are related to t, m, g, and d in the following manner: there are positive real numbers α, β, and γ such that

$$t' = \alpha t, \;\; d' = \beta d, \;\; g' = \frac{\beta g}{\alpha^2}, \;\; \text{and } m' = \gamma m \,.$$

Thus by substituting these equations into Equation 1.4 we get

$$\alpha t = F\left(\gamma m, \frac{\beta g}{\alpha^2}, \beta d, \theta\right) \,. \tag{1.5}$$

By dimensional invariance Equation 1.5 is true for all choices of positive reals α, β and γ. Since $\boldsymbol{\theta}$ is fixed and its measurement does not vary with changes of units, we can rewrite Equation 1.5 as

$$\alpha t = H\left(\gamma m, \frac{\beta g}{\alpha^2}, \beta d\right), \tag{1.6}$$

where for all positive real numbers s, x, y, and z,

$$s = F(x, y, z, \theta) \text{ iff } s = H(x, y, z).$$

Because in Equation 1.6 the first argument of H, γm, can take any positive real value (by an appropriate choice of γ) while leaving the value of H, αt, unchanged, it follows that H really does not depend on its first argument, and we can therefore rewrite Equation 1.6 as

$$\alpha t = K\left(\frac{\beta g}{\alpha^2}, \beta d\right). \tag{1.7}$$

By choosing units so that $\alpha = \sqrt{g}/\sqrt{d}$ and $\beta = 1/d$, it follows from Equation 1.7 that

$$t = \frac{\sqrt{d}}{\sqrt{g}} \cdot K(1,1). \tag{1.8}$$

It is easy to see by inspection that when the measurements of $\boldsymbol{t}$, $\boldsymbol{g}$, and $\boldsymbol{d}$ are changed to another coherent system of units that an equivalent form of Equation 1.8 with the same real constant $K(1, 1)$ is valid. If we assume that the measurement of $\boldsymbol{g}$ is known in one coherent set of units—and thus by Principle 1 in all coherent set of units—then Equation 1.8 gives a way of calculating for each length $\boldsymbol{d}$ the measurement of the period $\boldsymbol{t}$ in terms of the measurement of $\boldsymbol{d}$ and the real number $K(1, 1)$. The number $K(1, 1)$ cannot be determined by dimensional analysis alone. It can, however, be found by experiment; that is, a particular measured length can be chosen and the period measured for that particular length and $K(1, 1)$ can be computed by Equation 1.8.

In the previous argument we kept $\boldsymbol{\theta}$ fixed. As $\boldsymbol{\theta}$ varies so does Equation 1.8, but Equation 1.8 keeps the same form; that is, only the real number $K(1, 1)$ varies. In other words, $K(1, 1)$ is a function of $\boldsymbol{\theta}$ and therefore of θ. Thus for variable angle $\boldsymbol{\theta}$ we can rewrite Equation 1.8 as

$$t = \frac{\sqrt{d}}{\sqrt{g}} \Phi(\theta), \tag{1.9}$$

where θ is the (dimensionless) measurement of $\boldsymbol{\theta}$, and Φ is some particular real valued function. The usual law for the period of the pendulum using Newton's laws is basically the same as Equation 1.9 but with $\Phi(\theta)=\sin(\theta)$. Thus by using Newton's laws, $\Phi(\theta)$ can be completely determined without having to resort to experiment. However, by using the above dimensional analysis arguments, certain laws of pendulums can also be completely determined without having to resort to experiment. For example, if we have two pendulums of measured lengths d_1 and d_2 (measured of course in the same unit) that are released at the same angle with the vertical, and their resulting periods are t_1 and t_2 (measured in the same unit), then it immediately follows from Equation 1.9 that

$$\frac{t_1}{t_2} = \frac{\sqrt{d_1}}{\sqrt{d_2}}. \tag{1.10}$$

Note that Equation 1.10 does not depend on which units are used to measure distance and time and that the gravitational constant is not mentioned in the statement of the result or its conclusion, Equation 1.10.

Dimensional analysis is particularly useful in very complicated physical setting where exact solutions by purely mathematical methods are unknown or perhaps even impossible. It is also indispensable in those situations, where, for all practical purposes, it is impossible to give precise and detailed formulations of the fundamental equations from which the solution is sought. Such situations abound in the physical sciences, particularly in airplane and ship design and in engineering applications. Nevertheless, the philosophical justifications for the correctness of dimensional analysis have not been satisfactorily worked out in the literature despite the many words written about it. Many controversies have been generated by dimensional analysis. These usually center around the status of Principle 2, dimensional invariance. (Principle 1 has a rigorous mathematical/philosophical foundation and is considered noncontroversial.) There are several views in the literature about the nature of dimensional invariance, and some of the more prominent ones are discussed next.

The first is a view that was championed by the physicist P.W.Bridgman (1882–1961). It holds that the supposition that mathematical laws must satisfy dimensional invariance has never been demonstrated, and perhaps is even incorrect. But this does not matter in Bridgman's view, because the techniques of dimensional analysis really did not depend on physical laws being dimensionally invariant:

> Why is it that an equation which correctly describes a relation between various measurable physical quantities must in its form be independent of the size of the fundamental units? There does not seem to be any necessity for this in the nature of the measuring process itself. *(Bridgman, 1931, pg. 13)*
>
> The view presented here is one which apparently is increasingly accepted in the U.S. The crux of this point of view is that dimensional analysis is an analysis of an analysis: that is, an analysis of the implications of the fact that methods of analyzing experience have been found

> profitable which employ certain types of measuring process and certain methods of mathematical treatment of the results of the measurements. There is nothing absolute here, but a great deal of flexibility. *(Bridgman, 1968, pg. 449)*

The mathematician Garrett Birkhoff (1950) attempted to justify dimensional invariance by arguing that because all the fundamental equations used in physics are dimensionally invariant, all laws derivable from them must also be so. However, he did not formally describe what "derivability" is, and thus his theory is somewhat incomplete. His underlying idea is also somewhat flawed, for the derivations of laws in physics depended not only general laws which are dimensionally invariant but also on initial and boundary conditions, which in general are not dimensionally invariant, making it hard to see how a reasonable derivability concept could be easily formulated so its applications would yield only dimensionally invariant results. But even if we restrict our attention to those situations where the general laws and initial and boundary conditions are dimensionally invariant and believe that all proper methods of derivation preserve dimensional invariance, then there are still problems with Birkhoff's approach. First, it gives us no clue to why the general laws, initial conditions, and boundary conditions should be dimensionally invariant. Second, it cannot be used to justify the method of dimensional analysis for obtaining new general laws—it can only be used for obtaining special laws from already known general ones.

A different, widely held view is what Krantz et al. (1971) calls the "it couldn't be otherwise" view:

> An "it couldn't be otherwise" argument proceeds along lines something like these. Since the choice of units is a wholly arbitrary matter—the choice exists because of the way we choose to represent certain qualitative information numerically—any assertion that describes a natural phenomenon cannot depend upon such conventions. This is Sedov's view (1959, pg. 16), and it was described as a "principle of theory construction" by Luce (1959) who, however, backed away from this position (1962) in face of Rozeboom's (1962) criticisms. Many others have expressed more or less similar arguments. We suspect that many who hold this view are simply saying, as we did earlier, that if we knew how to formulate what we mean by a qualitative physical law, then we would find, as a purely logical consequence of our measurement assumptions, that the numerical representation of the law would be dimensionally invariant. Since this has not been proved, the position that "it couldn't be otherwise" is quite suspect and widely discredited. Perhaps we should add, however, that some arguments mounted against this position are not less suspect than the arguments for it. *(Krantz, Luce, Suppes, and Tversky, 1971, pp. 504–505)*

Still another widely held view for explaining why physical laws are dimensionally invariant is based on a theory of derived measurement and the concept of "physical similarity." This view will not be described here, because it has been superseded by a somewhat superior development due to Luce (1978), which is presented shortly. It should be noted, however, that its original proponents were physicists (e.g., Buckingham 1914, Campbell 1920, Lord Rayleigh 1915, and Tolman 1914), and their expositions on this matter were severely flawed by the fact that they never made explicit the empirical assumptions underlying their theories, leaving them open to all sorts of criticisms, including accusations of introducing metaphysical concepts into science. Later versions of this view that made explicit the underlying empirical assumptions were given by the philosopher Causey (1967, 1969) and the psychologist Luce (1971).

Luce (1978) presented an axiomatization of the algebra of physical qualities (briefly described in Section 5.10 of Chapter 5), and showed that it implied Principle 1. He further showed the following: If $\mathfrak{S}$ is the structure of physical qualities, $\boldsymbol{R}$ is a relation on physical qualities, and R is $\boldsymbol{R}$'s numerical representation under a particular proper system of measurements of individual physical qualities, then: $\boldsymbol{R}$ is invariant under the transformation group of $\mathfrak{S}$ iff R is dimensionally invariant. Thus Luce showed dimensional invariance corresponded to the concept of meaningfulness used by the Erlanger program. The problem of justifying the dimensional invariance of physical laws in this approach is essentially the same as justifying the concept of meaningfulness that results from invariance under transformation groups. Luce (1978) does not give such a justification; he only shows the equivalence of invariance under the transformation group of $\mathfrak{S}$ and dimensional invariance and leaves it at that. Of course all justifications must stop somewhere, and from some points of view showing invariance under a transformation group is a good stopping place. However, I find it somewhat a mysterious one, and have the feeling that something is left unsaid. One of the main themes of the book is finding other ways of saying "invariance under a transformation group" that are less mysterious and make more sense as meaningfulness concepts. General methods for accomplishing this are described in detail in Chapter 4, and these methods easily apply to the various axiomatizations of dimensional analysis presented in Section 5.10 of Chapter 5. In Section 6.4 of Chapter 6, it is argued that laws consist of more than meaningfulness, and consequently the invariance inherent in dimensionally invariant laws consists of more than invariance under the transformation group of $\mathfrak{S}$.

1.6 Eddington's "Method of Pure Numbers"

(The values of the experimentally determined physical constants used here—unless otherwise stated—are the ones given by E.Whittaker in his Tarner

Lectures presented at Cambridge University in 1947. The following brief exposition of Eddington's use of pure numbers is based on those lectures.)

The astronomer and physicist Arthur Eddington (1882–1944), like many before him, realized that the ratio of two physical quantities of the same physical dimension was a real number, that is, had no physical dimension. If these two physical quantities resulted from fundamental physical laws, then the resulting dimensionless number might have important physical significance, in much the same way that the ratio of the length of the circumference to the length of the diameter of a circle produces a dimensionless number, π, that has great geometrical importance. A physical example of this occurs in atomic physics. There are two relevant laws that describe interactions between an electron and a proton that are a distance d apart—Coulomb's law and Newton's gravitation law—and if the electric force given by Coulomb's law is divided by the gravitational force given by Newton's law, then a pure number results that is independent of the distance d and the units in which the relevant physical quantities are measured. This number, which is called the *force constant,* has been determined experimentally to be about 2.27×10^{39}.

Other constants can be similarly found. For example, the amount of time it would take (at the observed rate recession of the galaxies) for the universe to double its dimensions is 1.3×10^9 years. Another natural temporal constant is the time taken to transverse the length conventionally known as the radius of an electron with the velocity of light. This constant is about 10^{-23} seconds. The ratio of these two time intervals (measured in the same temporal units) is about 4×10^{39}, which is suggestive of the force constant.

Many pure numbers can be obtained in the above manner. If the convention is made that those less than 1 are replaced by their reciprocals, then these pure numbers, rather than being scattered about the number continuum, naturally cluster into three sizes of numbers, those (such as the ratio of the masses of a proton and an electron) that are less than 1900, those that are of order 10^{39}, and those that are of order 10^{79}. In this arrangement of pure numbers, Eddington found the evidence that gave rise to his theory of the physics of the universe.

Eddington, who did this theoretical work in the 1920s and 30s, assumed that the universe was composed of an equal number of electrons and protons. The sum of these two numbers, which represents the total number of particles in the universe, is a pure number, in fact necessarily *an integer,* which Eddington called "the cosmological number," and he denoted it by N. He assumed Einstein's cosmological model of the universe, and estimated N by doubling the estimated mass of the universe for Einstein's model. Its value, calculated from data available in 1930, was about $N=1.4 \times 10^{79}$, and thus too belonged to one of the three clusters of numbers mentioned previously. Whittaker makes the following comment about N and Eddington's calculation of it:

It is conspicuous by reason of two characteristics. First, since it represents a number of particles, it is necessarily a whole number, whereas the other pure numbers are (with one exception) non-integral: there is therefore a possibility of finding its value *exactly*. Secondly, Eddington suspected, and eventually proved, that when certain *qualitative* physical properties are assumed, N can be found without making any use of quantitative data derived from observation. The argument may be stated briefly thus. It is recognized in relativity theory that a measurement involves four entities, namely two to furnish an observable relation, and two to furnish the comparison relation in terms of which the first relation is measured: thus measured quantities are primitively associated with quadruple wave-functions. It is possible to calculate in space of constant positive curvature the number of independent wave-functions of this form with the necessary relativistic property, and the total number of elementary particles corresponds to this number. The value so deduced is

$$N = \frac{3}{2} \times 136 \times 2^{256},$$

which gives

$$N = 2.36216\ldots \times 10^{79}$$

in satisfactory agreement with the result previously derived from experimental data. *(Whittaker, 1958, pp. 193–194)*

Eddington used the number N as the basis for his cosmological theory that included subatomic relationships as well as extragalactic ones. He succeeded in qualitatively deriving values for many of the experimentally determined pure numbers previously discussed. In particular, the force constant was qualitatively derived to be

$$\frac{2}{3\pi} \cdot \left(\frac{136}{137}\right)^2 \cdot \sqrt{5N},$$

which to four decimal places is

$$2.2714 \times 10^{39},$$

which was very close to and easily within the error-range of the experimental determination of the constant.

One very interesting relationship that Eddington was able to derive was that

$$\frac{hc}{e^2},$$

the reciprocal of the fine-structure constant, where h is Planck's constant, c is the velocity of light, and e is the charge of an electron, must be the integer 137. At the time the derivation was published, the value of

$$\frac{hc}{e^2}$$

was 137.307 with probable error ±.048, and thus Eddington's theoretical value was outside the accepted range of experimentally determined values. However, due to new procedures of measurement, the accepted, experimentally determined value was changed a little later to 137.009, and remained there for a number of years, and Eddington's theoretical value was within the range of values permitted by measurement error. (Its current value according to the *CRC Handbook of Chemistry and Physics* is 137.0360 with probable error ±.0154.)

Thus Eddington was able to produce explanations for why the pure numbers derived from physical properties clustered around N, $\sqrt{N}$ or N^0, and showed that they were derivable qualitatively from the cosmic number N—a remarkable and somewhat amazing Pythagorean feat.

Pure numbers also play an important role in meaningfulness considerations outside of physics, and general meaningfulness issues regarding them are discussed in Subsection 5.10.8 of Chapter 5 and in Chapter 7.

1.7 Ideal Numbers

We now look at a different type of meaningfulness problem that arose from the introduction of algebraic techniques into mathematics. It involves distinguishing well-understood mathematical entities that corresponded to real magnitudes from others that have many of the formal characteristics of real magnitudes (such as additivity and multiplicativity) but lack interpretations as magnitudes. These latter were called "fictitious," and negative, imaginary, and infinitesimal numbers were the primary examples of such "nonexistent" entities. These particular examples were universally recognized as being enormously useful in understanding the real, existent entities. This presented a deep intellectual paradox: How could such nonexistent entities be so useful in discovering and explicating properties of existent ones? Various views about this issue were advanced, the most sophisticated being formalist-like positions that ultimately led in the latter part of the nineteenth century to George Boole's *Laws of Thought* and the discovery of symbolic logic.

Prior to the nineteenth century, mathematics was considered the science of quantities. As the use of algebraic techniques became increasingly popular during the postmedieval period, mathematicians of the time became inclined to use negative and imaginary numbers in their processes for solving algebraic equations. In 1629, the mathematician Girard explicitly wrote them down as part of the solution to an equation. He also clearly realized that the nonnegative and negative numbers taken together formed something like the modern concept of the number line: "The negative in geometry indicates a retrogression where the positive is an advance." The reason why he gave complex as well as real roots as solutions to an equation was that he believed that in doing so every equation would then have as many roots as its degree; he also realized that the use of complex roots allowed for general, convenient representations of equations.

Because the complex numbers had no proper mathematical interpretation, approaches like Girard's presented serious conceptual problems. This produced ambivalence about their use. Some mathematicians considered them wholly inappropriate. Others gave them shaky metaphysical interpretations. The great mathematician/philosopher G.Leibniz (1646–1716) likened them to a sort of amphibian that was halfway between existence and nonexistence. They gained much more acceptance when the mathematician F.Gauss (1777–1855) interpreted them as points on the plane, where the complex number $a+bi$ corresponded to the point (a, b) in Cartesian coordinates. This concrete interpretation quieted much of the uneasiness associated with using complex quantities in mathematical analyses. This is probably because it gave a means for establishing that the methods used for the manipulations of complex quantities would not lead to contradictions. However, it certainly did not solve the problem as to the true mathematical nature of complex quantities, as Gauss himself recognized. For him as well as others, it was a numerical problem which should have a numerical and not a geometrical solution. In 1831 he gave a different kind of construction that required no geometrical intuition. He identified the complex number $a+bi$ with the ordered pair of real numbers (a, b). He then defined algebraic operations on such ordered pairs by use of analogous operations on the reals. The result of this yielded what Gauss considered to be the "true metaphysics of $\sqrt{-1}$." Nevertheless, this solution was still unacceptable to many mathematicians of the day.

The basic objection to complex and negative numbers was not their lack of utility or interpretability, but their inability to be understood as objects of pure mathematics. "Mathematics," L.Euler (1707–1783) wrote, "is nothing other than the *science of magnitudes*" and his view was universally accepted until the latter part of the nineteenth century, where it continued to be a widely held, if not the dominant, view. The basic problem with the interpretations such as Gauss gave to complex and negative numbers was that they were seen as constructions that looked more like applied mathematical objects than descriptions of new quantities of pure mathematics. The following quote of William Frend, a contemporary of Gauss, expresses clearly this position:

> The ideas of number are the clearest and most distinct in the human mind; the acts of the mind upon them are equally simple and clear. There cannot be confusion in them... But numbers are divided into two sorts, positive and negative; and an attempt is made to explain the nature of negative number, by allusion to book-debts and other arts. Now, when a person cannot explain the principles of a science without a reference to

> metaphor, the probability is, that he has never thought accurately upon the subject. A number may be greater or less than another; it may be added to, taken from, multiplied into, and divided by another number; but in other respects it is intractable: though the whole world should be destroyed, one will be one, a three will be a three; and no art whatever can change their nature. You may put a mark before one, which it will obey: it submits to be taken away from another number greater than itself, but attempt to take it away from a number less than itself is ridiculous. Yet this is attempted by algebraists, who talk of a number less than nothing, of multiplying a negative number into a negative number and thus producing a positive number, of a number being imaginary. Hence they talk of two roots to every equation of the second degree ...they talk of solving an equation which requires two impossible roots to make it soluble: they can find out some impossible numbers, which, being multiplied together produce unity. This is all jargon, at which common sense recoils. (*William Frend,* Principles of Algebra, *1796, pp. x–xi)*

Infinitesimals were perceived as an entirely different kind of fictitious entity. There were many views about their nature. For example, Nicholas of Cusa associated them with Divine Thought; Pascal (1623–1662) saw them as mysteries, which man was supposed to admire but not understand; L'Hospital (1661–1704) believed them to be real quantities; and Leibniz thought of them as useful fictions that could be eliminated if need be by arguing in terms of quantities which are large enough or small enough to make the error smaller than any given number.

Infinitesimal analysis, which was the basis for calculus until the latter part of the nineteenth century, was founded upon principles that treated infinitesimal quantities formally like real numbers and considered curves to be infinitesimally close to a polygonal one made out of infinitesimally small line segments. However, unlike negative and complex numbers, infinitesimals disappeared from serious mathematical research in the late part of the nineteenth century. This is somewhat surprising, because they were even more useful than negative and complex numbers. The reason for their disappearance undoubtedly had to do with mathematicians' inability to interpret them in a manner similar to the negative and complex numbers, raising questions about their consistency.

From a philosophical perspective, one could ask why having an interpretation in terms of real elements should matter for the use of ideal ones. In fact, algebraists who were interested in similar questions argued that it should not. However, I believe that infinitesimals differ from the cases of negatives and complexes and many other algebraic examples in the following way: the infinitesimals cannot be defined in a reasonable way out of known quantities, that is, out of the reals. In other words, if a definability concept of meaningfulness were employed, then systems that included the infinitesimals of classical analysis would not be a meaningful extension of the reals. This includes the system of infinitesimals created in Robinson (1966) known as "nonstandard analysis." (More particularly, using results of Chapter 4, I believe it not difficult to give the following a rigorous formulation and proof: Robinson's system is not a meaningful extension of the real number system, if the real numbers are assumed to correspond to objects that form the science of magnitude.)

1.8 Actual Infinity and the Axiom of Choice

G.Cantor (1845–1918) worked out the beginnings of the mathematics of infinite sets in the 1870s. Before that, there was no coherent mathematical theory of actual infinity. In fact, most philosophers and mathematicians prior to Cantor thought that actual infinity was either a contradictory concept or one that was beyond the grasp of human intelligence, although there were a few notable exceptions who thought otherwise. Mathematicians were generally much more comfortable with the idea of *potential* infinity—that "infinite" mathematical processes were finite but indefinitely continuable. Uses of actual infinity were viewed with suspicion, if not outright hostility. This is clearly exemplified in the following quote from a celebrated letter from Gauss to Heinrich Schumacher:

> But concerning your proof, I protest above all against the use of an infinite quantity as a *completed* one, which in mathematics is never allowed. The infinite is only a *façon de parler,* in which one properly speaks of limits.

Thus it was natural that Cantor's bold theory of actual infinity would be vigorously attacked by many prominent mathematicians and philosophers of the time. Dauben writes:

> For example, among contemporary German philosophers opposed to the idea of completed infinities, Cantor singled out John Frederlck Herbart and Wilhelm Wundt as prime offenders. Their preoccupation with potential infinities precluded any satisfactory discussion of the actual infinite. In a letter to the Swedish mathematician and historian Gustav Eneström, Cantor summarized his opposition as follows:
>
> > All so-called proofs against the possibility of actually infinite numbers are faulty, as can be demonstrated in every particular case, and as can be concluded on general grounds as well. It is their πρῶτον ψευδος that from the outset they expect or even impose all the properties of finite numbers upon the numbers in question, while on the other hand the infinite numbers, if they are to be considered in any form at all, must (in their contrast to the finite numbers) constitute an entirely new kind of number, whose nature is entirely

> dependent upon the nature of things and is an object of research, but not of our arbitrariness or prejudices.
>
> Herbart was particularly open to Cantor's criticism. By defining the infinite in terms which admitted only potential forms of infinity, there was no way Herbart could have consistently allowed the idea of a completed or an actual infinity. Cantor believed that the actual infinite had to be studied without such arbitrariness or prejudice. *(Dauben, 1979, pp. 124–125)*

It should be noted that Cantor himself did not study infinity "without such arbitrariness or prejudice." In fact he often committed types of transgressions similar to those of which he accused Herbart and Wundt; in particular he assumed the existence of actual infinities in his arguments against potential infinities. He was also so entrapped by his concept of the infinite that he failed completely to understand other "infinite-like" concepts such as infinitesimals, and gave wholly confused and fallacious arguments against them.

Cantor's theory of sets was not generally accepted by mathematicians of the nineteenth century. It was seen as "metaphysical," its methods of proof were often vague, and some of its procedures of proof were very controversial and highly questionable. Today it is almost universally accepted among mathematicians, and it has become a necessary component in much of twentieth century mathematics. Most of the pertinent nineteenth century objections to it—that is, the ones that kept it initially from being accepted—were more keenly formulated in the early part of the twentieth century; and to this day they have not been adequately answered.

At the heart of the objections were meaningfulness issues about infinity and its uses. Mathematicians needed infinity for the type of mathematics they found most interesting, but they did not want to consider instances of it as "real," completed mathematical entities. Instead, they wanted to use it to discover truths about what they considered to be the real mathematical entities. Cantor provided them with a rich collection of techniques for using infinities. These techniques proved so fruitful that it became imperative to incorporate them into mainstream mathematics and find adequate foundations for their use so that, in the often quoted words of David Hilbert, "No one shall drive us out of the paradise which Cantor has created for us." Zermelo's 1908 axiomatization of set theory made considerable progress in this direction. However, it incorporated much more of Cantor's infinitistic theory than is needed for mathematical applications. It appeared that once mathematicians accepted some of Cantor's methods based on actual infinities, they were willing to accept all of them. This complete acceptance provided for very powerful methods of proof while avoiding the painful task of deciding which infinitistic procedures were valid and which were not. But it also created many meaningfulness problems about applying such infinitistic mathematics to empirical and qualitative phenomena. Many of these were discussed and debated in the early part of the twentieth century, but interest in such matters seems to have waned after that, probably owing to the combination of lack of progress in resolving the existing problems and the inability to produce new and interesting issues.

One of the most debated issues to come out of Cantor's theory of sets was the axiom of Choice. This principle, which was carefully and explicitly formulated by Zermelo in 1908, asserted that for each nonempty set S of nonempty sets there exists a set T that is formed by taking one element from each set in S. The use of this axiom produced several counter-intuitive mathematical results. A discussion of these and their relationship to meaningfulness issues are presented in Chapter 8.

2. Intuitive Theories of Meaningfulness

2.1 Overview

This chapter presents a review of some intuitive concepts of meaningfulness and measurement that have appeared in the literature. The adjective "intuitive" is used here because the concepts were designed primarily for providing new tools of immediate use to working scientists, and for the most part lacked sound philosophical and foundational justifications for their employment. For this context "intuitive" should not be construed to be "non-mathematical" or "non-formal," because most of the theories considered in this chapter are quite mathematical and formal. The dominant, most used foundational theory of meaningfulness and measurement in the literature is called the "representational theory," and it is discussed in detail in Chapter 5.

A domain X is measured by assigning numbers to elements of X. Usually this can be done in a variety of different proper ways. The proper ways constitute a set S of functions into the reals. The characterization of such sets of functions that arise from measurement is called the *theory of scale types.* The importance and need of a theory of scale types was first recognized in the seminal paper of S.S.Stevens (1946). A summary of the scale type portion of that paper is presented in Section 2.2. A more systematic presentation of scale types is given in Section 2.3, where relatively recent results about possible scale types are described. Stevens (1948) also introduced a concept of meaningfulness, and Section 2.4 discusses two intuitive examples of Stevens' meaningfulness concept.

Sections 2.5 through 2.7 and parts of Section 2.8 are concerned with developing methods similar to dimensional analysis in physics (Section 1.5) that are applicable to the social, behavioral, and economic sciences. Because these latter sciences encounter a wider range of scale types than in physics, they are often forced to make different kinds of assumptions about the relationships of variables of interest. Section 2.5 discusses Luce's influential 1959 paper, "On the possible psychophysical laws," which was the first systematic attempt to generalize principles inherent in dimensional analysis in ways that were applicable to non-physical sciences. Some of Luce's ideas in that paper were much criticized in the literature, primarily because they appeared to treat an inherently narrow situation as a rather general one. Sections 2.6 and 2.7 present ideas of Falmagne and Narens and of Roberts and Rosenbaum on how to link methods used by Luce to meaningfulness concepts and thereby produce mathematically sharper and philosophically more rigorous results. Section 2.8 presents applications of previous meaningfulness approaches.

Section 2.9 briefly summarizes the main strengths and weaknesses of the intuitive approaches discussed in the chapter.

Convention 2.1.1 In order to minimize the possible distortion of the views of the scholars presented in this book, frequent quotations from relevant papers are given about what they thought they were accomplishing and how they saw their research related to the literature of the time. To make these passages more readable in the present context, minor modifications are often made in notation, and so forth, so that the resulting passages conform more closely with the notation and conventions adopted throughout the book. For exact quotes the reader should consult the original papers.

2.2 Stevens' Theory of Scales and Meaningful Statistics

2.2.1 Stevens' Theory

Pressed by concerns that measurement in psychology was fundamentally different than in physics, S.S.Stevens, a psychologist, published in 1946 in the journal *Science* a conceptualization of measurement that was to dominate the thinking about measurement in the behavioral and social sciences for many years. This article was a response to the view held by many prominent scientists of the time that all strong forms of measurement—that is, those that consist of more than just counting or ranking—must in some foundational way incorporate methods of measurement similar to those of the physical sciences.

In his article, Stevens makes the following comment:

> For seven years a committee of the British Association for the Advancement of Science debated the problem of measurement. Appointed in 1932 to represent Section A (Mathematical and Physical Sciences) and Section J (Psychology), the committee was instructed to consider and report upon the possibility of "quantitative estimates of sensory events"—meaning simply: Is it possible to measure human sensation? Deliberation led only to disagreement, mainly about what is meant about the term measurement. An interim report in 1938 found one member complaining that his colleagues "came out by the same door as they went in," and in order to have another try at agreement, the committee begged to be continued for another year.
>
> For its final report (1940) the committee chose a common bone for its

> contentions, directing its arguments at a concrete example of a sensory scale. This was the Sone scale of loudness (S.S.Stevens and H.Davis, *Hearing*. New York: Wiley, 1938), which purports to measure the subjective magnitude of an auditory sensation against a scale having the formal properties of basic scales, such as those used to measure length and weight. Again 19 members of the committee came out by the routes they entered, and their view ranged widely between two extremes. One member submitted "that any law purporting to express a quantitative relation between sensation intensity and stimulus intensity is not merely false but is in fact meaningless unless and until a meaning can be given to the concept of addition as applied to sensation" (Final Report, p. 245).
>
> It is plain from this and from other statements by the committee that the real issue is the meaning of measurement. This, to be sure, is a semantic issue, but one susceptible of orderly discussion. Perhaps agreement can better be achieved if we recognize that measurement exists in a variety of forms and that scales of measurement fall into certain definite classes. These classes are determined both by the empirical operations invoked in the process of "measuring" and by the formal (mathematical) properties of scales. Furthermore—and this is of great concern to several of the sciences—the statistical manipulations that can legitimately be applied to empirical data depend upon the type of scale against which the data are ordered. *(Stevens, 1946, pg. 667)*

Stevens then goes on to define "measurement" and classify the kinds of scales that can result from measuring:

> Paraphrasing N.R.Campbell (Final Report, p. 340), we may say that measurement, in the broadest sense, is defined as the assignment of numerals to objects or events according to rules. The fact that numerals can be assigned under different rules leads to different kinds of scales and different kinds of measurement. The problem then becomes that of making explicit (a) the various rules for the assignment of numerals, (b) the mathematical properties (or group structure) of the resulting scales, and (c) the statistical operations applicable to measurements made with each type of scale.
>
> Scales are possible in the first place only because there is a certain isomorphism between what we can do with the aspects of objects and the properties of the numeral series. In dealing with the aspects of objects we invoke empirical operations for determining equality (classifying), for rank-ordering, and for determining when differences and when ratios between the aspects of objects are equal. The conventional series of numerals yields to analogous operations: We can identify the members of a numeral series and classify them. We know their order as given by

> convention. We can determine equal differences, as 8−6=4−2, and equal ratios, as 8/4=6/3. The isomorphism between these properties of the numeral series and certain empirical operations which we perform with objects permits the use of the series as a *model* to represent aspects of the empirical world.
>
> The type of scale achieved depends upon the character of the basic empirical operations performed. These operations are limited ordinarily by the nature of the thing being scaled and by our choice of procedures, but, once selected, the operations determine that there will eventuate one or another of the scales listed in Table[2.1]. *(Stevens, 1946, pp. 667–668)*

scale	basic empirical operations	mathematical group	permissible statistics (invariantive)
nominal	determination of equality	*permutation group* $x' = f(x)$ $f(x)$ means any one-to-one substitution	number of cases mode contingency correlation
ordinal	determination of greater or less	*isotonic group* $x' = f(x)$ $f(x)$ means any monotonic increasing function	median percentiles
interval	determination of equality of intervals or differences	*general linear group* $x' = ax + b$ $[a > 0,\ b \text{ real}]$	mean standard deviation rank–order correlation product–moment correlation
ratio	determination of equality of ratios	*similarity group* $x' = ax$ $[a > 0]$	coefficient of variation

Table 2.1. Stevens' classification of measurement scales

In a footnote, Stevens notes his indebtedness "to the late Prof. G.D. Birkhoff for a stimulating discussion which led to the completion of the table in essentially its present form."

In his paper, Stevens makes the following comments on how to interpre this table:

> In the column which records the group structure of each scale are listed the mathematical transformations which leave the scaleform invariant. Thus, any numeral, *x*, on a scale can be replaced by another numeral, *x*′, where *x*′

> is the function of x listed in this column. Each mathematical group in the column is contained in the group immediately above it.
>
> The last column presents examples of the type of statistical operations appropriate to each scale. This column is cumulative in that *all* statistics listed are admissible for data scaled against a ratio scale. The criterion for the appropriateness of a statistic is *invariance* under the transformations in Column 3. *(Stevens, 1946, pg. 678)*

He goes on to describe the various scales in the table and different kinds of circumstances in which some may appear.

He concludes his paper with:

> To the British committee, then, we may venture to suggest by way of conclusion that the most liberal and useful definition of measurement is, as one of its members advised, "the assignment of numerals to things so as to represent facts and conventions about them." The problem as to what is and is not measurement then reduces to the simple question: What are the rules, if any, under which numerals are assigned? If we can point to a consistent set of rules, we are obviously concerned with measurement of some sort, and we can then proceed to the more interesting question as to the kind of measurement it is. In most cases a formulation of the rules of assignment discloses directly the kind of measurement and hence the kind of scale involved. If there remains any ambiguity, we may seek the final and definitive answer in the mathematical group-structure of the scale form: In what ways can we transform its values and still have it serve all the functions previously fulfilled? We know that the values of all scales can be multiplied by a constant, which changes the size of the unit. If, in addition, a constant can be added (or a new zero point chosen), it is proof positive that we are not concerned with a ratio scale. Then, if the purpose of the scale is still served when its values are squared or cubed, it is not even an interval scale. And finally, if any two values may be interchanged at will, the ordinal scale is ruled out and the nominal scale is the sole remaining possibility.
>
> This proposed solution to the semantic problem is not meant to imply that all scales belonging to the same mathematical group are equally precise or accurate or useful or "fundamental." Measurement is never better than the empirical operations by which it is carried out, and operations range from bad to good. Any particular scale, sensory or physical, may be objected to on the grounds of bias, low precision, restricted generality, and other factors, but the objector should remember that these are relative and practical matters and that no scale used by mortals is perfectly free of their taint. *(Stevens, 1946, pg. 680)*

2.2.2 Comments

1. Stevens has no theory or results as to what kinds of scales are possible; he has only a list of the ones that he found in use at the time of his paper. Because of this, his theory of measurement is somewhat handicapped. In the next section, results are presented that show that much more can be done in classifying and specifying possible scale types.

2. Later, Stevens added two additional scale types to his list: *log-interval* scales, which have as their group of admissible transformations functions on the positive reals of the form $x \to bx^a$, where a and b are positive, and *absolute scales* which have the identity as their only admissible transformation. The log-interval scales get their name from the fact that they form interval scales when transformed by taking logarithms. Because of this, it is natural to consider log-interval scales and interval scales as variants of one another rather than two entirely distinct scale types. This is discussed in more detail in the next section.

3. Stevens, in his paper and other writings, is quite vague about what in general are proper rules for assigning numbers to empirical objects—that is, what should be proper methods of measurement. He gives many specific examples, but this is clearly inadequate for the rigorous founding of a theory of measurement.

2.3 A Formal Theory of Scales

Definition 2.3.1 Let *nominal, ordinal, interval, ratio, log-interval, and absolute scale* be as defined in Table 2.1 and Comment 2 of Subsection 2.2.2.

Let $\mathbb{R}$ denote the set of reals, $\mathbb{R}^+$ the set of positive reals, and $\mathbb{R}^-$ the set of negative reals.

There are six possibilities for the image of a ratio scale S: *(i)* $\mathbb{R}^+$, *(ii)* $\mathbb{R}^-$, *(iii)* $\mathbb{R}^+ \cup \{0\}$, *(iv)* $\mathbb{R}^- \cup \{0\}$, *(v)* $\mathbb{R}$, and *(vi)* $\mathbb{R}^+ \cup \mathbb{R}^-$. To avoid possible ambiguities about which of the six kinds of ratio scales is being referred to, the following convention is adopted:

Convention 2.3.1 Henceforth, unless explicitly stated otherwise, ratio scales as defined in Table 2.1 and Definition 2.3.1 are called *general ratio scales* and those general ratio scales with images $\mathbb{R}^+$ are called ratio scales.

As noted in the comments at the end of the previous section, Stevens' concept of scale is somewhat vague because it does not make clear which "rules" are acceptable for assigning numbers. Throughout the rest of the book, the following definition of "scale" is adopted.

Definition 2.3.2 S is said to be a *(real-valued) scale* on a nonempty set B if and only if S is a nonempty set of functions from B into $\mathbb{R}$. Nonempty subsets of a scale S are called *subscales* of S.

This definition of "scale" is quite general and would seem to include any clear instance of "scale" Stevens had in mind.

Definition 2.3.3 Let S be a scale. Elements of S are called *representations.* Stevens and others have used the term "scale" to stand for both "scale" as in Definition 2.3.2 and "representation" as in Definition 2.3.3. I prefer to explicitly distinguish these two rather different concepts. Thus throughout the rest of this book, I use expressions like "the gram representation of weight" instead of "the gram scale of weight."

The types of scales described by Stevens have special and useful properties that play an important role in recent developments in the theory of measurement. The more prominent of these are described in the following definitions, especially Definition 2.3.13.

Definition 2.3.4 Let S be a scale on B. By definition, the *image* of S is $\{f(x)\,|\,x \in B \text{ and } f \in S\}$.

Definition 2.3.5 Let f, g, and h be functions. Then by definition,

$$f * g = h$$

if and only if the domain of $h \subseteq$ domain of g and for all x in the domain of g, if $g(x)$ is in the domain of f then $f(g(x))=h(x)$. $*$ is called the operation of *functional composition.*

Definition 2.3.6 Let Z be a nonempty set. Then $\langle H, *\rangle$ is said to be a *group of functions* on Z if and only if for all f and g in H, *(i)* the identity function, ι, on Z is in H, *(ii)* if f and g is in H, then $f*g$ is in H, and *(iii)* if f is in H, then f^{-1} is in H. *By convention,* a group of functions is often referred to by its domain, for example, the group $\langle H, *\rangle$ is often denoted by H.

K is said to be a *subgroup* of H if and only if $\langle H, *\rangle$ and $\langle K, *\rangle$ are groups and $K \subseteq H$.

Let $\langle H, *\rangle$ and $\langle K*\rangle$ be groups of functions. Then $\langle H, *\rangle$ and $\langle K, *\rangle$ are said to be *isomorphic* if and only if there exists a one-to-one function γ from H onto K such that for all f and g in H, $\gamma(f*g)=\gamma(f)*\gamma(g)$.

Definition 2.3.7 Let S be a scale. S is said to be *regular* if and only if there exists a group H of functions from the image of S onto itself such that for some fixed f in S

$$S = \{\gamma * f \,|\, \gamma \in H\}\,. \tag{2.1}$$

Note that it follows from H being a group that for all f and g in S,

$$\{\gamma * f \mid \gamma \in H\} = \{\gamma * g \mid \gamma \in H\},$$

and so that the validity of Equation 2.1 is independent of the choice of f in S.

Suppose S is regular, all elements of S have the same range, H and K are groups of functions defined on the image of S, and ι is the identity function on the image of S. Let f be an arbitrary element of S, and suppose

$$\mathcal{S} = \{\gamma * f \mid \gamma \in H\} = \{\kappa * f \mid \kappa \in K\}.$$

Then it easily follows that for all γ in H, κ in K, and f in S that

$$\gamma * f = \kappa * f \text{ iff } \gamma^{-1} * \kappa = \iota,$$

that is, that H=K. This observation justifies the following definition: F is said to be the *scale group* of S if and only if F is a group on the image of S and for some f in S,

$$\mathcal{S} = \{\gamma * f \mid \gamma \in F\}.$$

Definition 2.3.8 Let S be a scale with scale group H. Then S is said to be *homogeneous* if and only if for each r and s in the image of S, there exists f in H such that $f(r)=s$.

As noted earlier, there are six possibilities for the image of a general ratio scale $\mathcal{S}$: (i) $\mathbb{R}^+$, (ii) $\mathbb{R}^-$, (iii) $\mathbb{R}^+ \cup \{0\}$, (iv) $\mathbb{R}^- \cup \{0\}$, (v) $\mathbb{R}$ and *(vi)* $\mathbb{R}^+ \cup \mathbb{R}^-$. For *(i)* and *(ii)*, S is homogeneous, and for the last four it is not; for *(iii)* and *(iv)* it is "homogeneous" except for a single element, 0; for *(v)* it is "homogeneous" on two pieces of its image, $\mathbb{R}^+$ and $\mathbb{R}^-$ and is distinctly "nonhomogeneous" on the remainder, which consists of a single element, 0; and for *(vi)* it is "homogeneous" on two pieces of its image, $\mathbb{R}^+$ and $\mathbb{R}^-$.

Note that interval, log-interval, ordinal, and nominal scales are homogeneous; absolute scales with images having more than one element are nonhomogeneous.

In summary, all scales considered by Stevens are regular, and all except the absolute ones and four of the six species of ratio scales are homogeneous, and the four nonhomogeneous species of ratio scales are "homogeneous" on large subparts of their images.

It is also worthwhile noting that all of the scales that Stevens considered have images that are infinite intervals of the reals.

Definition 2.3.9 Suppose S and T are regular scales with respective scale groups H and K, and suppose all representations σ in S and τ in T are one-to-one functions. Then S and T are said to be *u-conjugate* if and only if u is a one-to-one function from the image of S onto the image of T and

$$K = \{u * \gamma * u^{-1} \mid \gamma \in H\}.$$

S and T are said to be *conjugate* if and only if they are u-conjugate for some u.

It is easy to verify that conjugation is an equivalence relation on the class of regular scales and that conjugate scales have isomorphic scale groups. Because most essential measurement properties of regular scales (e.g., homogeneity) are defined in terms of their scale groups, it follows that conjugate scales share exactly all of these group-defined measurement properties, since they have isomorphic scale groups.

It is also easy to verify that log-interval and interval scales are u-conjugate for u=log.

u-conjugate scales of homogeneous ratio scales for u=log form another important class of measurement scales:

Definition 2.3.10 S is said to be a *translation scale* if and only if *(i)* S is a scale, *(ii)* for each φ and ψ in S there exists r in $\mathbb{R}$ such that $\psi=\varphi+r$, and *(iii)* $\eta+s$ is in S for each η in S and each s in $\mathbb{R}$.

Let S be a translation scale. Then elements of S are called *translation representations* or simply *translations.*

It is natural to ask what kinds of scales there are up to conjugation, and whether the ones described by Stevens are essentially the only interesting scales. Narens (1981a, b) developed a classification system to investigate this issue, and his results together with those of Alper (1983, 1987) indicate that Stevens' classification covers up to conjugation most—but not all—of the theoretically interesting and rich scale types. Their main result is stated in Theorem 2.3.1 below.

Definition 2.3.11 A regular scale S is said to be *ordered* if and only if its scale group consists of order preserving functions.

All of the scales considered by Stevens, except the nominal ones, are ordered. Note that the concept of "ordered scale" is not a scale type but a set of scale types. The following definition specifies a subset of ordered scale types.

Definition 2.3.12 S is said to be a *super-ratio scale* if and only if it is a regular scale and its scale group H satisfies the relation

$$F \subseteq H \subseteq K,$$

where F is the scale group for ratio scales and K is the scale group for loginterval scales. (Note by this definition ratio scales are super-ratio scales.)

It is easy to show that the scale group for a super-ratio scale consists of functions of the form

$$x \to bx^a,$$

where b ranges over the positive reals, and a ranges over a subset of positive reals R such that R is a group under multiplication. Two examples of R are the positive rationals and

$$\{r^n \mid r \in \mathbb{R}^+ \text{ and } n \in \mathbb{I}^+\},$$

where $\mathbb{I}^+$ is the set of positive integers.

The following concepts of homogeneity and uniqueness are vital for the general understanding of scale types.

Definition 2.3.13 A regular scale S is said to be *m-point homogeneous,* where m is a nonnegative integer, if and only if for all $r_1,\ldots, r_m$ and $s_1,\ldots, S_m$ in the image of S, if $r_1<\ldots< r_m$ and $s_1<\ldots<s_m$, then there exists γ in the scale group of S such that $\gamma(r_i)=s_i$ for $i=1,\ldots, m$. A regular scale is said to be *∞-point homogeneous* if and only if it is m-point homogeneous for each nonnegative integer m. A regular scale S is said to be *n-point unique,* where n is a nonnegative integer, if and only if for all γ and δ in the scale group of S, if γ and δ agree at n distinct points of the image of S, then $\gamma=\delta$. A regular scale is said to be *finitely point unique* if and only if it is n-point unique for some nonnegative integer n, and a regular scale is said to be *∞-point unique* if and only if it is not finitely point unique.

Note that 1-point homogeneity corresponds to homogeneity as given in Definition 2.3.8.[1] Also note that ratio scales are 1-point homogeneous but not 2-point homogeneous, interval scales are 1-point and 2-point homogeneous but not 3-point homogeneous, and ordinal and nominal scales are ∞-point homogeneous.

Let S be a regular, ordered scale. It easily follows that if S is n-point unique for some nonnegative integer n, then it is k-point unique for all nonnegative integers $k\geq n$. It is not true in general that $(m+1)$-point homogeneity implies m-point homogeneity. For example, if the image of S has exactly $m+1$ points, then it must be $(m+1)$-point homogeneous but not m-point homogeneous. However, it is easy to show that if the image of S has neither maximal nor minimal elements, then $(m+1)$-point homogeneity implies m-point homogeneity for each nonnegative integer m.

Let S be a regular, ordered scale. Let α be the largest number (possibly ∞) such that S is α-point homogeneous and β the smallest number (possibly ∞) such that S is β-point unique. Then it is easy to show that if the image of S has neither maximal nor minimal elements, then $\alpha\leq\beta$.

Members of the British Committee referred by Stevens (1946) and others have spoken of "strong" and "weak" forms of measurement—ratio and interval scales being exemplars of strong measurement, and ordinal and nominal scales of weak measurement. Using the above concepts of homogeneity and uniqueness, the following appears to me to be a reasonable way of generalizing the concepts of "strong" and "weak" beyond a few specific cases: All ∞-point homogeneous scales are "weak"; all finitely point unique scales are "strong." At the present

[1] 1-point homogeneity is similar to the concept of 1-transitivity used in the theory of permutation groups. However, the concept m-point homogeneity for positive integers $m>1$, does not correspond to the concept of m-transitivity in group theory.

time not enough is known about m-point homogeneous ($m<\infty$), ∞-point unique scales to give insight about their classifications into "strong" and "weak."

The following striking theorem is a direct consequence of results of Narens (1981a, b) and Alper (1983, 1987). The theorem and its implications for measurement theory are discussed in Subsection 5.4.8.

Theorem 2.3.1 Suppose S is a homogeneous, finitely point unique, ordered scale with image on a (possibly infinite) interval of the reals. Then S is conjugate to a super-ratio scale.

This theorem provides some insight into why so few scales types have appeared in science.

2.4 Intuitive Applications of Stevens' Meaningfulness Concept

The intuitive meaningfulness concept has been applied in various ways in the social, behavioral, and economic sciences. (See, for example, de Jong, 1967, for economic applications; Batchelder, 1987, 1992, for social network applications; Roberts, 1986, 1992, for respectively psychological and operation research applications; and Falmagne, 1985, for psychophysical applications.) In this section two elementary uses of the concept are presented. The first is a very simple application to propositions about physical temperature.

2.4.1 An Elementary Application

Consider the following: Suppose Monday's temperature is 10°C, Tuesday's 20°C, Wednesday's 15°C, and Thursday's 30°C. Then it is an empirically true statement that the ratio of Monday's temperature to Tuesday's is the same as the ratio of Wednesday's temperature to Thursday's, all temperatures being measured in Celsius, the ratio being $\frac{1}{2}$. It is also a true statement that the binary relation $\sim c$ defined on ordered pairs (x, y) of temperatures measured in Celsius by

$$(x, y) \sim_C (u, v) \text{ iff } \frac{x}{y} = \frac{u}{v}$$

is an equivalence relation. However, according to physical theory, $\sim c$ does not correspond to an equivalence relation on ordered pairs of qualitative, physical temperatures; that is, it does not correspond to an equivalence relation within physical theory. The reason is that physical theory demands that all qualitative equivalence relations within it, *when appropriately measured,* must be equivalence relations under measurement. Because clearly the ratio of Monday's temperature to Tuesday's is not in the same equivalence class as the ratio of Wednesday's temperature to Thursday's when measured in terms of Fahrenheit (that is, that these ratios, when measured in Fahrenheit, are not equal), it follows that in any physical theory in which Celsius and Fahrenheit are appropriate ways of quantitatively representing temperature, $\sim c$ is not a "physically relevant" equivalence relation on ordered pairs of temperatures. Stevens would say "$\sim c$ is meaningless."

The idea behind this very simple and elementary example, when properly employed, can lead to the rejection of sophisticated mathematical models, as is demonstrated in the psychological example given in Subsection 2.4.2.

Comment

In the above and subsequent examples, "meaningful" is used in a technical manner, and as such, it and its negation "meaningless" should not be given the ordinary English interpretations of "having meaning" and "having no meaning." As a technical term, it is always defined throughout the chapter with respect to some measurement scale. Thus the relation $\sim c$ in the above example is not meaningful with respect to a scale of measurement S that contains the Celsius and Fahrenheit representations, but is meaningful for the scale T that has the Celsius representation as its only representation. In theory and practice, both S and T can be used to properly measure temperature. Their differences are due to differing measurement processes, which supposedly result from different assumptions being made about temperature measurement, and thus for some applications, the choice of the assumptions that give rise to S may be appropriate, and for others, the choice of those that give rise to T. What is obviously lacking in the above and several other intuitive examples of the meaningfulness concept presented in this chapter and throughout the literature is a precise description of "measurement process" and its theoretical linkage to the above definition of "meaningfulness." Chapter 5 provides such a description and linkage for the measurement process that currently dominates most of the theoretical thinking about measurement; additional measurement processes and linkages are discussed in Chapter 7.

2.4.2 An Example Concerning Perceived Risk

Pollatsek and Tversky (1970) proposed a mathematical model of perceived risk, which resulted from a clever deduction from simple qualitative assumptions about how perceived risks might combine. For the purposes of this example, neither the details of the deduction nor motivation for the assumptions need be considered; only the mathematical model they imply is relevant, and the mathematical model needs to hold only for the special case of "simple gambles."

Simple gambles are represented by the notation (a, p, b), where a, b, and p are real numbers and $0<p<1$. Unless stated otherwise, the intended interpretation of the notation *(a, p, b)* is that the subject will receive *a* dollars with probability *p* and *b* dollars with probability $(1-p)$. Pollatsek and Tversky's model says that a subject will perceive gamble g_1 as having more risk than gamble g_2 if and only if $R(g_1)>R(g_2)$, where for simple gambles $g=(a, p, b)$, $R(g)$ is given by the formula,

$$R(g) = tV(g) - (1-t)E(g)\,, \tag{2.2}$$

where $E(g)$ is the expectation of *g*, that is,

$$E(g) = a \cdot p + b \cdot (1-p)\,,$$

and $V(g)$ is the variance of *g*, that is,

$$V(g) = (a - E(g))^2 + (b - E(g))^2\,,$$

and $0< t<1$ is an individual parameter that varies from subject to subject. The intuitive appeal of the model is that the perceived risk of simple gambles increase with variance and decline with positive expectation, and individuals vary in their risk perceptions according to how they trade-off increased variance with increase expectation, and this individual variability is captured by the single parameter *t*. Because of the latter, the size of *t* apparently has a psychological interpretation. However, as is shown next, this cannot be.

For the purpose of argument, let us suppose that Equation 2.2 holds empirically. That is, experiments have been conducted where subjects were asked to judge "Is there more risk in gamble *g* than gamble *h*?", and standard methods have been used to verify that subjects judged that there was more risk in *g* than *h* if and only if $R(g)>R(h)$. The "standard" methods are used to estimate each subject's value of *t* and verify the model across subjects. The issue to be investigated is: *Given these circumstances, does the model given in Equation 2.2 really measure perceived risk?*

In order to decide this, a "thought experiment" of Roskam (1989) is first considered. Let us relabel the simple gamble *(a,p,b)* as ($\$a$, *p*, $\$b$) in order to make explicit that the outcomes are paid in dollars. The notation *(f, c, g, f, d)* will stand for the gamble of receiving *c* guilders with probability *q* and *d* guilders with probability $(1-q)$. In our thought experiment we imagine a subject who preforms the experiment in accord with the model; that is, when presented $g=(\$a, p, \$b)$ and $g'=(\$a', p', \$b')$, the subject judges *g* as more risky than *g'* if and only if $R(g)>R(g')$, where $R(g)$ is as defined in Equation 2.2. It is assumed that the subject can convert from dollars into guilders without difficulty, and that he makes his risk judgments based upon the underlying probabilities and the *value* of the outcomes to him. Thus it is being assumed that the subject's perceived risk of a gamble *g* in dollars is the same as that of *g'*, where *g'* is the translation of *g* into guilders. This latter condition is a reasonable necessary condition for any perceived risk concept for those individuals who can translate from dollars into guilders without difficulty, and thus if this condition fails for such individuals, then it is highly questionable that *perceived risk* is being dealt with.

Assume the subject's value of t is .7, and the exchange rate is $f\,2.56$ to \$1. Consider the gambles given in Table 2.2. Note that at the exchange rate of $f\,2.56$ to \$1, g_2 is a translation of g_1, and g_4 is a translation of g_3, and that to the guilder amounts. Also note that for the guilder gambles g_2 and g_4, the the values $R(g_2)$ and $R(g_4)$ are computed by applying Equation 2.2 directly table uses the same value of t as for dollar gambles.

	gamble	$E(g)$	$V(g)$	t	$R(g)$
g_1	(\$1,.5,-\$.50)	+0.25	0.5625	.7	0.31875
g_2	(f 2.56,.5,-f 1.28)	+0.64	3.6864	.7	2.3885
g_3	(\$0,.5,-\$1.25)	-0.625	0.3906	.7	0.4609
g_4	(f 0.0,.5,-f 3.20)	-1.6	2.56	.7	2.272

Table 2.2. Roskam's gambles

Table 2.2 reveals an inconsistency: According to the model, the subject should perceive g_1 as less risky than g_3, because since $R(g_1)<R(g_3)$. Therefore, since g_2 is a translation of g_1 and g_4 is a translation of g_3, by assumption, the subject should perceive g_2 as less risky as g_4. However, in Table 2.2 $R(g_2)>R(g_4)$, which by the model yields that the subject should perceive g_2 as more risky than g_4. There is clearly an inconsistency here.

Roskam's thought experiment is based on the following five assumptions: *(i)* the subject perceived risk in accordance with Pollatsek and Tversky's model for gambles in dollars; *(ii)* the risk of the gamble is determined by the probabilities and the value to the subject of the outcomes; *(iii)* the subject can correctly translate gambles from dollars into guilders; *(iv)* Pollatsek and Tversky's model also holds for gambles in guilders; and *(v)* the individual parameter t is the same for gambles in dollars as for gambles in guilders. Let us suppose there are no problems with assumptions *(i), (ii),* and *(iii).* Because Pollatsek and Tversky derived their model from qualitative principles about risk that express exactly the same ideas for gambles in dollars as for gambles in guilders, their model should also apply to gambles in guilders, and thus there should be no problem with assumption *(iv).* This leaves assumption *(v)* as the prime suspect for the difficulty. The negation of *(v),* slightly generalized, says that the value of t varies with the unit of currency. In the context of *(iii),* this implies that the size of t for gambles in dollars does not have a psychological interpretation. Roskam (1989) shows that even if t were allowed to vary with unit of currency, the ratio of t values for a unit of currency of two subjects encounters similar difficulties. Roskam also considers and rejects other options.

The following, which in spirit is similar to Roskam (1989), reveals some of the difficulties inherent in Pollatsek and Tversky's model.

First we make assumption *(ii)* above more explicit: Let U be a function that converts money into its value to the subject. Thus by assumption *(iii)*, $U(\$x)=U(f\ 2.56x)$. To eliminate certain degenerate cases, we will also make the very weak assumption that $U(\$x)>U(\$y)$ for $x>y$. Applying Pollatsek and Tversky's theory to subjective values instead of monetary amounts then yields for $g=(a, p, b)$, where a and b are units of any currency and p is a probability,

$$R(g) = tV^{\star}(g) - (1-t)E^{\star}(g)\,, \tag{2.3}$$

where $R(g)$ is the perceived risk of g, $E^{*}(g)$ is the U-expectation of g, that

$$E^{\star}(g) = U(a)\cdot p + U(b)\cdot(1-p)\,,$$

and $V^{*}(g)$ is the U-variance of g, that is,

$$V^{\star}(g) = (U(a) - E^{\star}(g))^2 + (U(b) - E^{\star}(g))^2\,,$$

and t is an individual parameter that varies from subject to subject. This modified model, as given, does not have any of the problems discussed previously. However, a different kind of problem arises if one looks deeply into the nature of U.

The assigning of values to amounts of money is a measurement process, and it is natural to ask about the scale S to which U belongs. If it is an absolute scale, then U is its only representation, and there is no problem; the size of t can be given a psychological interpretation, and so forth. If, however, S is another scale type, then serious difficulties result, since it should be demanded that conclusions drawn using the representation U should remain the same for all representations in S. This is Stevens' theory of meaningfulness applied to this situation. If it is not to be met, then there should be some explicit reason for rejecting some element of S, and thus this reason can then be used to argue that a subscale S' of S is the proper scale to measure perceived risk on rather than S.

Let us assume for the moment that S is a ratio scale. Substituting rU, $r>0$, for U in Equation 2.3 yields

$$R(g) = t\cdot r^2V^{\star}(g) - (1-t)\cdot rE^{\star}(g)\,.$$

It is impossible for both this equation and Equation 2.3 to hold. The problem is that perceived risk should be viewed as being measured on a scale T and that the function R is just one of the representations in T. This can be taken into account by rewriting the previous equation as

$$R'(g) = t\cdot r^2V^{\star}(g) - (1-t)\cdot rE^{\star}(g)\,, \tag{2.4}$$

where R' is in T. However, this situation suffers the unwanted and unfortunate effect that the relation of equality of risks of gambles is not meaningful; that is, there are gambles g_1 and g_2 such that for R' as defined in Equation 2.4, $R'(g_1)=R'(g_2)$, but for another R'' in T, $R''(g_1)\neq R''(g_2)$. This is shown as follows:

Let g_1 be a gamble so that $V^*(g_1)=1/t$ and $E^*(g_1)=0$ and let g_2 be a gamble such that $V^*(g_2)=1$ and $E^*(g_2)=-1$. Then by Equation 2.4, $R'(g_1)=1$ and $R'(g_2)=1$. Let $U''=tU$. Then U'' is in S. Let R'' be defined by substituting tU for U in computing the variance and expectation in Equation 2.4. Then R'' is in T and $R''(g_1)=t^2$ and $R''(g_2)-t^3+t-t^2$, and because $0<t<1$, simple calculation shows that $R''(g_1)\neq R''(g_2)$.

Thus if the subject's value for money is appropriately measurable in terms of a ratio scale, then the model expressed in Equation 2.3, even if correct empirically at describing the subject's behavior in ranking simple gambles by "perceived risk" is not a correct psychological model of perceived risk. This is because the metatheory of "what correct psychological models should look like" should take into account meaningfulness considerations like those discussed above. For the above example when Equation 2.3 "explains the empirical data," this means that another kind of model is needed—one that "explains the empirical data *in terms of a model that is psychologically correct by the metatheory.*"

One way to obtain such a model is to let U be as above, and for each simple gamble g and each r in $\mathbb{R}$, let

$$R_r(g) = \frac{tV^\star(g)}{r} - E^\star(g)\,, \tag{2.5}$$

where $V^*(g)$ is the rU-variance of g and $E^*(g)$ is the rU-expectation of g. R_r in Equation 2.5 is to be interpreted as a representation in the scale T for perceived risk. It is easy to verify that for all simple gambles g and h and all r and s in $\mathbb{R}$,

$$R_r(g) \geq R_r(h) \text{ iff } R_s(g) \geq R_s(h)\,,$$

thus eliminating the above meaningfulness problems. In the model given by Equation 2.5, it is also easy to verify that for all r in $\mathbb{R}$, $R_r=r.\ R_1$, thus making consistent the assumption that T forms a ratio scale. The ratio scalability of T is not an assumption of Pollatsek and Tversky's model described above; it is, however, implied by the qualitative assumptions about perceived risk from which the model is derived.

2.5 Luce's Possible Psychophysical Laws

2.5.1 Luce's 1959 Paper

In a highly influential article, Luce (1959) proposed methods for determining functional relationships between subjective continua like brightness and loudness and the physical continua underlying them. His basic idea was that the functional relationship should obey a general "Principle of Theory Construction" that incorporated how the variables were measured as a central feature in determining the underlying law. He speculated that his Principle would constrain the psychophysical situation in much the same way as dimensional analysis constrained the possible physical relationships in physics. (See Section 1.5 for an elementary example for dimensional analysis.)

Luce viewed dimensional analysis as a special kind of "substantive theory" relating variables, and tried to generalize it by using a more general principle that is applicable to other cases, particularly psychophysical situations:

> A substantive theory relating two or more variables and the measurement theories for these variables should be that:
>
> 1. *(Consistency of substantive and measurement theories)* Admissible transformations of one or more of the independent variables shall lead, via the substantive theory, only to admissible transformations of the dependent variables.
>
> 2. *(Invariance of the substantive theory)* Except for the numerical values of parameters that reflect the effect on the dependent variables of admissible transformations of the independent variables, the mathematical structure of the substantive theory shall be independent of admissible transformations of the independent variables. *(Luce, 1959, pg. 85)*

In working out specific functional relationships, Luce restricted himself to the case of an unknown function of a single independent variable. He also assumed that both the dependent and independent variables of this relationship were measurable by some combination of ratio, interval, and log-interval scales, and that both variables range over continua. The following is typical of his method of reasoning about this type of situation:

> ...Let $x \geq 0$ denote a typical value of the independent variable and $u(x) \geq 0$ the corresponding value of the dependent variable, where u is the unknown functional law relating them. Suppose, first, that both variables form ratio scales. If the unit of the independent variable is changed by multiplying all values by a positive constant k, then according to the principle stated above only an admissible transformation of the dependent variable, namely multiplication by a positive constant, should result and the form of the functional law should be unaffected. That is to say, the changed unit of the dependent variable may depend upon k, but it shall not depend upon x, so we denote it by $K(k)$. Casting this into mathematical terms, we obtain the functional equation
>
> $$u(kx)=K(k)u(x)$$
>
> where $k>0$ and $K(k)>0$. *(Luce, 1959, pg. 86)*

By similar reasoning, functional equations for other combinations of scale types of variables are arrived at, and these are summarized in Table 2.3.

SCALE TYPES		Functional Equation	Comments
Independent Variable	*Dependent Variable*		
ratio	ratio	$u(kx) = K(k)u(x)$	$k > 0,\ K(k) > 0$
ratio	interval	$u(kx) = K(k)u(x) + C(k)$	$k > 0,\ K(k) > 0$
ratio	log-interval	$u(kx) = K(k)u(x)^{C(k)}$	$k > 0,\ K(k) > 0,\ C(k) > 0$
interval	ratio	$u(kx + c) = K(k,c)u(x)$	$k > 0,\ K(k,c) > 0$
interval	interval	$u(kx + c) = K(k,c)u(x) + C(k,c)$	$k > 0,\ K(k,c) > 0$
interval	log-interval	$u(kx + c) = K(k,c) \cdot u(x)^{C(k,c)}$	$k > 0,\ (k,c) > 0,\ C(k,c) > 0$
log-interval	ratio	$u(kx^c) = K(k,c)u(x)$	$k > 0,\ c > 0,\ K(k,c) > 0$
log-interval	interval	$u(kx^c) = K(k,c)u(x) + C(k,c)$	$k > 0,\ c > 0,\ K(k,c) > 0$
log-interval	log-interval	$u(kx^c) = K(k,c)u(x)^{C(k,c)}$	$k > 0,\ c > 0,\ K(k,c) > 0,\ C(k,c) > 0$

Table 2.3. The functional equations for the laws satisfying the Principle of Theory Construction

SCALE TYPES		Functional Equation	Comments
Independent Variable	*Dependent Variable*		
ratio	ratio	$u(x) = \alpha x^\beta$	$\beta/x; \beta/u$
ratio	interval	$u(x) = \alpha \log x + b,$ $u(x) = \alpha x^\beta + \delta$	α/x $\beta/x; \beta/u; \delta/x$
ratio	log-interval	$u(x) = \delta e^{\alpha x^\beta}$	$\alpha/u; \beta/x; \beta/u; \delta/x$
ratio	ratio	$u(x) = \alpha x^\beta$	$\beta/x; \beta/u$
interval	ratio	*impossible*	
interval	interval	$u(x) = \alpha x + \beta$	β/x
interval	log-interval	$u(x) = \alpha e^{\beta x}$	$\alpha/x; \beta/u$
log-interval	ratio	*impossible*	
log-interval	interval	$u(x) = \alpha \log x + \beta$	α/x
log-interval	log-interval	$u(x) = \alpha x^\beta$	$\beta/x; \beta/u$

Table 2.4. The possible laws satisfying the Principle of Theory Construction (The notation α/x means "α is independent of the unit of x.")

Luce showed that the solutions to these functional equations are highly constrained and yield the "possible laws" in Table 2.4.

2.5.2 Rozeboom's Criticism

Luce's Principle came under attack by Rozeboom (1962), who argued that it was "unsupported by reason" and that it should not be used as a general criterion for lawfulness, because it was inconsistent with certain types of physical laws, for example, the law of radioactive decay, which is given by the formula

$$q = ae^{-bt}, \tag{2.6}$$

where q is a quantity of mass, and t is a time after a specific time t_0, and a and b are empirically determined quantities.

Basically, Rozeboom argued that the only sensible interpretation of Luce's Principle is to say that a substantive theory about the variables x and y, x being taken as independent and y as dependent, is a mathematical function φ such that $y=\varphi(x)$ and for any admissible transformation U of the independent variable x there is a transformation T of the dependent variable y so that

$$T(y) = \varphi[U(x)]\,,$$

or because $y=\varphi(x)$,

$$T[\varphi(x)] = \varphi[U(x)]\,. \tag{2.7}$$

Rozeboom then argued that Equation 2.7 is far too restrictive for many lawful physical situations, and in particular for the law of radioactive decay as expressed in Equation 2.6.

2.5.3 Luce's Reply

In a reply to Rozeboom's criticisms, Luce (1962) agreed that Equation 2.7 essentially summarized his Principle and that it was too restrictive to capture all possible psychophysical laws. He still maintained, however, that the admissible transformations of measurements of variables are still an important part of a law relating them and must be accounted for in any proper, full statement of the law. For cases where Equation 2.7 fails, he argued that this must be done by including in the statement of the law "dimensional parameters" (constants whose values depend on the representations used to measure the variables). Luce comments:

> ... the problem remains whether or not the number of dimensional parameters involved in the statement of a law is of any inherent importance. Some physicists seem to feel that such parameters are undesirable and that their total number in a science should be held to a minimum. My earlier results strongly suggests that the minimum cannot possibly be zero. *(Luce, 1962, pg. 550)*

2.5.4 Some Observations

1. As Subsections 2.5.1 and 2.5.2 make clear, Luce's principle, as interpreted in Equation 2.7, is far too restrictive for determining *all possible* psychophysical laws. However, when Equation 2.7 does apply, Luce's method of analysis, particularly Table 2.3 and its mathematical transition to Table 2.4, can be quite useful, because in many situations Equation 2.7 follows from empirical and/or theoretical considerations.

2. Equation 2.7 is too restrictive because φ occurs on both sides. The most general way of modifying Luce's Principle would be to assume the functional relationship

$$y = \varphi(x)$$

and that if x is admissibly transformed by U, then y will be admissibly transformed by some T so that

$$T(y) = \varphi_{U,T}U(x),$$

which by assuming T is one-to-one and noting $y=\varphi(x)$ yields

$$\varphi(x) = T^{-1}\varphi_{U,T}U(x). \tag{2.8}$$

Luce's principle forces $\varphi=\varphi_{U,T}$.

Without some stronger restriction like Luce's Principle, it would appear that Equation 2.8 is quite useless, because it is satisfied for all functions φ. However, as we shall see, this is a quite misleading way of understanding the situation.

To make matters explicit, let X be the domain of the independent variable, Y be the domain of the dependent variable, P be the set of admissible transformations of the independent variable, T be the set of admissible transformations of the dependent variable, and φ be a function from X into Y. We suppose that the identity transformation ι is in P, and that for each U in P there is a $K(U)$ in T and a function φ_U from X into Y such that

$$K(U)[\varphi(x)] = \varphi_U[U(x)],$$

and $K(\iota)$ is the identity transformation on Y. Let

$$\mathcal{F} = \{\varphi_U \mid U \in \mathcal{P}\}.$$

Then $\varphi \in F$, because $L \in P$. F is called the *lawful family determined by* φ, P, *and* K.

Note that if φ satisfies Luce's principle (Equation 2.7), then $F=\{\varphi\}$.

Because in the above φ can be *any* function from X into Y, the condition of a function belonging to a lawful family puts—unlike Luce's Principle—no constraint on the function. However, in this more general context it is not the function that is the carrier of the law relating the independent and dependent variables, but the family. In fact, the same function φ can belong to quite different lawful families by appropriately choosing P and K; thus much of the information about the lawful relationship cannot in general be expressed by the function φ.

In the case where F is a singleton, $T=\{\varphi\}$, the "law" relating the independent and dependent variables, φ, does not depend on admissible transformations. This makes matters very convenient for communicating scientific information involving $F=\{\varphi\}$. When F is not a singleton, then scientific results involving F can still be communicated in a way that does not depend on admissible transformations of the variables, e.g., communicating by a predicate $\theta(w)$ for which

$$\theta(\gamma) \text{ iff } \theta(\lambda)$$

is true for all γ and λ in F. For example, for the case of radioactive decay, suppose

$$y = \varphi(x) \text{ iff } y = 5e^{-.6x}\,,$$

where y is measured in grams and x is measured in seconds. Transforming x to minutes, then yields the expression

$$y = 5e^{-.01x}\,.$$

Let $\theta(w)$ stand for

> w is a function from X into Y and there exists a positive real number b such that
>
> $$\forall x \forall y[y = w(x) \text{ iff } y = 5e^{-bx}]\,.$$

Then $\theta(\gamma)$ is true for all γ in F. By noting that

$$e^{-bx} = (e^{-b})^{x}\,,$$

θ says that the weight of a quantity (measured in grams) decays as a *particular* exponential function of $-x$ (where x is time) no matter how time is measured (within P). This is clearly a law. If y is measured in kilograms rather than grams, then a related law is true. The only difference between the gram and kilogram formulations is that 5 in the gram formulation is changed to 5/1000 in the kilogram formulation, and thus the weight of a quantity decays as a different exponential function of $-x$. The following law combines the two features: The weight of a quantity decays as an exponential function of $-x$, where x is time, no matter which elements of T and P are used to measure weight and time.

Note that this last version of the law violates the condition that K is a function and y must be measured on $K(U)$ for each U in P. This aspect of Luce's 1959 formulation builds in too many unnecessary and unwanted conventions, as results of the next two sections will show.

3. Luce's intention was to generalize the methods of dimensional analysis in physics, which is based upon ratio scalable independent and dependent variables, to cases of variables of strong scale types. He considered the strong types described by Stevens—ratio, interval, and log-interval. With these types of scales, he could formulate and solve the relevant functional equations.

In view of concepts and results of the previous section, we see that he was essentially on target: One needs homogeneity (or something very much like it) of the independent variable for the Principle to generate a workable functional equation, which will have solutions only when the dependent variable is also homogeneous, thus limiting the strong scales of interest to conjugates of super ratio scales.

4. There is clearly something amiss about Luce's results that lawful functional relationships exist for cases with a ratio scalable independent variable and an interval scalable dependent variable, but no lawful functional relationships exist for cases with an interval scalable independent variable and a ratio scalable dependent variable. After all, if the function f is lawful then so should be f^{-1}. This oddity is examined in Section 5.7, where it is argued using formal theories of meaningfulness and measurement that if f is "meaningful," then it is impossible for f to have a ratio scalable independent variable and an interval scalable dependent variable, thus removing the oddity.

5. A more penetrating analysis into the nature of "the possible psychophysical laws" is given in Subsection 6.4.2. Subsection 6.4.2 provides a qualitative description of the law of radioactive decay. That description, when compared to a formalization of Luce's possible psychophysical laws, shows that the law of radioactive decay and the possible psychophysical laws are functional relationships about very different kinds of measured objects. Subsection 6.4.2 also provides a summary and a discussion of Luce's 1991 reformulation of his theory of possible psychophysical laws.

2.5.5 Generalizations to Several Independent Variables

Luce's 1959 paper was concerned with the case of functions of a single independent variable. This section presents some direct extensions of the mathematical results contained in Table 2.4 to the case of n independent variables. We start with results of Luce (1964) concerning cases where the independent variables are either ratio or interval scalable.

Convention 2.5.1 Throughout the rest of this subsection, for $i=1,\dots, n+1$, let S_i be a scale on I_i, $C=I_1 \times\dots\times I_n$, and u be a function from C into I_{n+1}.

Luce (1964) assumes that for each $i=1,\dots, n+1$, I_i is either the interval of positive reals or the interval of reals, and if I_i is the interval of positive reals, then S_i is a ratio scale, and if I_i is the interval of reals, the S_i is an interval scale. He further assumes that there is a function K from $S_1\times\dots\times S_n$ into S_{n+1} such that

$$u[\varphi_1(x_1),\dots,\varphi_n(x_n)] = K(\varphi_1,\dots,\varphi_n) * u(x_1,\dots,x_n)$$

for all $(\varphi_1,\dots, \varphi_n)$ in $S_1\times\dots\times S_n$ and all $(x_1,\dots, x_n)$ in C. He notes that the above

> ... involves the assumption that the independent variables are indeed independent of each other in two strong senses. First, it is assumed that we may choose any combination of their values, i.e., no constraint

holds among their values. Second, it is assumed that we may choose any combination of their admissible transformations, i.e., no constraint holds among their representations. In physics, the second postulate is described by saying that they are a set of fundamental variables, and the first by saying that they are not related by a physical law. These two notions of independence are distinct. For example, suppose *m* and *v* denote mass and velocity, then if momentum is conserved, *mv*=constant, a constraint holds between the values of *m* and *v* but there is no constraint on the admissible measurement transformations since mass, length, and time have distinct dimensions. Conversely, in a system in which *m* and kinetic energy, $\frac{1}{2}mv^2$, are both treated as independent variables, clearly their values may be selected independently, but an admissible transformation on mass automatically forces one on the energy scale.

Within dimensional analysis, the nature of the constraints on physical laws when the independent variables are all ratio scales but they do not constitute a fundamental set of variables is covered by the so-called *Π*-theorem. So far as I am aware, its analogue for scales weaker than ratio ones has never been developed.

These remarks make it clear that the results to be given are of significance only when we are dealing with a set of fundamental independent variables that are in no way constrained by an empirical law. At present, this appears to mean that they are of significance to psychology only in psychophysical problems in which the independent variables are physical ones whose independence is determined by physical theory. Later, when we understand better what psychological variables are fundamental, our results should play a role in limiting the nature of psychological theories that involve non-ratio scale variables, just as dimensional analysis limits physical and other theories when the variables are on a ratio scale. *(Luce, 1964, Pg. 279)*

Luce (1964) shows the following theorem:

Theorem 2.5.1 Suppose *u* is continuous in and depends on each of its arguments. Then the following three statements are true:

1. If the independent variables and the dependent variable are ratio scales, then there exist $a>0$ and $b_i \neq 0$ such that

$$u(x_1, ..., x_n) = a \cdot x_1^{b_1} \cdots x_n^{b_n} .$$

2. If the independent variables are ratio scales and the dependent variable is an interval scale, then either there exist $a\neq0$, $b_i\neq0$ and *c* such that

$$u(x_1, \ldots, x_n) = a \cdot x_1^{b_1} \cdots x_n^{b_n} + c$$

or there exists $b_i\neq0$ and *c* such that

$$u(x_1, \ldots, x_n) = b_1 \log x_1 + \cdots + b_n \log x_n + c .$$

3. If one or more of the independent variables are interval scales and the remainder are ratio scales, and if the dependent variable is either a ratio scale or an interval scale, then it must be the case that $n=1$ and

$$u_1(x) = ax + b$$

for some $a \neq 0$ and b.

Osborne (1970) extends the above results of Luce (1964) by including cases where the independent variables or dependent variables are measured on log-interval or ordinal scales. Both Luce and Osborne assume that any combination of admissible transformations of the independent variables can be chosen. This restriction is not employed in a rather complete extension of the results of Luce and Osborne given in Aczél, Roberts, and Rosenbaum (1987), which also considers cases where u is not continuous or even measurable.

Luce (1964), Osborne (1970), and Aczél, Roberts, and Rosenbaum (1987) essentially employed principles very close to Luce's 1959 Principle of Theory Construction as the method for the theoretical justification of their methods. The next section presents a related development that weakens the underlying assumptions of Luce (1964) and attempts to provide justifications through intuitive meaningfulness considerations.

2.6 Falmagne's and Narens' Meaningful Quantitative Laws

Falmagne and Narens (1983) presented an approach similar in many respects to Luce's 1959 paper, but with the following six important differences: First, no claims were made about capturing all of the psychophysical laws—only certain specific types of laws were claimed to be captured. Second, the much more important and mathematically interesting case of functions of many variables was considered, rather than—as in Luce's 1959 paper— functions of single variables. Third, Luce's condition about the existence of functions relating admissible transformations of the dependent variable to admissible transformations of the independent variable was deleted. Being able to eliminate this condition is very important in practice, because often one knows directly and precisely the admissible transformations of the independent variables without having similar direct or precise knowledge of the scale type of the dependent variable. For example, psychophysicists often use physical independent variables with their usual physical transformations— multiplications by positive constants—as admissible transformations, and a psychological dependent variable for which they have little direct knowledge as to its admissible transformations. Fourth, the basic unit of analysis was lawful families of mathematical formulae rather than, as in Luce's case, a single mathematical formula that represents a law. Fifth, some justification was given for limiting the focus to cases that are conjugates (Definition 2.3.9) of ratio and interval scales. And finally sixth, there was an explicit realization that under certain types of theories of measurement, the "lawful relationships" considered correspond exactly to the Erlanger concept of meaningfulness.

Falmagne and Narens stated their aim as follows:

> There is a view of quantitative science which goes roughly as follows. In a first stage, data are collected. Next, these data are summarized and organized along the lines of a mathematical theory, which provides a temporary explanation. Such an explanation never fits the data perfectly. The discrepancies between theory and data suggest alternative theories, and further experiments. Science pursues its course towards an increasingly reliable description and explanation of the world.
>
> Such a view, even though certainly largely correct, only gives a simplified picture of the quantitative approach to science. In particular, there is a slow but steady recognition of the odd fact that the language itself which we use in our quantitative description of the world, conditions in a subtle way the image that we obtain. To establish with some hope of accuracy the relative importance in this image of our particular quantitative language, and of the data that it purports to explain, raises problems of considerable difficulty.
>
> Our paper aims at providing a specific contribution to that enterprise. It will be shown, through a detailed discussion of an important, exemplary case, that if we observe certain natural and plausible conventions regarding the interplay between changes of representations of the relevant empirical variables and the invariance of the (unknown) empirical law relating them, then the possible forms of the law are extremely limited. The precise formulation of this result suggests new ways of approaching empirical data and of formulating and testing mathematical models. In spirit at least, this program can be seen as bearing some resemblance to one enunciated by Luce in his influential 1959 paper, "On the Possible Psychophysical Laws." *(Falmagne and Narens, 1983, pg. 287)*

They considered a few specific examples and then concluded that

> … the following three concepts are intimately interrelated:
>
> *(i)* the admissible transformations of the variables (scales) entering into an equation describing some empirical law; *(ii)* the mathematical expression of this equation; *(iii)* the invariance of this equation under the admissible transformations of the variables.
>
> The approach taken in this paper is to assume that we are dealing with an empirical situation that is governed by an empirical law of which we know a little of its mathematical form and a little of its invariance properties, but a lot about the structure of the admissible transformations of its variables, and use this information to greatly delimit the possible equations that express this law. *(Falmagne and Narens, 1983, pg. 289)*

Falmagne and Narens considered empirical situations in which a (numerical) output quantity, q, is a function of two (numerical) input quantities b and x. They assumed that b and x range over (possibly infinite) intervals of real numbers, B and X respectively. They called groups of admissible transformations on b and x "scales." This usage is consistent with that of Definition 2.3.2 if one considers only regular scales on B and X with respective images B and X, which Falmagne and Narens did. In such cases, the scales can be identified with their scale groups, and in this section this practice is generally followed.

Let F and H be scales for b and x respectively. Let ι_B and ι_X be the identities of F and H respectively. Without loss of generality (since F and H are groups), Falmagne and Narens assume that the initial numerical codings of b and x are measured on ι_B and ι_X respectively. (Often, to simplify notation, the symbol "ι" will be used to stand for "ι_B" or "ι_X" or both.) The value of the output function, q, is in general not to be found (directly) in the data. Instead, it is assumed to be an unknown function of b and x. This function may of course change with the representations in F and H used to measure the input variables. In order to explicitly take this into account, the notation

$$M_{f,h}[f(b), h(x)]$$

will be used to denote the value of the function when b is measured by f and x by h. The function

$$M_{\iota,\iota}[\iota(b), \iota(x)] = M[b, x],$$

plays a special role throughout this section and will be denoted by

$$M(b, x).$$

It represents the unknown function in terms of the initial codings of the variables b and x.

Sometimes the variables b and x are interrelated so that only certain choices of f in F and h in H allow $M_{f,h}$ to be a valid computation of q. Two of the most important instances of this are *(i)* when a representation in F is a function of one in H (i.e., $f=K(h)$ for some function K) or *visa versa* or *(ii)* when they are totally independent of one another, that is, $M_{f,h}$ is allowable for all f in F and h in H. *(i)* occurs frequently when $B=X$ and $T=H$ and K is the identity function, for example, measuring areas of rectangles in terms of b, their length, and x, their width; and *(ii)* occurs frequently in physics, for example, momentum=mass×velocity.

In order to make the relationship between the representations f and h explicit, the notation $R(f, h)$ is used to denote it. The relation R is sometimes known beforehand; in other instances it has to be deduced.

It is important to realize that when R is a function, the situation is very different from Luce's Principle, because it is a function between representations of *independent variables* rather than, as in Luce's case, a function of representations of *the independent variable* yielding the representations of *the dependent variable.*

So far nothing has been said about the scale type of the dependent variable q. In Falmagne's and Narens' theory, its type need not be given initially, because the scale type of the dependent variable will be a consequence of other assumptions.

Suppose R has been determined (either explicitly or implicitly). Falmagne and Narens call $M_{f,h}$ such that $(f, h) \in R$ a *numerical code*, and

$$\mathcal{M} = \{M_{f,h} \mid (f, h) \in R\}$$

a *family of numerical codes*. M is said to be *homogeneous* if and only if F and H are homogeneous.

Falmagne and Narens investigated four kinds of invariance of M that they called *one-to-one meaningfulness, order meaningfulness, dimensional invariance, and isotonicity.* The first two are defined as follows:

Definition 2.6.1 Let $\mathcal{M} = \{M_{f,g} \mid (f, g) \in R\}$ be a family of numerical codes.

M is said to be *(one-to-one) meaningful* if and only if

$$M_{f,h}[f(b), h(x)] = M_{f,h}[f(c), h(y)] \text{ iff } M_{f',h'}[f'(b), h'(x)] = M_{f',h'}[f'(c), h'(y)]$$

for all b, c in B, x, y in X, f,f' in f', and h, h' in$H$.

M is said to be *order meaningful* if and only if

$$M_{f,h}[f(b), h(x)] \leq M_{f,h}[f(c), h(y)] \text{ iff } M_{f',h'}[f'(b), h'(x)] \leq M_{f',h'}[f'(c), h'(y)]$$

for all b, c in B, x, y in X, f, f' in F, and h, h' in H.

Convention 2.6.1 By convention, it will sometimes be said that a particular numerical code or a particular numerical law is *one-to-one meaningful (ordermeaningful)* to signify that the underlying family M of numerical codes is one-to-one meaningful (order-meaningful).

Because of this convention, the expression "one-to-one meaningful" is often shortened to "meaningful."

Falmagne and Narens made the following comments about this definition:

> We stress that, in one important respect, our definition of "meaningfulness" differs from that most frequently encountered in the measurement literature. A key feature of this definition is that it applies to a family of relations (the family of numerical codes), rather than to a single relation....
>
> Some reflection will probably convince that the concept of meaningfulness, as defined here, represents a rather minimal (yet essential) requirement for a family of numerical codes to be worthy of consideration

for scientific purposes. This may not be obvious. The following remarks may help the reader's examination of this notion.

Any numerical code $M_{f,h}$ is a translation, depending on the chosen scales f and h, of a collection of empirical facts. To be specific, to the Cartesian product $B \times X$ corresponds a set $B_0 \times X_0$ of empirical situations (inputs). Thus, $(\beta, \gamma) \in (B_0 \times X_0)$ is an empirical situation characterized by two aspects that are in general non-numerical. There is also an empirical set E_0 (of outputs), and a mapping

$$(\beta, \gamma) \rightarrow \rho(\beta, \gamma)$$

of $B_0 \times X_0$ onto E_0. The notation $\rho(\beta, \gamma)$ symbolizes the output in E_0 generated by the input (β, γ). We assume that some initial scaling has taken place, involving two real valued mappings

$$\beta \rightarrow \beta', \ \gamma \rightarrow \gamma'$$

respectively of B_0 onto B and X_0 onto X.

From an empirical viewpoint, the critical information is contained in the function ρ. This information should be preserved by any numerical translation. In particular, the fact that two inputs $(\beta, \gamma), (\alpha, \delta)$ generate the same output should be preserved no matter which numerical code $M_{f,h}$ is chosen. Symbolically,

$$\rho(\beta, \gamma) = \rho(\alpha, \delta) \text{ iff } M_{f,h}[f(\beta'), h(\gamma')] = M_{f,h}[f(\alpha'), h(\delta')].$$

Clearly, this leads to the definition of meaningfulness of families of numerical codes adopted here. The definition of order-meaningfulness arises when there is a natural ordering on the set E_0 of outputs. *(pp. 295–296)*

Definition 2.6.2 Let F and H be scale families on B and X respectively, $R \subseteq \mathcal{F} \times \mathcal{H}$ and $\mathcal{M} = \{M_{f,h} \mid (f,h) \in R\}$ be a family of numerical codes. Then M is said to be *dimensionally invariant* if and only if

$$M_{f,h}[f'(b), h'(x)] \leq M_{f,h}[f'(c), h'(y)]$$
$$\text{iff } M_{f,h}[f''(b), h''(x)] \leq M_{f,h}[f''(c), h''(y)] \tag{2.9}$$

for all b, c in B, x, y in X, f, f', f'' in F, and h, h', h'' in H.

Let $M_{f,h}$ be a numerical code in M. Then $M_{f,h}$ is said to be *dimensionally invariant* if and only if Equation 2.9 holds for all b, c in B, x, y in X, f', f'' in F and h', h'' in H.

Let F and H be scale families on B and X respectively, $R \subseteq \mathcal{F} \times \mathcal{H}$, and $\mathcal{M} = \{M_{f,h} \mid (f,h) \in R\}$ be a family of numerical codes and $M_{f,h}$ be in M. Assume the output quantity q is measured on a scale K that has only strictly monotonic representations, and assume that there is a function K on $F \times H$ such that for each $f' \in F$ and each $h' \in H$,

$$K(f', h') = k \in \mathcal{K} \text{ and } k(q) = M_{f,h}[f'(b), h'(x)] \tag{2.10}$$

for all $b \in B$ and $x \in X$. Then it easily follows that $M_{f,h}$ is dimensionally invariant.

Note that Equation 2.10, which generalizes the classical notion of "dimensionally invariant" used in dimensional analysis in physics (see, for example, Causey 1969, or Krantz et. al. 1971), is a two variable version of Luce's Principle discussed in Section 2.5.

Definition 2.6.3 Let F and H be scale families on B and X respectively, $R \subseteq \mathcal{F} \times \mathcal{H}$, and $\mathcal{M} = \{M_{f,h} \mid (f, h) \in R\}$ be a family of numerical codes. Then M is said to be *isotone* if and only if there exists a real valued function M' on $B \times X$ such that for all $(f, h) \in R$,

$$M_{f,h} = m_{f,h} * M'$$

for some strictly increasing, continuous function $m_{f,h}$ mapping the range of M' onto the range of $M_{f,h}$.

Notice that by Definition 2.6.3 there is no loss of generality in assuming $M' = M_{\iota,\iota} = M$, because by the definition of isotonicity,

$$M = M_{\iota,\iota} = m_{\iota,\iota} * M',$$

and thus,

$$M' = m_{\iota,\iota}^{-1} * M,$$

and therefore,

$$M_{f,h} = m_{f,h} * M' = m_{f,h} * (m_{\iota,\iota}^{-1} * M) = (m_{f,h} * m_{\iota,\iota}^{-1}) * M,$$

where $m_{f,h} * m_{\iota,\iota}^{-1}$ is a strictly increasing continuous function mapping the range of M onto the range of $M_{f,h}$.

With these considerations in mind, it is easy to see that isotonicity of M says that any numerical code in M can be obtained through some strictly increasing, continuous transformation of the initial code $M = M_{\iota,\iota}$. This is a particularly nice way of saying that all codes in M "have similar mathematical forms."

Order meaningfulness, dimensional invariance, and isotonicity are highly interrelated but quite distinct concepts, as the next three examples and theorem of Falmagne and Narens (1983) make clear.

The first example is about a family that satisfies isotonicity but fails to satisfy one-to-one meaningfulness, order meaningfulness, and dimensional invariance:

Example 2.6.1 (*Psychophysical choice* 1). Let x and y be numbers representing the light intensities of two visual stimuli. It is assumed that these numbers are measured on the same ratio scale. Let $P(x, y)$ be the probability that x is judged as brighter than y by a subject.

Suppose

$$P(x,y) = F\left[\frac{1+x}{1+y}\right], \tag{2.11}$$

where F is some strictly increasing, continuous function. Note that a critical piece of information is missing in the notation of Equation 2.11: the particular representation used to measure x and y is not indicated. Rewriting Equation 2.11 to explicitly take into account the representations used to measure x and y yields the equation

$$P_\lambda(x,y) = F_\lambda\left[\frac{1+x}{1+y}\right],$$

where $\lambda>0$ symbolizes the representation used to measure x and y, in the sense that $x=\lambda x'$, $y=\lambda y'$, where x', y' represent the initial numerical scaling of the intensities. It is easy to show that the family $\{P_\lambda\}$ of numerical codes is not one-to-one meaningful and is not order-meaningful, because it cannot be the case that

$$P_\lambda(\lambda x, \lambda y) = P_\lambda(\lambda z, \lambda w) \text{ iff } P_{\lambda'}(\lambda' x, \lambda' y) = P_{\lambda'}(\lambda' z, \lambda' w)$$

for all x, y, z, w, λ, and λ' in the interval $(0, \infty)$. It is also easy to check that the family $\{P_\lambda\}$ satisfies isotonicity but not dimensional invariance.

It is worthwhile noting that the expression

$$F_\lambda\left[\frac{1+x}{1+y}\right]$$

is very different from those encountered in physics because the dimensionless quantity 1 is being added to the dimensional quantities x and y. However, any example that violates dimensional invariance must in some way look different from expressions normally found in physics.

The following is an example of a family that satisfies order meaningfulness but fails to satisfy either isotonicity or dimensional invariance:

Example 2.6.2 (*Psychophysical choice* 2) With the probabilities $P_\lambda(x,y)$ having the same meaning as in Example 2.6.1, suppose

$$P_\lambda(x,y) = F_\lambda\left[\frac{\lambda K + x}{\lambda K + y}\right] \tag{2.12}$$

for all x, y, $\lambda>0$, where $K>0$ is a constant and F_λ is a strictly increasing, continuous function. Then Equation 2.12 is order-meaningful, because for all x, y, z, w, and $\lambda>0$,

$$P_\lambda(\lambda x, \lambda y) \leq P_\lambda(\lambda z, \lambda w) \text{ iff } F_\lambda\left[\frac{\lambda K + \lambda x}{\lambda K + \lambda y}\right] \leq F_\lambda\left[\frac{\lambda K + \lambda z}{\lambda K + \lambda w}\right]$$
$$\text{iff } \frac{K+x}{K+y} \leq \frac{K+z}{K+w}.$$

This last expression is independent of λ, from which order meaningfulness immediately follows. It is easy to verify that the family $\{F_\lambda|\lambda>0\}$ is neither isotone nor dimensionally invariant.

Example 2.6.3 For all positive *x, y,* and λ, let

$$M_\lambda = x + \lambda y\,,$$

and let $M=\{M_\lambda|\lambda>0\}$. The quantities x and y are measured on the same representation, denoted by λ, that comes from a ratio scale. Clearly the family *M.* is dimensionally invariant, but is neither isotone nor order meaningful.

The following theorem summarizes the logical relationships between ordermeaningfulness, isotonicity, and dimensional invariance:

Theorem 2.6.1 The properties of order-meaningfulness, isotonicity, and dimensional invariance are pairwise independent (none implies another). However, any two of these conditions implies the third.

Proof. Theorem 4 of Falmagne and Narens (1983).

Isotonicity often allows families of numerical codes to be written in simplified ways. For example, suppose *M* is a family of numerical codes that have multiplicative representations, i.e., each $M_{f,h}$ in *M.* can be represented by

$$M_{f,h}(b,x) = F'[\varphi(b)\psi(x)]\,, \tag{2.13}$$

where in general the functions F', φ, and ψ depend on the representations used to measure b and x. If, however, *M* is also isotone, then it may be assumed that φ and ψ are independent of the representations used to measure b and x, since by Equation 2.13 and the definition of isotonicity,

$$M_{f,h}(b,x) = m_{f,h}[M(b,x)] = (m_{f,h} * F)[\varphi(b)\psi(x)]\,,$$

where

$$M(b,x) = F[\varphi(b)\psi(x)]\,,$$

and thus

$$F' = m_{f,h} * F\,.$$

These remarks justify the following definition for isotone families:

Definition 2.6.4 Let F and H be scale families on B and X respectively, $R \subseteq \mathcal{F} \times \mathcal{H}$, and $\mathcal{M} = \{M_{f,h}\,|\,(f,h) \in R\}$ be an isotone family of numerical codes. Then (φ, ψ) is said to be a *multiplicative representation* of *M.* if and only if φ and ψ are continuous functions defined on B and X respectively which take values in the positive reals and are such that for all b in B and all x in X,

$$M(b,x) = F[\varphi(b)\psi(x)]\,,$$

where F is some strictly increasing, continuous function.

Falmagne and Narens show the following theorem:

Theorem 2.6.2 Let F and H be scale families on B and X respectively and $M = \{M_{f,h} | f \in F$ and $h \in H\}$ be an isotone family of numerical codes. Then any two of the following three statements imply the third:

1. M. is order-meaningful.
2. M has a multiplicative representation $(\varphi^{\Theta}, \psi^{\delta})$, where $\Theta > 0$ and $\delta \neq 0$ are constants and both φ and ψ are strictly increasing.
3. F and H are respectively φ-conjugate and ψ-conjugate to ratio scales.

Proof. Theorem 5 of Falmagne and Narens (1983).

The following corollary, which is an immediate consequence of Theorem 2.6.2, is quite useful:

Theorem 2.6.3 Let F and H be two ratio scales and suppose that $M = \{Mf,h | f \in f$ and $h \in H\}$ is an isotone, order-meaningful family of numerical codes. Then there exist constants $\Theta > 0$ and $\delta \neq 0$ such that for all $b \in B$ and all $x \in X$,

$$M(b, x) = K(b^{\Theta} x^{\delta}),$$

where K is a strictly increasing, continuous function.

Falmagne and Narens make the following comment about the last two theorems:

> The kinds of interconnections between variables captured in [Theorems 2.6.2 and 2.6.3] are fundamental throughout much of science. The characterizations presented here are immediate consequences of Section 4 of Narens (1981a), although the methods of proof developed here appear to us to be more straightforward and likely to lead to fruitful generalizations. Other characterizations have been presented in the literature (Narens and Luce 1976; Narens 1981a) for cases where naturally defined operations exist on the variables. *(Falmagne and Narens, 1983, pg. 304)*

Falmagne and Narens show many other results about isotone families of numerical codes. The most applicable are summarized in Table 2.5.

Falmagne and Narens do not see very much of a connection of their theory with that of Luce (1959):

> Another difference is that the bulk of Luce's results concerns a relation between one independent variable and one dependent variable, while, through our definition of a family of numerical codes, we are dealing here with a relation between two independent variables and one dependent variable. This may seem like a technical detail. However, a major focus of our paper…is on the understanding of the relationship between a multiplicative form of a numerical code

Case 1: F and H ratio scales and

$$\mathcal{M} = \{M_{f,h} \mid f \in \mathcal{F} \text{ and } h \in \mathcal{H}\}$$

is isotone. Then

$$M(b, x) = G(b^{\Theta} x^{\delta}), \quad \Theta > 0,\ \delta \neq 0\,.$$

Case 2: $M=\{M_{f,h} | f \in F\}$ is isotone, where F is one of the following scale types:

log-interval	$M(b,x) = F(b^{\Theta} x^{\delta})$	$\Theta > 0,\ \delta \neq 0$
ratio	$M(b,x) = F(\tau b^{\Theta} + \sigma x^{\Theta})$	$\tau, \Theta > 0,\ \sigma \neq 0$
	or $M(b,x) = F(b^{\Theta} x^{\delta})$	$\Theta > 0,\ \delta \neq 0$
interval	$M(b,x) = F(\tau b + \sigma x)$	$\tau > 0,\ \sigma \neq 0.$

Table 2.5. Representations of isotone families of numerical codes

and meaningfulness. For obvious reasons, such a topic has no place in Luce's 1959 paper. *(Falmagne and Narens, 1983, pg. 319)*

This view of Falmagne and Narens is somewhat in error. First, Luce (1959, 1962, 1964) considered his Principle to apply to cases of many independent variables. Indeed, he formulated his Principle so that the techniques of dimensional analysis of physics—which generally are applied to situations with several independent variables—could be extended to situations that did not rely on the ratio scalability of independent and dependent variables. Second, there are several ways of employing Falmagne's and Narens' results to improve upon of those of Luce (1959) for the case of a single independent variable. (See, for example, the communication from Luce to Roberts and Rosenbaum quoted at the end of Section 7).

Falmagne and Narens saw their theory more connected with the intuitive concept of meaningfulness prevalent in the literature. Their reasons were somewhat complicated:

They began with the following observation about the popular meaningfulness concept of the time:

> The most widely accepted usage of the word "meaningful" is that given by the following informal definition: "A statement involving (numerical) representations is meaningful if and only if its truth or falsity is unchanged under admissible transformations of all the representations in question" *(Roberts 1979, pg. 59)*
>
> The sentence,
>
> Φ: The ratio of Stendhal's weight to Jane Austen's on July 3, 1814 was 1.42,
>
> has been called "meaningful" since its truth value is the same for whatever representation is used to measure weight. A difficulty with the definition is that the expression "involving numerical representations" is unclear.

> The fact is that representations can be "involved" in more than one way in a numerical statement. In Φ for instance, a particular representation has been used to measure the weights of Jane Austen and Stendhal. However, that representation is not mentioned in the statement. Can a representation be "involved" without being mentioned? To illustrate the ambiguity, it is useful to contrast two interpretations of Φ, both of which make use of a ratio scale family F for the measurement of weight. *(Falmagne and Narens, 1983, pg. 319)*

Falmagne and Narens then proceed as follows: They first suppose that the initial scaling has been chosen so that the identity is the pound representation for weight. Then they give the following two interpretations:

First interpretation. The sentence Φ implicitly defines a numerical relation T such that:

$T(b, x)$ iff b is Stendhal's weight, x is Jane Austen's weight, and $b/x=1.42$.
Because for all $f \in F$ and all b and x in $\mathbb{R}^+$

$$T(b,x) \text{ iff } T[f(b), f(x)],$$

T is meaningful in the sense of Roberts (1979) mentioned in the above quote of Falmagne and Narens (1983).

Second interpretation. Consider the following family of relations, where $f \in F$:

$T_f(b,x)$ if and only if b and x are respectively Stendhal's and Jane Austen's weights, measured on representation f; moreover, $b/x = 1.42$.

If $T'(f, b, x)$ is written for $T_f(b, x)$, then it is clear that this latter relation is different from the T in the first interpretation, because T is a first order relation between numbers, whereas T' is a higher order relation between functions of numbers and pairs of numbers.

Since F is a ratio scale family, it follows that

$$T_f[f(b), f(x)] \text{ iff } T_f[h(b), h(x)] \tag{2.14}$$

for all b and x $\mathbb{R}$ in and all f and h in F. Note the strong resemblance between Equation 2.14 and the defining property of a meaningful family of numerical codes. Natural generalizations of the definitions of "families of numerical codes" and "(order-) meaningfulness" given below in Definition 2.6.5 will make "meaningful" the family of relations,

$$\mathcal{T} = \{T_f \mid f \in \mathcal{F}\}.$$

Under this interpretation, calling Φ "meaningful" is a harmless abuse of language.

Falmagne and Narens make the following comment about these two interpretations:

In this example, it does not matter which of the two interpretations is adopted since both lead to "meaningful" Φ with the same truth value. This might suggest that there is no essential discrepancy between the two concepts of "meaningfulness." Such a conclusion, however, would overlook an essential difference between the two interpretations, which in other situations could lead to serious misunderstandings. The definition below [Definition 2.6.5], which generalizes the notions of dimensional invariance, (order-) meaningfulness and isotonicity, emphasizes the distinction. *(Falmagne and Narens, 1983, pg. 320)*

Definition 2.6.5 For $1 \leq i \leq n$, let F_i be scales with domains B_i and $R \subseteq \mathcal{F}_1 \times \cdots \times \mathcal{F}_n$ For each $\alpha=(f_1,\ldots, f_n)$ in R, let T_α be an n-ary relation, and let

$$\mathcal{T} = \{T_\alpha \,|\, \alpha \in R\}\,.$$

Then T in T is said to be *meaningful in the first sense, or l-meaningful, (with respect to R)* if and only if

$$T(a_1,\ldots,a_n) \text{ iff } T[h_1(a_1),\ldots,h_n(a_n)]\,,$$

for all $(h_1,\ldots, h_n)$ in R and all ai in Bi, for $i=1,\ldots, n$. By extension, the family T is said to be *meaningful in the first sense, or* 1*-meaningful,* if and only if each of its relations are 1-meaningful.

T is said to be *meaningful in the second sense, or 2-meaningful,* if and only if for all $\alpha = (f_1,\ldots,f_n)$, $\beta = (h_1,\ldots,h_n)$ in R and all a_i in B_i for $i=1,\ldots, n$,

$$T_\alpha[f_1(a_1),\ldots,f_n(a_n)] \text{ iff } T_\beta[h_1(a_1),\ldots,h_n(a_n)]\,.$$

And T is said to be *degenerate* if and only if $T_\alpha=T_\beta$ for all α, β in R.

The following generalization of Theorem 2.6.1 shows the interrelationships of the meaningfulness conditions given in Definition 2.6.5.

Theorem 2.6.4 The three properties of 1-meaningfulness, 2-meaningfulness, and degeneracy are pairwise independent. However, any two of these three properties implies the remaining property.

Proof. Theorem 9 of Falmagne and Narens (1983).

Falmagne and Narens make the following comment about the above mentioned meaningfulness properties:

We suspect that most measurement theorists would choose 1-meaningfulness as the natural interpretation of Roberts' definition. This particular version of the concept of "meaningfulness" certainly dominates the literature. As an illustration, one important example is discussed below. *(Falmagne' and Narens, 1983, pg. 322)*

The important example referred to is an additional kind of meaningfulness condition, namely invariance of qualitative relations under automorphisms of the qualitative structure. Falmagne and Narens note than when the qualitative structure is "measured" by a real-valued scale, then this scale often has a scale group such that 1-meaningfulness of quantitative relations with respect to the scale group produces through isomorphism essentially the same meaningfulness concept as invariance of qualitative relations under automorphisms of the qualitative structure. They call 3-*meaningful* those qualitative relations that are invariant under automorphisms of the qualitative structure.

They comment:

> The importance of this concept—under the guise of 1-meaningfulness, 3-meaningfulness or dimensional invariance—is widely recognized by measurement theorists (see in particular Luce 1978) and will not be denied here. Whether it completely captures the notion of "meaningfulness" is less clear, and we shall argue that it does not.
>
> Two features of 2-meaningfulness as a general "meaningfulness" concept are especially worthy of note in this connection.
>
> 1. In the form of one-to-one or order-meaningfulness for families of numerical codes, it provides a constraint which is strong and yet very reasonable (in fact practically unescapable, as shown in our remarks ...).
>
> 2. In the same context, this concept permits the consideration of empirical laws of the form
>
> $$M_{\lambda,\mu}(a,x) = F_{\lambda,\mu}[(a+\lambda k_0)^{\beta}(x+\mu k_1)^{\delta}], \tag{2.15}$$
>
> in which $F_{\lambda,\mu}$ is a strictly increasing function, $a,x \in Re^+$ are variables, β, k_0, $k_1 \in Re^+$ and $\delta \neq 0$ are numerical constants, and λ, $\mu \in Re^+$ denote the representations....
>
> In particular, notice that Equation 2.15 does not satisfy dimensional invariance, nor any immediate natural generalization of this concept. As far as we know, this type of law has not been analyzed from a measurement viewpoint. Its main interest lies in the special role played by the additive "dimensional" constants k_0 and k_1 in the equation. A preliminary investigation of such laws, using and extending the techniques of this paper, indicates that a satisfactory analysis is possible. *(Falmagne and Narens, 1983, pg. 323)*

Falmagne and Narens conclude their paper with the following observation about the concept of "numerical law":

> To our knowledge, no one has provided a satisfying definition of "numerical law" for a general scientific context. There seems to be some agreement that laws should be "meaningful" in a sense close to the one used in this paper.

> Almost all the laws one encounters in science are instances of meaningful families of numerical codes. Thus, a natural strategy is to identify "law" with "meaningful families of numerical codes." We would feel uneasy about an uncritical endorsement of such strategy at this stage, since we do not fully understand the exact role of "meaningfulness" in scientific formulation, deduction and communication. Presumably, the strong liking for laws that are meaningful in the sense of this paper is due to a mixture of interrelated practical and theoretical reasons. On the practical side, it is clear that the adoption of a substantial number of nonmeaningful "laws," the form of which would depend on the representation(s) used, would quickly create a scientific Tower of Babel. On the theoretical side, meaningfulness appears to favor "coherent" systems (of quantitative notation of scientific facts) over "incoherent" ones. Here we are thinking about such aspects of a "scientific law" as those emphasized by Luce (1959) in his Principle of Theory Construction. *(pp. 323–324)*

2.6.1 Comments

1. Invariance under a group of automorphisms—3-meaningfulness in Falmagne's and Narens' terminology—is thoroughly investigated as a meaningfulness concept in Chapters 4 and 5.

2. In Chapter 6, a distinction is made between lawfulness and meaningfulness, with meaningfulness being a necessary but generally not a sufficient condition for lawfulness. It should be noted that in addition to meaningfulness concepts the relation R between F and H plays an important role in the derivation of laws through Falmagne's and Narens' methods.

3. Almost all the theory in the literature about the possible psychophysical laws and its applications have formulated laws in terms of individual functions. Falmagne and Narens (1983), which takes a family of functions as the basic concept, is a notable exception to this practice. In Subsection 6.4.5, a generalization of Luce's Principle of Theory Construction (Section 2.5) is formulated in terms a new meaningfulness-like concept and a family of functions.

2.7 Roberts' and Rosenbaum's Possible Psychophysical Laws

Roberts and Rosenbaum (1986) gave an updated and improved version of Luce's 1959 possible psychophysical laws. In their version, Luce's Principle of Theory Construction was replaced with the requirement that certain kinds of assertions about the dependent variable were meaningful. This also allowed for the elimination of scale type assumptions about the dependent variable.

Roberts and Rosenbaum assume that measurement scales arise from the representational approach, which is discussed in depth in Chapter 5. Essentially, in Roberts' and Rosenbaum's version of this approach an empirical or qualitative situation is a structure of the form $\mathfrak{X} = \langle X, R_j \rangle_{j \in J}$ where X is a nonempty set and R_j, $j \in J$, are $n(j)$-ary relations on X, and a "scale" S for $\mathfrak{X}$ is a regular scale (in the sense of Definition 3.4) such that all representations in S are homomorphisms of $\mathfrak{X}$ into some numerical structure $\mathfrak{N} = \langle N, S_j \rangle_{j \in J}$. (A *homomorphism* of $\mathfrak{X}$ into $\mathfrak{N}$ is a function φ from X into N such that for all $j \in J$ and all $x_1, \ldots, x_n(j)$ in X,

$$R_j[x_1, \ldots, x_{n(j)}] \text{ iff } S_j[\varphi(x_1), \ldots, \varphi(x_{n(j)})]\,.)$$

Elements of the scale group for S are called *admissible transformations* (of S), and they are used to define meaningfulness as following: A statement involving a representation f of S is *meaningful* if and only if its truth value is unchanged under admissible transformations of S.

In their analysis of the possible psychophysical laws, Roberts and Rosenbaum employ the scale types listed in Table 2.6. Note that this list is broader than the one used in Luce (1959).

Scale type	*Class of admissible transformations: all functions $\varphi : f(A) \to B$ so that*
Absolute	$\varphi(x) = x$, all x in $f(A)$
Ratio	for some $\gamma > 0$, $\varphi(x) = \gamma x$, all x in $f(A)$
Interval	for some γ, δ real, $\gamma > 0$, $\varphi(x) = \gamma x + \delta$, all x in $f(A)$
Weak interval	for some γ, δ real, $\gamma \neq 0$, $\varphi(x) = \gamma x + \delta$, all x in $f(A)$
Ordinal	φ is strictly increasing on $f(A)$
Nominal	φ is one-to-one
Difference	for some δ real, $\varphi(x) = x + \delta$, all x in $f(A)$
Log-interval	for some $\gamma > 0$ and $\delta > 0$, $\varphi(x) = \gamma x^\delta$, all x in $f(A)$
Weak log-interval	for some $\gamma > 0$, $\delta \neq 0$, $\varphi(x) = \gamma x^\delta$, all x in $f(A)$

Table 2.6. Scale Types

Roberts and Rosenbaum gave the following description of their research objectives:

> Luce's (1959) paper is concerned with a situation where we know one representation f (a physical representation), and we want to relate a second representation g (a psychological representation) to f, by a functional relationship Ψ (the psychophysical function). That is, $g=\Psi(f)$. Luce argues that if we know the scale type of f and the scale type of g, and we make some reasonable continuity assumptions about Ψ then we can derive a general form for Ψ. The basic principle Luce uses, which he calls the *principle of theory construction,* is that an admissible transformation of f should lead to an admissible transformation of g. In response to criticisms

by Rozeboom (1962), Luce (1962) modified the principle to apply only to situations in which the relation between *f* and *g* involves no dimensional parameters whose values can change with admissible transformations. (He also agreed that the 'principle' was not really broadly applicable enough to be called a 'principle'.) The modified principle was applied to more general situations by Luce (1964) and by Osborne (1970). See also the comments by Rozeboom (1962b).

Luce's approach is a reasonable one in the situation where both *f* and *g* are defined, say, by physical or psychological requirements. However, imagine a situation in which we are given one representation *f*, and we want to define a new representation *g* from *f*. We can define *g* in any way we want, by writing down an explicit formula for it. However, we might want certain comparisons using *g* to be meaningful, for instance that $g(a)>g(b)$, or that $g(a)=2g(b)$. This situation might arise, for instance, if we have an economic or psychological measure and we want to define an index number from it. This is a situation of *derived measurement:* the representation *g* is defined in terms of a known representation *f*. We shall observe that knowing the scale type of *f* and requiring that a certain statement involving *g* is meaningful places restrictions on the possible functions Ψ which we can use to define *g* from *f*. *(Roberts and Rosenbaum, 1986, pg. 79)*

The following example illustrates the kind of analysis Roberts and Rosenbaum had in mind. Suppose *S* is a ratio scale on *X* with image $\mathbb{R}^+$ $f\in S$, and $g=\Psi(f)$ is a function onto $\mathbb{R}^+$ and we want

$$g(a) = \alpha g(b) \tag{2.16}$$

to be meaningful for all *a* and *b* in *X* and all positive α. The problem is to determine what possible functions Ψ can be. Using $g=\Psi(f)$, Equation 2.16 becomes

$$\Psi[f(a)] = \alpha\Psi[f(b)]\,. \tag{2.17}$$

The meaningfulness of Equation 2.17 says that if an admissible transformation is applied to *f*, the truth value of Equation 2.17 should not change. Thus, since *S* is a ratio scale, it follows that for all positive *k*,

$$\Psi[f(a)] = \alpha\Psi[f(b)] \text{ iff } \Psi[kf(a)] = \alpha\Psi[kf(b)]\,.$$

Since *f* is onto $\mathbb{R}^+$ this logical equivalence says that

$$\Psi(x) = \alpha\Psi(y) \text{ iff } \Psi(kx) = \alpha\Psi(ky)\,, \tag{2.18}$$

and this latter logical equivalence is true for all positive *x, y, k*, and α. Let $K(k)=\Psi(k)/\Psi(1)$. Then $K(k)$ is positive, since $\Psi>0$. Letting $\alpha=\Psi(kx)/\Psi(k)$ and

y=1 in Equation 2.18, which is permissible since Equation 2.18 is true for all positive x, y, k, and α, we get

$$\Psi(x) = \frac{\Psi(kx)}{\Psi(k)}\Psi(1) \text{ iff } \Psi(kx) = \frac{\Psi(kx)}{\Psi(k)}\Psi(k),$$

or in other words,

$$K(k)\Psi(x) = \Psi(kx) \text{ iff } \Psi(kx) = \Psi(kx),$$

that is,

$$\Psi(kx) = K(k)\Psi(x), \tag{2.19}$$

for all positive k and x. This last equation is one of the functional equations that Luce solves in his 1959 paper under the additional assumption that Ψ is continuous. With this additional assumption,

$$\Psi(x) = \lambda x^{\gamma},$$

where $\lambda>0$. Conversely, it is easy to show that

$$g(a) = \alpha g(b)$$

is meaningful if Ψ has this form and f is a ratio scale.

Roberts and Rosenbaum analyzed situations like the above as follows:

> The main point is that Luce's assumption about the scale type of g is replaced by the assumption that a certain statement involving g is meaningful. This also avoids the 'principle of theory construction', which some authors have considered questionable. Similar results can be obtained in other situations. For instance, if f is a ratio scale, f is onto $\mathbb{R}^+$, and
>
> $$g(a) - g(b) = \alpha[g(c) - g(d)]$$
>
> is meaningful for all a, b, c, d in A, all $\alpha>0$, and
>
> $$g(a) > g(b)$$
>
> is meaningful for all a, b in A, then we can derive the same functional equation which arises in Luce's case of f ratio, g interval. We shall show this below. What these observations led to was the conclusion that the meaningfulness of certain statements involving scales (like g) is essentially equivalent to the assumption that these scales have certain scale types. *(Roberts and Rosenbaum, 1986, pg. 80)*

Let S and T be regular scales. Then by definition, S is said to be *stronger than T* if and only if the scale group of S is a subgroup of the scale group of T. A scale type Δ is said to be stronger than a scale type Γ if and only if each scale in Δ is *stronger than* each scale in Γ.

Roberts and Rosenbaum commented that

> ... An analysis of the proofs in Luce (1959) shows that his results do not require the full assumption about the scale of type *g*, only that *g* has the scale type assumed *or a stronger one.* For instance, in the case where *f* is a ratio scale and *g* is a ratio scale, we can replace the latter assumption by the assumption that *g* is a ratio scale or stronger. To see why, note that the assumption that *g* is a ratio scale is used by Luce to derive equation [2.19] which is then solved by methods of functional equations. In deriving equation [2.19], Luce in fact uses the following assumptions: every admissible transformation of *f* leads to an admissible transformation of *g*; multiplication by a positive constant is always an admissible transformation of *f*; and every admissible transformation of *g* arises by multiplication by a positive constant. He does not use the assumption that every multiplication of *g* by a positive constant is admissible.

Roberts and Rosenbaum then worked out the equivalences between the scale types given in Table 2.6 and sets of meaningful statements. These equivalences are given in Table 2.7[2].

Scale (or stronger)	*Statement*	*Meaningful for*
Absolute	$g(a) = \alpha$	all α
Ratio	$g(a) = \alpha g(b)$	all $\alpha > 0$
Interval	$g(a) - g(b) = \alpha[g(c) - g(d)]$ and $g(a) > g(b)$	all $\alpha > 0$
Weak interval	$g(a) - g(b) = \alpha[g(c) - g(d)]$	all $\alpha > 0$
Ordinal	$g(a) > g(b)$	
Difference	$g(a) - g(b) = \alpha$	all $\alpha > 0$
Log-interval	$\log g(a) - \log g(b) = \alpha[\log g(c) - \log g(d)]$ and $g(a) > g(b)$	all $\alpha > 0$
Weak log-interval	$\log g(a) - \log g(b) = \alpha[\log g(c) - \log g(d)]$	all $\alpha > 0$
Nominal	$g(a) = g(b)$	

Table 2.7. Equivalence between scale types and meaningful statements

Roberts and Rosenbaum made the following remark about the possible connection of their results with those of Falmagne and Narens (1983):

2 *Note:* The table summarizes the relationships between scale types and meaningful statements, i.e., *g* is *a*...scale or stronger if and only if the corresponding statement is meaningful for all *a, b, c, d* in *A* and all a satisfying the restriction given. It is assumed that $g: A \to \mathbb{R}$ in all cases, except $g: A \to \mathbb{R}^+$ in the ratio, log-interval, and weak log-interval scale cases. There is no need to assume *g* is onto.

> …R.D.Luce (personal communication, 22 February 1985) has observed that Corollary 1.1 [i.e., the ratio scale entry in Table 2.7] follows, under some reasonable assumptions, from the work of Falmagne and Narens (1983). Specifically, assume that g is onto $\mathbb{R}^+$, and assume that all admissible transformations are continuous, increasing automorphisms (one-to-one, onto homomorphisms). Suppose $[g(a)=\alpha g(b)]$ is meaningful for all a, b in A and all $\alpha>0$. Define $M_{g,\iota}[g(a),\alpha]$ to be $\alpha g(a)$. Then, to use the terminology of Falmagne and Narens, Luce shows that the family of such functions generated by admissible transformations on g is multiplicative, isotone, and order meaningful. That g is a ratio scale or stronger follows by invoking Theorem 5 of Falmagne and Narens.

2.8 Applications of the Lawfulness/Meaningfulness Concept

In this section, three applications of the lawfulness/meaningfulness concept are presented. The first concerns psychophysical functions obtained through magnitude estimation and a testable assumption that makes them into "psychophysical laws."

2.8.1 Magnitude Estimation

"Magnitude estimation" is a term used to describe numerical representations that result from having a subject give numerical estimations of the magnitudes of stimuli. For example, in a psychological experiment, a subject may be asked to estimate numerically the brightness of various lights of a fixed wavelength. Let the notation $\psi(x)=r$ stand for x is a stimulus item and r is the subject's numerical estimation of x. Let φ be one of the usual representations that physicists use for measuring physical intensity of light. Note in this case that the subject gives her response to the stimulus x; she does not know or care which numerical representation the experimenter is using to record the physical intensity of the lights.

It is natural to ask, "What is the relationship between φ and ψ?" To formulate this more precisely, define the function Ψ as follows:

$$\Psi(s) = r \text{ iff } \exists x[\varphi(x) = s \text{ and } \psi(x) = r].$$

Then Ψ is the "relationship" between φ and ψ. It is called the "psychophysical function" relating measurements of physical intensity with measurements of psychological brightness. As an idealization, it is supposed that the lights range over all theoretically possible physical intensities, and that the subject has a (theoretical) response for each of these intensities. In this context, the experimental data represent a finite sample of these responses, generally with "errors." With these conventions in mind, Ψ is a function from the positive reals into the positive reals. Throughout this section—unless by context or explicitly stated otherwise—it will be assumed that Ψ is strictly increasing and continuous.

The most common way of describing Ψ from finite data is called "the curve fitting" approach. This approach hypothesizes Ψ coming from some specific well-known set of functions, and uses standard statistical techniques to determine if the experimental data is consistent with the hypothesis. A different method is to try and understand the relationship between φ and ψ algebraically and use functional equation techniques to determine the form of Ψ:

Suppose that Ψ were conjectured to be a power function. Then if *x, y, u,* and *v* are lights such that

$$\frac{\varphi(x)}{\varphi(y)} = \frac{\varphi(u)}{\varphi(v)}, \tag{2.20}$$

then it immediately follows that

$$\frac{\psi(x)}{\psi(y)} = \frac{\psi(u)}{\psi(v)}. \tag{2.21}$$

This implication is testable experimentally, because the experimenter can select stimuli *x, y, u,* and *v* that satisfy Equation 2.20 and see if the subject's responses are consistent with Equation 2.21. Thus if the collected data systematically violated the above implication, then the conjecture of Ψ being a power law could be rejected. The logical equivalence of Equations 2.20 and 2.21 is also testable experimentally. This equivalence implies, through functional equation considerations, that Ψ is a power law:

For all positive β and r, since

$$\frac{r}{1} = \frac{\beta r}{\beta},$$

it follows from the equivalence of Equations 2.20 and 2.21 and the definition of Ψ that

$$\frac{\Psi(r)}{\Psi(1)} = \frac{\Psi(\beta r)}{\Psi(\beta)}.$$

Letting

$$K(\beta) = \frac{\Psi(\beta)}{\Psi(1)},$$

the above equation becomes

$$K(\beta)\Psi(r) = \Psi(\beta r)$$

a well-known functional equation with solutions $\Psi(r)=\lambda r^{\gamma}$ for positive λ and γ.

In practice, the data are quite "noisy," and statistical methods are needed to determine whether or not Ψ should be considered to be a particular kind of function, for example, a power function. Since the "curve fitting" and "functional equation" approaches operate on different kinds of data (values of Ψ versus number of times specific assertions involving Ψ are true), they often use different kinds of data collection techniques and/or statistical analyses. Because of this, for some situations one approach may be easier or more feasible than the other.

In the above paradigm, any function ψ from lights into positive real numbers is theoretically possible. This means that theoretically any possible psychophysical function Ψ could result from the subject's responses. The unrestricted calling of all such functions "laws" seems premature. If not all psychophysical functions are to be considered lawful, then any claim of lawfulness for a particular function should be justified. For the special case of Ψ being a power function, one is tempted to do this by saying its "lawfulness" consists of the logical equivalence of Equations 2.20 and 2.21.1 find this "justification" unconvincing: For what is so special about Equations 2.20 and 2.21? Or put another way, there are many possible numerical statement pairs (or more generally, pairs of sets of numerical statements) involving φ and ψ. Do we want to call "lawful" all the Ψ whose forms are specified by logical equivalence of some pair?

Luce (1959), Falmagne and Narens (1983), Roberts and Rosenbaum (1986), and many others have identified lawfulness with invariance. In the derivation of the power law form of Ψ from the equivalence of Equations 2.20 and 2.21, the scales S and T to which φ and ψ respectively belong were not used. If S is assumed to be a ratio scale—as it is in physics—then Equation 2.20 is meaningful in the sense that its truth value is the same no matter what representation in S is used to measure physical intensity. It is easy to verify that Equation 2.21 is meaningful if and only if it is a subscale of a log-interval scale. If Equations 2.20 and 2.21 are meaningful, then it makes sense to say that the determination of the power law form of Ψ was done in a meaningful way, i.e., by using facts about the physical stimulus (Equation 2.20) that are physically significant, and facts about the psychological stimulus (Equation 2.21) that are psychologically significant.

A similar conclusion is reached by following—at least in spirit—the methods of Falmagne and Narens (1983). They would write $\Psi_{\varphi,\psi}$ for Ψ, since the function depends on the particular representations φ and ψ selected from S and T, and search for conditions that would imply that the family of numer ical codes

$$\{\Psi_{\varphi,\psi} \mid \varphi \in S \text{ and } \psi \in T\}$$

is a set of power functions. As the reader can easily show by the above discussion, such conditions would imply that S and T are subscales of loginterval scales and that Equations 2.20 and 2.21 are meaningful with respect to S and T respectively.

The above lines of reasoning suggest that the "lawfulness" of a power function Ψ consists of the restrictions imposed on it by the meaningfulness and equivalence of Equations 2.20 and 2.21. Note that this is very different from saying that the relationship is "lawful" because Ψ is a, "simple" or "wellbehaved" function.

In the above example, the scale S is known from physics, and the logical equivalence of Equations 2.20 and 2.21 is either confirmed or disconfirmed by the data. The scale T, however, is not known: It is only known that ψ belongs to T. Thus, with the current collected data, it cannot be determined whether or not Ψ is lawful—even for the case of it being a power function. In other words, the determination of the lawfulness of Ψ (from this data) requires an additional *theoretical* assumption involving in one way or another the scale T. For the case where the nature of the lawfulness is captured by the equivalence of Equations 2.20 and 2.21, this additional assumption can be that T is a subscale of a log-interval scale, or equivalently, that Equation 2.21 is meaningful.

Of course, it is highly desirable to convert, when feasible, important theoretical concepts into observational ones. This and other issues involving magnitude estimation and the psychophysical power law are considered in depth in Section 5.8 and Subsection 6.4.4.

2.8.2 Meaningful Averaging of Rating Data

The behavioral and social sciences, as well as athletics. the arts, and other organized social activities, often use data from raters, judges, and so forth, as the basis for ranking objects, performances, or individuals. There are many methods in current use for doing this. Most use some form of arithmetic averaging of individual ratings; but such processes present serious theoretical difficulties. To see this, consider the following example:

Example 2.8.1 Suppose raters R_1, R_2, and R_3 are rating two performances, P_1 and P_2. Further, suppose each rater Ri measures performance on a ratio scale Si and uses representation φ_i from this scale to report his rating. Because of this, $\varphi i(P_j)$ is called "R_i*'s rating* of Pj" Suppose performance P is judged better than performance P'—in symbols $P \succ P'$—if and only if the arithmetic mean of the ratings of P is greater than the arithmetic mean of the ratings of P'. Further, suppose the following data have been collected:

$$\begin{array}{lll} \varphi_1(P_1) = 5 & \varphi_2(P_1) = 5 & \varphi_3(P_1) = 5 \\ \varphi_1(P_2) = 8 & \varphi_2(P_2) = 9 & \varphi_3(P_2) = 1\,. \end{array}$$

Because the arithmetic means of P_1's and P_2's ratings are respectively 5 and 6, it follows by the ranking rule stated above that $P_2 \succ P_1$. However to justify this conclusion—and the rule from which it is derived—more needs to be said about how the φ_i were selected from S_i. This is where the theoretical problems with arithmetic averaging enter. If, for example, it is additionally assumed—as is often done in the social and behavioral sciences—that for $i \neq j$ the selection of φ_i from S_i is independent from the selection of φ_j from S_j, then, in general, the arithmetic mean is a bad choice for determining overall rankings. For example, in the above case it would then follow that R_1 could have used the same representation, φ_1, in giving his ratings, while R_2 could have used representation $2\varphi_2$ instead of φ_2 in giving his. Then the arithmetic mean ranking rule with these new representations would yield $P_1 \succ P_2$, contradicting the previous overall rankings by this rule using the old representations. This sort of inconsistency makes the following conclusion unavoidable: *If the arithmetic mean ranking rule is to be valid in the above situation with the S_i being ratio scales, then the selection of representations from the S_i must be coordinated.*

The coordination of rating representations is a variation of an important problem studied in economics called "the interpersonal comparison of utility." Many prominent economic theorists have concluded that value scales between individuals cannot be validly compared. This idea applied to rating situations says that systematic coordinations of rating representations do not produce—except perhaps fortuitously—valid comparison of values underlying those ratings. This issue is discussed in some detail in Narens and Luce (1983).

The "problem of interpersonal comparisons" is not that individual ratings for some activity can not be coordinated—for they obviously can, for example, in ranking diving performances judges can be encouraged to choose representations for which the "perfect dive" is given the value 10. The problem with using such coordinations as methods for intercomparing values is that they appear to depend on some arbitrary choice; for example, with the above scheme for coordinating diving preferences, there is no guarantee that the (qualitative) magnitude of one judge's value of a perfect dive is the same as another. In fact some philosophers would argue strongly that it could not make any sense to say they are the same. In other words, the coordination of the representations is based on an apparently arbitrary feature—having all judges agree to assign 10 to a perfect dive—rather than an intercomparison of the judged value of dives.

Without validly being able to intercompare values, the arithmetic mean looses most of its intuitive appeal as a plausible statistic for determining overall rankings. This suggests looking at other rules for producing overall rankings for cases where the ranking representations are uncoordinated.

Roberts (1986) and others have suggested the geometric mean as a method of producing overall rankings for rating representations selected from ratio scales, that is, using the rule, $P \succ P'$ if and only if the geometric mean of the ratings for P is greater than that for P'. Since for all positive integers n and positive real numbers $a_1, \ldots, a_n$, $b_1, \ldots, b_n$, $r_1, \ldots, r_n$,

$$\frac{a_1 \cdots a_n}{n} \geq \frac{b_1 \cdots b_n}{n} \text{ iff } \frac{r_1 a_1 \cdots r_n a_n}{n} \geq \frac{r_1 b_1 \cdots r_n b_n}{n},$$

it is easy to see that the geometric mean rule is *meaningful* in the sense that it produces the same overall ranking no matter which individual rating representations are chosen from the ratio scales.

It is of natural interest to ask, "What other ranking procedures are meaningful for uncoordinated rating representations?" The somewhat surprising answer is, "Essentially none for ratings taken from ratio scales, and only a highly specific variant of the procedure of comparing geometric means for ratings taken from homogeneous scales." To demonstrate this, the concept of an uncoordinated ranking function needs to be introduced:

Definition 2.8.1 F is said to be an *uncoordinated ranking function* for scales $S_1, \ldots, S_n$ based on B with images $I_1, \ldots, I_n$ respectively, if and only if F is a strictly increasing, real valued function on $I_1 \times \ldots \times I_n$ such that for all b and c in B, if

$$F[\varphi_1(b), \ldots, \varphi_n(b)] > F[\varphi_1(c), \ldots, \varphi_n(c)]$$

for some $(\varphi_1, \ldots, \varphi_n)$ in $\mathcal{S}_1 \times \cdots \times \mathcal{S}_n$, then

$$F[\psi_1(b), \ldots, \psi_n(b)] > F[\psi_1(c), \ldots, \psi_n(c)]$$

for all $(\psi_1, \ldots, \psi_n)$ in $\mathcal{S}_1 \times \cdots \times \mathcal{S}_n$.

Note that uncoordinated ranking functions satisfy the definition of "dimensional invariance" given in Definition 2.6.2, when the latter is naturally generalized to numerical codes with n independent variables.

Definition 2.8.2 Let F be an uncoordinated ranking function for scales $S_1, \ldots, S_n$ based on B with images $I_i, \ldots, I_n$ respectively. Then F is said to be *solvable* if and only if for all i=1,…, n and all

$r_1, \ldots, r_{i-1}, r, r_{i+1}, \ldots, r_n$ and $r_1, \ldots, r_{i-1}, s, r_{i+1}, \ldots, r_n$ and $t_1, \ldots, t_n$ in $I_i \times \ldots \times I_n$, if

$$F(r_1, \ldots, r_{i-1}, r, r_{i+1}, \ldots, r_n) > F(t_1, \ldots, t_n) > F(r_1, \ldots, r_{i-1}, s, r_{i+1}, \ldots, r_n),$$

then there exists t in I_i such that $r > t > s$ and

$$F(r_1, \ldots, r_{i-1}, t, r_{i+1}, \ldots, r_n) = F(t_1, \ldots, t_n).$$

Solvability is a natural and desirable condition for ranking functions: For if judge i giving rating r to object a causes a to be ranked above b, and judge i giving rating s to a causes b to be ranked above a, then it is natural to consider that judge i could give a rating t between r and s that causes a to be ranked the same as b.

Lemma 2.8.1 Suppose $n \geq 2$ and F is a solvable, uncoordinated ranking function for scales $S_1, \ldots, S_n$ based on B with images $I_1, \ldots, I_n$ respectively that are open (possibly infinite) intervals of real numbers. Then there exists a continuous function F' that is also a solvable, uncoordinated ranking function for scales $S_1, \ldots, S_n$ based on B with images $I_1, \ldots, I_n$ respectively such that for all $(r_1, \ldots, r_n)$ and $(s_1, \ldots, s_n)$ in $I_i \times \ldots \times I_n$,

$$F(r_1, \ldots, r_n) > F(s_1, \ldots, s_n) \text{ iff } F'(r_1, \ldots, r_n) > F'(s_1, \ldots, s_n).$$

Lemma 2.8.2 Suppose $n \geq 2$ and F is a solvable, uncoordinated ranking function for scales $S_1, \ldots, S_n$ based on B with images $I_1, \ldots, I_n$ respectively. Also suppose for $i=1, \ldots, n$, S_i is regular and homogeneous and that for all φ and ψ in S_i, $\varphi(B)=\psi(B)=$an open interval of reals. Then the S_i are conjugate to ratio scales for $i=1, \ldots, n$.

The proofs of Lemmas 2.8.1 and 2.8.2 are straightforward and will not be given here.

Lemma 2.8.3 Suppose $n \geq 2$ and F is a solvable, uncoordinated ranking function for scales $S_1, \ldots, S_n$ based on B with images $I_1, \ldots, I_n$ respectively. Also suppose for $i=1, \ldots, n$, S_i is regular and homogeneous and that for all φ and ψ in S_i, $\varphi(B)=\psi(B)=$an open interval of reals. Then there exist functions $u_1, \ldots, u_n$ from B onto $\mathbb{R}^+$ such that the S_i are u_i-conjugate to ratio scales and for all b and c in B and all $(\varphi_1, \ldots, \varphi_n)$ in $S_1 \times \ldots \times S_n$,

$$\begin{aligned} &F[\varphi_1(b), \ldots, \varphi_n(b)] > F[\varphi_1(c), \ldots, \varphi_n(c)] \\ &\quad \text{iff } (u_1 * \varphi_1(b_1))^{\theta(1)} \cdots (u_n * \varphi_n(b_n))^{\theta(n)} \\ &\quad\quad > (u_1 * \varphi_1(c_1))^{\theta(1)} \cdots (u_n * \varphi_n(c_n))^{\theta(n)}, \end{aligned}$$

for some positive real numbers $\theta(1), \ldots, \theta(n)$.

Proof. The lemma follows from the monotonicity and increasingness of F, Lemmas 2.8.1 and 2.8.2, and extending Theorem 2.6.2 to functions of n variables, which is easy to do by modifying the proof of the corresponding theorem *(Theorem 5, pp. 303–308)* of Falmagne and Narens (1983).

Definition 2.8.3 Let F be an uncoordinated ranking function for scales $S_1, \ldots, S_n$ based on B with images $I_1, \ldots, I_n$ respectively. Then F is said to be *unbiased* if and only if for all permutations π on $1, \ldots, n$ and all $r1, \ldots, r_n$ in $I_1 \times \ldots \times I_n$,

$$F(r_1, \ldots, r_n) = F(r_{\pi(1)}, \ldots, r_{\pi(n)}).$$

Unbiasedness is a desirable property of uncoordinated rating functions. Its inclusion into the hypotheses of Lemma 2.8.3 yields the following theorem.

Theorem 2.8.1 Suppose $n \geq 2$ and F is a solvable and unbiased uncoordinated ranking function for scales $S_1,\ldots, S_n$. Also suppose for $i=1,\ldots, n$, S_i is regular and homogeneous and that for all φ and ψ in S_i, $\varphi(B)=\psi(B)$ =an open interval of reals. Then for some function u, the S_i are u-conjugate to ratio scales, and for all b and c in B and all $(\varphi_1,\ldots, \varphi_n)$ in $S_1\times\ldots\times S_n$,

$$F[\varphi_1(b_1),\ldots,\varphi_n(b_n)] > F[\varphi_1(c_1),\ldots,\varphi_n(c_n)]$$
$$\text{iff } [u * \varphi_1(b_1) \cdots u * \varphi_n(b_n)]^{1/n} > [u * \varphi_1(c_1) \cdots * \varphi_n(c_n)]^{1/n}.$$

Note that S_i being conjugate to ratio scales is a conclusion of Theorem 2.8.1; the hypotheses of the theorem do not even assume that they are of the same scale type. Because of this and the rest of its conclusion, Theorem 2.8.1 becomes a quite powerful tool for delimiting possible ranking rules.

The above discussion and theorem demonstrate the power of the intuitive meaningfulness concept in uncovering serious conceptual flaws in certain types of ranking rules. It also shows how the use of this kind of meaningfulness concept can sometimes lead to a unique solution, given plausible assumptions about the interrelationships of the scales of individual variables.

Meaningfulness considerations involving the merging of individual's scales have been examined in depth in Aczél and Roberts (1989), where very general and comprehensive results for a variety of merging situations are obtained. Theorem 2.8.1 and the above proof was known to Falmagne and Narens in 1981. The first published proof I know of a result like it is in Aczél and Roberts (1989).

2.8.3 A Psychophysical Application

The example in this subsection is more complicated and is presented in a more detailed manner than the other intuitive examples of the chapter. It demonstrates the effectiveness of Falmagne's and Narens' approach to meaningfulness discussed in Section 2.6 in analyzing an actual data set. Because the results and methods of this subsection are not used elsewhere in the book, they may be skipped without loss of continuity.

Table 2.5, which summarizes several meaningfulness results of Falmagne and Narens (1983), can be used to narrow down the class of numerical functions which are candidates for a description of a body of data. Such an application was carried out in detail in Falmagne and Narens (1983), where experiments and empirical data of Pavel (1980) and Iverson and Pavel (1981) are discussed. The following is a description of Falmagne's and Narens' approach to this data set.

The data were from a psychoacoustic experiment in which a subject was required to detect a stimulus—a faint click—embedded in or preceded by a burst of white noise. Three independent variables were considered: the intensities of the click and the noise, and the delay τ between the end of the noise and the click (thus, τ can be negative). A basic notion for analyzing the data is the probability

$$P(x, n, \tau)$$

that a click of intensity x is detected over a masking noise of intensity n, with a delay τ. The data supported the assumption that, over some intervals, $P(x, n, \tau)$ is strictly increasing in x and τ, and is strictly decreasing in n. The data were also consistent with the condition

$$P(x, n, \tau) \leq P(x', n', \tau) \text{ iff } P(\lambda x, \mu n, \tau) \leq P(\lambda x', \mu n', \tau), \tag{2.22}$$

for all positive λ and μ.

In order to place this situation in the context developed in Section 2.6, τ will be (temporarily) fixed, and it will be assumed that $P(x, n, \tau)$ (an abbreviation of $P_{\iota,\iota}(x, n, \tau)$) is the initial code of a family F of numerical codes. The numbers x and n represent measurements of physical intensities on some representation belonging to a ratio scale.

Equation 2.22 says that the initial code is dimensionally invariant (Definition 2.6.2). Making also the reasonable assumption that F is ordermeaningful, it follows by Theorem 2.6.1 that F is isotone. Because of the ratio scale character of the physical intensities, the possible forms for $P(x, n, \tau)$ are highly restricted.

Following Falmagne and Narens (1983), two cases are considered:

Case 1. The intensities of the click and the noise are measured by distinct ratio scales.

In this case, the admissible transformations of the numbers x and n in $P(x, n, \tau)$ are unrelated to one another. By Table 2.5 (or Theorem 2.6.2), the only possible form is

$$P(x, n, \tau) = F_\tau[x/n^{\delta(\tau)}], \tag{2.23}$$

where F_τ is continuous and strictly increasing. Interestingly, Pavel's analysis of his data led him to postulate a form identical to Equation 2.23 but with F not depending on τ, namely,

$$P(x, n, \tau) = F[x/n^{\delta(\tau)}]. \tag{2.24}$$

It will first be shown that Equation 2.24 and the assumption that P varies with τ contradicts order meaningfulness. Fix x and consider the two admissible transformations $n \rightarrow \lambda n$, $\tau \rightarrow \theta\tau$. These transform Equation 2.24 into

$$P_{\lambda,\theta}(x, n, \tau) = F_{\lambda,\theta}[x/n^{\delta(\lambda,\theta(\tau))}],$$

where λ and θ are the scales used to measure n and τ respectively. It then follows from order-meaningfulness and Theorem 2.6.2 that

$$P_{\lambda,\theta}(x, \lambda n, \theta\tau) = (H_{\lambda,\theta} * F)[x/n^{\delta(\tau)}] = F_{\lambda,\theta}[x/(\lambda n)^{\delta(\theta\tau)}]$$

for some strictly increasing, continuous function $H_{\lambda,\theta}$. Letting $x=1$ and

$$G_{\lambda,\theta}(s) = [F_{\lambda,\theta}^{-1} * H_{\lambda,\theta} * F)(1/s)]^{-1},$$

yields

$$(\lambda n)^{\delta(\theta\tau)} = G_{\lambda,\theta}[n^{\delta(\tau)}]. \tag{2.25}$$

Setting $n=1$ in Equation 2.25 then gives

$$\lambda^{\delta(\theta\tau)} = G_{\lambda,\theta}(1). \tag{2.26}$$

Since the right side of Equation 2.26 does not depend upon τ it follows that δ is a constant function. This conclusion, however, is not supported by the data (see Figure 2.1).

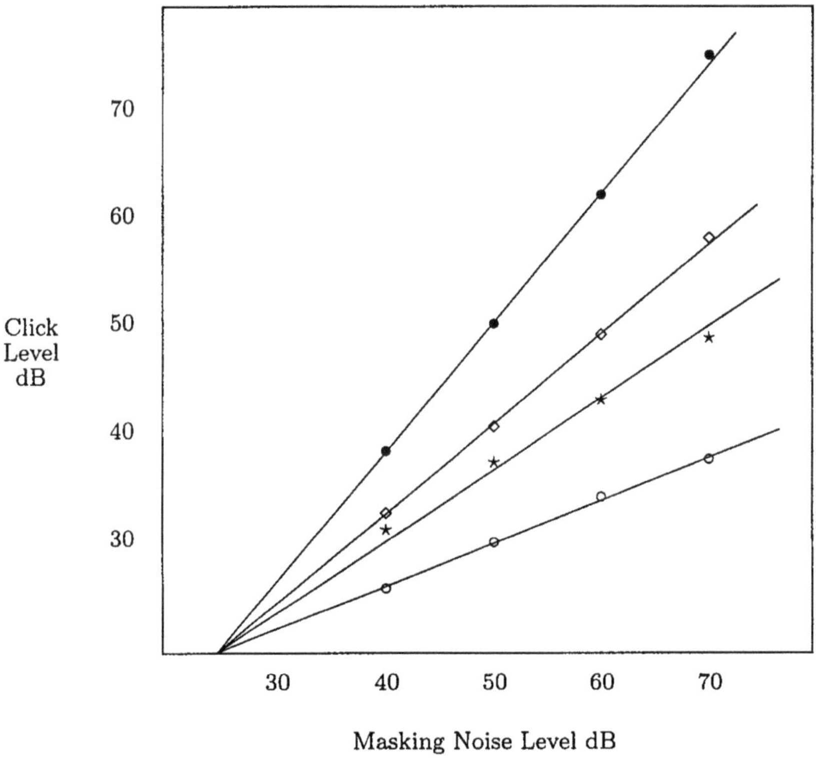

Fig. 2.1. Pavel's data. The straight lines are least square fit of Equation 2.23, plotted in decibel units.

Falmagne and Narens conclude from this that if an order-meaningful expression is to be chosen to describe Pavel's empirical results, it must be of the form given in Equation 2.23, with the function F_τ necessarily depending on τ. In fact,

Equation 2.23 is consistent with the data, a part of which is displayed in Figure 2.1. This is demonstrated as follows:

Applying F_τ^{-1} to both sides of Equation 2.23 yields

$$F_\tau^{-1}[P(x,n,\tau)] = x/n^{\delta(\tau)}\,. \tag{2.27}$$

Setting

$$P(x,n,\tau) = \rho \text{ and } \gamma(\tau,\rho) = F_\tau^{-1}(\rho)$$

and solving for x in Equation 2.27 and noting that x is a function, α, of n, τ, and ρ then yields

$$x = \alpha(n,\tau,\rho) = \gamma(\tau,\rho)n^{\delta(\tau)}\,, \tag{2.28}$$

which by taking logarithms of both sides yields (with obvious notation) in decibel units

$$x' = \alpha'(n,\tau,\rho) = \gamma'(\tau,\rho) + \delta(\tau)n'\,. \tag{2.29}$$

Thus, for any fixed values of ρ and τ, a linear relationship is predicted between the intensity of the click and that of the masking noise—when both are evaluated in decibel units. As indicated by the data in Figure 2.1, this is exactly what was observed by Pavel. Moreover, a fixed point property is apparent in Figure 2.1, which is expressed by the equation

$$\alpha[n_0(\rho),\tau,\rho] = \alpha[n_0(\rho),\tau^*,\rho] \tag{2.30}$$

for some $n_0(\rho)$. In words: n_0 (ρ) is the intensity value of the noise at which the delay τ has no effect on the intensity of the stimulus. This indicates that the parameters of the linear equation (2.30) are linked by a constraint of the form

$$\gamma'(\tau,\rho) = K'(\rho) - \delta(\tau)n_0'(\rho)\,,$$

for some constants $n'_0(\rho)$ and $K'(\rho)$ independent of τ. In terms of the initial units, this gives

$$F_\tau^{-1}(\rho) = K(\rho)[n_0(\rho)^{\delta(\tau)}]^{-1} = \gamma(\tau,\rho)\,,$$

yielding, as a special case of Equation 2.28,

$$\alpha(n,\tau,\rho) = K(\rho)[n/n_0(\rho)]^{\delta(\tau)}\,. \tag{2.31}$$

Note that for order-meaningfulness to hold, $n_0(\rho)$ must vary with ρ. Otherwise—as is easy to check—Equation 2.31 becomes equivalent to Equation 2.23, which is not order-meaningful.

Case 2. The quantities x and n are physical quantities measured on the same representation of a ratio scale.

In this case, the admissible transformations $x \to \gamma x, n \to \theta n$ are linked by $\gamma = \theta$.

By using arguments similar to those in Case 1, one can conclude that there are only the following two possible forms:

$$P(x, n, \tau) = F_\tau[x/n^{\delta(\tau)}] \tag{2.32}$$

and

$$P(x, n, \tau) = F_\tau[x^{\delta(\tau)} + \beta(\tau)n^{\delta(\tau)}]\,. \tag{2.33}$$

Equation 2.32 was analyzed in Case 1. With, as before,

$$\gamma(\tau, \rho) = F_\tau^{-1}(\rho)\,,$$

Equation 2.33 gives

$$x = \alpha(n, \tau, \rho) = [\gamma(\tau, \rho) - \beta(\tau)n^{\delta(\tau)}]^{\frac{1}{\delta(\tau)}}\,,$$

a result for which the linearity in the data of Figure 2.1 will create difficulties.

2.9 Conclusions

Stevens' (1948) intuitive theory of measurement and meaningful statistics and Luce's (1959) intuitive Principle of Theory Construction can be refined and more deeply developed to produce an interesting array of methods that are of direct applicability to scientists of a number of disciplines. Some of the main strengths of the intuitive approach discussed in the chapter are: *(i)* it provides powerful methods of drawing conclusions from the interrelationships of representations of the underlying variables; *(ii)* given that scales were selected correctly, it gives clear and desirable criteria for rejecting certain concepts as relevant; and *(iii)* it formalizes many long-lasting scientific practices and conventions. Two obvious weaknesses of the intuitive approach are: *(i')* it does not explain clearly enough how scales are constructed or scale types are determined; and *(ii')* the definition of meaningfulness upon which it relies is unjustified; for example, no reason is given as to why empirically significant relations should be meaningful. These two weaknesses are addressed in detail Chapters 4 and 5.

3. Axiomatic Set Theory

3.1 Introduction

Axiomatic set theory provides a very powerful means for the precise formulation of mathematical and scientific concepts. Because it is formulated in terms of simple primitive concepts and has a simple axiomatization, it is relatively easy to prove certain results about it that are relevant to meaningfulness issues, especially how definability and invariance concepts are related. It also has enough power to describe all ordinary mathematics and science, making it an ideal vehicle for the formal investigation of the role of mathematics in science.

There are several axiomatic versions of set theory. The one used as a basis for formulating the theories of meaningfulness in this book is called *axiom system* ZFA. It is a variant of the set theories of Zermelo (1908) and Fraenkel (1922b). (In the expression ZFA, "ZF" is for "Zermelo-Fraenkel" and "A" is for "atoms." The axiom system ZFA includes Zermelo's axiom of Choice.)

Within ZFA, concepts of "set" and "nonset" are formulable. The nonsets are called *atoms,* and the collection of atoms A forms a set within the theory. As is customary in set theory, the symbol $\in$ denotes the binary relation of set membership and Ø denotes the empty set. The theory of sets can be formulated entirely in terms of $\in$, A, and Ø, and thus can be thought of as an ordered 4-tuple $\langle V, \in, A, \varnothing\rangle$, where V is the collection of entities with which the theory is concerned. To axiomatize this theory of sets, an appropriate first order language is needed; this is provided in the next section.

3.2 A Language for the Theory of Sets

Definition 3.2.1 The *language* **L(∈, A, Ø)** is the first-order language with equality that is based on the binary relation $\in$ and the individual constant symbols $\boldsymbol{A}$ and **Ø**, That is, **L(∈, A, Ø)** is built from following set of symbols:

(i) individual constant symbols: $\boldsymbol{A}$ and **Ø**;
(ii) variables: $x_1, \ldots, x_i, \ldots$;
(iii) a 2-place nonlogical predicate symbol: $\in$;

(iv) logical symbols: connectives ¬ and Λ; a quantifier ∃; and a 2-place predicate symbol=; and

(v) parentheses: (and).

The nonlogical predicate symbol, ∈, and individual constant symbols, ***A*** and **Ø,** are called the *primitives* of **L(∈, *A*, Ø)**.

The *atomic formulas* of **L(∈, *A*, Ø)** are defined as follows: If *u* is a variable or individual constant symbol and *υ* is a variable or individual constant symbol, then *(u=υ)* and *(u∈ v)* are atomic formulas of **L(∈, *A*, Ø)**, and all atomic formulas of **L(∈, *A*, Ø)** have these forms.

The *formulas* of **L(∈, *A*, Ø)** are defined inductively as follows:

1. Atomic formulas of **L(∈, *A*, Ø)** are formulas of **L(∈, *A*, Ø)**.
2. If Φ and Ψ are formulas of **L(∈, *A*, Ø)**, then ¬ Φ and $(\Phi \Lambda \Psi)$ are formulas of **L(∈, *A*, Ø)**.
3. If Φ is a formula of **L(∈, *A*, Ø)**, and x is a variable, then $(\exists x)\Phi$ is a formula of **L(∈, *A*, Ø)**.
4. All formulas of **L(∈, *A*, Ø)** can be obtained by repeated application of rules 1–3 above.

In interpreting **L(∈, *A*, Ø)**, ¬ is to be interpreted as negation or "not," Λ as conjunction or "and," ∃ as the existential quantifier "there exists" or "for some," and=as the identity relation. Other logical connectives and quantifiers are definable in terms of these. In particular, implication, →, ("if …then…") can be defined by

$$(\Phi \rightarrow \Psi) \text{ iff } \neg(\Phi \wedge \neg\Psi)\,,$$

logical equivalence, ↔, ("if and only if") and disjunction, V, ("or") by

$$(\Phi \leftrightarrow \Psi) \text{ iff } ((\Phi \rightarrow \Psi) \wedge (\Psi \rightarrow \Phi))\,,$$

and

$$(\Phi \vee \Psi) \text{ iff } \neg(\neg\Phi \wedge \neg\Psi)\,,$$

and the universal quantifier, ∀ ("for all") by

$$(\forall x)\Phi \text{ iff } \neg(\exists x)\neg\Phi\,.$$

A precise inductive definition can be given for "the occurrence of a variable x in the formula Φ is free" and for "the occurrence of a variable x in the formula is bound," and it is assumed that the reader is familiar with these fundamental concepts of formal logic. In the formula

$$(((\exists x_1)(x_1 = x_2) \vee (\forall x_1)(\forall x_2)(x_2 \in x_1)) \wedge (x_3 = x_4))$$

all occurrences of x_1 are bound, the first occurrence of x_2 (reading left to right) is free whereas the second occurrence is bound, and the occurrences of x_3 and x_4 are free. A variable x in a formula Φ is said to be *free* (or more precisely, a *free variable* of Φ) if and only if there is an occurrence of the variable x in Φ that is free.

If Φ is a formula of **L(∈, *A*, Ø)** and $y_1,\ldots, y_n$ are variables of **L(∈, *A*, Ø)**, then the notation $\Phi(y_1,\ldots, y_n)$ indicates that the free variables of Φ are *among* the variables $y_1,\ldots, y_n$ (in that order). Note that $\Phi(y_1,\ldots, y_n)$ does not say that $y_1,\ldots, y_n$ *are* the free variables of Φ, but merely that the free variables of Φ form a subset $y_1,\ldots, y_n$. By this convention, $\Psi(y_1, y_2)$ and $\Psi(y_2, y_1)$ say slightly different things; they both say that the set of free variables occurring in Φ is a subset $\{y_1, y_2\}$, but they mention these free variables in different orders.

The formulas of **L(∈, *A*, Ø)** that have no free variables are called *sentences,* and once an interpretation is given to the nonlogical symbols **∈**, ***A***, and **Ø**, each sentence through this interpretation is either true or false. To be a little more precise, a *model* of **L(∈, *A*, Ø)** is an ordered 4-tuple $\langle V, \in, A, \varnothing\rangle$. V is called the *universe of discourse* and is thought of as a collection of objects. The elements of V are called *entities.* **∈** is some binary relation on V, and A and Ø are elements of V. The quantifiers $\exists$ and $\forall$ are assumed to range over the entities, so that $\exists x$ is read "for some entity x" and $\forall x$ "for all entities x." **∈** is interpreted as ∈, ***A*** as A, and **Ø** as Ø. The logical connectives are interpreted by their usual meanings, for example, ¬ as negation, Λ as conjunction, → as implication, V as disjunction, and ↔ as "if and only if." The logical predicate symbol, **=**, is interpreted as the identity relation on the entities. Under such interpretations, each sentence of **L(∈, *A*, Ø)** is a statement about $\langle V, \in, A, \varnothing\rangle$ that is either true or false. Nevertheless, some formulas of **L(∈, *A*, Ø)** are not statements about $\langle V, \in, A, \varnothing\rangle$, for example, x_1**∈ A**. However, for each entity a in V, the expression a**∈*A*** can be interpreted as a statement about $\langle V, \in, A, \varnothing\rangle$, a true statement if $a \in A$ and a false statement if not $a \in A$. In general, if $y_1,\ldots, y_n$ are variables, $\Phi(y_1,\ldots, y_n)$ is a formula of **L(∈, *A*, Ø)**, and $a_1,\ldots, a_n$ are entities of V, then $\Psi(a_1,\ldots, a_n)$ is a statement about $\langle V, \in, A, \varnothing\rangle$. (A word should be said about the case where a variable, say y_1 does not occur in Φ. Suppose that $\Phi(y_2,\ldots, y_n)$ is a formula of **L(∈, *A*, Ø)** and $a_1,\ldots, a_n$ are entities. Then by our conventions, we can also write $\Phi(y_1,\ldots,y_n)$. To avoid confusion, let $\Phi'(y_1,\ldots,y_n)$ stand for $\Phi(y_1,\ldots, y_n)$. Then by definition, $\Phi'(a_1,\ldots, a_n)$ is true if and only if $\Phi(a_2,\ldots, a_n)$ is true.)

This book follows the usual mathematical practices of simplifying notation. In particular, bold face symbols are often not used; for example, x_1**∈*A*** is often written as $x_1 \in A$ and **L(∈, *A*, Ø)** as L(∈, A, Ø). Additional symbols for variables are used to avoid the proliferation of subscripts. Brackets [and] are used in addition to parentheses to improve readability. Parentheses are often omitted; for example, $(\forall x)(x=x)$ is written $\forall x(x=x)$, $(\Phi \Lambda \Psi)$ is often written $\Phi \Lambda \Psi$. The usual practice that → and ↔ dominate the other logical symbols in the interpretation of formulas is observed; for example, $\Phi\Lambda\Psi\rightarrow\Theta$ is read as $(\Phi\Lambda\Psi)\rightarrow\Theta$, and $\Phi\Lambda\Psi\leftrightarrow\Psi\Lambda\Theta$ as $(\Phi\Lambda\Psi)\leftrightarrow(\Psi\Lambda\Theta)$. In a similar way, connectives dominate quantifiers; for example, $\forall x\Phi(x)\Lambda\Psi(x, y)$ is read as $[\forall x\Phi(x)]\Lambda\Psi(x, y)$. The word "and" is often used in place of Λ, "or" in place of V, "for all" in place of $\forall$, and so forth. Furthermore, the logical symbols of **L(∈, *A*, Ø)** are often used in expressions outside the language **L(∈, *A*, Ø)**.

3.3 The Axiom System ZFA

The axioms for ZFA can be explicitly and entirely formulated in L(∈, *A*, Ø). In this section, an informal but systematic presentation of the axioms is provided.

Definition 3.3.1 $x \notin y$ stands for $\neg x \in y$, and $x \neq y$ stands for $\neg x = y$. We will often say "*x* is an element of *y*" or "*x* is a member of *y*" in place of $x \in y$.

Definition 3.3.2 Elements of the universe of discourse are called *entities.* An entity *a* is said to be an *atom* if and only if $a \in A$. An entity *a* is said to be a *set* if and only if $a \notin A$.

Definition 3.3.3 *Axiom of Atoms:* The following three statements are true:

1. Ø is not an atom.
2. $\exists x(x \in A)$.
3. $\forall x[x \text{ is an atom} \rightarrow \forall y(y \notin x)]$

Theorem 3.3.1 *A* and Ø are sets.

Proof. Immediate from Definitions 3.3.2 and 3.3.3

Definition 3.3.4 *A* is called the *set of atoms* and Ø is called the *empty set.*

Definition 3.3.5 *Axiom of the Empty Set:* $\forall x(x \notin \emptyset)$.

Theorem 3.3.2 $A \neq \emptyset$.

Proof. Immediate from the axiom of Atoms and the axiom of the Empty Set.

Definition 3.3.6 Axiom of Extensionality:

$$\forall x \forall y[x \text{ and } y \text{ are sets } \rightarrow (\forall z(z \in x \leftrightarrow z \in y) \rightarrow x = y)]\,. \tag{3.1}$$

The axiom of Extensionality says that if two sets have the same elements, then they are identical. (Of course by the definition of identity used in logic, identical sets have the same elements.)

Theorem 3.3.3 $\forall x(x \text{ is a set and } \forall y(y \notin x) \rightarrow x = \emptyset)$.

Proof. Immediate from the axiom of the Empty Set, the axiom of Extensionality, and Theorem 3.3.1.

Definition 3.3.7 *Axiom of Pairs:*

$$\forall x \forall y \exists z \forall w (w \in z \leftrightarrow w = x \vee w = y) .$$

Definition 3.3.8 Let a and b be entities. Then, by definition, $\{a, b\}$ is the set c such that c has as its only elements a and b. c exists by the axioms of Pairs and Extensionality. By definition, let $\{a\}$ be $\{a, a\}$. $\{a, b\}$ is called the *unordered pair* of a and b, and $\{a\}$ is called the *singleton set* containing a.

Definition 3.3.9 Let a and b be entities. Then, by definition,

$$(a, b) = \{\{a\}, \{a, b\}\} .$$

(a, b) is called the *ordered pair* of a and b.

Theorem 3.3.4 Let a, b, c, d be entities. If $(a, b)=(c, d)$ then $a=c$ and $b=d$.

Proof. Suppose $(a, b)=(c, d)$.

Case 1. $a=b$. Then $(a, b)=\{\{a\}\}$, and thus (a, b) has only one element, $\{a\}$. Since $(a, b)=(c, d)$, it follows that (c, d) has $\{a\}$ as its only element. Since by the definition of ordered pair $\{c\} \in (c, d)$, it follows that $\{a\}=\{c\}$. But this implies that $a=c$. Since

$$(c, d) = \{\{c\}, \{c, d\}\} = \{\{c\}\} ,$$

$\{\{c, d\}\}=\{\{c\}\}$, and therefore $\{c,d\}=\{c\}$, that is, $c=d$. Thus $b=a= c=d$.

Case 2. $a \neq b$. Then (a, b) is not a singleton, and therefore (c, d) is not a singleton. Therefore $c \neq d$. Now, because

$$\{\{a\}, \{a, b\}\}=\{\{c\}, \{c, d\}\},$$

either *(i)* $\{a\}=\{c,d\}$ and $\{a, b\}=\{c\}$, or *(ii)* $\{a\}=\{c\}$ and $\{a, b\}= \{c, d\}$. *(i)* is not possible since $\{a\}$ is a singleton and $\{c, d\}$ is not a singleton. Therefore *(ii)* holds, and thus $\{a\}=\{c\}$, which implies $a=c$, which together with $\{a, b\}=\{c, d\}$ implies $b=d$.

Definition 3.3.10 *Union axiom:*

$$\forall x (x \text{ is a set } \rightarrow \exists y \forall z [z \in y \leftrightarrow \exists w (w \in x \wedge z \in w)]) .$$

Definition 3.3.11 For each set a, let $\bigcup a$ be the set b such that

$$\forall x [x \in b \leftrightarrow \exists y (y \in a \wedge x \in y)] .$$

Such a set b exists by the Union Axiom and the axiom of Extensionality.

Suppose c, d, e are arbitrary entities. Then by the axiom of Pairs $f=\{\{c, d\}, \{e\}\}$ is a set, and by the Union axiom, $g=\bigcup f$ is a set. It is easy to show that g has exactly c, d and e as its elements, and in this case we write $g=\{c, d, e\}$. In general, for a positive integer $n \geq 3$, any n entities (not necessarily distinct)

$a_1,\ldots, a_n$ can be collected together to form a set h that has exactly $a_1,\ldots, a_n$ as elements, and we use the notation $h=\{a_1,\ldots, a_n\}$ to describe this situation. (This easily follows by induction by noting that if $\{a_1,\ldots, a_k\}$ is a set then

$$\{a_1,\ldots,a_{k+1}\} = \bigcup\{\{a_1,\ldots,a_k\},\{a_{k+1}\}\}.)$$

Let a and b be arbitrary sets. By definition, $a \cup b$ is $\bigcup\{a,b\}$. $a \cup b$ is called the *union* of a and b

Definition 3.3.12 For all entities a and b $a \subseteq b$ is defined as follows:

$$a \subseteq b \text{ iff } a \text{ and } b \text{ are sets and } \forall x(x \in a \rightarrow x \in b).$$

For all sets c and d, c is said to be a *subset* of d if and only if $c \subseteq d$.

Definition 3.3.13 *Power-set axiom:*

$$\forall x[x \text{ is a set } \rightarrow \exists y \forall z(z \in y \leftrightarrow z \subseteq x)].$$

Definition 3.3.14 For each set a let $p(a)$ be defined as follows:

$$\forall y(y \in \mathfrak{p}(a) \leftrightarrow y \subseteq a).$$

$p(a)$ is well-defined by the axioms of Power-set and Extensionality. $p(a)$ is called the *power-set* of a.

Definition 3.3.15 *Axiom of Infinity:*

$$\exists x[\varnothing \in x \wedge \forall y(y \in x \rightarrow \{y\} \in x)].$$

Definition 3.3.16 *Axiom of Comprehension:* For each formula $\varphi(x, x_1,\ldots, x_n)$ of $L(\in, A, \varnothing)$ (where $x, x_1,\ldots, x_n$ are variables of $L(\in, A, \varnothing)$),

$$\forall x_1 \cdots \forall x_n \forall y \exists z \forall x[z \text{ is a set and } (x \in z \leftrightarrow \varphi(x,x_1,\ldots,x_n) \wedge x \in y)].$$

Definition 3.3.17 Let $\varphi(x, x_1,\ldots, x_n)$ be a formula of $L(\in, A, \varnothing)$, and let $a_1,\ldots, a_n$ be entities. If b is an entity such that

$$\forall x[x \in b \leftrightarrow \varphi(x,a_1,\ldots,a_n)],$$

then we write

$$b = \{x \mid \varphi(x,a_1,\ldots,a_n)\}$$

and read this as "b is the set of all x such that $\varphi(x, a_1,\ldots, a_n)$."

We often extend this notation and often write $\{(x, y) \mid \Theta(x, y)\}$ for

$$\{z \mid z = (x,y) \text{ and } \Theta(x,y)\},$$

and so forth.

The axiom of Comprehension together with the axiom of Extensionality allows us to freely define sets. The axiom of Comprehension is a sufficiently strong principle for the kinds of constructions and proofs used in ordinary mathematics. Fraenkel (1922b) introduced a stronger principle called the "axiom of Replacement." This principle implies the axiom of Comprehension as well as the existence of very large infinite sets, and is currently accepted as a standard axiom of set theory. Intuitively, the axiom of Replacement is formulated as follows: Let $\varphi(x, y, u_1, \ldots, u_n)$ be a formula of L(∈, A, Ø) and $a_1, \ldots, a_n$ be entities. Let F be the binary relation that is defined on the collection of entities as follows: for all entities a and b,

$$F(a, b) \text{ iff } \varphi(a, b, a_1, \ldots, a_n)\,.$$

Suppose F is a function, c is a set, and H is the function that is defined as follows: the domain of H is the intersection of the domain of F and c, and for all entities a and b, if a is in the domain of H, then

$$H(a) = b \text{ iff } F(a, b)\,.$$

Then the axiom of Replacement (as given in the next definition) will guarantee that the range of H is a set. The relation F given above corresponds to a formula of L(∈, A, Ø) and is a function. Such relations definable in L(∈, A, Ø) are often called *functional relations.*

Definition 3.3.18 *Axiom of Replacement:* For each formula

$$\varphi(x, y, u_1, \ldots, u_n)$$

of L(∈, A, Ø), the following statement is true:

$$\begin{aligned}&\forall u_1 \cdots \forall u_n [\forall x \forall y \forall z [\varphi(x, y, u_1, \ldots, u_n) \wedge \varphi(x, z, u_1, \ldots, u_n) \rightarrow y = z] \\ &\quad \rightarrow \forall t \exists w \forall v [w \text{ is a set and } v \in w \leftrightarrow \exists s (s \in t \wedge \varphi(s, v, u_1, \ldots, u_n))]]\,.\end{aligned}$$

Theorem 3.3.5 The axiom of Replacement implies the axiom of Comprehension.

Proof. Let $\varphi(s, u_1, \ldots, u_n)$ be a formula of L(∈, A, Ø) and $a, a_1, \ldots, a_n$ be entities. Let x and y be variables different from $u_1, \ldots, u_n$, and let $\psi(x, y, u_1, \ldots, u_n)$ be the formula $x = y \wedge \varphi(x, u_1, \ldots, u_n)$. If a is an atom, then

$$\begin{aligned}\varnothing &= \{y \,|\, \exists x \psi(x, y, a_1, \ldots, a_n) \wedge x \in a\} \\ &= \{x \,|\, \varphi(x, a_1, \ldots, a_n) \wedge x \in a\}\end{aligned}$$

because a has no elements, and thus in this instance the axiom of Comprehension holds. So suppose a is not an atom. $\psi(x, y, a_1, \ldots, a_n)$ defines a function $F(x, y)$ on a subcollection of entities, and the axiom of Replacement implies that when this function is restricted to the intersection of its domain and a, the range of the resulting restricted function is a set b, that is,

$$\begin{aligned} b &= \{y \mid \exists x(\psi(x, y, a_1, \ldots, a_n) \wedge x \in a)\} \\ &= \{y \mid \exists x[x = y \wedge \varphi(x, a_1, \ldots, a_n) \wedge x \in a\} \\ &= \{x \mid \varphi(x, a_1, \ldots, a_n) \wedge x \in a\}\,. \end{aligned}$$

The axiom of Comprehension does not imply the axiom of Replacement, but we will not give the proof of this here.

Definition 3.3.19 Let a and b be arbitrary sets. By the axiom of Comprehension, let c be the set such that $c = \{x \mid x \in a \wedge x \in b\}$. Then, by definition, we say that c is the *intersection* of a and b and write $a \cap b = c$.

Definition 3.3.20 *Axiom of Foundation:*

$$\forall x[x \text{ is a set and } x \neq \varnothing \rightarrow \exists y(y \in x \text{ and either } y \in A \text{ or } y \cap x = \varnothing)]\,.$$

The axiom of Foundation rules out the possibility of having a set a such that $a \in a$, sets a_1 and a_2 such that $a_1 \in a_2$ and $a_2 \in a_1$, or sets a_1, a_2, and as such that $a_1 \in a_2$, $a_2 \in a_3$, and $a_3 \in a_1$, etc. The next theorem shows the first of the cases.

Theorem 3.3.6 $\forall x(x \notin x)$.

Proof. Let a be an arbitrary entity.

Case 1. a is an atom. Then by the axiom of Atoms $a \notin a$.

Case 2. a is a set. Then $\{a\}$ is a set and $\{a\} \neq \varnothing$. Thus applying the axiom of Foundation to $\{a\}$, let $y \in \{a\}$ be such that *(i)* either $y \in A$ or *(ii)* $y \cap \{a\} = \varnothing$. Then $y = a$. *(i)* is impossible since by hypothesis a is a set and is therefore not an element of A. Thus *(ii)* must hold. Therefore $a \cap \{a\} = \varnothing$, and thus $a \notin a$.

Definition 3.3.21 *Axiom of Choice:* Suppose a is an arbitrary nonempty set such that each element of a is also a nonempty set. Then there exists a set b such that

$$\forall x[x \in a \rightarrow \exists y(x \cap b = \{y\})]\,.$$

Definition 3.3.22 The axiom system ZFA consists of the following axioms: Atoms, Empty Set, Extensionality, Pairs, Union, Power Set, Infinity, Comprehension, Replacement, Foundation, and Choice.

The axiom system ZFA is very close to other Zermelo-Fraenkel-like axiomatizations of set theory except for the following minor alterations: *(i)* it has a primitive set of atoms A; *(ii)* the empty set $\varnothing$ is taken as a primitive concept rather than as a derived one; and *(iii)* some of the axioms are consequences of other axioms.

3.4 Consequences of the Axiom System ZFA

In this section, some of the familiar theorems of set theory are stated for the system ZFA. Almost any elementary book on set theory has theorems whose proofs can be modified so that they become proofs of the corresponding theorems presented here of ZFA.

3.4.1 Elementary Algebra of Sets

Let a and b be sets. The subset relation $a \subseteq b$, the (finite) union operation $a \cup b$, and (finite) intersection operation $a \cap b$ have already been defined. By definition, let $a - b = \{x \mid x \in a \wedge x \notin b\}$. All the familiar properties of these set operations and relations are consequences of ZFA. The set union operation $\bigcup$ was defined in Definition 3.3.11. The alternate notation

$$\bigcup_{x \in a} x$$

is sometimes used for $\bigcup a$. By definition, for each nonempty set c, let

$$\bigcap c = \{x \mid \forall y(y \in c \to x \in y)\} .$$

For $a \neq \emptyset$, the alternate notation

$$\bigcap_{x \in a} x$$

is sometimes used for $\bigcap a$. The familiar properties of $\bigcup$ and $\bigcap$ also follow from ZFA.

3.4.2 Relations, Functions, and Cartesian Products

The definition for ordered pairs of entities has already been given. By induction, for all positive integers $k>2$ and all entities $a_1, \ldots, a_k$, define the *ordered k-tuple* $(a_1, \ldots, a_k)$ as follows:

$$(a_1, \ldots, a_k) = ((a_1, \ldots, a_{k-1}), a_k) .$$

For each entity a, define the *ordered* 1-*tuple (a)* by *(a)=a*. Let n be a positive integer. a is said to be a *n-ary relation* if and only if a is a set such that each element of a is an ordered n-tuple. If a is a n-ary relation, $(b_1, \ldots, b_n) \in a$ will often be expressed by writing $a(b_1, \ldots, b_n)$. A 2-*ary relation* will often be called a binary relation.

Suppose R is a binary relation. By definition,

(i) $R^{-1} = \{(y, x) \mid R(x, y)\}$,

(ii) the *domain* of R is $\{x|\exists yR(x, y)\}$, and
(iii) the *range* of R is $\{y|\exists xR(x, y)\}$.

f is said to be a *function* if and only if f is a binary relation and

$$\forall x\forall y\forall z[(x,y) \in f \wedge (x,z) \in f \rightarrow y = z].$$

Suppose f is a function and $x \in$ domain of f. Then by definition, $f(x)$ is the entity y such that $(x, y) \in f$.

Suppose f is a function. f is said to be *on a* (or *from a*) if and only if a is domain of f. f is said to be *into b* if and only if range of f is a subset of b. f is said to be *onto b* if and only if the range of f is b. f is said to be *one-to-one* if and only if for all x and y in the domain of f, if $f(x)=f(y)$ then $x=y$. It is easy to show that f is one-to-one if and only if f^{-1} is a function.

Suppose f is a function and a is a subset of the domain of f. Then *the restriction of f to a* is the function g with domain a such that for each x in a, $f(x)=g(x)$.

Suppose a and b are sets. By definition,

$$a \times b = \{(x,y)\,|\,x \in a \wedge y \in b\}.$$

Suppose $a_1, \ldots, a_k$ are sets and k is an integer ≥ 3. Define $a_1 \times \ldots a_k$ inductively as follows:

$$a_1 \times \cdots \times a_k = (a_1 \times \cdots \times a_{k-1}) \times a_k.$$

3.4.3 Ordering Relations

Let R be a binary relation. aRb is often used as an alternate notation for $R(a, b)$. By definition,

1. R is *transitive* if and only if $\forall x\forall y\forall z(xRy \wedge yRz \rightarrow xRz)$,
2. R is *connected* if and only if

$$\forall x\forall y(\{x,y\} \subseteq (\text{domain of } R) \cup (\text{ range of } R) \rightarrow \text{either } xRy \text{ or } yRx),$$

3. R is *reflexive* if and only if $\forall x\ (x \in (\text{domain of } R) \cup (\text{range of } R) \rightarrow xRx)$,
4. R is *symmetric* if and only if $\forall x\forall y(xRy \rightarrow yRx)$,
5. R is *antisymmetric* if and only if $\forall x\forall y(xRy \wedge yRx \rightarrow x=y)$.

Note that if R is connected then R is reflexive.

R is said to be an *equivalence relation* if and only if it is a binary relation that is reflexive, transitive, and symmetric. Equivalence relations are often written as $\equiv$ or $\sim$.

R is said to be a *total ordering* if and only if it is a binary relation that is transitive, connected, and antisymmetric. Total orderings are often written either as $\geq$ or $\succeq$ and are to be read as “at least as great as.”

Suppose $\geq$ is a total ordering. Then, $>$, $\leq$, and $<$ have their usual definitions. (Note that $\leq$ is also a total ordering.) a is said to be the least element of $\geq$ if and only if $x \geq a$ for all $x \in$ domain of $\geq$. Since $\geq$ is reflexive, it follows that the domain of $\geq$ is identical with the range of $\geq$.

$\geq$ is said to be a *well-ordering* if and only if $\geq$ is a total ordering and

$$\forall s[s \subseteq (\text{domain of } \geq) \wedge s \neq \varnothing \rightarrow \exists y(y \in s \wedge \forall x(x \in s \rightarrow x \geq y))].$$

3.4.4 Cardinal Numbers

Let a and b be sets. Then a and b are said to be *equinumerous,* in symbols, $a \equiv b$, if and only if there exists a one-to-one function from a onto b. The following theorem is easy to show:

Theorem 3.4.1 For all sets a, b, and c, the following three statements are true:

1. $a \equiv a$.
2. $a \equiv b \wedge b \equiv c \rightarrow a \equiv c$.
3. $a \equiv b \rightarrow b \equiv a$.

Theorem 3.4.1 says that $\equiv$ is an equivalence relation as is normally defined in mathematics. However, even though $\equiv$ is definable through the language L($\in$, A, $\varnothing$), it does not necessarily correspond to a set within the theory ZFA and thus not to a relation *within* that theory. In fact, it is not difficult to show that it cannot correspond to a relation *within* that theory. Because $\equiv$ is an equivalence relation on the collection of sets, it partitions that collection into equivalence classes. These equivalence classes will be called *cardinal numbers,* and for each set a, $\bar{\bar{a}}$ denotes the equivalence class to which a belongs. $\bar{\bar{a}}$ is called the *cardinal number* of a.

Let a and b be sets. By definition, $\bar{\bar{a}} \geq \bar{\bar{b}}$ if and only if there exists a one-to-one function from b into a. It is easy to show the following theorem:

Theorem 3.4.2 For all sets a, b and c, if $\bar{\bar{a}} \geq \bar{\bar{b}}$ and $\bar{\bar{b}} \geq \bar{\bar{c}}$, then $\bar{\bar{a}} \geq \bar{\bar{c}}$.

The following is more difficult to prove:

Theorem 3.4.3 (The Cantor-Schröder-Bernstein Theorem.) For all sets a and b, if $\bar{\bar{a}} \geq \bar{\bar{b}}$ and $\bar{\bar{b}} \geq \bar{\bar{a}}$, then $\bar{\bar{a}} = \bar{\bar{b}}$.

Theorem 3.4.3 was first proved by G.Cantor. His proof used the axiom of Choice. Later the theorem was proved independently by E.Schröder and F.Bernstein without the use of the axiom of Choice.

Theorem 3.4.4 For all sets a and b , either $\bar{\bar{a}} \geq \bar{\bar{b}}$ or $\bar{\bar{b}} \geq \bar{\bar{a}}$.

The proof of Theorem 3.4.4 necessarily requires the axiom of Choice. In fact,

Theorem 3.4.4 together with the axioms of ZFA minus the axiom of Choice implies the axiom of Choice.

Theorems 3.4.2, 3.4.3, and 3.4.4 immediately imply the following:

Theorem 3.4.5 $\geq$ is a total ordering on the cardinal numbers.

Theorem 3.4.6 Let a be a set and $b=p(a)$, where p is the power-set operation. Then $\bar{\bar{b}} > \bar{\bar{a}}$.

3.4.5 Ordinal Numbers

α is said to be an *ordinal* if and only if the following four conditions hold:

(i) α is a set;
(ii) all elements of α are sets;
(iii) $\forall x(x\in\alpha\rightarrow x\subseteq\alpha)$;
(iv) $\forall x\forall y[(x\in\alpha\wedge y\in\alpha\wedge x\neq y)\rightarrow$ either $x\in y$ or $y\in x]$.

The empty set Ø is an example of an ordinal. So is the set $\{\varnothing\}$. It is easy to show that if a is an ordinal then so is $\alpha\cup\{\alpha\}$.

Define the ordering relation $\geq$ on the collection of ordinals as follows: for all ordinals α and β,

$$\alpha \geq \beta \text{ iff } \beta \in \alpha .$$

Although $\geq$ is definable by a formula of L($\in$, A, Ø), it does not correspond to a set of ordered pairs within ZFA.

Theorem 3.4.7 $\geq$ is a well-ordering on the collection of ordinals, that is, if a is a set of ordinals then a has a least element with respect to $\geq$.

Theorem 3.4.8 Suppose α and β are ordinals. Then the following four statements are true:

1. $\alpha\leq\beta$ iff $\alpha\subseteq\beta$.
2. $\alpha=\{x|x$ is an ordinal and $x<\alpha\}$.
3. $\bigcup\alpha$ is an ordinal (and in fact is the $\geq$-least upper bound of the set of ordinals, α).
4. $\bigcap\alpha$ is an ordinal (and is in fact the least element of α).

Theorem 3.4.9 $\forall x[x$ is a set$\rightarrow\exists\alpha(\alpha$ is an ordinal and $\alpha\equiv x)]$.

The proof of Theorem 3.4.9 necessarily requires the axiom of Choice, and in fact, in the presence of the other axioms of ZFA Theorem 3.4.9 is logically equivalent to the axiom of Choice.

Theorem 3.4.9 is often called the *Well-Ordering Theorem,* because it in conjunction with Theorem 3.4.7 says that every set can be well-ordered.

When considered as an ordinal Ø is often written as 0. If α is an ordinal, by definition, $\alpha^{+}=\alpha\cup\{\alpha\}$. The notation $\alpha+1$ is also be used for α^{+}.

Theorem 3.4.10 The following four statements are true for all ordinals α:

1. 0 is an ordinal and $0\leq\alpha$.
2. α is an ordinal and $\alpha<\alpha+$.
3. $\forall x$(x is an ordinal and $\alpha\leq x\leq\alpha+\rightarrow$ either $\alpha=x$ or $\alpha^{+}=x$).
4. There exists a smallest ordinal ω such that $0<\omega$ and for all ordinals β, if $\beta<\omega$ then $\beta^{+}<\omega$.

The ω that exists by Statement 4 of Theorem 3.4.10 is called the *set of finite ordinals* and will be denoted from now on by the symbol ω. Elements of ω will be called *finite ordinals.* A set α is said to be *finite* if and only if there exists a finite ordinal α such that $\alpha\equiv\alpha$. Non-finite sets are called *infinite.* It is easy to show that ω is an infinite set.

α is said to be a *limit ordinal* if and only if α is an ordinal and there does not exist an ordinal β such that $\alpha=\beta^{+}$. 0 and ω are examples of limit ordinals.

Statements involving ordinals can often be proved by a procedure known as *transfinite induction:* Suppose $\varphi(x)$ is a formula of L($\in$, *A*, Ø). If $\varphi(0)$ is true, and for all ordinals β, $\varphi(\beta)\rightarrow(\varphi(\beta^{+})$, and for all limit ordinals γ,

$$\forall\delta[(\delta \text{ is an ordinal and } \delta<\gamma \text{ and } \varphi(\delta))]\rightarrow\varphi(\gamma)\,,$$

then $\forall x\varphi(x)$ is true.

Certain functions on the collection of ordinals can be defined in terms of L($\in$, *A*, Ø) by a procedure known as *transfinite recursion.* In particular, suppose *F* is a function on the collection of ordinals that is defined by:

(i) $F(0)=a$;
(ii) for each ordinal α, $F(\alpha^{+})=\psi[F(\alpha)]$; and
(iii) for each limit ordinal $\beta>0$,

$$F(\beta)=\bigcup_{\gamma<\beta}F(\gamma)\,,$$

where a is an entity for which there exists a formula $\varphi(x)$ of L($\in$, *A*, Ø) such that $\forall x[\varphi(x)\leftrightarrow x=a]$, and $\psi(x, y)$ is a formula of L($\in$, *A*, Ø) that is a functional relation on the collection of ordinals (and hence is informally written as $y=\psi(x)$). Then the Recursion Theorem for set theory implies that there exists a formula $\Theta(x, y)$ of L($\in$, *A*, Ø) such that Θ defines the function *F*, that is, for all entities u and v

$$\Theta(u,v) \leftrightarrow F(u) = v\,,$$

that is, *F* is a functional relation.

3.5 The Rank Function

Definition 3.5.1 For each ordinal α, define the sets V_α and P_α by transfinite recursion as follows:

(i) $V_0 = A$ and $P_0 = \emptyset$,
(ii) if α is an ordinal, then,

$$V_{\alpha+1} = p(V_\alpha) \cup V_\alpha \text{ and } P_{\alpha+1} = p(P_\alpha) \cup P_\alpha,$$

where *p* is the power-set function, and
(iii) if β is a limit ordinal >0, then

$$V_\beta = \bigcup_{\gamma<\beta} V_\gamma \text{ and } P_\beta = \bigcup_{\gamma<\beta} P_\gamma\,.$$

Then by the Recursion Theorem there exist formulas $\varphi(x, y)$ and $\Psi(x, y)$ of L($\in$, *A*, $\emptyset$) such that for all entities *a* and *b*,

$$\varphi(a, b) \text{ iff } a \text{ is an ordinal and } b = V_a, \tag{3.2}$$

and

$$\psi(a, b) \text{ iff } a \text{ is an ordinal and } b = P_a. \tag{3.3}$$

Let φ and ψ be as in Equations 3.2 and 3.3, and let $V(x)$ and $P(x)$ be the following formulas of L($\in$, *A*, $\emptyset$):

$$V(x)\text{: } \exists y \exists z(\varphi(y, z) \wedge x \in z)$$

and

$$P(x)\text{: } \exists y \exists z(\psi(y,z) \wedge x \in z)$$

Then $V(a)$ is true for an entity *a* if and only if $a \in V_\alpha$ for some ordinal α. And similarly, $P(b)$ is true for an entity *b* if and only if $b \in P_\alpha$ for some ordinal α. Notation will often be confused a little by letting the symbol *V* stand also for the collection of all entities a such that $V(a)$ and *P* stand also for the collection of all entities *b* such that $P(b)$. Theorem 3.5.8 will show that $\forall x V(x)$, and because of this *V* is called the *collection of entities*. For similar reasons, *P* is called the *collection of pure sets*.

Throughout this book, *A* is often interpreted as a set of qualitative entities. When this is done, it is natural to consider elements of V_ω as qualitative. This approach to qualitativeness is discussed in detail in Chapter 7. Also, throughout the book, *P* is often identified with pure mathematics. (This identification is explained in Section 4.1.)

Theorem 3.5.1 For each ordinal α, $P_\alpha \subseteq V_\alpha$.

Proof by transfinite induction. $P_0 \subseteq V_0$. Suppose α is an ordinal and $P_\alpha \subseteq V_\alpha$. Then

$$P_{\alpha+1} = \mathfrak{p}(P_\alpha) \cup P_\alpha \subseteq \mathfrak{p}(V_\alpha) \cup V_\alpha = V_{\alpha+1}\,.$$

Suppose β is a limit ordinal >0 and for all $\gamma<\beta$, $P_\gamma<V_\gamma$. Then

$$P_\beta = \bigcup_{\gamma<\beta} P_\gamma \subseteq \bigcup_{\gamma<\beta} V_\gamma = V_\beta\,.$$

Theorem 3.5.2 $P \subseteq V$.

Proof. Immediate from Theorem 3.5.1.

Theorem 3.5.3 For all ordinals α and β, if $\alpha \leq \beta$, then $V_\alpha \subseteq V_\beta$.

Proof. Suppose the theorem is false. Let α and β be ordinals such that $\alpha \leq \beta$ and V_α is not a subset of V_β. Then it must be the case that $\alpha<\beta$. Since the ordinals are well-ordered, let γ be the least ordinal such that for some $\kappa' < \gamma$, V_κ' is not a subset of V_γ. Let $\kappa<\gamma$ be such that V_κ is not a subset of V_γ.

Case 1. γ is a limit ordinal. But since

$$V_\gamma = \bigcup_{\delta<\gamma} V_\delta$$

and $\kappa<\gamma$, it follows that $V_\kappa \subseteq V_\gamma$, contrary to the choice of γ and κ.

Case 2. γ is not a limit ordinal. Let δ be the ordinal such that $\gamma=\delta+1$.

Then since $\kappa<\gamma$, it follows that $\kappa\leq\delta$. If $\kappa=\delta$, then

$$V_\kappa \subseteq \mathfrak{p}(V_\kappa) \cup V_\kappa \subseteq \mathfrak{p}(V_\delta) \cup V_\delta = V_{\delta+1} = V_\gamma\,,$$

contrary to the choice of γ and κ. If $\kappa<\delta$, then since $\delta<\gamma$, it follows from the choice of γ that $V_\kappa \subseteq V_\delta$, and thus

$$V_\kappa \subseteq V_\delta \subseteq \mathfrak{p}(V_\delta) \cup V_\delta = V_\gamma\,,$$

again contrary to the choice of γ and κ. Thus Case 2 is also impossible, and the theorem follows by contradiction.

Definition 3.5.2 For each entity *a* in *V*, the *rank* of *a* is the least ordinal α such that $a \in V_\alpha$. We often write "α=rank*(a)*" for "α is the rank of *a*."

Theorem 3.5.4 Let *a* be an entity in *V*. Then the following two statements are true:

1. rank*(a)*=0 if and only if *a* is an atom.
2. rank*(a)*=β+1 for some ordinal β if and only if *a* is a set.

Proof. Statement 1 follows immediately from Definitions 3.5.1 and 3.5.2. Statement 2 follows from the observation that rank*(a)* can never be a limit ordinal γ>0, for if

$$a \in \bigcup_{\delta<\gamma} V_\delta ,$$

then $a \in V_\delta$ for some $\delta<\gamma$.

Theorem 3.5.5 Suppose a is a set. Then

$$a \in V \text{ iff } \forall x(x \in a \rightarrow x \in V).$$

Proof. Suppose $a \in V$. Since by hypothesis a is a set, by Theorem 3.5.4 let β be the ordinal such that rank$(a)=\beta+1$. Then $a \in V_{\beta+1}$ and $a \notin V_\beta$. Thus $a \subseteq V_\beta$, and therefore all elements of a are in V_β and therefore are also in V.

Suppose $\forall x(x \in a \rightarrow x \in V)$. Let $F(x,y)$ stand for rank$(x)=y$ Then $F(x,y)$ can be formulated as a formula of L($\in$, A, Ø), and as such represents a function, $y=F(x)$. Therefore, since a is a set, by the axiom of Replacement $\{F(x) \mid x \in a\}$ is a set of ordinals b. By Theorem 3.4.7, let γ be the least ordinal such that $\gamma>\alpha$ for each α in b. Then $F(x)<\gamma$ for each x in a. Therefore, for each x in a, $x \in V_{F(x)}$ and (by Theorem 3.5.3) $x \in V_\gamma$. Thus $a \in V_{\gamma+1}$, and therefore $a \in V$.

Theorem 3.5.6 Suppose a is a set, a is in V, and rank$(a)=\alpha$. Then

$$\forall x(x \in a \rightarrow \text{rank}(x) < \alpha).$$

Proof. By Theorem 3.5.4, let β be an ordinal such that $\alpha=\beta+1$. Since rank$(a)=\beta+1$, it follows that $a \in V_{\beta+1}$ and $a \notin V_\beta$. Since $V_{\beta+1}=p(V_\beta) \cup V_\beta$, it then must be the case that $a \in p(V_\beta)$, that is, $a \subseteq V_\beta$. Thus each element of a is in V_β and therefore must have rank$\leq\beta$, and thus rank$<\alpha$.

Theorem 3.5.7 The following six statements are true:

1. $\forall x(x \in P \rightarrow x$ is a set).
2. $\forall x \forall y$ (x and y are ordinals and $x \leq y \rightarrow P_x \subseteq P_y$).
3. $\forall x \forall y \forall \alpha$ (α is an ordinal, $y \in x$, and $x \in P_\alpha \rightarrow y \in P_\alpha$).
4. $\forall x \forall y$ ($x \in P \wedge y \in x \rightarrow y \in P$).
5. $\forall x$(x is an ordinal$\rightarrow x \in P$).
6. $\forall x$(x is an ordinal$\rightarrow$rank$(x)=x+1$).

Proof. 1. Statement 1 immediately follows from the definitions of P_α and P (Definition 3.5.1).

2. The proof of Statement 2 is almost identical to the proof of Theorem 3.5.3.

3. Suppose α is an ordinal, $y \in x$ and $x \in P_\alpha$. We need only show $y \in P_\alpha$. Let β be the least ordinal such that $x \in P_\beta$. There are two cases to consider. *Case* 1. β is a limit ordinal. $\beta \neq 0$ since $P_0=$Ø and $x \in P_\beta$. Thus $x \in P_\gamma$ for some $\gamma<\beta$, and this contradicts the choice of β. *Case* 2. γ is an ordinal and $\beta=\gamma+1$. Then $P_\beta=p(P_\gamma) \cup P_\gamma$.

$x \notin P_\gamma$ by the choice of β. Therefore $x \in p(P_\gamma)$, i.e., $x \subseteq P_\gamma$. But then $y \in P_\gamma$. Since by Statement 2 and the choice of β, $P_\gamma \subseteq P_\beta \subseteq P_\alpha$, it follows that $y \in P_\alpha$.

4. Statement 4 follows immediately from Statement 3 and the definition of P (Definition 3.5.1).

5. Statement 5 follows by transfinite induction: $0 \in P_1$ (and thus $0 \in P$). Suppose α is an ordinal and $\alpha \in P_{\alpha+1}$ (and thus $\alpha \in P$). Then $\{\alpha\} \subseteq P_{\alpha+1}$, and therefore

$$\alpha + 1 = \alpha \cup \{\alpha\} \in \mathfrak{p}(P_{\alpha+1}) \cup P_{\alpha+1} = P_{(\alpha+1)+1}$$

(and thus $\alpha+1 \in P$). Suppose $\beta>0$ is a limit ordinal and for all $\gamma \in \beta$, $\gamma \in P_{\gamma+1}$. Then

$$\beta = \bigcup_{\gamma<\beta} \gamma \subseteq \bigcup_{\gamma<\beta} P_{\gamma+1} = P_\beta\,,$$

and therefore $\beta \in P_{\beta+1}$ and (thus $\beta \in P$).

6. To show Statement 6, by Theorem 3.5.3 we need only show that for all ordinals α, $\alpha \in V_{\alpha+1}$ and $\alpha \notin V_\alpha$. Suppose this were not the case. A contradiction will be shown. Let β be the least ordinal such that either $\beta \notin V_{\beta+1}$ or $\beta \in V_\beta$. *Case* 1. $\beta \notin V_{\beta+1}$ $\beta \neq 0$, since $0 = \emptyset \in V_1$. Now, since β is an ordinal, each element of β is an ordinal, and thus by the choice of β, for each γ in β, $\gamma \in V_{\gamma+1}$, and therefore

$$\beta \subseteq \bigcup_{\gamma<\beta} V_{\gamma+1}\,. \tag{3.4}$$

But by Theorem 3.5.3, $V_\gamma \subseteq V_\beta$ for each $\gamma<\beta$. Thus Equation 3.4 yields $\beta \subseteq V_\beta$, from which it follows that $\beta \in V_{\beta+1}$, contrary to hypothesis. *Case* 2. $\beta \in V_\beta$. Since β is an ordinal, β is a set. By Theorem 3.5.4, let γ be an ordinal such that $\beta \in V_{\gamma+1}$ and $\beta \notin V_\gamma$. Since $\beta \in V_\beta$, it follows from Theorem 3.5.3 that $\beta \geq \gamma+1$. Since $\beta \in V_{\gamma+1}$ and $\beta \notin V_\gamma$, it follows that $\beta \subseteq V\gamma$. Since $\beta>\gamma$ and β and γ are ordinals, $\gamma \in \beta$. Therefore $\gamma \in V_\gamma$, contradicting the choice of β.

Theorem 3.5.8 $\forall x V(x)$.

Proof. Suppose a is an entity and $a \notin V$. A contradiction will be shown. a is not an atom, since each atom is in V_0. Thus a is a set. Since $\emptyset \in V_1$, $a \neq \emptyset$. Define the sequence a_α by transfinite recursion as follows: $a_0 = a$, $a_{\alpha+1} = \bigcup a_\alpha$, and for limit ordinals $\beta>0$,

$$a_\beta = \bigcup_{\gamma<\beta} a_\gamma\,.$$

Consider a_ω (where ω is the set of finite ordinals). a_ω is a set and a_ω satisfies the following condition:

$$\forall x \forall y[x \in a_\omega \wedge y \in x \rightarrow y \in a_\omega]\,. \tag{3.5}$$

(To show Equation 3.5, suppose $x \in a_\omega$ and $y \in x$. Let α be the first ordinal for which $x \in a_\alpha$. Then $y \in \bigcup a_\alpha$. Because $\bigcup a_\alpha$ is an ordinal $\leq a_\alpha$, it follows that $y \in a_\omega$.) Let

$$b = \{x \mid x \in a_\omega \wedge \neg V(x)\}. \tag{3.6}$$

Then b is a set. Since $a \notin V$, it follows from Theorem 3.5.5 that a has an element c such that $c \notin V$. Since $a = a_0 \subseteq a_\omega$, it then follows by Equation 3.6 that $c \in b$. Thus $b \neq \emptyset$. $c \notin A$ since $A = V_0$. Therefore, by the axiom of Foundation, let $d \in b$ be such that either $d \in A$ or $d \cap b = \emptyset$. $d \notin A$ since $A = V_0 \subseteq V$, and by Equation 3.6 $d \notin V$. $d \neq \emptyset$, since $\emptyset \in V$ and therefore by Equation 3.6 $\emptyset \notin b$. Therefore by Theorem 3.5.5, let $e \in d$ be such that $e \notin V$. Since $b \subseteq a_\omega$, it then follows from Equation 3.5 that $e \in a_\omega$. Since it is also the case that $e \notin V$, it follows by Equation 3.6 that $e \in b$. Thus $e \in d \cap b$, and this contradicts $d \cap b = \emptyset$.

3.6 Permutations of Atoms

Transformations from a set onto itself have an important role in mathematics and science. In algebra, such transformations are called permutations. Permutations on the set of atoms were introduced into the theory of sets by Fraenkel (1922a, 1935), who used them to construct models of set theory in which the axiom of Choice is false but the other axioms of ZFA are true. They still play an important role in the metamathematics of set theory (e.g., see Jech, 1973). In Chapter 4, permutations of atoms are used to provide a formal basis for the systematic description of invariance concepts.

Definition 3.6.1 f is said to be a *permutation* on A if and only if f is a one-to-one function from A onto A. The symbol $\imath$ is used to denote the identity function on A. It is immediate that $\imath$ ι is a permutation on A. $*$ denotes the composition of functions, that is, $(f * g)(x) = f[g(x)]$. It is easy to show that if f and g are permutations on A then so are f^{-1} and $f * g$.

Let f be a function from A into A. Suppose $a \subseteq A$. f can be extended naturally to yield a value at a as follows:

$$f(a) = \{f(y) \mid y \in a\},$$

and in general if f has been extended to all elements of some set b, then f can be extended to yield a value at b by

$$f(b) = \{f(x) \mid x \in b\}.$$

In this way f can be extended to the entire collection of entities. This is now done formally in the following definition.

Definition 3.6.2 Let f be a function from A into A. By transfinite recursion define the function f_α on V_α as follows: $f_0 = f$. Suppose α is an ordinal and the function f_α on V_α has been defined. Define $f_{\alpha+1}$ on $V_{\alpha+1}$ as follows: for each $a \in V_{\alpha+1}$, if $a \in V_\alpha$

then $f_{\alpha+1}(a)=f_\alpha(a)$, and if $a\in V_{\alpha+1}-V_\alpha$ then $f_{\alpha+1}(a)=\{f(x)|x\in a\}$. Suppose $\beta>0$ is a limit ordinal and f_γ has been defined for each $\gamma<\beta$. Define

$$f_\beta = \bigcup_{\gamma<\beta} f_\gamma .$$

From this definition it easily follows that for all ordinals α and β, f is a function from V_α into V_α and if $\alpha\leq\beta$ then $f_\alpha\subseteq f_\beta$. By the Recursion Theorem, for each ordinal α, f_α is an entity (within ZFA). Let f' be the functional relation on V that is defined by: for each $a\in V$, $f'(a)=f_\alpha(a)$, where α=rank(a). (It can be shown that f' cannot be considered as an entity within ZFA. However, it is a functional relation definable within L(∈, A, Ø).) Since f' is a natural extension of f, we will confuse notation a little and write f for f'. If there is need for distinguishing between these two uses of the symbol f, we will do so by explicitly indicating which usage is intended, for example, "the permutation f," "the functional relation f," "the function f from A into A," and so forth.

Theorem 3.6.1 The following seven statements are true for all permutations f on A:

1. For each set a, $f(a) = \{f(x)\,|\,x \in a\}$..
2. For each set a, $\iota(a)=a$, $f^{-1}[f(a)]=a$, and for all permutations g on A, $(f*g)(a)=f[g(a)]$.
3. For each ordinal α, $f(V_\alpha)=V_\alpha$.
4. For each entity a, rank(a)=rank$(f(a))$.
5. $\forall x\forall y[x \in y \leftrightarrow f(x) \in f(y)]$.
6. For all entities a and b if $a\neq b$ then $f(a)\neq f(b)$.
7. For all permutations g and h on A, if $f=g*h$ then for each entity a, $f(a)=g[h(a)]$.

Proof. 1. Statement 1 follows immediately from Definition 3.6.2.

2. Statement 2 follows easily from Statement 1 by transfinite induction on the rank of a.

3. Let α be an ordinal. It immediately follows from Definition 3.6.2 that $g(V_\alpha)\subseteq V_\alpha$ for each permutation g on A. In particular, $f(V_\alpha)\subseteq V_\alpha$ and $f^{-1}(V_\alpha)\subseteq V_\alpha$,. Now, by Statement 2 and $f^{-1}(V_\alpha)\subseteq V_\alpha$, we can conclude that

$$V_\alpha = \iota(V_\alpha) = f * f^{-1}(V_\alpha) = f[f^{-1}(V_\alpha)] \subseteq f(V_\alpha)\,,$$

and thus $V_\alpha=f(V_\alpha)$.

4. Statement 4 immediately follows from Statement 3.

5. $\forall x\forall y[x \in y \rightarrow f(x) \in f(y)]$ immediately follows from Statement 1.

We will now show $\forall x\forall y[f(x) \in f(y) \rightarrow x \in y]$. Suppose a and b are entities such that $f(a) \in f(b)$. Then by what we have already shown, $f^{-1}[f(a)] \in f^{-1}[f(b)]$, which by Statement 2 yields $a \in b$.

6. Statement 6 will be shown by contradiction. Suppose Statement 6 were not true. Let α be the least ordinal such that there exist entities x and y such that α=rank(x), $x \neq y$, and $f(x)=f(y)$, and let a and b be entities such that α=rank(a), $a \neq b$, and $f(a)=f(b)$. $a \neq 0$, for if $a=0$, then a is either Ø or an atom, and in either case it immediately follows from $f(a)=f(b)$ that $a=b$, a contradiction. Suppose $a>0$. Then a is a set, and thus $f(a)$ and $f(b)$ are sets, from which it follows that b is a set. Since $a \neq b$, either $a-b \neq Ø$ or $b-a \neq Ø$. *Case* 1. $a-b \neq Ø$. Let $c \in a-b$. Then by Statement 1, $f(b) = \{f(x) \,|\, x \in b\}$. From this it follows that we can select d in b such that $f(c)=f(d)$. Since $a-b \neq Ø$ and $c \in a-b$ and $d \in b$, it follows that $c \neq d$. By Theorem 3.5.6, rank(c)<rank$(a)=\alpha$. Thus rank$(c)<\alpha$, $c \neq d$, and $f(c)=f(d)$, and these three assertions taken together contradict the choice of α. *Case* 2. $b-a \neq Ø$. Similar to Case 1.

7. Suppose g and h are permutations on A and $f=g*h$. Statement 7 will be shown by contradiction. Suppose a is an entity of smallest rank α such that $f(a) \neq g[h(a)]$. Since by hypothesis if $a \in A$ then $f(a)=g[h(a)]$, it follows that $\alpha \neq 0$. Therefore a is a set and by Theorem 3.5.6 each element of a has rank<α. Thus by the choice of α, $\forall x(x \in a \rightarrow f(x) = g[h(x)])$, which by Statement 1 yields

$$f(a) = \{f(x) \,|\, x \in a\} = \{g[h(x)] \,|\, x \in a\} = g(\{h(x) \,|\, x \in a\}) = g[h(a)].$$

Theorem 3.6.2 Let f be a permutation on A. Then

$$\forall x[x \in P \rightarrow f(x) = x].$$

Proof. Suppose it were not the case that $\forall x(x \in P \rightarrow f(x) = x)$. A contradiction will be shown. Let a be an entity of least rank, α, such that $a \in P$ and $f(a) \neq a$. By Theorem 3.5.7, a is a set since $a \in P$. $a \neq Ø$, since $f(Ø)=Ø$. Now by Theorems 3.5.7 and 3.5.6, for each entity b, if $b \in a$ then $b \in P$ and rank$(b)<\alpha$. Thus by the choice of α, $f(b)=b$ for all $b \in \alpha$. Therefore,

$$f(a) = \{f(b) \,|\, b \in a\} = \{b \,|\, b \in a\} = a,$$

and this contradicting $f(a) \neq a$.

3.7 Pure Set Theory and Axiom System ZF

As mentioned earlier, the collection of pure sets is identified with pure mathematics. For this identification to make sense, the pure sets have to have the correct kind of structure for formulating mathematical concepts. This section shows that the pure sets together with the $\in$-relation form a model of set theory (without atoms) that is very similar to ZFA in form and richness.

Definition 3.7.1 Let L(∈) be the first order language that has a 2-place nonlogical predicate symbol ∈ and no individual constant symbols. (The formal definition of L(∈) is similar to the definition of **L(∈, *A*, Ø)** given in Definition 3.2.1.)

In the literature, the most prevalent forms of axiomatic set theories are formulated in the language L(∈) and have axioms that prohibit the existence atoms. The system ZF *(Zermelo-Fraenkel)* is one of the most popular of these. It is very similar to the system ZFA and its exact description will be given shortly.

Definition 3.7.2 Assume ZFA. Let ∈′ be the restriction of ∈ to the collection of pure sets *P*. The structure $\langle P, \in' \rangle$ is called *pure set theory*.

Each sentence of L(∈) can be interpreted either as a statement about the structure $\langle P, \in' \rangle$ or, because L(∈) is a sublanguage of **L(∈, *A*, Ø)**, as a statement about $\langle V, \in, A, \varnothing \rangle$. In this section, we are primarily interested in interpreting sentences and other expressions based on L(∈) in terms of the structure $\langle P, \in' \rangle$. In carrying out such interpretations, the interpretations of the quantifiers and free variables must be restricted to *P*. The same result can be obtained within ZFA by taking an expression Θ of L(∈) and modifying it to an expression Θ' of **L(∈, *A*, Ø)** that is its relativization to *P*. For example, the sentence

$$\forall x \exists y [x \in y]$$

of L(∈) interpreted in $\langle P, \in' \rangle$ says that for every pure set *x* there is a pure set *y* such that $x \in' y$. But this says exactly the same thing as its relativization to *P*,

$$\forall x[P(x) \rightarrow \exists y(P(y) \wedge x \in y)],$$

formulated in the language **L(∈, *A*, Ø)** and interpreted in $\langle V, \in, A, \varnothing \rangle$. Similarly, if $\varphi(x, y, u_1, u_2)$ is a formula of L(∈), a_1 and a_2 are elements of *P*, and in $\langle P, \in' \rangle$ $\varphi(x, y, a_1, a_2)$ is a functional relation—which we will write as $y=F(x)$—then in the language **L(∈, *A*, Ø)** a relativization $\varphi'(x, y, u, v)$ of φ can easily be given, so that in the structure $\langle V, \in, A, \varnothing \rangle$, $\varphi'(x, y, a_1, a_2)$ defines exactly the same function as determined by the functional relation *F*.

The following is a more precise definition of "relativization":

Definition 3.7.3 Assume ZFA. The relativization (in **L(∈, *A*, Ø)**) of a formula of L(∈) to *P* is defined inductively as follows:

1. If φ is an atomic formula of L(∈) of the form $u \in v$, then its relativization to *P* is the formula $u \in v \Lambda P(u) \backslash P(v)$; if φ is an atomic formula of L(∈) of the form $u=v$, then its relativization to *P* is the formula $v=v \Lambda P(u) \Lambda P(v)$.
2. If Ψ' and Θ' are the relativization of the formulas Ψ and Θ of L(∈) to *P*, then $\Psi' \Lambda \Theta'$ is the relativization of $\Psi \Lambda \Theta$ to *P*.
3. If $\Theta'(u_1, \ldots, u_n)$ is the relativization of the formula $\Theta(u_1, \ldots, u_n)$ of L(∈) to *P*, then $\neg\Theta'(u_1, \ldots, u_n) \wedge P(u_1) \wedge \cdots \wedge P(u_n)$ is the relativization of $\neg\Theta$ to *P*.

4. If $\Theta'(u, u_1, \ldots, u_n)$ is the relativization of the formula $\Theta(u, u_1, \ldots, u_n)$ of L(∈) to P, then $\exists u[\Theta'(u, u_1, \ldots, u_n) \wedge P(u)]$ is the relativization of $\exists u\Theta$ to P.
5. All relativizations of formulas of L(∈) to P can be obtained by repeated applications of rules 1–5 above.

Note that in the definition of relativization given above some extra subformulas might appear, for example, because $x \in y \wedge P(x) \wedge P(y)$ is the relativization of $x \in y$, the formula $\exists x[x \in y \wedge P(x) \wedge P(y) \wedge P(x)]$ is the relativization of $\exists x(x \in y)$ that is, there is an "extra" $P(x)$-subformula occurring in the relativization of $\exists x(x \in y)$. These "extra" subformulas do no harm and allow for a slightly simpler formulation.

We now present the axioms of ZF and show that they are true in $\langle P, \in' \rangle$. Each axiom of ZF is prefaced by the expression "ZF-" to distinguish it from a similar axiom in ZFA. In showing that the ZF-axioms are true in $\langle P, \in' \rangle$, various properties of pure sets given in Section 3.5 (as well as some easy deductions from them) are used freely.

ZF-axiom of the Empty Set: $\exists y \forall x(x \notin y)$. This axiom is true in $\langle P, \in' \rangle$ since $\varnothing \in P$.

ZF-axiom of Extensionality: $\forall x \forall y[\forall z(z \in x \leftrightarrow z \in y) \rightarrow x = y]$. This axiom is true in $\langle P, \in' \rangle$ because the axiom of extensionality is true in $\langle V, \in, A, \varnothing \rangle$ and because every element of a pure set is a pure set.

ZF-axiom of Pairs: $\forall x \forall y \exists z \forall w(w \in z \leftrightarrow w = x \vee w = y)$. This axiom is true in $\langle P, \in' \rangle$ by the properties of pure sets given in Section 3.5.

ZF-Union axiom: $\forall x \exists y \forall z[z \in y \leftrightarrow \exists w(w \in x \wedge z \in w)]$. This axiom is true in $\langle P, \in' \rangle$ by the properties of pure sets given in Section 3.5.

ZF-Power-set axiom: $\forall x \exists y \forall z(z \in y \leftrightarrow z \subseteq x)$. This axiom is true in $\langle P, \in' \rangle$ by the properties of pure sets given in Section 3.5.

ZF-axiom of Infinity: $\exists u \exists v[\forall x(x \notin u) \wedge u \in v \wedge \forall y(y \in v \rightarrow \{y\} \in v)]$. This axiom is true in $\langle P, \in' \rangle$ because $\varnothing \in P$ and for each a in $P, \{a\}$ is in P.

ZF-axiom of Foundation: $\forall x[\exists u(u \in x) \rightarrow \exists y(y \in x \wedge \neg \exists z(z \in y \cap x))]$. This axiom is true in $\langle P, \in' \rangle$ by the properties of pure sets given in Section 3.5.

ZF-axiom of Choice: $\forall x[\exists y(y \in x) \wedge \forall z(z \in x \rightarrow \exists w(w \in z)) \rightarrow \exists u \forall v(v \in x \rightarrow \exists t(v \cap u = \{t\}))]$. This axiom is true in $\langle P, \in' \rangle$ by the properties of pure sets given in Section 3.5 and the axiom of Choice in $\langle V, \in, A, \varnothing \rangle$.

ZF-axiom of Replacement: For each formula $\varphi(x, y, u_1, \ldots, u_n)$ of L(∈)

$$\forall u_1 \cdots \forall u_n(\forall x \forall y \forall z[\varphi(x, y, u_1, \ldots, u_n) \wedge \varphi(x, z, u_1, \ldots, u_n) \rightarrow y = z]$$
$$\rightarrow \forall t \exists w \forall v[v \in w \leftrightarrow \exists s(s \in t \wedge \varphi(s, v, u_1, \ldots, u_n))]) .$$

To show that the ZF-axiom of Replacement is true in $\langle P, \in' \rangle$, suppose that $\varphi(x, y, u_1, \ldots, u_n)$ is a formula of L(∈), $a, a_1, \ldots, a_n$ are elements of P, and $\varphi(x, y, a_1, \ldots, a_n)$ is a functional relation in $\langle P, \in' \rangle$. Let $y=F(x)$ be the function that is defined by $\varphi(x, y, a_1,\ldots, a_n)$. (Note that F is not necessarily an element of P.) Then the range of F is a subcollection of P. Let $\varphi'(x, y, a_1,\ldots, a_n)$ be the formula of **L(∈, *A*, Ø)** that is the relativization of $\varphi(x, y, a_1,\ldots, a_n)$ to P. Then $\varphi'(x, y, a_1,\ldots, a_n)$ is a functional relation on $\langle V, \in, A, \varnothing\rangle$ and $F=F'$, where F' is the function defined by $\varphi'(x, y, a_1,\ldots, a_n)$. Therefore by the axiom of Replacement in $\langle V, \in, A, \varnothing\rangle$,

$$b = \{y \mid \exists x(x \in a \wedge y = F'(x))\}$$

is a set that is an element of V. But since the range of $F'=F$ is a subcollection of P, each element of b is a pure set. Thus by the properties of pure sets given in Section 3.5, b is a pure set. Therefore in $\langle P, \in'\rangle$

$$b = \{y \mid \exists x(x \in a \wedge y = F(x))\} = \{y \mid \exists x(x \in a \wedge \varphi(x, y, a_1, \ldots, a_n))\}$$

is an element of P, and we have shown that the ZF-axiom of Replacement is true in $\langle P, \in'\rangle$.

ZF-axiom of Comprehension: For each formula $\varphi(x, x_1,\ldots, x_n)$ of L(∈)

$$\forall y \forall x_1 \cdots \forall x_n \exists z \forall x[x \in z \leftrightarrow \varphi(x, x_1, \ldots, x_n) \wedge x \in y].$$

By very slightly modifying the proof of Theorem 3.3.5, it follows that the ZFaxiom of Replacement implies the ZF-axiom of Comprehension. Thus because the ZF-axiom of Replacement is true in $\langle P, \in'\rangle$, it follows that the ZF-axiom of Comprehension is also true in $\langle P, \in'\rangle$.

Definition 3.7.4 The *axiom system* ZF is formulated in the language L(∈) and has as its axioms the ZF-axioms of the Empty Set, Extensionality, Pairs, Union, Power-set, Infinity, Foundation, Choice, Replacement, and Comprehension.

Theorem 3.7.1 Assume ZFA. Let $\in'$ be the restriction of $\in$ to the collection of pure sets P. Then $\langle P, \in'\rangle$ satisfies the axiom system ZF.

Proof. By Definition 3.7.4 and the above arguments.

4. Axiomatic Generalizations of the Erlanger Program

4.1 A Formal Approach to Meaningfulness

4.1.1 Introduction

Klein's Erlanger Program (Section 1.4) is one approach to a specific theory of meaningfulness. In this chapter, it and some of its natural generalizations are formalized, and some of their implications for the foundations of science are explored. Other approaches to meaningfulness based on measurementtheoretic considerations are discussed in Chapter 5, and approaches to the related concepts of lawfulness and qualitativeness are discussed in Chapters 6 and 7.

In order to increase the readability of the chapter, many of the proofs are given in Section 4.10 at the end of the chapter.

4.1.2 The Language L($\in$, *A*, Ø, *M*)

In this chapter, both definitional and transformational approaches are used to develop formal, axiomatic theories of meaningfulness. These axiomatic formulations are heavily influenced by concepts, techniques, and results of the axiom system ZFA. As remarked earlier, ZFA is very powerful, and—according to a view widely held among philosophers of science—it is sufficiently rich in structure and concepts to provide the necessary mathematical and conceptual base for all of mathematical science.

Throughout this chapter, axiom system ZFA is used for the formulation of mathematical and scientific ideas, and, unless otherwise indicated by context or explicit reference, the following three identifications are made: *(i)* Pure sets and concepts based solely upon them are identified with pure mathematics. *(ii)* The set of atoms, *A,* are identified with the set of objects that constitute the domain of current interest. This domain and its elements provide the basis for the development of all applied mathematical concepts. The set *A* is considered a qualitative entity. In general, its choice varies from one application to another.

For example, for the applied mathematics of space-time, A might be selected as the set of space-time points; and for the applied mathematics of the measurement of mass, A might be selected as the set of objects with mass. *(iii)* Once A has been selected, the applied mathematics based on it will be identified with concepts that are definable in terms of entities of V and the language L($\in$, A, Ø) or some slight extension of this language.

Note that by this usage, applied mathematics includes pure mathematics. From a theoretical point of view, this is reasonable, because we do not want to exclude on a priori grounds any part of pure mathematics as being irrelevant for scientific applications. More is said about this later.

Although each proposition of applied mathematics is either true or false, it does not follow that each such proposition has anything of significance to say about the particular applied situation under consideration. The concepts of "significant applied mathematical relation" and "significant applied mathematical proposition" play a central role throughout the book, and in this chapter various axiomatic formulations of "significant applied mathematical relation" are given. However, instead of the word "significant," the word "meaningful"—which in the measurement literature has come to describe a closely related, if not identical, concept—is used throughout this chapter.

"Meaningful" is taken as an undefined term to be axiomatized in terms of some of its more recognizable properties. The main thrust of the axiomatizations is to consider entities that are properly defined in terms of meaningful entities are themselves meaningful. Of course, the major difficulty in successfully bringing this idea to fruition is providing a precise description of "properly defined" and justifying it as an appropriate basis for a concept of meaningfulness.

In axiomatizing meaningfulness formally, it is desirable from a number of perspectives to do it in a language that is more powerful than L($\in$, A, Ø). The one that is used throughout this chapter adds a 1-place predicate symbol, M, to L($\in$, A, Ø):

Definition 4.1.1 The language L($\in$, A, Ø, M) is defined in almost exactly the same manner as L($\in$, A, Ø) but with two exceptions: *(i)* the nonlogical symbols of L($\in$, A, Ø, M) include an additional 1-place predicate symbol, M, and *(ii)* the definition of "atomic formula" includes the following three additional clauses: "$M(u)$ is an atomic formula of L($\in$, A, Ø, M) for each variable u of L($\in$, A, Ø, M)," "$M(A)$ is an atomic formula of L($\in$, A, Ø, M)," and "M(Ø) is an atomic formula of L($\in$, A, Ø, M)."

$M(x)$ is often written as "x is meaningful."

Note that because L($\in$, A, Ø) is a sublanguage of L($\in$, A, Ø, M), the axiom system ZFA can be formulated in L($\in$, A, Ø, M). However, when this is done, care should be taken that the axioms of Comprehension and Replacement are restricted to the sublanguage L($\in$, A, Ø); that is, the axioms of Comprehension and Replacement are to be applied only to formulas of L($\in$, A, Ø, M) not involving the predicate M.

As mentioned previously, the basic approach to meaningfulness taken in this chapter is an axiomatic one that takes meaningfulness as a undefined term. In general, I find for the specification of slippery concepts this kind of approach superior to those that try to capture them through explicit definitions, because to justify the latter one needs to have a fairly complete understanding of the concept, whereas to justify the former one needs only to justify the axioms, which are often formulated to capture some well-understood, integral part of the concept rather than the entire concept. Also, the axiomatic approach can apply to situations for which no explicit definition is possible. An example of this for meaningfulness is given in Section 4.4, where a structure is produced in which M is not directly expressible in terms of L($\in$, A, $\emptyset$). Nevertheless, as various theorems of Sections 4.2, 4.3, and 4.4 make clear, it is often the case that various natural axiomatizations of meaningfulness formulated in the richer language L($\in$, A, $\emptyset$, M) are reducible to ones formulated in L($\in$, A, $\emptyset$). However, it should be noted that in most of the interesting cases such reductions are not easy and depend on complicated series of theorems.

4.1.3 The Axiom System ZFA

Embodied in almost all of the axiom systems of the chapter is the axiom system ZFA. It provides a well-accepted axiomatization of set theory. Within it, the collection of pure sets P is definable (Definition 3.5.1), and P to gether with the restriction of the membership relation $\in$ to it form a model of a closely related and much used axiomatization of set theory, ZF (Theorem 3.7.1).

By use of ZF alone, it is easy to show that the finite ordinals, ω, (which, of course, are themselves pure sets) together with the ordinal successor function $a^+ = a \cup \{a\}$ satisfy G.Peano's well-known axiom system for the natural numbers. Then, by using the well-known constructions of Peano and Dedekind, it is easy to show that the operations of addition and multiplication are definable (within ZF) on ω and that the resulting algebraic structure—which we call the *set-theoretic natural numbers*—is isomorphic to the natural numbers with its usual operations of addition and multiplication. By taking ordered pairs of set-theoretic natural numbers and (in terms of the set-theoretic arithmetical operations of addition and multiplication) defining the appropriate operations of addition and multiplication on them (which is easy to do within ZF), another algebraic structure results—which we call the *set-theoretic integers*—that is isomorphic to the usual system of integers with its usual operations of addition and multiplication. Similarly, by taking ordered pairs of the set-theoretic integers and defining for them the appropriate operations of addition and multiplication, the structure of *set-theoretic rational numbers* can be constructed, and it is easy to show that it is isomorphic to the rational numbers under addition and multiplication. By using similar, but little more involved constructions, the structure of *set-theoretic real numbers* can be constructed out of the set-theoretic rational numbers, and it can be shown that it is isomorphic to the structure of real numbers with its usual operations of addition and multiplication. And of course these latter constructions can also be carried out in the axiom system ZF.

Today it is well-known that the structures used in ordinary mathematics have isomorphic models constructed (within ZF) from the set-theoretic natural, rational, and real numbers. For example, Euclidean plane geometry is isomorphic to the *set-theoretic Euclidean plane geometry,* which is a structure based on ordered pairs of set-theoretic real numbers (which are its "points") and has as "lines" sets of "points" (x, y) that satisfy an equation of the form $ax+by=c$, where a, b, and c are set-theoretic real numbers such that a and b are not both 0, and has as "circles" sets of "points" (x, y) that satisfy equations of the form $(x-a)^2+(y-b)^2=c^2$, where a, b, and c are set-theoretic real numbers and $c>0$.

In the previous development, I have taken great care to distinguish mathematical structures from their isomorphic, set-theoretic counterparts. There are various reasons for this. First, I did not want to embrace the philosophical doctrine of logicism, which identifies mathematical structures with their set-theoretic counterparts. Second, I thought it was important to stress that a great deal is obtained by understanding mathematical and scientific structures axiomatically, which is most efficiently accomplished by providing categorical sets of axioms and isomorphic set-theoretic models. And finally, I wanted to allow for theories of meaningfulness for mathematical as well as scientific structures, for example, for systems of geometry, and I felt that mathematical situations were best dealt with by distinguishing them from their set-theoretic counterparts. Such distinctions would allow, for example, for interesting mappings from qualitative mathematical structures into numerically based ones—where for the purposes of this chapter "numerically based" can be taken to be "set-theoretic." Traditionally, these kinds of mappings have played a central role in mathematics—particularly in geometry— where they were the focus of many meaningfulness issues, for example, the construction and use of canonical systems of coordinates.

Within ZFA, the domain of interest—be it mathematical or scientific—is represented by the set of atoms, A, and the relevant mathematical or scientific structure is represented by a structure of concepts based on A, e.g, by an ordered set of relations on A . In this book, the mathematical or scientific structure are often called the *qualitative structure,* and its isomorphic counterpart—which of course has a pure set as its domain and pure sets as its relations, etc.—is often called the (isomorphic) *quantitative structure.* It is not difficult to show the existence of such isomorphic quantitative counter-parts within ZFA. (However, it can be shown that any general proof of this necessarily involves the axiom of Choice.) Also, various kinds of relationships between qualitative and quantitative structures are easily formulated within ZFA. Because of the aforementioned and the enormous richness and flexibility of ZFA, I consider it to be an ideal basis for the formulation of general definitions of meaningfulness.

Some might object to this, since ZFA brings with it too much "metaphysics" with its assumptions of actual infinities and the axiom of Choice. Although I am in sympathy with this point of view, I see no way to effectively avoid such infinitistic approaches if the mathematics used in science is to be incorporated into the theory, because such "metaphysical commitments" are already an integral part of modern, ordinary mathematics. Of course, one could do with a weaker fragment of set theory—for example, with one that did not assume the axiom of Replacement or the existence of "large" infinities—and still be able to capture almost all of "ordinary" mathematics. But such modifications would not resolve the basic problem of "too much metaphysics."

My limited approach to this problem is to consider some axiomatic systems of meaningfulness that require meaningful entities not to be "too meta-physical." More will be said about this throughout the book.

4.1.4 Pure versus Applied Mathematics

In discussing the role of mathematics in science, it is useful to have a clear distinction between pure and applied mathematics. The usual use of "applied mathematics" appears to me to be quite muddled and of little value in theoretical discussions about the nature of science. In this chapter, a different concept of applied mathematics is adopted, one that is clearly stated and has important ramifications for theories of meaningfulness: *Applied mathematics* is concepts and true statements formulated in the system ZFA; *pure mathematics* is concepts and true statements formulated in the system ZF.

By these definitions, applied mathematics includes all of pure mathematics. This is reasonable if one wants to say that any part of pure mathematics is potentially applicable in some scientific or qualitative mathematical setting. In any case, history has taught us that it is hazardous to rule out parts of pure mathematics as nonapplicable. Also by this definition, applied mathematics has concepts that are not part of pure mathematics, for example, the qualitative set of entities, A, and concepts that are partly based on pure mathematics, for example, certain functions from A into the (set-theoretic) real numbers, which in scientific applications are often called "measurements" of A.

I realize that others may prefer a different kind of distinction between applied and pure mathematics, and for other applications such distinctions may make more sense than the one I have presented, which was designed to clarify the role pure mathematics plays in applied mathematics, particularly in the meaningfulness considerations that I perceive to be inherent in the Erlanger Program.

4.1.5 The Intended Use and Scope of the Meaningfulness Concept

Various definitions and uses of meaningfulness have appeared in the literature. In this book, four of these are singled out and given precise roles and technical formulations: *empiricalness, qualitativeness, intrinsicness,* and *meaningfulness.* The intended interpretations of the first two are the obvious ones corresponding to their names. Intrinsicness is a generalization of an intuitive concept of lawfulness. The concept of meaningfulness is intended to capture in a very general way the topic or subject matter that the piece of science or applied mathematics under consideration is about.

As an example of the latter, consider the case where A is the set of points of a Euclidean plane. A can be used as a basis for many geometries, including Euclidean geometries of higher dimension as well as noneuclidean ones. This is because no structure has been given to A other than it being a set of atoms. A is made into a geometry by adding extra structure in terms of special relations and concepts based on A. Suppose that this is the case for $A, R_1, \ldots, R_n$; specifically, suppose $A, R_1, \ldots, R_n$ is one of the ways of specifying Euclidean plane geometry in terms of higher order relations, (e.g., A is the set of points, R_1 is the set of lines, where each line is a set of points, etc.) Then *for this geometry* the meaningful relations and concepts based on A are the Euclidean ones, that is, the relations and concepts that belong to the topic of this Euclidean plane geometry based on A. There are other ways of specifying the topic of this geometry, for example, by $A, S_1, \ldots, S_m$, where some of the S_i are sufficiently different from the R_j. The nonmeaningful relations based on A are those relations based on A that do not belong to the topic of this Euclidean geometry based on A. This does not mean that such relations are not "geometric" in all senses. Indeed, it may be the case that some of them belong to topics of other geometries based on A.

As another example, consider Einstein's Special Theory of Relativity. Special relativity is concerned with observations made from a specific set of frames of reference. There are two obvious topics to associate with this situation. The first is about relationships that hold for every frame of reference. The second is about relationships and the frames of reference for which they hold. While both topics are valid, only the first is used in powerful ways in the special theory of relativity. The second, although feasible as a topic for physical matters, fails to isolate the interesting phenomena that is the core of special relativity. This is why the first and not the second is chosen to be the "topic of special relativity." Note that the phenomena considered by the second topic includes all the phenomena considered by the first topic.

In general, throughout this book, meaningfulness is a relative concept, often formulated in terms of specific relations that determine a specific topic, and often crafted with specific objectives in mind. Thus it is fair to debate *(i)* whether the specific relations capture the correct topic, or *(ii)* whether the particular meaningfulness concept is correct for the objectives in mind. It is not fair to challenge such relative meaningfulness concepts by *(a)* providing examples with different objectives in mind or *(b)* providing examples about different topics; that is, it is not fair to treat a specifically designed relative concept as an absolute one.

In geometry and physics, invariance under a group of automorphisms has been traditionally identified with a topic. This is consistent with the above view of meaningfulness when meaningfulness is defined to be invariance under a group of automorphisms, as in Section 4.2. The following quotation from Veblen and Young (1946) illustrates for geometry the identification of meaningfulness (as invariance) with a topic, and shows how this identification is used in geometry to transfer results and concepts from one geometric structure to another:

> At each step we have helped ourselves forward by transferring the results of one geometry to another, combining these with easily obtained theorems of the second geometry, and thus extending our knowledge of both. This is one of the characteristic methods of modern geometry. It was perhaps first used with a clear understanding by O.Hesse [*Gesammelte Werke,* p. 531], and was formulated as a definite geometrical principle (Ubertragungsprinzip) by F.Klein [1872].
>
> This principle of transference or of carrying over the results of one geometry to another may be stated as follows: *Given a set of elements [e] and a group G of permutations of these elements, and a set of theorems [T] which state relations left invariant by G. Let [e'] be another set of elements, and G' a group of permutations of [e']. If there is a one-to-one reciprocal correspondence between [e] and [e'] in which G is simply isomorphic with G', the set of theorems [T] determines by a mere change of terminology a set of theorems [T'] which state relations among elements e' which are left invariant by G'.*
>
> This principle becomes effective when the method by which *[e]* and *G* are defined is such as to make it easy to derive theorems which are not so easily seen for *[e']* and *G'*. This has been abundantly illustrated in the present chapter…
>
> DEFINITION. Given a set of elements *[e]* and a group *G* of permutations of [e], the set of theorems *[T]* which state relations among the elements of *[e]* which are left invariant by *G* and are not left invariant by any group of transformations containing *G* is called a *generalized geometry* or *a branch of mathematics.*
>
> This is, of course, a generalization of the definition of a geometry [by Klein, 1872]… *(Veblen and Young, 1946, pp. 284–285)*

Throughout this book, invariance is taken as a necessary condition for meaningfulness. For some axiomatic systems, it is also a sufficient condition. Because in this book empiricalness is taken as a concept separate from meaningfulness and invariance, invariance is in general neither a necessary nor a sufficient condition for empiricalness. However, because invariance is a necessary condition for meaningfulness, it follows that invariance is also a necessary condition for the concept "empirical and meaningful." I believe that this latter concept, which singles out the empirical relationships and concepts that belong to a topic, is the concept that many authors of meaningfulness papers in the literature had in mind when they identified meaningfulness (as invariance) with empiricalness.

4.2 The Erlanger Program: Transformational Meaningfulness

In various parts of mathematics and science, meaningfulness concepts have been identified with invariance, the concept of "geometric" of Felix Klein's Erlanger Program being one of several prominent examples. In this section, a precise way of making this identification is given, and some implications of the identification are explored.

The following important concepts are used throughout the rest of the book.

Definition 4.2.1 Assume L(∈, *A*, Ø, *M*) and the *axiom system* ZFA. Let

$$\Pi = \{f \mid f \text{ is a permutation on } A\}.$$

Let *G* be the collection of all elements *f* of *Π* such that for each entity *x*, if *M(x)*, then *f(x)=x*. *G* is called the *transformation group* of *M*.

Note that in Definition 4.2.1, *Π* is a *set* that is defined within ZFA through a use of the axiom of Comprehension, whereas *G* is defined to be a *collection* of elements of *Π* and is not assumed to be a set within ZFA. We are not able to conclude that *G* is a set, because it is defined in terms of a concept, namely *M(x)*, that is not part of the language L(∈, *A*, Ø) and thus is not an appropriate concept for an application of the axiom of Comprehension of ZFA.

Convention 4.2.1 Throughout the remainder of this book, unless explicitly stated otherwise, *Π* will denote the set of permutations on *A* and *G* will denote the transformation group of *M*.

If there happens to be an entity *H* such that for all entities *f*, *f*∈*H* if and only if *f* is in *G*, then *G* will be identified with *H*. Such situations will usually be noted by saying "*G* is an entity" or "*G* is a set," with the understanding that in the latter, "set" means "an entity in ZFA that is a set." The more general situation, where the collection *G* is not necessarily a set, will usually be noted by stating explicitly that *G* is a collection. In most of the axiomatic systems considered in this Chapter, axioms are added to ZFA that ensure that *G* is a set.

In ZFA a collection of entities that is not a set has already been encountered, namely, the collection of all entities, *V*. (This follows from the axiom of Regularity and Theorem 3.5.8, which states that all entities are in *V*.) Collections like *V* and the set of all ordinals, which are specifiable through formulae of L(∈, *A*, Ø) but are "too large" to be sets of the theory, are often in the literature called "classes," and they play an important role in some axiomatic versions of set theory. When the transformation group of M is not a set, it plays a different role than the one that has been traditionally reserved for "classes that are not sets," since in such situations G cannot be considered to be "too large," because it is a subcollection of a set, namely the set Π.

Theorem 4.2.1 G (under function composition) is a group of permutations on A.

Proof. By Definition 4.2.1, each element of G is a permutation on A. It is immediate from Definition 4.2.1 that the identity permutation on A, ι, is in G. Suppose f and g are arbitrary elements of G and x is an arbitrary element of A such that $M(x)$. Then by Definition 4.2.1, $f(x)=x$ and $g(x)=x$. Thus

$$f^{-1}(x) = x \text{ and } f * g(x) = f[g(x)] = f(x) = x\,.$$

Since x is an arbitrary element of A such that $M(x)$, it then follows from Definition 4.2.1 that f^{-1} and $f{*}g$ are in G. Thus G is a group.

Definition 4.2.2 Let H be a collection of permutations on A and x be an entity. Then x is said to be *H-invariant,* in symbols, $I_H(x)$, if and only if for each f in H, $f(x)=x$.

Invariance under groups of permutations plays a critical role throughout applied mathematical science. Definition 4.2.1 provides a key link between invariance and meaningfulness. The interrelationships between these two concepts will be examined in greater detail throughout the chapter. The Erlanger Program, as captured in the following definition, identifies the two concepts.

Definition 4.2.3 The *axiom system* TM has as its language L($\in$, A, Ø, M). The axioms of TM consist of the following:

1. ZFA (formulated in L($\in$, A, Ø));
2. *TM (Transformational Meaningfulness):*

$$\exists H(H \subseteq \Pi \text{ and } \forall x[M(x) \leftrightarrow I_H(x)])\,.$$

Theorem 4.2.2 Assume TM. Then G is an entity and $\forall x[M(x) \leftrightarrow I_G(x)]$..

Proof. By Definition 4.2.3, let H be a subset of Π such that

$$\forall x[M(x) \leftrightarrow I_H(x)].$$

Then

$$G = \{f \mid f \in \Pi \wedge \forall x[I_H(x) \to f(x) = x]\}\,,$$

and thus G is an entity by the axiom of Comprehension of ZFA, and $\forall x\ [M(x) ® I_G(x)]$. Suppose a is an arbitrary entity such that $I_G(a)$. To show the theorem, we need only to show $M(a)$. Since $\forall x[M(x) \leftrightarrow I_H(x)]$, to show $M(a)$ we need only show $I_H(a)$. By Definition 4.2.1, G is the largest collection of all f in Π such that

$$\forall x[M(x) \rightarrow f(x) = x]. \tag{4.1}$$

Let f be an arbitrary entity such that $f \in H$. It then follows from the choice of H and Equation 4.1 that f is in G. From this and $I_G(a)$, it then follows that $I_H(a)$.

The following two lemmas give some useful consequences of TM.

Lemma 4.2.1 Assume TM. Then A and $\emptyset$ are meaningful.

Proof. Since G is a permutation group on A, it immediately follows that $I_G(A)$ and $I_G(\emptyset)$. Thus by Theorem 4.2.2, $M(A)$ and $M(\emptyset)$.

Lemma 4.2.2 Assume TM. Suppose a is a set such that for each $f \in G$, $f(a) \subseteq a$. Then a is meaningful.

Proof. If $a=\emptyset$, then $M(a)$ by Lemma 4.2.1. So suppose $a \neq \emptyset$. Let f be an arbitrary element of G. We will show $f(a)=a$. $f(a) \subseteq a$ by hypothesis. Let y be an arbitrary element of a. Since by Theorem 4.2.1 G is a group, $f^{-1} \in G$. Thus by hypothesis, $f^{-1}(a) \subseteq a$. Thus in particular, $f^{-1}(y) \in a$. Then by Theorem 3.6.1,

$$y = f[f^{-1}(y)] \in f(a).$$

Since y is an arbitrary element of a, we have shown $a \subseteq f(a)$. Thus $f(a)=a$. Since f is an arbitrary element of G, $I_G(a)$ follows. Thus by Theorem 4.2.2, $M(a)$.

Axiom system TM will imply other axiomatic systems of meaningfulness introduced throughout in the chapter. Many of the important axioms used in these systems are discussed in this section.

Definition 4.2.4 Assume L($\in$,A,$\emptyset$,M) and axiom system ZFA. Then *axiom MP (Meaningful Pure Sets)* is said to hold if and only if for each pure set x, $M(x)$ is true.

Theorem 4.2.3 TM→*MP*.

Proof. By Definition 4.2.3, let H be a subset of permutations on A such that

$$\forall x[M(x) \leftrightarrow I_H(x)]. \tag{4.2}$$

Let a be an arbitrary pure set. Then by Theorem 3.6.2, $f(a)=a$ for each permutation f on A. Thus in particular, $I_H(a)$. Therefore by Equation 4.2, $M(a)$. Since a is an arbitrary pure set, axiom *MP* has been shown.

The axiom of Meaningful Pure Sets, which by Theorem 4.2.3 is an immediate consequence of the Erlanger Program (as axiomatized by TM), at first sight appears to be a rather puzzling consequence. The intent of the meaningfulness concept is to capture significant mathematical applied relations and propositions, and pure sets are good examples, in the context of ZFA, of things that belong to pure rather than applied mathematics. Nevertheless, I want to have the opportunity to include the pure sets or a portion of them among the "meaningful" entities for the following four reasons: First, Axiom *MP* is a straightforward consequence of the Erlanger Program. Second, to exclude *MP* or weakened versions of it as meaningfulness principles would introduce a degree of inelegance to many of the axiomatizations and proofs given in this chapter, because extra clauses and steps that accomplish very little would have to be routinely added. For example, one of the important meaningfulness concepts presented in this chapter is a definability one which states that an entity that is appropriately defined in terms of meaningful entities is itself meaningful. Without the axiom of Meaningful Pure Sets, this would have to be reworded as follows: "An entity that is not a pure set that is appropriately defined in terms of pure sets and/or meaningful entities is itself meaningful." Third, because it is easy to exclude the meaningfulness of pure sets by introducing additional, inelegant clauses and extra steps in proofs, their inclusion as meaningful entities can be treated a harmless extension of the meaningfulness concept. By this I mean that the reader could, in a relatively mechanical way, make changes throughout this chapter so that the axiom of Meaningful Pure Sets needs not be assumed or used, while at the same time preserve the substance of all of the important results that make use of it. And fourth, the axiom of Meaningful Pure Sets together with the identification of pure set theory with pure mathematics reflect the commitment to treat the role of pure mathematics in science in a manner similar to that of logic in (pure) mathematics.

The axiom of Meaningful Pure Sets, particularly in the presence of certain definability axioms, turns out to be a very powerful principle, and it might be argued that a weakened form of it should be used, for example, only certain "constructible" pure sets should be meaningful. I see this as a reasonable option, but one that will not be pursued in any systematic way in this book. This option is briefly discussed in Section 4.7.

Definition 4.2.5 Assume ZFA and L(∈, *A*, Ø, *M*). Then *axiom MI (Meaningful Inheritability)* holds if and only if for each set *x*, if all elements of *x* are meaningful, then *x* is meaningful.

Theorem 4.2.4 TM→*MI*.

Proof. Assume TM. To show axiom *MI*, suppose *x* is a set such that

$$\forall y[y \in x \rightarrow M(y)]\,.$$

Then by Theorem 4.2.2

$$\forall y[y \in x \rightarrow I_G(y)]\,,$$

which by the definition of I_G yields

$$\forall y \forall f(y \in x \wedge f \in G \rightarrow f(y) = y)\,,$$

and from this immediately follows that for each $f \in G$,

$$f(x) = \{f(y) \mid y \in x\} = \{y \mid y \in x\} = x\,,$$

that is, that $I_G(x)$. Thus by Theorem 4.2.2, $M(x)$, and the axiom of Meaningful Inheritability has been shown.

Definition 4.2.6 Assume ZFA and L(∈, *A*, Ø, *M*). Then axiom DM* *(Definable Meaningfulness*)* is said to hold if and only if there exists a formula $\Phi(y, u_1,\ldots, u_n)$ of L(∈, *A*, Ø) such that

$$\exists u_1 \cdots \exists u_n [M(u_1) \wedge \cdots \wedge M(u_n) \wedge \forall y (M(y) \leftrightarrow \Phi(y, u_1, \ldots, u_n))]\,.$$

Lemma 4.2.3 Assume ZFA. Suppose *f* is a permutation on *A*, (x, y) an ordered pair of entities, and $(x_1,\ldots, x_n)$ an ordered *n*-tuple of elements, for $n>2$. Then

(i) $f[(x, y)]=(f(x), f(y))$, and
(ii) $f[(x_1,\ldots, x_n)]=(f(x_1),\ldots,f(x_n))$.

Proof. *(i)* By the definition of "ordered pair,"

$$(x, y) = \{\{x\}, \{x, y\}\}\,.$$

Thus

$$\begin{aligned} f[(x,y)] = f[\{\{x\},\{x,y\}\}] &= \{f[\{x\}], f[\{x,y\}]\} \\ &= \{\{f(x)\},\{f(x),f(y)\}\} \\ &= (f(x), f(y))\,. \end{aligned}$$

(ii) will be shown by induction. Suppose $2<k<n$ and

$$f[(x_1,\ldots,x_k)] = (f(x_1),\ldots,f(x_k))\,.$$

Now

$$(x_1,\ldots,x_k,x_{k+1}) = ((x_1,\ldots,x_k),x_{k+1})\,.$$

Thus by Part *(i)*, and the induction hypothesis,

$$\begin{aligned} f[(x_1,\ldots,x_k,x_{k+1})] &= f[((x_1,\ldots,x_k),x_{k+1})] \\ &= (f[(x_1,\ldots,x_k)], f[x_{k+1}]) \\ &= ((f(x_1),\ldots,f(x_k)), f(x_{k+1})) \\ &= (f(x_1),\ldots,f(x_k), f(x_{k+1}))\,. \end{aligned}$$

Lemma 4.2.4 TM→*M(G)*.

Proof. Assume TM. *G* is an entity by Theorem 4.2.2. We will show *G* is meaningful. Let *g* and *f* be arbitrary elements of *G*. Then

$$g = \{(x, g(x)) \mid x \in A\}.$$

Since *f* is onto *A*, it then follows that

$$g = \{(f^{-1}(x), g[f^{-1}(x)]) \mid x \in A\}. \tag{4.3}$$

Since by Lemma 4.2.3,

$$f[(f^{-1}(x), g * f^{-1}(x))] = (f * f^{-1}(x), f * g * f^{-1}(x)) = (x, f * g * f^{-1}(x)),$$

it follows from Equation 4.3 that

$$f(g) = \{(x, f * g * f^{-1}(x)) \mid x \in A\} = f * g * f^{-1},$$

and since *G* is a group, it follows from this that $f(g) \in G$. Because *g* is an arbitrary element of *G*, we have shown that *f(G)* is a subset of *G*. Since *f* is an arbitrary element of *G*, it then follows from Lemma 4.2.2 that *G* is meaningful.

Theorem 4.2.5 TM→*DM**.

Proof. Assume TM. By Lemma 4.2.4, *G* is meaningful. Therefore by Theorem 4.2.2 and Definition 4.2.6, axiom *DM** is true.

Theorem 4.2.6 Assume ZFA and L($\in$,*A*,Ø,*M*). Suppose axiom *DM**. Then

$$[M(\varnothing) \wedge MI] \rightarrow MP.$$

Proof. Assume *M*(Ø) and *MI*. We need only show that the axiom of Meaningful Pure Sets, *MP*, is true. Suppose *MP* is false. *A* contradiction will be shown. Let *a* be an element of *P* that has least rank α. such that *a* is not meaningful. (Note that axiom *DM** is needed to show the existence of *a*. Two other axioms that will be given later—*MC* and *DM*—will also show the existence of a, and therefore can be used in place of *DM** for the purposes of this theorem.) $a \neq Ø$ since Ø is meaningful. Since *a* is a pure set, each element of *a* is a pure set (Statement 4 of Theorem 3.5.7). Since each element of *a* has rank$<\alpha$ (Theorem 3.5.6), it follows from the choice of α that each element of *a* is meaningful. Thus by axiom *MI a* is meaningful, a contradiction.

The axioms of Meaningful Inheritability and Definable Meaningfulness* are consequences of the Erlanger Program, at least when it is formulated as axiom system TM. They together with *M*(Ø) implies axiom *MP* (Theorem 4.2.6). Although at first sight axiom *MI* looks rather innocuous, it is in fact a very powerful assumption, as Theorem 4.2.6 and other theorems of this chapter will show. For meaningfulness, it is not a true definability or constructibility principle, because if *a* is a set of meaningful entities, then in general there need not be a rule that specifies *a* in terms of finitely many of its elements. Indeed, a can even result by an application of the axiom of Choice to a well-specified set of sets of meaningful entities.

The following axiom provides the basis for a definitional treatment of meaningfulness that will be given in the next section:

Definition 4.2.7 Assume ZFA and L(∈, *A*, Ø, *M*). Then *axiom MC' (Meaningful Comprehension']* is said to hold if and only if for all formulas

$$\Phi(x, u_1, \ldots, u_n)$$

of L(∈, *A*, Ø) and all entities $a, a_1, \ldots, a_n$, if $a_1, \ldots, a_n$ are meaningful and

$$a = \{x \,|\, \Phi(x, a_1, \ldots, a_n)\},$$

then *a* is meaningful.

Note that in Definition 4.2.7, Φ is a formula of L(∈, *A*, Ø) and therefore does not contain the predicate *M*.

The axiom of Meaningful Comprehension′ allows one to "construct" or "define" new meaningful entities in terms of already known ones. It is a principle very similar to the axiom of Comprehension of ZFA. Because of this, it is useful to remark briefly about the introduction of the axiom of Comprehension into mathematics.

In 1908, Zermelo axiomatized set theory. He used the following version of the axiom of Comprehension (which he called the "axiom of Separation"): If a propositional function *P(x)* is definite *(definit)* for a set *S*, then there is a set *T* containing precisely those elements *x* of *S* for which *P(x)* is true. Moore (1982) writes the following about this definition: "For Zermelo, a propositional function *P(x)* was *definit* for a set *S* if the membership relation on *B* [i.e., on the collection of entities] and the laws of logic determined whether *P(x)* held for each *x* in *S*."

In the 1920s, A.Fraenkel and T.Skolem gave specific suggestions as to what *definit* should be, and these essentially amounted to versions of the axioms of Comprehension and Replacement. Zermelo was unhappy with these suggestions. He believed that *definit* should be allowed to include more powerful principles than first-order definability (which was the basis for Fraenkel's and Skolem's treatments of *definit*). Zermelo, however, was never able to provide a completely satisfactory account as to what these more powerful definability concepts should be.

The initial reaction to Zermelo's axiom of Comprehension was mixed, with many prominent mathematicians and philosophers considering it to be either vague or completely wrong. A good deal of the criticism directed at it was really about more general issues than the axiom of Comprehension—such as whether mathematics should be constructible—and does not apply directly to meaningfulness issues of the current context. Most of the remaining criticism concerned the vague way that *definit* was originally presented, but the later work of Fraenkel and Skolem basically took care of this concern.

Axiom *MC'* creates no new entities; it only describes a procedure by which already known entities can be determined to be meaningful. The basic motivation for axiom *MC'* is that entities that are "definable" from meaningful entities should themselves be meaningful. Because axiom *MC'* is clearly a "definability" principle, the issue about its acceptability reduces to basically this: Does its use produce some meaningful sets that are really not "meaningful"? That is, is *MC'* too strong a definability principle for meaningfulness concerns? This is difficult to answer. As we shall see, it is certainly much weaker than the principles behind the Erlanger Program's concept of meaningfulness, and it is certainly a natural concept of "definability" given that one is working in the context of ZFA, which already contains a closely related concept of "definability," namely the axiom of Comprehension. It seems to me that this is the type of question that is most likely to be answered by observing the consequences of *MC'* in the contexts of other meaningfulness axioms, and deciding whether or not such consequences make sense in such contexts.

Theorem 4.2.7 TM→*MC'*.

Proof. The following concepts and results of model theory will be used: A *model* of a first-order language $L(\mathbf{P}_1,\ldots, \mathbf{P}_n, \mathbf{c}_1,\ldots, \mathbf{c}_m)$, where the $\mathbf{P}_i$ are *k(i)*-ary predicate symbols and the c_j are individual constant symbols, is a structure of the form $\mathfrak{X} = \langle X, R_1, \ldots, R_n, a_1, \ldots, a_m\rangle$, where X is a nonempty set, the R_i are *k(i)*-ary relations on X, and the a_j are elements of X. Sentences of $L(\mathbf{P}_1,\ldots, \mathbf{P}_n, \mathbf{c}_1,\ldots, \mathbf{c}_m)$ are interpreted in $\mathfrak{X}$ by interpreting $\mathbf{P}_i$ by R_i, c_j by a_j, and giving the logical symbols and parentheses their usual interpretations, e.g., $\forall$ as "for all," $\neg$ as "not," and = as =. Suppose $\mathfrak{Y}$ is another model of $L(\mathbf{P}_1,\ldots, \mathbf{P}_n, \mathbf{c}_1,\ldots, \mathbf{c}_m)$ that is isomorphic to $\mathfrak{X}$ by an isomorphism *f*. Then it is a theorem of model theory that *(i)* for each sentence Σ of $L(\mathbf{P}_1,\ldots, \mathbf{P}_n, \mathbf{c}_1,\ldots, \mathbf{c}_m)$, Σ is true about $\mathfrak{X}$ if and only if it is a true statement about $\mathfrak{Y}$, and *(ii)* for each formula $\Phi(x)$ of $L(\mathbf{P}_1,\ldots, \mathbf{P}_n, \mathbf{c}_1,\ldots, \mathbf{c}_m)$, if a in X is such that $\Phi(a)$ is true about $\mathfrak{X}$, then $\Phi(f(a))$ is true about $\mathfrak{Y}$. A consequence of the latter, which plays a central role in this proof, is that isomorphisms of $\mathfrak{X}$ onto $\mathfrak{X}$ (i.e., automorphisms of $\mathfrak{X}$) preserve relations that are defined through first-order definitions in terms of the primitives of $\mathfrak{X}$. (For example, see Corollary 220 on page 93 of Enderton, 1972.) Properties *(i)* and *(ii)* are natural and intuitive properties about isomorphisms of models. A structure $\mathfrak{Z}$ is said to be a *model of a collection of sentences* Γ of $L(\mathbf{P}_1,\ldots, \mathbf{P}_n, \mathbf{c}_1,\ldots, \mathbf{c}_m)$ if and only if $\mathfrak{Z}$ is a model of $L(\mathbf{P}_1,\ldots, \mathbf{P}_n, \mathbf{c}_1,\ldots, \mathbf{c}_m)$ and each sentence in Γ is true about $\mathfrak{Z}$.

Now for the proof of Theorem 4.2.7. Assume axiom system TM. It is formulated in the first-order language $L(\in, A, \emptyset, M)$. To show the theorem, it is sufficient to show the following: MC' is true about $\mathfrak{V}'$ for each model $\mathfrak{V}'$ of TM. So let $\mathfrak{V}' = \langle V', \in', A', \varnothing', M' \rangle$ be an arbitrary model of TM. Note that $\mathfrak{V}'$ is an ordinary mathematical structure; that is V' is a nonempty set, $\in'$ 'is a binary relation on $\mathfrak{V}'$, A' and $\varnothing'$ are elements of V', and M' is a subset of V'. It is not assumed that any of the elements of V' (as ordinary mathematical objects) are sets, nor is it assumed that $\in'$ (as an ordinary mathematical relation) is the membership relation between sets in ordinary mathematics: $\in'$ is just a binary relation between elements of V', and $\mathfrak{V}'$ is just one of infinitely many quite different structures of ordinary mathematics that satisfies the axiom system TM.

Since $\mathfrak{V}'$ is a model of TM, every consequence of TM (within the language L($\in$, A, $\varnothing$, M)) is also a true statement about $\mathfrak{V}'$. Thus in $\mathfrak{V}'$, let Π' be the element of V' such that the statement

Π' is the set of one-to-one functions from A onto A

is a true statement about $\mathfrak{V}'$. Let f be an arbitrary element of $\mathfrak{V}'$ such that

$$f \in' \Pi' .$$

A function $\tilde{f}$ (of ordinary mathematics) from V' onto V' is associated with f in the following manner: For all elements u, v of V', let $(u, v)'$ denoted the element w of V' such that

$$w = (u, v)$$

is a true statement about $\mathfrak{V}'$. Thus within the model $\mathfrak{V}'$, $(u, v)'$ is the "ordered pair of elements u and v." By Definition 3.6.2, f can be extended to a functional relation $\hat{f}$ on $\mathfrak{V}'$ (which is expressible in L($\in$, A, $\varnothing$, M)), which by convention is also denoted by f. By Theorem 3.6.1, this functional relation has the following important property: For all a, b in V',

$$a \in' b \leftrightarrow f(a) \in' f(b) .$$

Define $\tilde{f}$ on V' as follows: for all a, b in V',

$$\tilde{f}(a) = b \text{ iff } (a, b)' \in' f .$$

Using facts about f that were established in Section 3.6 of Chapter 3 (and are expressible in L($\in$, A, $\varnothing$, M)), it is easy to establish the following:

(1) $\tilde{f}$ is a one-to-one function from V' onto V'.
(2) $\tilde{f}(A') = A'$ and $\tilde{f}(\varnothing') = \varnothing'$..
(3) For all a and b in V', $a \in' b$ if and only if $\tilde{f}(a) \in' \tilde{f}(b)$..

(For example, the one-to-oneness of $\tilde{f}$ follows from the fact that

$$\forall x \forall y \forall z[(x, y) \in f \wedge (z, y) \in f \rightarrow x = z]$$

is a true statement about $\mathfrak{V}'$.) (1), (2), and (3) say that $\tilde{f}$ is an isomorphism of $\langle V', \in', A', \varnothing' \rangle$ onto itself, that is, is an automorphism of $\langle V', \in', A', \varnothing' \rangle$.

Let $\Phi(x, u_1, \ldots, u_n)$ be an arbitrary formula of L($\in$, A, Ø) and $a, a_1, \ldots, a_n$ be arbitrary elements of V' such that

$$M'(a_1) \wedge \cdots \wedge M'(a_n) \wedge \forall x(x \in' a \leftrightarrow \Phi(x, a_1, \ldots, a_n))$$

is a true statement about $\mathfrak{V}'$. To show MC', it is sufficient to that $M'(a)$ is a true statement about $\mathfrak{V}'$. Since axiom system TM is true about $\mathfrak{V}'$, let G' in V' be such that

G' is the transformation group for M'

is a true statement about $\mathfrak{V}'$. Let g be an arbitrary element of $\mathfrak{V}'$ such that $g' \in G'$. $\tilde{g}$ Let be defined in a manner analogous to $\tilde{f}$ above. Because axiom system TM is true about $\mathfrak{V}'$, to show $M'(a)$ it is sufficient to show that

$$(a, a)' \in' g \,.$$

This is equivalent to showing $\tilde{g}(a) = a$. Since by hypothesis $M'(a_1) \wedge \cdots \wedge M'(a_n)$, it follows from axiom system TM that

$$(a_1, a_1)' \in' g, \ldots, (a_n, a_n)' \in' g \,,$$

and thus that

$$(4)\ \tilde{g}(a_1) = a_1 \text{ and } \cdots \text{ and } \tilde{g}(a_n) = a_n \,.$$

Consider the structure

$$\mathfrak{S} = \langle V', \in', A', \varnothing', M', a_1, \ldots, a_n \rangle \,.$$

By (4) and using (1), (2), and (3) above with $\tilde{f} = \tilde{g}$, it follows that $\tilde{g}$ is an automorphism of $\mathfrak{S}$ Since $\Phi(x, \mathrm{a}_1, \ldots, \mathrm{a}_n)$ is a formula of

$$\mathsf{L}(\in, \boldsymbol{A}, \varnothing, \boldsymbol{M}, \mathbf{a}_1, \ldots, \mathbf{a}_n)$$

(of $\mathfrak{S}$ which is a model) and

$$\forall x(x \in' a \leftrightarrow \Phi(x, a_1, \ldots, a_n))$$

is a true statement about $\mathfrak{S}$, it follows from the theorem of model theory about automorphisms preserving definitions mentioned at the beginning of this proof that $\tilde{g}(a) = a$.

Axiom MC' says that if a *set* is definable through a formula of L($\in$, A, Ø) and meaningful entities, then it is meaningful. This axiom does not apply to atoms that are definable through formulas of L($\in$, A, Ø). (MC' applied to this latter case would yield the singleton sets of such atoms as meaningful.) The following principle is a particularly simple axiom that when combined with *MC′* yields the meaningfulness of appropriately defined atoms:

Definition 4.2.8 Assume ZFA and L(∈, *A*, Ø, *M*). Then *axiom AL (Atomic Legacy)* is said to hold if and only if for each *a* in *A*, if *{a}* is meaningful, then *a* is meaningful.

Theorem 4.2.8 TM→*AL*.

Proof. Assume TM. To show *AL*, suppose *a* is an element of *A* and *M({a})*. Then by Theorem 4.2.2, $I_G(\{a\})$. Let *f* be an arbitrary element of *G*. Then *f({a})={a}*. But *f({a})={f(a)}*. Therefore *{f(a)}={a}*, i.e., *f(a)=a*. Since *f* is an arbitrary element of *G*, we have shown $I_G(a)$. Therefore by Theorem 4.2.2, *M(a)*, and thus the axiom of Atomic Legacy is true.

Theorem 4.2.9 Assume ZFA and *MC'* and *AL*. Then

$$\forall x[M(x) \leftrightarrow M(\{x\})].$$

Proof. Suppose *a* is an arbitrary entity and *M(a)*. Let

$$b = \{y \mid y = a\}.$$

Then *b={a}*, and by axiom *MC′*, *M(b)*.

Now suppose *a* is an arbitrary entity and *M[{a}]*. If *a*∈*A*, then *M(a)* by axiom *AL*. If *a*∉*A*, then *a* is a set and

$$a = \{y \mid y \in a\} = \{y \mid \exists z(y \in z \wedge z \in \{a\})\},$$

which by axiom *MC'* yields *M(a)*.

Within the context of ZFA with a definability approach to meaningfulness, the axiom of Atomic Legacy is a very reasonable principle, since clearly an atom a should be able to be "defined in a meaningful way" out of the set {a}.

The previous theorems show that axiom system TM imply axioms *MP*, *MI*, *DM**, *MC′*, and *AL*. In addition, axiom system ZFA is part of TM, so TM also implies ZFA. The following theorem shows that the conjunction of these consequences are in fact logically equivalent to TM:

Theorem 4.2.10 TM↔[ZFA ∧ *MP* ∧ *MI* ∧ *DM** ∧ *MC′* ∧ *AL*].

Proof. TM→[ZFA ∧ *MP* ∧ *MI* ∧ *DM** ∧ *MC′* ∧ *AL*] by Theorems 4.2.3, 4.2.4, 4.2.5, 4.2.7, and 4.2.8. [ZFA ∧ *MP* ∧ *MI* ∧ *DM** ∧ *MC′* ∧ *AL*]→TM by Theorems 4.10.3 and 4.2.11 (below), and Definition 4.2.10.

There are other and more elegant ways of stating logical equivalences of TM that stress definability concepts of meaningfulness. One such employs the following axiom:

Definition 4.2.9 Assume ZFA and L(∈, *A*, Ø, *M*). Then *axiom MC (Meaningful Comprehension)* is said to hold if and only if for all formulas

$$\Phi(x, u_1, \ldots, u_n)$$

of L(∈, A, Ø, M) and all entities a, $a_1, \ldots, a_n$, if a, $a_1, \ldots, a_n$ are meaningful and a is a set, then there exists a meaningful set b such that

$$b = \{x \mid \Phi(x, a_1, \ldots, a_n) \wedge x \in a\}.$$

Note that unlike *MC′* axiom *MC* applies to all formulas of L(∈, A, Ø, M) and not just to those of L(∈, A, Ø). Also note that because of this, an extra clause is required to guarantee the existence of the set [b in Definition 4.2.9] whose elements satisfy the relevant formula of L(∈, A, Ø, M) [$\Phi(x, a_1, \ldots, a_n) \wedge x \in$ in Definition 4.2.9].

Clearly *MC*→*MC′*. The following theorem shows that in certain situations the converse is also true:

Theorem 4.2.11 Assume ZFA. Then

$$(DM^\star \text{ and } MC') \rightarrow MC.$$

Proof. Assume *DM** and *MC′*. By *DM**, let $\Psi(y, u_1, \ldots, u_n)$ be a formula of L(∈, A, Ø) and $a_1, \ldots, a_n$ be entities such that $M(a_1), \ldots, M(a_n)$ and

$$\forall y[M(y) \leftrightarrow \Psi(y, a_1, \ldots, a_n)]. \tag{4.4}$$

To show the theorem we need only show that *MC* is true. This will immediately follow from the axiom of Comprehension of ZFA and axiom *MC′* if we can show that for each formula $\Phi(x, v_1, \ldots, v_k)$ of L(∈, A, Ø, M) there exists a formula $\Phi'(x, v_1, \ldots, v_{k+n})$ of L(∈, A, Ø) such that

$$\forall x \forall v_1 \cdots \forall v_k[\Phi(x, v_1, \ldots, v_k) \leftrightarrow \Phi'(x, v_1, \ldots, v_k, a_1, \ldots, a_n)]. \tag{4.5}$$

The following shows how to construct Φ': If an expression of the form "$M(w)$" occurs in Φ, replace it by the expression "$\Psi(w, a_1, \ldots, a_n)$" and call the formula resulting from all such replacements Φ'. Then Equation 4.5 follows from Equation 4.4.

Axiom *MC* appears to be a weaker principle than the combination of *DM** and *MC′*, However, as Theorem 4.2.12 shows, in the presence of ZFA, *AL*, and *MI* this is not the case.

Definition 4.2.10 *Axiom system M* is formulated in the language

$$\mathsf{L}(\in, A, \varnothing, M)$$

and consists of the following axioms: ZFA, *MC*, *AL*, and *MI*.

Theorem 4.2.12 TM↔M.

Proof. Theorem 4.10.3.

TM is the most powerful axiom system considered in the chapter, and because it represents a formalization of meaningfulness for the Erlanger Program, each of its equivalences can be considered as an alternative formulation of the Erlanger Program, and each of its implications that is not an equivalence as a proper generalization of the Erlanger Program. Its most important equivalence is axiom system M, which attempts to capture the Erlanger Program as much as possible in terms of definability concepts. I believe that for philosophical considerations the formulation given by M is far superior to that given by TM for the following reason:

In TM, a group of transformations H on the atoms is selected and meaningfulness is defined in terms of invariance under it. As noted before, it is no means obvious why such a group should play a central role in meaningfulness considerations. Axiom system M, however, is formulated in terms of conditions that are far more understandable as meaningfulness principles, and because of this, it presents far more clearly just what the Erlanger Program is assuming about meaningfulness.

The next two sections generalize the Erlanger Program in two ways, each corresponding to a side of the equivalence TM $\leftrightarrow$ M: *(i)* by weakening some of the axioms that make up M; and *(ii)* by reformulating TM using a weaker form of invariance. Interestingly enough, when these generalizations are done in the most natural ways, parallels to Theorem 4.2.11 result; that is, meaningfulness based on definitional generalizations of M correspond to meaningfulness based on the generalized ideas of invariance.

4.3 Definitional Generalizations of the Erlanger Program

In this section, several definitional systems of meaningfulness are presented axiomatically. They have in common the idea that meaningful entities are characterized by the fact that they are explicitly defined in terms of other specified entities. They differ in the kinds of formulae allowed and the kinds of entities specified.

4.3.1 Axiom System D′

The first axiom system considered, D′, is one of the weakest of the chapter. It provides a core of ideas that is either implied or directly assumed by most of the other systems.

Definition 4.3.1 *Axiom system* D′ is formulated in L($\in$, A, $\emptyset$, M) and consists of the following four axioms:

(i) ZFA;
(ii) *MP* (Definition 4.2.4);

(iii) *AL* (Definition 4.2.8); and
(iv) *MC′* (Definition 4.2.7).

By Theorems 4.2.3, 4.2.8, and 4.2.7, axiom system D′ is implied by TM. Later it is shown that D′ does not imply TM. Consequently, D′ is a proper generalization of TM.

Throughout this book, axiom system D′ and its extensions are often referred to as *definitional systems of meaningfulness.* Strictly speaking, axiom system D′ is not a form of *logical* definability, because it assumes the axiom of Meaningful Pure Sets. However, it appears to me to capture a form of "definability" that is commonly used in science, if it is accepted that pure sets corresponds to pure mathematics. For example, in science qualitative relations are often "defined" in terms of other qualitative objects and real numbers. Axiom system D′ is intended to capture this and similar practices of *scientific* definability. At various places in the book, the kind of "definability" inherent in axiom system D′ is called "set-theoretic definability," in part, to distinguish it from "logical definability."

The following theorem presents two simple consequences of D′:

Theorem 4.3.1 Assume ZFA, *MC′,* and *AL.* Then the following two statements are true:

1. $M(A)$ and $M(\varnothing)$.
2. $\forall x[M(x) \leftrightarrow M(\{x\})]$.

Proof. Statement 1 follows from *MC′*, since $A = \{x \mid x \in A\}$ and $\varnothing = \{x \mid x \neq x \wedge x \in A\}$ and "$x \in A$" and "$x \neq x \wedge x \in A$ are formulas of L($\in$, A, $\varnothing$). Statement 2 follows from Theorem 4.2.9.

Similarly, the following is easily established:

Theorem 4.3.2 Assume D′. Then $M(\Pi)$, and if $*'$ is composition of elements of Π, then $M(*')$.

Proof. Theorem 4.10.1.

The next theorem is useful in many proofs:

Theorem 4.3.3 Assume D′ and G is an entity. Suppose H is an entity such that *(i)* each element of H is a function from A into A, and *(ii)* for each f in H and each set x, if $M(x)$ then $f(x) \subseteq x$. Then $H \subseteq \Pi$. Furthermore, if H is a group, then $H \subseteq G$.

Proof. Immediately follows from Lemmas 4.10.1, 4.10.3, and 4.10.4.

An interesting consequence of the proof of Theorem 4.3.3 is that any system of meaningfulness based on the following concept of invariance is incompatible with axiom system D′: Let H be an entity that is a nonempty set of functions from A into A and is such that at least one of its elements is either not one-to-one or not onto A. Define the predicate $M(x)$ by:

$$\forall x[M(x) \leftrightarrow \forall f(f \in H \rightarrow f(x) \subseteq x)]\,.$$

Then axiom system D′ is false about the structure $\langle V, \in, A, \varnothing, M\rangle$ In fact, only a very small portion of axiom system D′ is needed to obtain this result. For example, if an element *f* of *H* is not onto *A*, then *M({A})* cannot be true, since if it were, then (by adapting the relevant part of the proof of Theorem 4.3.3) *f* would be an element of *G* and thus onto *A*, contrary to the choice of *f*.

Axiom system D′ can be strengthened in several ways. The first considered replaces *MC′* by the more powerful condition *MC*.

4.3.2 Axiom System D

Definition 4.3.2 *Axiom system* D is formulated in language L(∈, *A*, Ø, *M*) and consists of the following four axioms:

(i) ZFA;
(ii) *MP* (Definition 4.2.4);
(iii) *AL* (Definition 4.2.8); and
(iv) *MC* (Definition 4.2.9).

With axiom system D, the predicate *M* can appear in formulas used to establish the meaningfulness of entities. In particular, it can be used to show that the transformation group, *G*, of *M* is meaningful:

Theorem 4.3.4 Assume D. Then *G* is an entity and *M(G)*. Furthermore, if *□″ represents the composition of elements of *G*, then *″ is an entity and *M*(*″).

Proof. $G = \{f \mid f \in \Pi \wedge \forall x[M(x) \rightarrow f(x) = x\}$, and thus *G* is a meaningful entity by Theorem 4.3.2 and axiom *MC* of D. Let *′ be the operation of function composition of elements of *Π*. Then

$$*'' = \{(x, y, z) \mid (x, y, z) \in *' \wedge x \in G \wedge y \in G \wedge z \in G\}\,,$$

is a meaningful entity by Theorem 4.3.2, the meaningfulness of *G*, and axiom *MC* of D.

In Theorem 4.2.11, it was shown that the conjunction of *MC'* and *DM** implied axiom *MC*. This suggests the following strengthening of D′.

4.3.3 Axiom System D*

Definition 4.3.3 *Axiom system* D* is formulated in language L(∈, *A*, Ø, *M*) and consists of the following five axioms:

(i) ZFA;
(ii) *MP* (Definition 4.2.4);
(iii) *AL* (Definition 4.2.8);
(iv) *MC'* (Definition 4.2.7); and
(v) *DM** (Definition 4.2.6).

By Theorem 4.2.11, D* → D.

The basic idea behind axiom systems D′ and D is that entities that are appropriately defined in terms of meaningful entities are themselves meaningful. They do not directly say anything about nonmeaningful entities. Axiom system D*, however, contains axiom *DM**, which makes a strong statement about what is nonmeaningful. (Note, however, that D* is not specific about the kinds of formulae of L(∈, *A*, Ø) that can be used to specify *M*.)

Axiom *DM** says that the meaningfulness predicate *M* is definable in terms of a formula Φ of L(∈, *A*, Ø) and finitely many meaningful entities $a_1,\ldots, a_n$. This idea is weakened in the next axiom, which does not assume the meaningfulness of $a_1,\ldots, a_n$:

Definition 4.3.4 Assume ZFA and language L(∈, *A*, Ø, *M*). Then axiom *DM (Definable Meaningfulness)* is said to hold if and only if there exists a formula $\Phi(x, u_1,\ldots, u_n)$ of L(∈, *A*, Ø) and entities $a_1,\ldots, a_n$ such that

$$\forall x[\Phi(x, a_1, \ldots, a_n) \leftrightarrow M(x)]\,.$$

The next axiom system is a weakened version of D* that results from replacing *DM** with *DM*.

4.3.4 Axiom System $\overline{\mathsf{D}}$

Definition 4.3.5 Axiom system $\overline{\mathsf{D}}$ is formulated in language L(∈, *A*, Ø, *M*) and consists of the following five axioms:

(i) ZFA;
(ii) *MP* (Definition 4.2.4);
(iii) *AL* (Definition 4.2.8);
(iv) *MC′* (Definition 4.2.7); and
(v) *DM* (Definition 4.3.4).

Clearly $\mathsf{D}^* \rightarrow \overline{\mathsf{D}}$ and $\overline{\mathsf{D}} \rightarrow \mathsf{D}'$.

Note that like D*, axiom $\overline{\mathsf{D}}$ system does not say anything directly about the form of any particular defining formula for the meaningfulness predicate *M*.

The next axiom system makes an explicit statement about the meaningfulness/nonmeaningfulness of entities.

4.3.5 Axiom System D″*(a)*

Definition 4.3.6 Assume $\langle V, \in, A, \varnothing, M\rangle$ is a model of L(∈, *A*, Ø, *M*). Let *a* be a subcollection of *V*. Then *axiom system* D″*(a)* consists of the following two axioms:

1. ZFA.
2. *DC(a)* (*Definable Closure* of *a*): For all entities *b*, *M(b)* if and only if there exists a formula $\Phi(x, u_1,\ldots, u_n)$ of L(∈, *A*, Ø) and entities $a_1,\ldots, a_n$ such that *(i)* for each i=1,..., *n*, either a_i is in *a* or a_i is a pure set, and *(ii)*

$$\{b\} = \{x \,|\, \Phi(x, a_1, \ldots, a_n)\}\,.$$

As shown below, Definition 4.3.6 avoids the introduction of an additional clause corresponding to the axiom of Atomic Legacy by having the meaningfulness of *b* be defined in terms of the definability of *{b}*.

Theorem 4.3.5 Let $\mathfrak{V} = \langle V, \in, A, \varnothing, M\rangle$ be a model of L(∈, *A*, Ø, *M*) and *a* be a subcollection of *V*. Then D″*(a)*→ D′.

Proof. Assume axiom system D″*(a)*.

1. Axiom system ZFA is true since D″*(a)* is true.
2. Suppose *b* is an arbitrary pure set. Then *M(b)* is true since

$$\{b\}=\{x|x=b\}$$

and "*x*=*u*," where *u* is a free variable, is a formula of L(∈, *A*, Ø) and axiom system D″*(a)* is true. Thus, since *b* is an arbitrary pure set, axiom *MP* is true.

3. Suppose *b* is an arbitrary atom such that *M({b})*. By axiom system D″*(a)*, let $\Phi(x, u_1,\ldots,u_n)$ be a formula of L(∈, *A*, Ø) and $a_1,\ldots,a_n$ be entities such that *(i)* for each i=1,..., *n*, either $a_i \in a$ or a_i is a pure set and *(ii)*

$$\{\{b\}\}=\{x|\Phi(x,a_1,\ldots,a_n)\}.$$

Then, since

$$\{b\}=\{y|\exists x[y\in x\wedge\Phi(x, a_1,\ldots,a_n)]\},$$

it follows from axiom system D″*(a)* that *M(b)*. Thus axiom *AL* is true.

4. Suppose *b* is a set, $\Phi(x, u_1,\ldots,u_n)$ is a formula of L(∈, *A*, Ø), $b_1,\ldots, b_n$ are entities such that $M(b_1),\ldots, M(b_n)$, and

$$b=\{x|\Phi(x,b_1,\ldots,b_n)\}.$$

It will be shown that *M(b)*. Let *i* be an integer such that $1 \le i \le n$. Since $M(b_i)$ is true, let $\Phi_i(x, u_1,\ldots,u_{m(i)})$ be a formula of L(∈, *A*, Ø) and $a^i_1, \ldots, a^i_{m(i)}$ be entities such that for each *j*, *j*=1,..., *m(i)*, either $a^i_j \in a$ or a^i_j is a pure set and such that

$$\{b_i\} = \{x \mid \Phi_i(x, a_1^i, \ldots, a_{m(i)}^i)\}\,.$$

Let $\Theta_i(u)$, where u is a free variable, be the expression $\Phi_i(u, a_1^i, ..., a_{m(i)}^i)$.

Then it follows that

$$\forall u[\Theta_i(u) \leftrightarrow u = b_i]$$

is a true statement. Then, since

$$\begin{aligned}\{b\} &= \{y \mid \forall z(z \in y \leftrightarrow \forall w(w \in z \leftrightarrow \Phi(w, b_1, \ldots, b_n))\} \\ &= \{y \mid \forall z[z \in y \leftrightarrow \forall w(w \in z \leftrightarrow \exists x_1 \ldots \exists x_n \\ &\qquad \Phi(w, x_1, \ldots, x_n) \wedge \Theta_1(x_1) \wedge \ldots \wedge \Theta_n(x_n))]\}\,,\end{aligned}$$

it follows from axiom system D″*(a)* that b is meaningful. Thus axiom *MC′* is true.

Let a be a collection of entities. Axiom system D″*(a)* gives a specific description of meaningfulness: Those elements that are definable in terms of elements of a and pure sets are meaningful; those that are not so definable are nonmeaningful. In applications, a is usually taken to be an entity whose elements are considered to be the "primitive" relations that "characterize" the empirical or qualitative situation under consideration. Under D″*(a)*, entities are meaningful if and only if they are appropriately definable in terms of primitives and pure mathematics.

Convention 4.3.1 Assume $\mathfrak{V} = \langle V, \in, A, \varnothing\rangle$ is a model of the axiom system ZFA. The term "set," unless otherwise made clear by context, will be reserved for those elements a of V such that "$\neg[a \in A]$" is a true statement about $\mathfrak{V}$ As noted before, V itself is not a set in this sense of "set." Instead, V is called a "collection." Associated with each set a in V is a subcollection of V denoted by $\hat{a}$ that is defined as follows: $\hat{a}$ consists of all those entities b of V such that it is a true statement about $\mathfrak{V}$ that $b \in a$. To make statements of certain ideas more uniform, "$\hat{a}$" will often be written as "a" for sets a. Note that for the set a the statement "y is an element of the subcollection $\hat{a}$" says exactly the same thing as the expression "$y \in a$" of L($\in$, A, Ø).

Axiom system D″*(a)*, where a is a *finite subcollection* of V, has certain special properties. But before going into these, it is best to make explicit the differences between "finiteness" of a subcollection and a set.

Definition 4.3.7 It is well-known in logic that axiom system ZFA is not categorical; that is, ZFA has many non-isomorphic models. Thus it is not unexpected that some of these may have unusual properties. For example, let $\mathfrak{V} = \langle V, \in, A, \varnothing\rangle$ be a model of ZFA. Then in logic it is well-known that the set of finite ordinals of $\mathfrak{V}$, ω, may be such that it has an element α that when considered as a subcollection of V—i.e., when considered as $\hat{\alpha}$ (see Convention 4.3.1 above)—is an infinite subcollection of V. This is often described as follows: as an internal element of $\mathfrak{V}$ (that is, as a set), α is "finite," because the statement "α is a finite set" is a true statement [in the language L(∈, A, Ø)] about $\mathfrak{V}$; as an external element of $\mathfrak{V}$ (that is, as the subcollection $\hat{\alpha}$), α is "infinite," because $\hat{\alpha}$ is an infinite subcollection of V. These two cases are distinguished by the following definitions: *A* set *a* is said to be *internally finite* if and only if the statement "*a* is finite" of L(∈, A, Ø) is a true statement about the model $\mathfrak{V}$. A set *a* is said to be externally finite if and only if the subcollection $\hat{\alpha}$ is finite.

Note that if *b* is any finite subcollection of *V*, then there exists an externally finite set a such that $b=\hat{a}$. For example, if *b* has exactly the entities a_1, a_2, and a_3 as elements, then by theorems of ZFA a set *a* in *V* can be found so that

$$a_1 \in a \wedge a_2 \in a \wedge a_3 \in a \wedge \forall z(z \in a \leftrightarrow z = a_1 \vee z = a_2 \vee z = a_3)$$

is a true statement about $\mathfrak{V}$. But then, by the definition of $\hat{a}$, $\hat{a}=b$. Thus *the concepts of "finite subcollection of* $\mathfrak{V}$*" and "external finite set" coincide.*

Theorem 4.3.6 Suppose *a* is an externally finite set and D″*(a)*. Then axiom system TM is true and $G = \{f \mid f \in \Pi \wedge \forall x[x \in a \rightarrow f(x) = x]\}$.

Proof. Theorem 4.10.4.

Theorem 4.3.6 is quite surprising and has several important consequences that are discussed next.

Assume D″*(a)*, where a is an externally finite set. Then by Theorem 4.3.6, axiom system TM is true. Thus *G* is a meaningful entity and *M* is definable in terms of *G* through a formula of L(∈, *A*, Ø), namely the formula $I_G(x)$. In the next section, a related method is given for defining *M* when *D″(b)* holds for an arbitrary set *b* that may or may not be externally finite.

Recall the major flaw of the Erlanger Program was its perceived inability to deal effectively with cases where the transformation group consisted only of the identity transformation. In such situations all entities are meaningful. Part of the motivation for developing the above definitional theories of meaningfulness was to avoid this disturbing state of affairs by considering concepts of meaningfulness that were proper generalizations of the Erlanger Program. Unfortunately, there are many important situations that can be axiomatized in terms of finitely many primitives and have trivial transformation groups. If in such situations axiom system D″*(a)* for an externally finite set *a* applies, then by Theorem 4.3.6 the same plaguing problem of having all entities meaningful results. Using the weaker axiom system D′ for these situations does not help, as the following example shows.

Suppose the situation of interest is characterized by the structure $\mathfrak{S} = \langle A, R_1, \ldots, R_{15}\rangle$, where $R_1, \ldots, R_{15}$ are fifteen relations on *A*. Also suppose $\mathfrak{S}$. has the identity as its only automorphism. Suppose axiom system D′ and $M(R_1), \ldots, M(R_{15})$. Because the identity is the only automorphism of $\mathfrak{S}$, it easily follows that the transformation group for *M* is trivial. Because the collection that consists of exactly the fifteen relations $R_1, \ldots, R_{15}$ is finite, there is a set *a* that is externally finite and is such that it has exactly $R_1, \ldots, R_{15}$ as elements. Consider D″*(a)*. Let *M*″ be its meaningfulness predicate. By the way axiom system D″*(a)* was defined (Definition 4.3.6), it is easy to show (by methods similar to those of the proof of Theorem 4.3.6) that $M'' \subseteq M$. However, because *a* is externally finite, it follows from Theorem 4.3.6 that $M''=V$. Thus $M=V$.

Historically, the first critical challenge to the Erlanger Program—and one it could not effectively meet—was Einstein's general theory of relativity, which had a trivial transformation group. It seemed unnatural that every relation and concept about physical space-time should be meaningful, and so the application of the Erlanger Program to this situation was highly suspect. However, it is highly unlikely that physical space-time, as conceptualized in the general theory of relativity, can be captured by finitely many entities, and thus the reasons for doubting the applicability of axiom system TM in this case does not extend to axiom system D′.

One important place in science where finitely specifiable structures with trivial transformation groups occur repeatedly is for situations that can be formulated as a finite set of atoms with a qualitative total ordering relation. For such situations, it does not immediately seem unreasonable that all *n*-ary relations on the set of atoms are meaningful, and in general that all entities are meaningful. Thus at first glance such finite structures do not appear to present any unusual difficulty for the definitional theories of meaningfulness presented previously.

The type of ordered infinite structure most cited in the measurement literature as being problematic meaningfulness theories because it has a trivial transformation group is an extensive structure with a maximal element. This structure arises in a natural way in probability contexts. It is defined in terms of a total ordering relation $\succeq$ and a ternary relation $C(x, y, z)$ which allows one to concatenate elements *x*, *y* of *A* that "are not too large" to form another element *z* of *A*. A good example of such a structure is to take *A* as the half-open set of real numbers (0,1] (considered as qualitative entities), $\succeq$ as the usual ordering $\geq$ on (0, 1], and *C* as the relation that is defined by

$$C(x, y, z) \text{ iff } x, y, z \text{ are in } (0,1] \text{ and } x + y = z\,.$$

Then $\langle A, \geq, C\rangle$ is an extensive structure with maximal element 1. It can be shown that this structure has the identity as its only automorphism. Thus if $\geq$ and *C* are assumed to be meaningful, then it follows from a remark following Theorem 4.3.6 that under axiom system D′ every entity is meaningful.

Extensive structures with maximal elements occur in probability theory in the following way: Suppose ε is an algebra of events and $\succsim$ is an ordering relation on ε, called *the at least as* likely as relation. Then $\langle \mathcal{E}, \succsim\rangle$ is said to be a *qualitative probability structure* if and only if there exists a function *P* from ε into the closed interval of real numbers [0, 1] such that for all *B* and *C* in ε,

(i) $P(X)=1$ and $P(\emptyset)=0$, where X is the sure event in ε;
(ii) $B \succsim C$ if and only if $P(B) \geq P(C)$; and
(iii) if $B \cap C=\emptyset$, then $P(B \cup C)=P(B) + P(C)$.

A function P that satisfies conditions *(i)*, *(ii)*, and *(iii)* above is called a *probability representation.*

Suppose ε is a qualitative probability structure. Define the binary relation $\equiv$ on ε as follows: for all B and C in ε, $B=C$ if and only if for all probability representations P of ε, $P(B)=P(C)$. It is easy to show that $\equiv$ is an equivalence relation on ε. Let ε' be the set of $\equiv$-equivalence classes of ε, and define $\succsim'$ on ε' as follows: for all Γ and Σ in ε', $\succsim' \Sigma$ if and only if there exist B in Γ and C in Σ such that $B \succsim' C$. Let ε'' be the set that consists of those equivalence classes in ε' that have elements B such that $P(B)>0$ for all probability representations of $\langle \mathcal{E}, \succsim \rangle$ Intuitively, ε'' consists of equivalence classes of events that have "positive likelihood." Let $\succsim''$ be the restriction of $\succsim'$ to ε''. Let $\oplus$ be defined on ε'' as follows: For all Γ, Σ, and Δ in ε'', $\Gamma \oplus \Sigma = \Delta$ if and only if there exist B in Γ, C in Σ, and D in Δ such that $B \cap C=\emptyset$ $B \cup C=D$. Then if $\langle \mathcal{E}, \succsim \rangle$ sat isfies certain plausible conditions (which hold in the most important probability structures used in mathematics), then $\langle \mathcal{E}'', \succsim'', \oplus \rangle$ is an extensive structure with a maximal element, and from this it is easy to show that $\langle \mathcal{E}, \succsim \rangle$ has a unique probability representation. (See pp. 95–97 of Narens, 1985, or pp. 211–214 of Krantz et al., 1971, for details.)

Let us assume that $\langle \mathcal{E}, \succsim \rangle$ satisfies the plausible conditions mentioned above. Then $\langle \mathcal{E}'', \succsim'', \oplus \rangle$ is an extensive structure with a maximal element, and, as mentioned before, it therefore has the identity as its only automorphism. This does not mean that the structure $\langle \mathcal{E}, \succsim \rangle$ has the identity as its only automorphism; indeed, in most of the important mathematical cases, $\langle \mathcal{E}, \succsim \rangle$ has a great abundance of automorphisms, but these all have the property of taking elements of ε into $\equiv$-equivalent ones. The type of symmetry produced by such automorphisms is useful in some meaningfulness contexts. It is because of the lack of other kinds of automorphisms that $\langle \mathcal{E}, \succsim \rangle$ has a unique probability representation and that $\langle \mathcal{E}'', \succsim'', \oplus \rangle$ has a trivial group of automorphisms. And thus, this lack is the principal reason behind the doubt about the appropriateness of TM as a theory of meaningfulness for $\langle \mathcal{E}, \succsim \rangle$.

Some (Fine, 1973, pp. 55–56; Luce and Narens, 1978, pp. 232–33) have argued that Condition *(i)* of the standard notion of probability representation given above—which states that the probability of the sure event is 1—is unjustified and perhaps incorrect. Their arguments are based on the idea that the choice of the number 1 is arbitrary and does not correspond to anything of qualitative significance. They recommend that the notion of "probability representation" should be changed to accommodate this. Luce and Narens (1978) suggests Condition *(i)* above be changed to read "$P(X)=r$, where r is a positive real number." When this is done, situations that had unique probability representations will now have a set of probability representations that forms a ratio scale. Also, this new definition of probability still allows for the formulation of all important probabilistic concepts. For example, B and C are "independent" if and only if

$$P(A \cap B) = P(A) \cdot \frac{P(B)}{P(X)},$$

where X is the sure event and P is an arbitrary probability representation in the scale. One of their prime reasons for wanting probability to be a ratio scale was that then it would be

> ...just like the other basic extensive measures of physics. In particular, probability can be incorporated into the dimensional structure of physics (if it is found to interrelate with other dimensions via distribution laws) and meaningfulness can be studied as with other dimensions. *(Luce and Narens, 1978, pg. 233)*

They also point out that the different probability representations in the ratio scale had qualitative significance in the sense that they were related to one and another through homomorphisms (or their inverses) of the qualitative probability structure into itself. These homomorphisms also correspond to the choices of the probabilities assigned to the sure event. Luce and Narens also recommended that one should seriously consider invariance under endomorphisms (i.e., under homomorphisms of the qualitative structure into itself) as the basis for meaningfulness considerations in probabilistic contexts. (Meaningfulness as invariance under endomorphisms first made it appearance in measurement literature in Pfanzagl, 1968).

I do not find persuasive the above argument of Luce and Narens (1978) against the use of TM as a theory of meaningfulness for the structure $\langle \mathcal{E}, \succsim \rangle$ It seems to me that for the type of applications that Luce and Narens (1978) refer to, the appropriate "theory of meaningfulness" required would probably have to be somewhat custom designed and may not apply to other "meaningfulness" applications.

There are four basic options for dealing with meaningfulness situations like qualitative probability above. The first is to consider that there is no problem; that is, to consider that it is alright to have every entity meaningful in situations like the qualitative probability structure $\langle \mathcal{E}'', \succsim'' \rangle$ above, or perhaps even in each situation which has a trivial transformation group. Such a view could support D′ or TM as reasonable theories of meaningfulness for all situations. It would also hold that it is not automatic that the meaningfulness of every entity should be avoided, and would note that reasons for such an avoidance have never been satisfactorily given in the literature. The second option is to broaden the theories of meaningfulness above. Because of Theorem 4.3.6, this would mean that either the axiom of Atomic Legacy or the axiom of Meaningful Comprehension' or the axiom of Meaningful Pure Sets should be weakened. The latter two axioms are weakened in the axiom system for qualitativeness discussed in Chapter 7. The third option is to require meaningful entities to satisfy additional definitional or invariance conditions. This option is explored in Chapter 6. Of the above options, options 2 and 3 appear to lead to sound theories of meaningfulness for cases of structures with trivial automorphism groups. A fourth option is to base meaningfulness on more general invariance conditions. This option is explored in Sections 4.7, 6.2, and 6.3.

4.3.6 Interrelationships Among the Definitional Meaningfulness Concepts

In the preceding, several definability systems of meaningfulness have been presented. The following theorems describe their interrelationships.

Theorem 4.3.7 The following two statements are true (where, as usual, $\Phi \rightarrow \Psi \rightarrow \Theta$ stands for $\Phi \rightarrow \Psi$ and $\Psi \rightarrow \Theta$):

1. $\mathsf{M} \rightarrow \mathsf{D}^* \rightarrow \overline{\mathsf{D}} \rightarrow \mathsf{D}'$.
2. D′ does not logically imply $\overline{\mathsf{D}}$; $\overline{\mathsf{D}}$ does not logically imply D*; and D* does not logically imply M.

Proof. Statement 1 follows by Theorems 4.2.10, 4.2.12, 4.2.11, and Definitions 4.3.3, 4.3.2, and 4.3.1. Statement 2 follows by Theorems 4.10.24, 4.10.23, and 4.10.22.

Theorem 4.3.7 shows that D′, $\overline{\mathsf{D}}$, D*, and M form a sequence of increasingly stronger axiomatic systems of meaningfulness. Just where D and D″*(a)* for various choices of *a* fit into this sequence are interesting questions. Their answers—as given in this book—use an additional method of inference in set theory that I call "External Finite Induction."

If it is assumed that set theory is about some sort of reality, then the collection of sentences *T* of L(∈, *A*, Ø) that is true about that reality is denumerable, since L(∈, *A*, Ø) is a denumerable language. The axioms of ZFA and their consequences through first-order logic belong to *T*. But because of a classical incompleteness result of Gödel (1931), they do not form all of *T*.

It is well-known in mathematical logic that *T* has many non-isomorphic models. Because every sentence of L(∈, *A*, Ø) or its negation is in *T*, L(∈, *A*, Ø) cannot distinguish between models of *T*. It will be argued that the principle of External Finite Induction is valid for all models of *T*. Because of this, it too can not distinguish between models of *T*. However, it can be shown that this principle is not valid for all models of ZFA. Therefore, it can sometimes distinguish between a model of *T* and a model of ZFA that is not a model of *T*. This is why it is a useful principle.

Definition 4.3.8 *EFI (External Finite Induction):* Assume ZFA. For all formulas

$$\Theta(x, u_1, \ldots, u_n)$$

of L(∈, *A*, Ø), if $a_1, \ldots, a_n$ are entities and $\Theta(\alpha, a_1, \ldots, a_n)$ for each externally finite ordinal α (i.e., for each finite ordinal α that is an external finite set), then $\forall x[x \in \omega \rightarrow \Theta(x, a_1, \ldots, a_n)]$.

Note that in some model $\mathfrak{V}'$ of ZFA, $\alpha \in w$ may be true for some α that is not externally finite, i.e., in $\mathfrak{V}'$, the set of finite ordinals *w* may have "nonstandard elements." It can be shown through mathematical logic that such "nonstandard" models exist, even for the collection *T* of all true settheoretic sentences of L(∈, *A*, Ø).

The intuitive argument for the validity of *EFI* goes as follows: Suppose $\mathfrak{V}'$ is a model of the collection *T* of true set-theoretic sentences of L(∈, *A*, Ø). Suppose *EEI* is false about $\mathfrak{V}'$. A contradiction will be shown, thus establishing the validity of *EFI.* By Definition 4.3.8, let $\Theta(x, u_1, \ldots, u_n)$ be a formula of L(∈, *A*, Ø) and $a_1, \ldots,$ an be entities such that $\Theta(\alpha, a_1, \ldots, a_n)$ is true about $\mathfrak{V}'$ for each externally finite ordinal a, and such that $\forall x[x \in \omega \rightarrow \Theta(x, a_1, \ldots, a_n)]$ is false about $\mathfrak{V}$ But in *the reality that set theory is about, each element of w is externally finite. Thus* $\forall x[x \in \omega \rightarrow \Theta(x, a_1, \ldots, a_n)]$ *is true about this reality.* Therefore, $\mathfrak{V}'$ cannot be a model of *T,* contrary to hypothesis.

The following definition extends *EFI* to the language L(∈, *A*, Ø, *M*). Its justification as a valid principle is similar to the one just given for JEFJ.

Definition 4.3.9 *EFI* (External Finite Induction*):* Assume ZFA. For all formulas $\Theta(x, u_1, \ldots, u_n)$ of L(∈, *A*, Ø, *M*) and all entities $a_1, \ldots, a_n$, if $\Theta(\alpha, a_1, \ldots, a_n)$ for each external finite ordinal *a* (that is, for each finite ordinal a that is an external finite set), then $\forall x[x \in \omega \rightarrow \Theta(x, a_1, \ldots, a_n)]$

Theorem 4.3.8 The following three statements are true:

1. Assume *EFI**. Then D*↔D.
2. Assume *EFI.* Then $\overline{\mathsf{D}} \leftrightarrow \exists x[x \text{ is a set and } \mathsf{D}''(x)]$.
3. Assume *EFI.* Then D*↔∃*x*[*x* is a meaningful set and D″*(x)*].

Proof. Theorems 4.10.19 and 4.10.21.

Statement 1 of Theorem 4.3.8 is somewhat of a surprise, since the axioms of D do not give criteria for the nonmeaningfulness of entities. Statements 2 and 3 are also somewhat surprising, since D″*(a)* assumes a complicated axiom of closure to specify the meaningfulness predicate.

In view of Theorem 4.3.8, the main thing missing about the above descriptions of systems D*, $\overline{\mathsf{D}}$ and D″*(a),* where a is an entity, is a statement about the kinds of formulae of L(∈, *A*, Ø) that can be used to define the meaningfulness predicate, *M.* This gap is filled in the next section.

4.4 Transformational Generalizations of the Erlanger Program

In the Erlanger Program, meaningfulness was identified with invariance under a permutation group. Theorem 4.3.3 and its proof shows that identifying meaningfulness with invariance under a more general set of transformations produces violations of meaningfulness principles like *MC′*. A different way of generalization of the Erlanger Program is to identify meaningfulness with invariance under a *family of transformation groups,* so that an entity is meaningful if and only if it is invariant under some group that belongs to the family. In this section, these kinds of generalizations are developed. The principal result is that they lead naturally to transformational equivalents of axiom systems D′, $\overline{\text{D}}$ and D*.

Invariance under a family of transformation groups was, to my knowledge, first used in Fraenkel (1922a, 1935) to show the consistency of the negation of the axiom of Choice with the remaining axioms of ZFA. It still continues to be a useful concept in the metamathematics of set theory—see, for example, Jech (1973). Its use in set theory is related to but is different from its use in this book.

Definition 4.4.1 Assume ZFA and L(∈, *A*, Ø, *M*). Then Δ is said to *generate M by invariance* if and only if the following three conditions hold:

(i) Δ is a nonempty collection of sets;
(ii) each element of Δ is a subgroup of Π under function composition; and
(iii) for all entities *x*, *M(x)* if and only if $I_K(x)$ for some *K* in Δ.

Note that in Definition 4.4.1, all the elements of Δ are entities, but Δ itself need not be an entity.

Definition 4.4.2 Assume ZFA and L(∈, *A*, Ø, *M*). Δ is said to be a *transformational family* for *M* if and only if the following two conditions hold:

(i) Δ generates M by invariance; and
(ii) for each *F* and *H* in Δ, there exists *K* in Δ such that $K \subseteq F \cap H$.

Theorem 4.4.1 below shows that the existence of a transformational family for M is necessary and sufficient for axiom system D′ to hold.

Definition 4.4.3 The *axiom system* GTM′ *(Generalized Transformational Meaningfulness′)* is formulated in the language L(∈, *A*, Ø, *M*) and consists of the following two axioms: (1) ZFA and (2) there exists a transformational family for *M.*

Theorem 4.4.1 GTM′ ↔ D′.
Proof. Theorem 4.10.16.

To obtain the stronger axiom system $\overline{\mathsf{D}}$, an additional condition on Δ is needed.

Definition 4.4.4 The *axiom system* GTM *Generalized Transformational Meaningfulness*) is formulated in the language L(∈, *A*, Ø, *M*) and consists of the following two axioms: (1) ZFA and (2) there exists an entity Δ such that Δ is a transformational family for *M*.

Note that the only difference between axiom systems GTM′ and GTM is that GTM requires that there exists a transformational family for *M* that is an *entity,* GTM′ only requires that there exists a transformational family for *M* that is a *collection* (which, of course, may be an entity).

Theorem 4.4.2 The following two statements are true:

1. GTM → $\overline{\mathsf{D}}$.
2. Assume EFI. Then GTM ↔ $\overline{\mathsf{D}}$.

Proof. Theorem 4.10.18.

To obtain axiom system D*, we need to only additionally require that Δ is meaningful:

Definition 4.4.5 The axiom system GTM* (Generalized Transformational Meaningfulness*) is formulated in the language L(∈, *A*, Ø, *M*) and consists of the following two axioms: (1) ZFA and (2) there exists a meaningful entity Δ such that Δ is a transformational family for *M.*

Theorem 4.4.3 The following three statements are true:

1. GTM* → D*.
2. Assume *EFI.* Then GTM* ↔ D*.
3. Assume *EFI**. Then GTM* ↔ D* ↔ D.

Proof. Theorem 4.10.19.

Theorems 4.4.1, 4.4.2, and 4.4.3 give transformational equivalents of axiom systems D′, $\overline{\mathsf{D}}$ and D*. Although $\overline{\mathsf{D}}$ and D* have axioms that state the meaningfulness predicate M is definable through finitely many entities and a formula of L(∈, *A*, Ø), they do not specify forms of particular formulae that accomplish this. By using transformational families of groups of permutations, such formulae can be specified for $\overline{\mathsf{D}}$ and D*. The idea is as follows:

Assume the relatively weak system D′. For each set a of meaningful entities, let

$$G_a = \{f \mid f \in \Pi \text{ and } f(x) = x \text{ for all } x \text{ in } a\} .$$

Let Δ be the collection of all G_a such that *a* is a meaningful finite set. Δ is not necessarily a set, because axiom system D′ is assumed and Δ is partly defined in terms of *M.* However, if either of the stronger axiom systems $\overline{\mathsf{D}}$ or D* are assumed, then Δ will be a set. Under the additional assumption of *EFI,* it can be shown (Theorem 4.4.2) that Δ is a transformational family for *M.* Thus by Theorems 4.4.2 and 4.4.3, the formula of L(∈, *A*, Ø),

$$M(x) \leftrightarrow \exists H(H \in \Delta \wedge I_H(x)) \ ,$$

provides a definition in L(∈, *A*, Ø) of *M* for situations where axiom systems $\overline{\mathsf{D}}$ or D* hold.

For the case where *M* is axiomatized by a set *a* of primitives through the axiom system D″*(a)*, the same idea works, but this time with

Δ={*H*|*H* is a subgroup of *Π* and there exists a finite subset *b* of *a* such that *f(x)*=*x* for all *x*∈*b* and all *f*∈*H*}.

The following independence results are established through Theorems 4.4.1, 4.4.2, and 4.4.3 and the appropriate choices of families of subgroups of *Π*:

Theorem 4.4.4 D′ ∧ *MI* does not imply either axiom *DM* or axiom *MC*.
Proof. Theorem 4.10.24.

Theorem 4.4.5 Then $\overline{\mathsf{D}}$ does not imply either axiom *DM** or axiom *MC*.
Proof. Theorem 4.10.23.

Theorem 4.4.6 D* does not imply axiom *MI*.
Proof. Theorem 4.10.22.

Theorem 4.4.6 uses the following construction: Assume ZFA, and suppose *A* is infinite. Within ZFA, let

Δ={*H*|*H* is a subgroup of *Π* and there exists a nonempty finite subset *a* of *A* such that *f(x)*=*x* for all *x*∈*a* and all *f*∈*H*}.

Define *M* as follows: for all entities *b*,

$$b \text{ is in } M \text{ if and only if } \exists H[H \in \Delta \wedge I_H(b)].$$

Then it can be shown that $\mathfrak{V} = \langle V, \in, A, \varnothing, M \rangle$ satisfies axiom system D* but not axiom *MI.*

Theorem 4.4.5 is shown by constructing within ZFA a particular nonempty set *Γ* of subgroups of *Π* with the following two properties: *(i)* for each *F* and *H* in *Γ*, there exists *K in Γ* such that $K \subseteq F \cap H$; and *(ii)* for each *H* in *Γ*, $\neg I_H(\Gamma)$. Define *M* as follows: for all entities *b*,

$$b \text{ is in } M \text{ if and only if } \exists H[H \in \Gamma \wedge I_H(b)].$$

Since *Γ* is a set, axiom $\overline{\mathsf{D}}$ system is true by Theorem 4.4.2. Suppose that *DM** were also true. A contradiction is shown as follows: By Theorem 4.4.3 there is a meaningful set Δ that is a transformational family for *M*. By the way *Γ* was selected, it turns out that it must be the case that for all *H* in *Γ*, $\neg I_H(\Delta)$, that is, Δ is not meaningful, a contradiction.

The proof of Theorem 4.4.4, which uses the Löwenheim-Skolem theorem of mathematical logic, is more complicated.

4.5 Consequences of D′, D*, and TM

This section presents a sampling of some of the kinds of results and concepts that are easily developed within the axiomatic theories of measurement of the previous sections; it is not meant to be a thorough or exhausted study. It has been included primarily to show that the meaningfulness concepts have rich and interesting mathematical structures associated with them.

4.5.1 Some Basic Concepts

Lemma 4.5.1 Assume D′. Suppose H is a meaningful function. Then the following two statements are true:

1. The domain of H and the range of H are meaningful.
2. If a is a meaningful subset of the domain of H, then the restriction of H to a is meaningful.

Proof. Lemma 4.10.5.

Lemma 4.5.2 Assume D′. Then the following two statement are true:

1. V_α is meaningful for each ordinal α.
2. For all f in Π and all entities a and b, $f[(a,b)]=(f(a),f(b))$.

Proof. Lemma 4.10.6.

Lemma 4.5.3 Assume D′. Suppose H is a function that is G-invariant. Then the following two statements are true:

1. The domain and range of H are G-invariant.
2. If a is G-invariant and a is a subset of the domain of H, then the restriction of H to a is G-invariant.

Proof. Lemma 4.10.7.

Definition 4.5.1 Assume D′. A set is said to be *homomeaningful* if and only if all of its elements are meaningful. A set is said to be *homomeaningless* if and only if all of its elements are nonmeaningful. If a is a set, then by definition

$$^{+}a = \{x \,|\, x \in a \wedge M(a)\}$$

and

$$^{-}a = \{x \,|\, x \in a \wedge \neg M(a)\}$$

^{+}a is called the *meaningful part* of a and ^{-}a is called the *meaningless part* of a.

Note that by Definition 4.5.1, Ø is both a homomeaningful and homomeaningless set.

Lemma 4.5.4 Assume M. Then the following two statements are true for all sets *a* and *b*:

1. If *a* is homomeaningful, then *a* is meaningful.
2. If *b* is meaningful, then ^{+}b and ^{-}b are meaningful.

Proof. Lemma 4.10.8.

Definition 4.5.2 Assume D′. Let *y* be an entity. Then *x* is said to be the *minimal meaningful set containing y as an element* if and only if $y \in x$, *x* is meaningful, and for all sets *z*, if *z* is meaningful and $y \in z$ then $x \subseteq z$.

It is immediate from Definition 4.5.2 that for each entity *y*, there is at most one minimal meaningful subset containing *y* as an element.

Lemma 4.5.5 Assume M. Then for each entity *y* there exists a minimal meaningful set containing *y* as an element.

Proof. Lemma 4.10.9.

Definition 4.5.3 Assume D′. A set *x* is said to be *minimeaningful* if and only if it is meaningful and each of its nonempty proper subsets is nonmeaningful.

Lemma 4.5.6 Assume D′. Suppose *x* is the minimal meaningful set containing *y* as an element. Then *x* is minimeaningful.

Proof. Lemma 4.10.10.

Definition 4.5.4 Assume D′ and *G* is an entity. Then, by definition, for each entity *a*, the *orbit* of *a* is $\{f(a) \mid f \in G\}$.

Lemma 4.5.7 Assume D′ and *G* is an entity. Suppose *a* and *b* are entities with respective orbits $\hat{a}$ and $\hat{b}$. Then the following two statements are true:

1. $\hat{a}$ is *G*-invariant.
2. Either $\hat{a}=\hat{b}$ or $\hat{a} \cap \hat{b} = \varnothing$.

Proof. Lemma 4.10.11.

Lemma 4.5.8 Assume TM. Suppose *a* is an entity and $\hat{a}$ is its orbit. Then $\hat{a}$ is minimeaningful.

Proof. Lemma 4.10.12.

4.5.2 Homogeneity

In mathematics, a relational structure $\mathfrak{X} = \langle X, R_0, R_1, \ldots \rangle$ is said to be "homogeneous" if and only if for each *x* and *y* in *X*, there exists an automorphism α of $\mathfrak{X}$ such that $\alpha(x)=y$. This form of "homogeneity" is also very important in science, and it plays a major role in the theory of measurement (see Chapter 5). In this subsection, a generalization of it is presented that is based on the concept of minimeaningfulness.

The following definition, which uses the transformation group G instead of "automorphisms", generalizes the mathematical concept of homogeneity to arbitrary sets:

Definition 4.5.5 Assume D′. Let a be a set. a is said to be *homogeneous* if and only if for each x and y in a, there exists f in G such that $f(x)=y$. a is said to be *nearly homogeneous* if and only if for each x and y in ^{-}a there exists f in G such that $f(x)=y$.

The usual mathematical concept of homogeneity results by applying Definition 4.5.5 with $a=A$.

Theorem 4.5.1 Assume D′. Suppose a is a meaningful set. Then if a is homogeneous, then it is minimeaningful.

Proof. Statement 1 of Theorem 4.10.5.

Theorem 4.5.1 shows that under axiom system D′, minimeaningfulness is a generalization of homogeneity. The next theorem shows that under axiom system TM they are equivalent:

Theorem 4.5.2 Assume TM. Suppose a is a meaningful set. Then the following two statements are true:

1. a is homogeneous if and only if a is minimeaningful.
2. a is nearly homogeneous if and only if ^{-}a is minimeaningful.

Proof. Theorem 4.10.6.

Theorems 4.5.1 and 4.5.2 applied to the special case of $a=A$ is worth singling out because of its strong link to the usual mathematical concept of homogeneity:

Theorem 4.5.3 The following two statements are true:

1. If D′ is true and A is homogeneous, then A is minimeaningful.
2. If TM: then A is homogeneous if and only if it is minimeaningful.

Proof. Theorem 4.10.7.

4.5.3 Meaningful Cardinals

In Chapter 3, the concept of two sets, a and b, being equinumerous, $a\equiv b$, was defined and used as the basis for the concept of cardinality. In this subsection, corresponding concepts based on meaningfulness considerations are developed. (As briefly discussed in Section 8.2, the concepts of this subsection bear some relationship to ideas of F. Bernstein concerning the axiom of Choice.)

Definition 4.5.6 Assume D. Define $\equiv_M$ as follows: $a\equiv_M b$ if and only if a and b

are sets and there exists a meaningful one-to-one function from a onto b. $a \equiv_M b$ is to be read as "a and b are meaningfully equinumerous."

Theorem 4.5.4 Assume D. Then for all sets a and b,

$$a \equiv_M b \rightarrow a = b.$$

Proof. Immediate from Definition 4.5.6 and axiom system D.

Lemma 4.5.9 Assume D′. Then the following three statements are true:

1. Suppose a is an arbitrary meaningful set and $\imath$ is the identity function on a, that is, $\imath(x)=x$ for all $x \in a$. Then $\imath$ is a meaningful one-to-one function.
2. Suppose f is a meaningful one-to-one function. Then f^{-1} is a meaningful one-to-one function.
3. Suppose f and g are meaningful one-to-one functions. Then $f{*}g$ is a meaningful one-to-one function.

Proof. Lemma 4.10.18.

Theorem 4.5.5 Assume D. Then $\equiv_M$ is an equivalence relation on the collection of meaningful sets.

Proof. Theorem 4.10.8.

The next theorem shows that for pure sets, equinumerosity and meaningful equinumerosity coincide:

Theorem 4.5.6 Assume D. Then

$$\forall x \forall y [x \in P \wedge y \in P \rightarrow (x \equiv_M y \leftrightarrow x \equiv y)] .$$

Proof. Theorem 4.10.9.

Theorem 4.5.7 Assume M. Then

$$\exists x (x \in P \wedge x \equiv_M A) \rightarrow \forall y M(y) .$$

Proof. Theorem 4.10.10.

Definition 4.5.7 Assume D. By Theorem 4.5.5, $\equiv_M$ is an equivalence relation on the collection of meaningful sets. For each meaningful set a, let $\overline{a}$ be the $\equiv_M$ equivalence class of meaningful sets that contains a. For each meaningful set a, $\overline{a}$ is called the meaningful cardinal of a.

Theorem 4.5.8 For each meaningful set a, $\overline{a}$ is a subcollection of $\overline{\overline{a}}$ (where as usual $\overline{\overline{a}}$ denotes the cardinal of a).

Proof. Immediate from Theorem 4.5.4.

Definition 4.5.8 Assume D. Let a and b be arbitrary meaningful sets. Then by definition, $\overline{a} \leq \overline{b}$ if and only if there exists a meaningful one-to-one function from a into b.

Theorem 4.5.9 Assume D. Suppose a and b are meaningful sets, $c \in \overline{a}$ and $d \in \overline{b}$. Then

$$\overline{a} \leq \overline{b} \text{ iff } \overline{c} \leq \overline{d} .$$

Proof. Theorem 4.10.11.

Theorem 4.5.10 Assume D. Suppose *a, b* and *c* are meaningful sets, $\overline{a} \leq \overline{b}$ and $\overline{b} \leq \overline{c}$. Then $\overline{a} \leq \overline{c}$.

Proof. Theorem 4.10.12.

The next theorem is a version of the Cantor-Schröder-Bernstein Theorem for meaningful cardinals.

Theorem 4.5.11 Assume D.Suppose *a* and *b* are meaningful sets and $\overline{a} \leq \overline{b}$ and $\overline{b} \leq \overline{a}$. Then $\overline{a} = \overline{b}$.

Proof. Theorem 4.10.13.

Theorems 4.5.10 and 4.5.11 show that $\leq$ is a transitive and antisymmetric relation on the collection of meaningful cardinals. Unlike the case for cardinals, it is in general not a connected relation, not even under the assumption of TM.

Theorem 4.5.12 Assume M. Suppose *a* is a homomeaningful set. Then

$$\exists x (x \in P \wedge a \equiv_M x) .$$

Proof. Theorem 4.10.14.

4.5.4 Meaningful "Set Theory"

In many ways the collection of meaningful entities resemble a model of ZFA. In this section this connection is investigated by examining how well the axioms of ZFA describe the class of meaningful entities.

Convention 4.5.1 Throughout this subsection the axiom system D* is assumed.

Definition 4.5.9 Let V_M be the collection of all meaningful entities. Elements of V_M are called *M-entities.*

x is said to be a *M-atom* if and only if x is a M-entity, $x \neq \emptyset$, and for all M-entities y, $y \notin x$.

x is said to be a *M-set* if and only if x is a M-entity, x is not a *M*-atom, and $\forall y (y \in x \rightarrow y$ is a *M*-entity).

x is said to be a *M-mix* if and only if x is a *M*-entity, x is not a *M*-atom, and x is not a *M*-set.

Note that a set $x \neq \emptyset$ is homomeaningless if and only if it is a *M*-atom. Thus we may have M-atoms x and y such that $x \subseteq y$. Also note that a set z is a M-set if and only if it is homomeaningful.

In addition, note that the theory of meaningful cardinals presented in the previous subsection was based upon correspondences which may be M-atoms as well as M-mixes or M-sets. If M-functions are defined as M-sets of ordered pairs of M-entities and M-cardinals are defined in terms of one-to-one M-functions, then a theory of cardinals will result that is different from the one in the previous subsection. This new theory of "M-cardinals" is an example of a concept developed in what I call "meaningful set theory"–that is, the "set theory" that results from developing set-theoretic concepts in terms of V_M.

Meaningful "set theory" will now be investigated in terms of how well the axioms of ZFA hold for it:

Axiom of Atoms

The axiom of Atoms (Definition 3.3.3) fails in meaningful set theory in two important ways: First, not every M-entity is a M-set or M-atom. And second, the M-atoms do not in general form a set and therefore in particular not a M-set. To see an example of the latter, suppose $A=\{a, b\}$, where $a \neq b$, and suppose axiom system TM holds with $G=II$. It will be shown by contradiction that the M-atoms do not form a set. Suppose s is the set of M-atoms. Let a=rank(s). Since rank $(A)=1$ and A is a M-atom, it follows that $\alpha>1$. By Theorem 3.5.6, each element of s has rank $<a$. Consider the set $x=a \cup \{a\}$. By construction, rank$(x)>a$. We will show that x is not meaningful. Let f be the element of G such that $f(a)=b$. Then $f(x)=\alpha \cup \{b\}$ and therefore $f(x) \neq x$. By TM, x is not meaningful. Similarly $\alpha \cup \{b\}$ is not meaningful. Consider

$$y = \{\alpha \cup \{a\}, \alpha \cup \{b\}\} .$$

Then y is meaningful by TM. However, no element of y is meaningful. Thus y is a M-atom. But since rank $(y)>\alpha$ and rank$(s)=\alpha$, it cannot be that $y \in s$. This contradicts that s is the set of M-atoms.

Axiom of Extensionality

The axiom of Extensionality (Definition 3.3.6) holds for M-sets but fails for M-mixes. (Two M-mixes may have as elements the same M-entities and different non-M-entities.)

Axiom of Pairs

The axiom of Pairs (Definition 3.3.7) holds for M-entities, that is, for all M-entities a and b there exists a M-set, namely $\{a, b\}$, that has a and b and only a and b as elements.

Union Axiom

The Union axiom (Definition 3.3.10) holds for *M*-entities: for each *M*-set or *M*-mix *x* define within the axiom system D (which by Definition 4.3.3 and Theorem 4.2.11 is implied by D*) the *M-union u* of *x* as follows:

$$u = \{z \mid M(z) \wedge \exists y[M(y) \wedge y \in x \wedge z \in y]\} .$$

By the axiom of Meaningful Comprehension, *u* is a *M*-set.

Power Set Axiom

The Power Set axiom (Definition 3.3.13) holds for *M*-entities: First define *M*-subsets as follows: for each *M*-set and *M*-mix *x*, *y* is said to be a *M-subset* of *x* if and only if *y* is a *M*-set and $y \sqsubseteq x$. Define for each *M*-set or *M*-mix *x* the *M-power set* of *x* to be the set of all *M*-subsets of *x*. It then follows by the axiom of meaningful Comprehension (which by Theorem 4.4.3 is implied by D*) that the *M*-power set of a *M*-set or *M*-mix is a *M*-set.

Axiom of Infinity

The axiom of Infinity (Definition 3.3.15) holds for *M*-entities, since by the axiom of Pure Sets, the finite ordinals and the ordinal *w* are *M*-entities.

The following definition is useful:

Definition 4.5.10 The *M-formulas* of L(∈, *A,* Ø, *M*) are formulas Φ of L(∈, *A,* Ø, *M*) that are formed by using the following four rules:

1. Φ is of the form

   $$\Psi(u_1, \ldots, u_n) \wedge M(u_1) \wedge \ldots \wedge M(u_n),$$

   where $1 \leq n$, $u_1, \ldots, u_n$ are variables of L(∈, *A,* Ø), and $\Psi(u_1, \ldots, u_n)$ is a formula of L(∈, *A,* Ø) that has no quantifiers.
2. Φ is of the form $\Psi \wedge$ Ø, where Ψ and Θ are *M*-formulas.
3. Φ is of the form $\neg\Psi(u_1,\ldots,u_n) \wedge M(u_1) \wedge \ldots \wedge M(u_n)$, where $u_1,\ldots, u_n$ are variables of L(∈, *A,* Ø) and $\Psi(u_1,\ldots, u_n)$ is a *M*-formula.
4. Φ is of the form $\exists u[M(u) \wedge \Psi(u)]$ where $\Psi(u)$ is a *M*-formula.

Axiom of Comprehension

The axiom of Comprehension holds for *M*-entities, that is, it is a consequence of D* that for all *M*-formulas $\Phi(x,u_1,\ldots, u_n)$ of L(∈, *A,* Ø, *M*) and all *M*-entities $a_1,\ldots, a_n$, *b,* there exists a *M*-set *a* such that for all *M*-entities *x*,

$$x \in a \leftrightarrow [\Phi(x, a_1, \ldots, a_n) \wedge x \in b] .$$

Axiom of Replacement

We will show that the axiom of Replacement holds for *M*-entities.

Let $\Gamma(x, y, u_1, \ldots, u_n)$ be an arbitrary *M*-formula and $a_1, \ldots, a_n$ be arbitrary *M*-entities such that

$$\forall x \forall y \forall z[M(x) \wedge M(y) \wedge M(z) \wedge \Gamma(x, y, a_1, \ldots, a_n) \wedge \Gamma(x, z, a_1, \ldots, a_n) \rightarrow y = z]. \quad (4.6)$$

Since we suppose D*, by *DM** let

$$\Psi(x, v_1, \ldots, v_k) \text{ and } \Theta(x, y, u_1, \ldots, u_n, v_1, \ldots, v_k)$$

be formulas of L(∈, *A*, Ø) and $b_1, \ldots, b_k$ be entities such that $M(b_1), \ldots, M(b_k)$ and

$$\forall x(\Psi(x, b_1, \ldots, b_k) \leftrightarrow M(x))$$

and

$$\forall x \forall y \forall u_1 \ldots \forall u_n[\Gamma(x, y, u_1, \ldots, u_n) \leftrightarrow \Theta(x, y, u_1, \ldots, u_n, b_1, \ldots, b_k)] .$$

Let $\Phi(x, y, u_1, \ldots, u_n, v_1, \ldots, v_k)$ be the formula

$$\Psi(x) \wedge \Psi(y) \wedge \Theta(x, y, u_1, \ldots, u_n, v_1, \ldots, v_k)$$

of L(∈, *A*, Ø). Then from Equation 4.6 it follows that

$$\forall x \forall y[(\Phi(x, y, a_1, \ldots, a_n, b_1, \ldots, b_k) \wedge \Phi(x, z, a_1, \ldots, a_n, b_1, \ldots, b_k)) \rightarrow y = z] \quad (4.7)$$

holds. Let a be an arbitrary *M*-set. By applying the axiom of Replacement of ZFA (Definition 3.3.18) to *a* and Equation 4.7, we obtain a function *f* with domain *a* such that

$$\forall x[x \in a \rightarrow \Phi(x, f(x), a_1, \ldots, a_n, b_1, \ldots, b_k)]$$

holds. Thus to show the axiom of Replacement for *M*-entities, we need only show the following:

(1) for each *M*-entity *x* in a, *f(x)* is a *M*-entity;
(2) *f* is a *M*-set; and
(3) range*(f)* is a *M*-set.

(1) Suppose *x* is a *M*-entity that is in *a.* Then

$$\{f(x)\} = \{y \,|\, \Phi(x, y, a_1, \ldots, a_n, b_1, \ldots, b_k)\} .$$

By MC, *{f(x)}* is meaningful. Therefore by Theorem 4.3.1, *f(x)* is a *M*-entity.

(2) Since

$$f = \{z \mid z = (x, y) \wedge x \in a \wedge \Phi(x, y, a_1, \ldots, a_n, b_1, \ldots, b_k)\},$$

f is a *M*-entity by *MC*. We will show that *f* is a *M*-set. Since *f* is a function, it is a set. Thus to show that *f* is a *M*-set, we need only show that each element of it is a *M*-entity. Let *(x, y)* be an arbitrary element of *f*. Then $x \in a$. Since by assumption *a* is a *M*-set, it follows that *x* is a *M*-entity. Thus by (1), *y* is a *M*-entity. From this it easily follows by *MC* that *(x,y)* is a *M*-entity.

(3) Since

$$\text{range}(f) = \{y \mid \exists x[x \in a \wedge \Phi(x, y, a_1, \ldots, a_n, b_1, \ldots, b_k)]\},$$

range *(f)* is a *M*-entity by *MC*. Since range *(f)* is also a set and by (1) all of its elements are *M*-entities, it follows that range *(f)* is a *M*-set.

Axiom of Foundation

The axiom of Foundation for *M*-entities takes the following form:

$\forall x[M(x)$ and x is not a *M*-atom and $x \neq \varnothing \rightarrow \exists y(M(y)$ and
$y \in x$ and either y is a *M*-atom or $\{z \mid M(z) \wedge z \in y \cap x\} = \varnothing)]$.

We will now show that it is true. Suppose *x* is a *M*-entity, *x* is not a *M*-atom, and $x \neq \varnothing$. Let $w = \{y \mid y \in x \wedge M(y)\}$. Then by hypothesis and the axiom of Comprehension, *w* exists, is a set, is $\neq \varnothing$, and by *MC*, *w* is meaningful. Let *y* be an element of *w* that has least rank. Then $y \in x$ and *M(y)*. Suppose *y* is not a *M*-atom. We need only show that

$$\{z \mid M(z) \wedge z \in y \cap x\} = \varnothing. \tag{4.8}$$

Suppose Equation 4.8 is false, that is, suppose *z* is an entity such that *M(z)* and $z \in y \cap x$. A contradiction will be shown. Since $z \in x$ and *M(z)*, it follows that $z \in w$. Since $z \in y$, it follows by Theorem 3.5.6 that *z* has smaller rank than *y*, which contradicts the choice of *y*.

Axiom of Choice

We will show that under the assumption of the axiom system M the axiom of Choice holds for *M*-sets, that is, that

$$\begin{aligned} &\forall x[x \text{ is a } M\text{-set and } x \neq \varnothing \\ &\quad \text{and } \forall y(y \in x \rightarrow y \text{ is a } M\text{-set and } y \neq \varnothing) \\ &\quad \rightarrow \exists z(z \text{ is a } M\text{-set and } \forall y(y \in x \rightarrow \exists w(y \cap z = \{w\})))]. \end{aligned} \tag{4.9}$$

Suppose axiom system M. Let *x* be a nonempty *M*-set such that for each *y* in *x*, *y* is a *M*-set and $y \neq \varnothing$. Then by the axiom of Choice of ZFA (Definition 3.3.21), let z' be a set such that

$$\forall y[y \in x \rightarrow \exists w(y \cap z' = \{w\})].$$

Let

$$z = \{w \mid \exists y(y \in x \wedge w \in y \cap z')\}. \tag{4.10}$$

Since by hypothesis each y in x is a M-set, each element w in z is an M-entity. Thus by *MI*, z is meaningful and therefore is a M-set. It then follows from Equation 4.10 that the axiom of Choice, that is, Equation 4.9, must hold.

4.6 Formulations in Second-Order Languages

4.6.1 Introduction

In the foundations of science, it is customary to keep track of the degree of abstraction of the formulation of scientific concepts, with minimal abstraction being the preferred condition. There is also a strong tendency towards lesser complexity. The desire for minimal abstraction usually wins out against the desire for lesser complexity when the two are in conflict.

The axiomatic systems of meaningfulness of the previous sections, which assume the entirety of set theory as part of their bases, are by any reasonable measure highly abstract. Thus their use as bases for formulating foundational issues in science runs counter to the trend towards minimal abstraction.

Interesting enough, within ZFA there is a natural measure of the "abstractness" of a concept, namely its rank. However, as a measure, this has some unfortunate consequences: the power set of the set of atoms, A, has rank 2, while any binary relation on A has rank 3, and in general any n-ary relation on A, where $n \geq 2$, has rank greater than n. In empirical applications, binary, ternary, 4-ary, and so forth, relations are often directly observable, and so are not thought to be very abstract. However, for infinite A, the power set of A—the set of all 1-ary relations on A—is not directly observable and is thought to be more abstract than any particular finitary relation on A. From a scientific perspective, the difference between 2-ary and 3-ary relations is not one of abstractness but one of complexity.

Because of this, most researchers involved in the qualitative foundations of science have shied away from set-theoretic approaches and focused instead on finitary relations on some set A of qualitative objects; that is, on m-ary relations on A, where m is some nonnegative integer. (A 0-*ary relation* on A is, by definition, an element of A.) Scientific qualitative situations are usually formulated as a (first-order) relational structure, $\mathfrak{S} = \langle A, R_1, \ldots, R_n \rangle$, where the R_i are finitary relations on A that capture the scientific qualitative situation. The R_i are called the *primitives* of $\mathfrak{S}$ They are usually chosen so that they are in sorne sense "observable." Whatever higher-order concepts are needed are stated in terms of A, $R_1, \ldots, R_n$. For example, if R_1 is a total ordering on A, the higher-order concept of Dedekind completeness of $\langle A, R_1 \rangle$ (that is, every nonempty

R_1-bounded subset of A has a $R1$-least upper bound) is stated in terms of A and R_1 using the higher-order quantifier "for all subsets." When meaningfulness issues are relevant, the primitives are taken as meaningful, and it is asked, "What other finitary relations on A are meaningful?" In the literature, it is usually conceded that relations that are first-order definable in terms of the primitives are meaningful. But there are at most denumerably many such relations, and they do not appear rich or diverse enough for many applications of meaningfulness, as for example in the uses of dimensional analysis in physics.

The kinds of suggestions for meaningfulness put forth in the literature have almost always been based on some notion of invariance. In those cases where meaningfulness considerations seem to produce the best results, the kind of invariance used essentially reduces to invariance under automorphisms of $\mathfrak{S}$, that is, becomes the Erlanger Program restricted to finitary relations on A.

Invariance approaches to meaningfulness must use some abstract, nonfirst order concept. For example, the Erlanger Program uses the group of automorphisms of $\mathfrak{S} = \langle A, R_1, \ldots, R_n \rangle$ to define which first-order relations on A are meaningful. Because of this, the problem of justifying the Erlanger Program for first-order relations is essentially the same as justifying it for higher-order relations.

In this section, a definitional characterization for meaningfulness for firstorder relations on A is given that is very similar to axiom system D″(a), where $a = R_1, \ldots, R_n$, and it is shown that this characterization agrees with the Erlanger Program for finitary relations on A.

4.6.2 The Language L^2

Because the formal description of the second-order language

$$L^2(\boldsymbol{A}, \boldsymbol{P}_1, \ldots, \boldsymbol{Pn}, \boldsymbol{M}, \boldsymbol{N}),$$

which in this section is often abbreviated to L^2, is much more complicated than the first-order language L($\in$, A, Ø, M), it is described by giving various aspects of its syntax along with the intended interpretations of its symbols.

The *non-logical predicate symbols* of L^2 are $\boldsymbol{A}$, $\boldsymbol{P}_1, \ldots, \boldsymbol{Pn}$, $\boldsymbol{M}$, and $\boldsymbol{N}$. **A**, $\boldsymbol{M}$ and $\boldsymbol{N}$ are 1-place predicate symbols and $\boldsymbol{Pi}$ is a $m(i)$-ary predicate symbol for $i=1, \ldots, n$. ($m(i)$ may be 0, in which case $\boldsymbol{Pi}$ may be called an *individual constant symbol*).

$\boldsymbol{A}$ is to be interpreted as the qualitative domain of interest, which is denoted by A. A plays the same role as the set of atoms in the axiomatic theories of meaningfulness of the previous sections.

$\boldsymbol{P}_1, \ldots, \boldsymbol{Pn}$ are interpreted as the relations $P1, \ldots, P_n$ on A (with the interpretations of 0-ary predicate symbols being elements of A.)

$\boldsymbol{M}$ is interpreted as the meaningfulness predicate, M.

N is interpreted as a set of "numbers," N. These may be real numbers, complex numbers, or any collection of entities of pure mathematics.

Let

$$D=\{R|R \text{ is a finitary relation on } A\cup N\}.$$

D denotes the domain over which the logical predicate symbols and quantifiers apply. Note that A, N, and Pi, for $i=1,\ldots, n$, are in D.

The *logical predicate symbols* of L^2 are=, $\boldsymbol{\pi}_0$, $\boldsymbol{\pi}_1,\ldots$, $\boldsymbol{\pi k},\ldots$, where k runs through the nonnegative integers.

=is interpreted as the identity relation on D, which is denoted by=.

For $k=0, 1,\ldots$, $\boldsymbol{\pi k}$ is a k-ary predicate symbol. It is interpreted as the set of k-ary relations in D, which is denoted by π_k.

$\Lambda, V, \neg, \rightarrow$, and $\leftrightarrow$ have their usual interpretations.

L^2 has an infinite list of free variables $x, y, z, x_1, y_1, z_1, \ldots$.

The quantifiers $\forall x$ and $\exists x$ should be interpreted as "for all x in D" and "there exists x in D" respectively.

Let $\mathfrak{M} = \langle D, A, P_1, \ldots, P_n \rangle$, $\mathfrak{M}$ is called a L^2-*measurement model.*

Expressions like

$$\boldsymbol{\pi}_3(x) \wedge x(u, v, z), \tag{4.11}$$

where x, u, v, z are free variables are well formed: Equation 4.11 says x is a 3-ary predicate and $x(u, v, z)$. Thus

$$\exists x\exists u\exists v\exists z[\pi 3(x) \wedge x(u, v, z)]$$

is a true statement about $\mathfrak{M} = \langle D, A, P_1, \ldots, P_n \rangle$ if $A\neq\emptyset$, because it says there is a (nonempty) 3-ary relation on A. Expressions like

$$\pi_2(A) \wedge A(x)$$

and

$$\exists x[\pi_2(A) \wedge A(x)]$$

are well-formed with the latter being false about $\mathfrak{M}$.

Sentences of L^2 are well-formed expressions that have no free variables. Precise descriptions of "well-formed formula of L^2" and "sentence of L^2" are complicated and are not given here. However, it is easy to write wellformed formulas and sentences of L^2, and to understand the results and proofs discussed in this section, this is all that is needed.

4.6.3 Axiom System D^2

Definition 4.6.1 Suppose $\mathfrak{E} = \langle E, A, P_1, \ldots, P_n, M, N \rangle$ is a measurement model for L^2. Then $\mathfrak{E}$ is said to satisfy axiom system D^2 if and only if the following four axioms are true about $\mathfrak{E}$:

1. $A \neq \emptyset$ and $A \cap N = \emptyset$. (In L^2 this is formulated as follows:

$$\exists x \boldsymbol{A}(x) \wedge \forall x \neg [\boldsymbol{A}(x) \wedge \boldsymbol{N}(x)] \,.)$$

2. There exists a one-to-one function from A onto N. (In L^2 this is formulated as follows: $\exists x \Theta_1(x)$, where $\Theta_1(x)$ is the following formula of L^2:

$$\begin{aligned} \boldsymbol{\pi_2}(x) &\wedge \forall z \forall u \forall v [x(z,u) \wedge x(z,v) \rightarrow u = v] \\ &\wedge \forall z \forall u \forall v [x(u,z) \wedge x(v,z) \rightarrow u = v] \\ &\wedge \forall u (\boldsymbol{A}(u) \leftrightarrow (\exists v) x(u,v)) \\ &\wedge \forall v (\boldsymbol{N}(v) \leftrightarrow (\exists u) x(u,v)) \,. \end{aligned}$$

($\Theta_1(x)$ says that x is a one-to-one function from A onto N. Thus $\exists x \Theta_1(x)$ says there exists a one-to-one function from A onto N. Such a function "measures" A by assigning numbers [elements of N] to elements of A. Hence the use of the terminology "measurement model" for describing $\mathfrak{E}$.)

3. Axiom AL^2: For each a in A, $M(a)$ if and only if $M(\{a\})$. (In L^2, this is formulated as follows:

$$\forall x [\boldsymbol{A}(x) \rightarrow [\boldsymbol{M}(x) \leftrightarrow \exists y [\boldsymbol{\pi_1}(y) \wedge \boldsymbol{M}(y) \wedge y(x) \wedge \forall z [y(z) \leftrightarrow z = x]]]] \,.)$$

4. Axiom D^2: For each k-ary relation a in E, $k \geq 1$, $M(a)$ if and only if there exists well-formed formula $\Phi(x_1, \ldots, x_k, u_1, \ldots, u_m)$ of L^2 and relations (possibly 0-ary) $a_1, \ldots, a_m$ of N such that

$$\forall x_1 \ldots \forall x_k [a(x_1, \ldots, x_k) \leftrightarrow \Phi(x_1, \ldots, x_n, a_1, \ldots, a_m)] \,.$$

Although axiom D^2 uses L^2 in its formulation, it cannot be completely formulated in L^2, because it quantifies over formulae of L^2. This is similar to the problem we faced with $L(\in, A, \emptyset, M)$ and axiom systems D^* and $D''(a)$. It turns out that for showing the main result of this section (Theorem 4.6.1) the form of Φ in Definition 4.6.1 can be severely limited, since the proof of that theorem use only formulas of L^2 that are constructed in a very special way. (The proof of Theorem 4.6.1 is not given in this book. Its main ideas are similar to those of the proof of Theorem 4.3.6.)

Theorem 4.6.1 Assume $\mathfrak{E} = \langle E, A, P_1, \ldots, P_n \rangle$ is a L^2-measurement model, and assume axiom system D^2 is true about $\mathfrak{E}$. Let G be the group of automorphisms of $\mathfrak{S} = \langle A, P_1, \ldots, P_n \rangle$. Then for each a in A,

$$M(a) \text{ iff } f(a) = a \text{ for all } f \text{ in } G,$$

and for each k-ary relation b on A,

$$M(b) \text{ iff for all } f \text{ in } G \text{ and all } a_1, \ldots, a_k \text{ in } A,$$
$$b(a_1, \ldots, a_k) \textit{ iff } b[f(a_1), \ldots, f(a_k)].$$

4.6.4 What Has Been Accomplished

Axiom system D^2 essentially says two things: First, elements of A can be assigned "numbers"; and second, a finitary relation on A is meaningful if and only if it is definable through a well-formed formula of L^2 in terms of A, $P_1, \ldots, P_n$, N and finitary relations on N. Theorem 4.6.1 shows that axiom system D^2 yields the same meaningfulness concept on finitary relations on A as the Erlanger Program.

The motivation for developing the axiom system D^2 was to demonstrate that many of the concepts and results of Section 4.3 could be adapted to situations that did not assume the immensely rich collection of higher order concepts that exists in set theory. Theorem 4.6.1 is only one example of a translation of a result of Section 4.3 (i.e., the result that for externally finite a, $D''(a)$ and TM are equivalent) into a relatively weak version of a second-order language; other results of that section can be similarly translated.

4.7 Additional Approaches to Invariance and Definability

I believe that the axiomatic approaches to meaningfulness of this chapter provide a rich set of concepts and techniques that work reasonably well for the kind of issues considered in this book. Obviously, these approaches can be refined further, and other approaches based on different concepts might be as productive. This section very briefly discusses a few alternatives to the axiomatic theories of meaningfulness of this chapter.

4.7.1 Weakening Axiom System ZFA

One way to make the ontologies of mathematics and science more coordinate is to weaken axiom system ZFA. In the nineteenth century mathematics and science were on a parallel course of trying to cleanse their disciplines of metaphysics. In mathematics, this process resulted in the "arithmetization of analysis" and an emphasis on constructive, non-infinitistic methods. In the twentieth century, mathematics and science diverged in their foundational objectives, with twentieth century mathematics abandoning the non-infinitistic constructive approach in favor of Cantorian set theory, which provided the kinds of concepts and techniques needed for modern mathematics, especially for the fields of analysis and topology. However, it later became recognized that much of the infinitistic structure inherent in the Cantorian approach is not needed for ordinary mathematics. This is particularly the case for the highly applicable areas of analysis and topology. For example, the typical proofs of several important theorems of analysis and topology from axiom system ZFA require the use of the axiom of Choice (Definition 3.3.21). For most of these, other proofs can be formulated that use only very weakened forms of the axiom of Choice. Thus most of analysis and topology can be formulated in a axiomatic set theory that is like ZFA except that the axiom of Choice is replaced by a weaker version. It also can be shown, that such weakened versions of ZFA often eliminate counter-intuitive implications of the axiom of Choice, for example, the existence of Lebesgue non-measurable subsets of the reals. (See Chapter 8 for a discussion of the non-intuitive consequences of the axiom of Choice.) Other axioms of ZFA can be similarly weakened. For example, the axiom of Replacement implies the existence of very large cardinals that are not used in ordinary mathematics or science. This axiom can be weakened to yield a set theory that has a much smaller metaphysical commitment to the existence of very large, infinite sets.

4.7.2 Changing Axiom System ZFA

It appears to be advantageous in the modeling of certain ideas in the behavioral sciences to invoke specific principles that, in the presence of the axioms of ZFA other than Choice, imply the negation of the axiom of Choice. (E.g., see Brunner and Mihara, 2000, for a discussion and an example involving social welfare functions.) Sometimes the scientific applications of such principles may also be obtainable through meaningfulness considerations similar to those discussed in this Chapter, but discussions of this interesting issue is outside the intended scope of this book.

4.7.3 Weakening Axioms *MC′* and *MC*

Axioms *MC′* and *MC* provide powerful ways of defining meaningful entities in terms of other meaningful entities. Obviously these methods of definability can be weakened and still provide for interesting meaningfulness-like concepts. However, for intended use of the concept of meaningfulness as discussed in Subsection 4.1.5, I see no reason for such an action. For different intended uses, for example, for capturing certain ideas about qualitativeness or empiricalness, one may want to weaken axioms *MC′* and *MC*.

4.7.4 Weakening Axiom *MP*

Axiom *MP* allows pure mathematics to be employed freely in the defining of meaningful entities through applications of axiom *MC′* (or axiom *MC*). The weakening of axiom *MP* therefore restricts the amount of pure mathematics employable in this manner. One obvious way to accomplish this is to restrict the use of pure mathematics by methods similar to those discussed in Section 4.7.1. By this strategy, axiom system ZFA is assumed, but the pure mathematics available for use in defining meaningful entities consists of a weakened form of axiom system ZF applied to collection of pure sets (Section 3.7).

Implementation of this strategy would make the metaphysics inherent in the collection of meaningful entities more coordinate with the kind of metaphysics underlying infinitistic scientific applications that are encountered in physics and other highly mathematical sciences. A much more severe limiting of axiom *MP* is described in Chapter 7.

4.7.5 Weakening Axiom *TM*

One approach for weakening axiom *TM* is to define meaningfulness in terms in of invariance under a more general set of transformations than a group of permutations. Historically, this kind of consideration occurred with attempts to generalize the Erlanger's Program concept of "geometric" to situations in differential geometry which had trivial transformations groups. More recently, a similar program has been undertaken in the theory of measurement for a similar reason (see Section 5.5). The ones in geometry had only limited success, and the ones for the theory of measurement appear, to me, to be also of limited use, because they work only in highly specialized circumstances. In Chapter 6 a different way of generalizing axiom *TM* is given, one that uses the idea of invariance across equivalent structures.

4.7.6 Using Infinitary Languages and Logics

Infinitary languages and logics generalizes the concept of first-order language and logic to situations that allow for infinitely long formulas and sentences. Although such languages and logics are sometimes useful in providing simple formulations of various infinitary concepts, including some invariance concepts, I am very doubtful about their value as bases for general theories of meaningfulness. First, they produce a very artificial variant of mathematics, one that appears, to me, to be quite unsuited for investigating the foundations of science. Second, such languages and logics place much of their metaphysical commitment at the meta-level, for example, the existence of some kinds of formulas through uses of the axiom of Choice. For meaningfulness issues, I find it preferable to have as much as possible this kind of metaphysical commitment be at the object level, where it can be explicitly described as meaningfulness axioms. Third, to carry out mathematics and science in the most natural way, one needs higher-order concepts, and only a very restricted set of higher-order concepts can be captured by the traditional infinitary languages. And fourth, a comprehensive framework for meaningfulness needs to account for the purely mathematical, the qualitative-like concepts based upon a qualitative domain, and the concepts based on mappings between the qualitative domain and a purely mathematical domain, and such a framework is outside the traditional treatments of infinitary languages and logics.

4.8 Conclusions

As discussed in Chapter 1, the need to separate the meaningful from the nonmeaningful occurred repeatedly throughout the histories of mathematics and science. In my view, this resulted from two different but related concerns, which I call *ontological* and *epistemological* after the philosophical usage. The ontological need arises naturally when the metaphysical entities in mathematics have a radically different logical form from those occurring in science, as for example in the case of mathematical entities sometimes having very high orders of infinity. The epistemological need arises primarily when material reality possesses inherent properties that when combined with material (empirical) facts allow for inferences that do not follow from the facts by logic alone. Of course, these two needs are highly interrelated, and in some instances are very difficult to separate. In terms of some of the historical examples of Chapter 1, I consider Plato's distinction between pure and applied geometry to be primarily an ontological one; Kepler's distinction between primary and secondary qualities to be a mixture of ontological and epistemological concerns; and dimensional analysis in physics to be primarily an epistemological matter.

Ontological meaningfulness concerns become prominent when either mathematics or science make radical shifts in the kinds of entities they are willing to consider, or when their ontological commitments appear to be incompatible. Both kinds of situations occur in modern mathematical science.

Recall that in the nineteenth century, mathematics made a concerted effort to rid itself of as many metaphysical concepts as possible. (This foundational cleansing—which became known as "the arithmetization of analysis"— is discussed in Section 1.2.) At about the same time, there were similar trends in the philosophical foundations of science, positivism today being the best known example. Thus in the nineteenth century there was a convergence—in attitude at least—in the philosophical trends occurring in mathematics and science, making the ontological commitments of the two fields more coordinate. This trend was abruptly disrupted at the turn of the twentieth century by the infusion of Cantorian set theory into ordinary mathematics. The result of this is that today mathematical science has a highly platonic, highly infinitistic, mathematical side with a strong tendency for making its empirical side highly positivistic and highly constructive. This discordance obviously poses a wide range of ontological and epistemological difficulties, especially if the full range of mathematics is to be available for scientific application.

The book's method for dealing with these difficulties is to adopt an extended set-theoretic framework that can simultaneously handle qualitative and quantitative concepts, their interactions, and their separation. The basis for the set-theoretic framework is axiom system ZFA. Pure mathematics is identified with the set theory of pure sets, and (purely) qualitative entities are identified with atoms, relations on atoms, relations of relations on atoms, and so forth. In axiom system ZFA, the qualitative, the purely mathematical, and mappings between them are highly coordinate ontologically. This is one of the main reason for the selection of ZFA as the basis for the theory. The book's method for accounting for the ontological discordance between scientifically based concepts and (purely) mathematical ones is by adding meaningfulness, qualitativeness, and empiricalness predicates to L(∈, *A*, Ø) and adding additional axioms involving these predicates to ZFA. Although, in the resulting axiomatic systems there exist ontological correlates based on the set of atoms for almost all of the (purely) mathematical concepts, only some of these correlates are meaningful, qualitative, or empirical. This method clearly mitigates some of the difficulties of discordant ontologies, but it does not completely eliminate them—especially for the meaningfulness concepts of this chapter.

The fundamental property of meaningfulness stressed throughout this chapter is that entities that are definable in terms of meaningful entities are themselves meaningful. This is captured by the axioms of Meaningful Comprehension' and Atomic Legacy. Although these principles allow for highly abstract entities to be meaningful, they in themselves are generally not responsible for the infusion of highly infinitistic concepts from the mathematical domain into the qualitative domain: the latter usually results from applications of the axiom of Meaningful Pure Sets. ZFA and the abovementioned axioms imply the existence of a *meaningful set S* of one-to-one functions from the set of atoms *A* onto a pure set *T*. Because all (pure) concepts based on *T* are meaningful by the axiom of Meaningful Pure Sets, the functions in *S* can be used to transfer (in a meaningful way) much (but not all) of the ontology of pure mathematics to meaningful concepts based on *A*. The obvious way to prevent this kind of transfer is to weaken the axiom of Meaningful Pure Sets. This option is pursued in Chapter 7 for the concepts of qualitativeness and empiricalness.

Interestingly enough, the axiom of Meaningful Pure Sets was also diagnosed in Section 4.3 as the cause for the apparent failure of the axiomatic theories of meaningfulness in certain situations with trivial transformation groups for *M*.

Except for these difficulties linked to the axiom of Meaningful Pure Sets, the axiomatic generalizations of the Erlanger Program presented in this chapter function quite well. In particular, the following have been accomplished: (1) The concept of meaningfulness inherent in the Erlanger Program is now far better understood because an equivalent of it (axiom system M) was produced that stressed definability concepts. (2) This equivalent naturally generalizes into various concepts of meaningfulness (D′, D, $\overline{\text{D}}$.D*, D″(a)) that avoid a strongly nondefinitional aspect of the Erlanger Program, namely axiom *MI.* (3) These generalizations turn out to be equivalents to weaken invariance forms of the Erlanger concept (GTM′, GTM, GTM*), providing interesting links between definability and invariance.

As noted in Section 4.6, the set-theoretic approaches to meaningfulness of the chapter produce some awkwardnesses, for example, *n*-ary relations on the set of atoms being highly abstract for $n>1$. I consider these to be relatively minor and easily dealt with in the set-theoretic approach. In fact, the methods of Section 4.6 show that all of the axiomatic results can be effectively carried over to higher-order model-theoretic formulations. While such modeltheoretic formulations are more appropriate for various foundational aspects of science, I still prefer the set-theoretic ones, because they apply to mathematical as well as physical situations, and one does not have to change the order of the higher-order language with applications. Also, the set-theoretic approaches are simpler and more elegant to present.

The main argument against the axiomatic approaches to meaningfulness of this chapter is that they allow some possibly irrelevant entities to be meaningful, and these may influence the meaningfulness of some relevant ones, that is, they are not conservative in what they allow to be meaningful. *But one should realize that this argument is really double-edged: by not being conservative in what they allow to be meaningful, they are very conservative in what they allow to be nonmeaningful, and in many meaningfulness techniques— such as dimensional analysis—it is the excluding of the nonmeaningful that drives the technique, and thus for such cases these axiomatic approaches are very conservative.*

4.9 Summary of Axioms and Axiom Systems

This section summarizes the various axiomatic generalizations of the Erlanger Program given in the chapter and the theorems interrelating them.

The definitions of the set-theoretic concepts ZFA, "externally finite," *EFI,* and *EFI** are given in Definitions 3.3.22, 4.3.7, 4.3.8, and 4.3.9.

4.9.1 Axioms

1. *MP (Meaningful Pure Sets):* Each pure set is meaningful. (Definition 4.2.4)
2. *AL (Atomic Legacy):* If a is an atom and $\{a\}$ is meaningful, then a is meaningful. (Definition 4.2.8)
3. *MC' (Meaningful Comprehension'}:* A set that is definable in terms of meaningful entities through a formula of L(∈, *A*, Ø) is itself meaningful. (Definition 4.2.7)
4. *MC (Meaningful Comprehension):* If $\Phi(x, u1, \ldots, u_n)$ is a formula of L(∈, *A*, Ø, *M*) and $b, b_1, \ldots, b_n$, are meaningful entities, then there exists a meaningful set a such that
$$a = \{x \mid \Phi(x, b_1, \ldots, b_n) \wedge x \in b\}.$$
(Definition 4.2.9)
5. *DM (Definable Meaningfulness):* M is definable in terms of entities and a formula of L(∈, *A*,Ø). (Definition 4.3.4)

6. *DM* (Definable Meaningfulness*): M* is definable in terms of meaningful entities and a formula of L(∈, *A*, Ø). (Definition 4.2.6)
7. *DC(a)* (*Definable Closure* of *a*): *a* is a collection of entities and for all entities *b, b* is meaningful if and only if *{b}* is definable in terms of elements of *a* and pure sets through a formula of L(∈, *A*, Ø). (Definition 4.3.6)
8. *MI (Meaningful Inheritability):* If each element of a set is meaningful, then the set is meaningful. (Definition 4.2.5)
9. *TM (Transformational Meaningfulness):* There exists an entity *H* that is a set of permutations on *A* and is such that for each entity *x, x* is meaningful if and only if for each *f* in *H, f(x)=x.* (Definition 4.2.3)

4.9.2 Axiom Systems

1. D′: ZFA and axioms *MP, AL,* and *MC′* are true. (Definition 4.3.1)
2. D: ZFA and axioms *MP, AL,* and *MC* are true. (Definition 4.3.2)
3. $\overline{\mathrm{D}}$: Axiom system D′ and axiom *DM* are true. (Definition 4.3.5)
4. D*: Axiom system *D′* and axiom *DM** are true. (Definition 4.3.3)
5. D″(a): ZFA and axiom *DC(a)* are true. (Definition 4.3.6)
6. M: ZFA, axioms *AL, MC,* and *MI* are true. (Definition 4.2.10)
7. TM *(Transformational Meaningfulness):* ZFA and axiom *TM* are true. (Definition 4.2.3)
8. GTM’ *(Generalized Transformational Meaningfulness'): Γ* is said to be a *transformational family* for *M* if and only if *Γ* is a nonempty collection of entities such that each element of *Γ* is a group of permutations of *A* and such that *(i)* for each entity *x, M(x)* if and only if there exists *K* in *Γ* such that for all *f* in *K, f(x)=x,* and *(ii)* for each *F* and *H* in *Γ* there exists *K* in *Γ* such that *K* is a subset of *F*∩*H.* Axiom system GTM’ is said to hold if and only if ZFA and there exists a transformational family for *M.* (Definitions 4.4.1, 4.4.2, and 4.4.3)
9. GTM *(Generalized Transformational Meaningfulness):* Axiom system GTM is said to hold if and only if ZFA and there exists a transformational family for *M* (see GTM′ above) that is an entity. (Definition 4.4.4)
10. GTM* *(Generalized Transformational Meaningfulness*):* Axiom system GTM* is said to hold if and only if ZFA and there exists a transformational family for *M* (see GTM′ above) that is a meaningful entity. (Definition 4.4.5)

4.9.3 Theorems Interrelating the Axiom Systems

1. GTM′↔D′. (Theorem 4.4.1)
2. GTM→D. (Theorem 4.4.2)
3. Assume *EFI.* Then GTM↔$\overline{\mathrm{D}}$↔ D″*(a)* for some set *a*. (Theorems 4.3.8 and 4.4.2)

4. GTM*→D*. (Theorem 4.4.3)
5. Assume *EFI*. Then GTM*↔D*↔D″*(a)* for some meaningful set *a*. (Theorems 4.4.3 and 4.3.8)
6. Assume *EFI**. Then GTM*↔D*↔D. (Theorem 4.4.3)
7. M↔TM. (Theorem 4.2.12)
8. If *a* is an externally finite set and D″*(a)*, then TM. (Theorem 4.3.6)
9. D′ ∧ *MI* does not imply either axiom *DM* or axiom *MC*. (Theorem 4.4.4)
10. $\overline{\mathrm{D}}$ does not imply either axiom *DM** or axiom *MC*. (Theorem 4.4.5)
11. D* does not imply axiom *MI*. (Theorem 4.4.6)

4.10 Additional Proofs and Results

4.10.1 Introduction

In this section, proofs of various theorems stated in the chapter will now be proved. The order of these proofs will not in general follow the order in which the theorems appeared in previous sections. Because of this, such theorems in this section will have two labeling numbers associated with them: one from a previous section and one from this section. (The previous section's number will be in parentheses just before the statement of the theorem.)

4.10.2 Basic Concepts I

Theorem 4.10.1 (Theorem 4.3.2) Assume D′. Then $M(\Pi)$, and if $*'$ is composition of elements of Π, then M($*'$).

Proof. Since

$$\Pi=\{f|f \text{ is a one-to-one function from } A \text{ onto } A\}$$

and

$$*' = \{(f,g,h)\,|\, f \in \Pi \wedge g \in \Pi \wedge h \in \Pi \wedge \forall x[x \in A \rightarrow h(x) = f(g(x))]\},$$

the theorem follows by axiom *MC′*.

Lemma 4.10.1 Assume D′. Suppose *H* is an entity such that *(i)* each element of *H* is a function from *A* into A, and *(ii)* for each *f* in *H* and for each set *x*, if *M(x)* then $f(x)\subseteq x$. Then each element of *H* is a function that is onto *A*.

Proof. Suppose $f\in H$. Then *M(A)* by Theorem 4.3.1. Thus also by Theorem 4.3.1, $M[\{A\}]$. Therefore, by the definition of *H*, $f[\{A\}]\subseteq\{A\}$. However,

$$f[\{A\}] \subseteq \{A\} \text{ iff } f(A)=A \text{ iff } f \text{ is onto } A.$$

Lemma 4.10.2 Assume D′. Let

$$D = \{x \mid \exists u \exists v (u \in A \wedge v \in A \wedge u \neq v \wedge x = \{u, v\})\}.$$

Then *M(D)*.

Proof. By Theorem 4.3.1, *M(A)*. *M(D)* then follows by axiom *MC′*.

Lemma 4.10.3 Assume D′. Suppose *H* is an entity such that *(i)* each element of *H* is a function from *A* into *A*, and *(ii)* for each *f* in *H* and for each set *x*, if *M(x)* then $f(x) \subseteq x$. Then each element of *H* is a one-to-one function.

Proof. Let $f \in H$. Suppose *f* is not a one-to-one function. A contradiction will be shown. Let *a* and *b* in *A* be such that $a \neq b$ and $f(a)=f(b)$. Then $\{a,b\} \in D$, where *D* is as defined in Lemma 4.10.2. Since $f \in H$ and $f(a)=f(b)$, it follows that

$$f[\{a, b\}] = \{f(a), f(b)\} = \{f(a)\} \notin D.$$

But since *M(D)* by Lemma 4.10.2, it follows from the definition of *H* that $f(D) \subseteq D$. Since $\{a, b\} \in D$, it then follows that $f[\{a,b\}] \in D$, which contradicts $f[\{a,b\}] \notin D$.

Lemma 4.10.4 Assume ZFA. Suppose *H* is a group of permutations and *a* is a set such that $f(a) \subseteq a$ for each *f* in *H*. Then $I_H(a)$.

Proof. Let *f* be an arbitrary element of *H*. It need only be shown that $f(a)=a$. Let *x* be an arbitrary element of a. It need only be shown that $f(y)=x$ for some $y \in a$. Since *H* is a group, $f^{-1} \in H$. Let $y=f^{-1}(x)$. Then $y \in a$ since $f_{-}^{1}(a) \subseteq a$. By Statement 7 of Theorem 3.6.1,

$$f(y) = f[f^{-1}(x)] = f * f^{-1}(x) = x.$$

4.10.3 Basic Concepts II

Lemma 4.10.5 (Lemma 4.5.1) Assume D′. Suppose *H* is a meaningful function. Then the following two statements are true:

1. The domain of *H* and the range of *H* are meaningful.
2. If *a* is a meaningful subset of the domain of *H*, then the restriction of *H* to a is meaningful.

Proof. 1. The domain of *H* is $\{x \mid \exists y[(x, y) \in H]\}$ and the range of *H* is $\{y \mid \exists x[(x, y) \in H]\}$, and consequently the domain and range of *H* are meaningful by the meaningfulness of *H* and *MC′*.

2. Suppose *a* is a meaningful subset of the domain of *H*. Then the restriction of *H* to a is

$$\{z \mid \exists x \exists y (x \in a \wedge (x, y) \in H \wedge z = (x, y))\}$$

and this is meaningful by *MC′*.

Lemma 4.10.6 (Lemma 4.5.2) Assume D′. Then the following two statements are true:

1. V_α is meaningful for each ordinal α.
2. For all f in II and all entities a and b, $f[(a, b)]=(f(a), f(b))$.

Proof. 1. It follows from the Recursion Theorem and Definition 3.5.1 that there is a formula $\Phi(u,v)$ of L($\in$, A, Ø) such that for all entities a and b,

$$\Phi(a, b) \text{ iff } a \text{ is an ordinal and } b=V_\alpha.$$

Then

$$V_\alpha=\{x \mid \exists y(\Phi(\alpha, y) \text{ and } x\in y\},$$

and thus it immediately follows from Theorem 3.5.7 and axioms *MP* and *MC′* and Theorem 3.3.1 that V_a is meaningful for each ordinal α.

2. Suppose $f\in G$ and a and b are entities. Then $f[\{a\}]=\{f(a)\}$ and $f[\{_o, 6\}]=\{f(a),f(b)\}$. Thus

$$\begin{aligned} f[(a,b)] &= f[\{\{a\},\{a,b\}\}] = \{f[\{a\}], f[\{a,b\}]\} \\ &= \{\{f(a)\},\{f(a),f(b)\}\} = (f(a),f(b)). \end{aligned}$$

Lemma 4.10.7 (Lemma 4.5.3) Assume D′. Suppose H is a function that is G-invariant. Then the following two statements are true:

1. The domain and range of H are G-invariant.
2. If a is G-invariant and a is a subset of the domain of H, then the restriction of H to a is G-invariant.

Proof. 1. Let d be the domain of H. Let f be an arbitrary element of G and x be an arbitrary element of d. Let y be such that $(x, y)\in H$. Since H is G-invariant, it follows by Lemma 4.10.6 that $f[(x,y)]=(f(x), f(y))$ is in H. Therefore $f(x)$ is in d. Since x is an arbitrary element of d, it has been shown that $f(d)\subseteq d$. Since f is an arbitrary element of G, d is G-invariant by Lemma 4.10.4. A similar argument shows that the range of H is G-invariant.

Statement 2 follows by an argument similar to the proof of Statement 1.

Lemma 4.10.8 (Lemma 4.5.4) Assume M. Then the following two statements are true for all sets a and b:

1. If a is homomeaningful, then a is meaningful.
2. If b is meaningful, then ${}^{+}b$ and ${}^{-}b$ are meaningful.

Proof. Statement 1 is an immediate consequence of *MI*. To show Statement 2, suppose b is a meaningful set. Then ${}^{+}b$ is meaningful by Statement 1. Since b and ${}^{+}b$ are meaningful and ${}^{-}b=b-{}^{+}b$, it follows by *MC'* that ${}^{-}b$ is meaningful.

Lemma 4.10.9 (Lemma 4.5.5) Assume M. Then for each entity y there exists a minimal meaningful set containing y as an element.

Proof. Let y be an arbitrary entity and α be its rank. Let

$$W = \{w \mid M(w) \wedge y \in w \wedge w \subseteq V_\alpha\}.$$

$W \neq \emptyset$, since V_α is meaningful (Lemma 4.10.6), $y \in V_\alpha$, and consequently $V_\alpha \in W$. Since each element of W is meaningful, it follows by *MI* that W is meaningful. Therefore by *MC*, $x = \bigcap W$ is meaningful. We will show that x is the minimal meaningful set containing y as an element. $y \in x$, since y is a member of each element of W. Suppose z is another meaningful set such that $y \in z$. Then $y \in z \cap V_\alpha \subseteq V_\alpha$, and $z \cap V\alpha$ is meaningful by *MC*. Therefore $z \cap V_\alpha$ is in W and thus $x \subseteq z \cap V_\alpha \subseteq z$.

Lemma 4.10.10 (Lemma 4.5.6) Assume D′. Suppose x is the minimal meaningful set containing y as an element. Then x is minimeaningful.

Proof. Since x is the minimal meaningful set containing y as an element, it is meaningful. Suppose it is not minimeaningful. A contradiction will be shown. Since it is not minimeaningful, let z be a nonempty proper subset of it such that z is meaningful. Then by *MC′*, $x—z$ is meaningful. $x—z$ is also a proper subset of x. Now, either $y \in z$ or $y \in x—z$. In either case, y is an element of a meaningful set that is a proper subset x, which is impossible, since x is the minimal meaningful set containing y as an element.

Lemma 4.10.11 (Lemma 4.5.7) Assume D′ and G is an entity. Suppose a and b are entities with respective orbits $\hat{a}$ and $\hat{b}$. Then the following two statements are true:

1. $\hat{a}$ is G-invariant.
2. Either $\hat{a}=\hat{b}$ or $\hat{a} \cap \hat{b} = \emptyset$.

Proof. 1. Let f be an arbitrary element of G and x be an arbitrary element of $\hat{a}$. By the Lemma 4.10.4, it needs to only be shown that $f(x) \in \hat{a}$. Since $x \in \hat{a}$, let g in G be such that $x=g(a)$. Then by Statement 7 of Theorem 3.6.1,

$$f(x)=f[g(a)]=f{*}g(a),$$

and since $f{*}g$ is in G, it then follows from Definition 4.5.4 that $f{*}g(a) \in \hat{a}$, i.e., that $f(x) \in \hat{a}$.

2. Suppose $\hat{a} \cap \hat{b} \neq \emptyset$. Let y be an element of $\hat{a} \cap \Theta$. By Definition 4.5.4, let f and g in G be such that $f(a)=y$ and $g(b)=y$. Then it follows from Statement 7 of Theorem 3.6.1 and the group properties of G that

$$\begin{aligned} \hat{a} &= \{h(a) \mid h \in G\} = \{h[f^{-1}(y)] \mid h \in G\} \\ &= \{h * f^{-1}(y) \mid h \in G\} = \{h(y) \mid h \in G\} \\ &= \{h[g(b)] \mid h \in G\} = \{h * g(b) \mid h \in G\} \\ &= \{h(b) \mid h \in G\} = \hat{b}\,. \end{aligned}$$

Lemma 4.10.12 (Lemma 4.5.8) Assume TM. Suppose a is an entity and $\hat{a}$ is its orbit. Then $\hat{a}$ is minimeaningful.

Proof. By Theorems 4.2.4, 4.2.5, 4.2.7, 4.2.8, and 4.2.11, it follows that TM implies M and TM implies D′. Thus by Lemmas 4.10.9 and 4.10.10, let b be the minimeaningful set such that $a \in b$. Since by Lemma 4.10.11 $I_G(\hat{a})$, it follows from TM that $\hat{a}$ is meaningful, and since $a \in \hat{a}$ and $a \in b$ and b is minimeaningful, it then follows that $\hat{a} \supseteq b$. Now since b is meaningful, it follows from TM that $I_G(b)$. Therefore, since $a \in b$ and $I_G(b)$,

$$\hat{a} = \{f(a) \mid f \in G\} \subseteq b.$$

Thus $\hat{a}=b$, and therefore $\hat{a}$ is minimeaningful since b is.

4.10.4 Sufficient Conditions for TM

In this subsection it will be shown that the axiom systems M and TM are logically equivalent, and that axiom system D″(a), where a is an externally finite set, implies axiom system TM. In order to establish these results, some preliminary lemmas about extending certain types of functions on A are needed:

4.10.5 Preliminary Lemmas From Set Theory

Definition 4.10.1 Assume ZFA. Suppose q is a function on A. Then q can be extended to all elements of V by transfinite induction as follows: Let q_0 be the function q on $V_0=A$. Suppose β is an ordinal and q_β has been defined on $V\beta$. Define $q_{\beta+1}$ on $V_{\beta+1}$ as follows: if $x \in V_\beta$ then $q_{\beta+1}(x)=q\beta(x)$, and if $x \subseteq V_\beta$ then $q_{\beta+1}(x) = \{q_\beta(y) \mid y \in x\}$. Suppose $\gamma>0$ is a limit ordinal and q_δ has been defined for all $\delta<\gamma$. Let $q_\gamma = \bigcup_{\delta<\gamma} q_\delta$ Then it is easy to show by using the Recursion Theorem that there exists a functional relation $\hat{q}$ on V such that for each x in V, $\hat{q}(x)=q_\delta(x)$ where δ is the rank of x. We will often confuse notation and write $\hat{q}$ as q. When this is done, the context will always make clear whether the symbol "q" stands for the function on A or the functional relation on V.

The following lemma easily follows from Definition 4.10.1 and transfinite induction:

Lemma 4.10.13 Assume ZFA. Suppose q is a function from A onto a pure set. Then for each x in V, $q(x)$ is a pure set.

Lemma 4.10.14 For each ordinal δ there exists a one-to-one function q from A onto a pure set such that the restriction of q to V_δ is a one-to-one function.

Proof. Let δ be an arbitrary ordinal. Let (by Theorem 3.4.9) β be the least ordinal such that β is equinumerous with V_δ. It is not difficult to show by using elementary properties of cardinals and ordinals that there exists an ordinal $\gamma'>\beta$

and a one-to-one function q' from A into γ' such that for each x in $V\delta$: *(i)* if $x \in V_\delta — A$ then $q'(x) \in \beta$, and *(ii)* if $x \in A$ then $q'(x) \notin \beta$. So let γ and q be such that

(1) γ is an ordinal and $\gamma > \beta$,
(2) q is a one-to-one function from A into γ,
(3) for each $x \in V_\delta — A$, $q(x) \in \beta$, and
(4) for each x in A, $q(x) \notin \beta$.

Let $\alpha = \{q(x) \mid x \in A\}$. Since $\alpha \subseteq \gamma$ and γ is an ordinal, it follows from Theorem 3.5.7 that α is a pure set. Thus by Lemma 4.10.13, the range of the restriction of q to V_δ is a pure set. Since the restriction of q to V_δ is a function on $V\delta$, to show the lemma we need only establish that the restriction of q to V_δ is one-to-one. Suppose it is not. A contradiction will be shown. Let μ be the smallest ordinal such that there exist x and y in V_δ such that $x \neq y$, $q(x)=q(y)$, and μ=the minimum of rank(x) and rank(y). Without loss of generality, let a and b be elements of V_δ such that $a \neq b$, $q(a)=q(b)$, and μ=rank(a). Then by the choice of μ, rank$(b) \geq \mu$. There are two cases to consider.

Case 1. a is a set. Since a is a set, rank$(a)>0$. Since rank$(b)>$rank$(a)=\mu$, b is also a set. Thus since $a \neq b$, either $a—b \neq \emptyset$ or $b—a \neq \emptyset$. Subcase *(i)*. $a—b \neq \emptyset$. Let $c \in a—b$. Then since a and b are sets and $q(a)=q(b)$, it follows that we can find d in b such that $q(c)=q(d)$. Since $c \in a—b$ and $d \in b$, it follows that $c \neq d$. But since $c \in a$, rank$(c)<\mu$, and this together with $c \neq d$ d and $q(c)=q(d)$ contradicts the choice of μ. Subcase *(ii)*. $b—a \neq \emptyset$. Let $d \in b—a$. Since $q(b)=q(a)$, we can find c in a such that $q(d)=q(c)$. Then rank$(c)<\mu$, $c \neq d$, and $q(c)=q(d)$, contradicting the choice of μ.

Case 2. a is an atom. Since q was chosen so that it is one-to-one on A and $a \neq b$ and $q(a)=q(b)$, it follows that b is not an atom. Thus b is a set. Therefore $q(b) \in \beta$. Since a is an atom, $q(a) \notin \beta$. Thus $q(a)=q(b)$ is contradicted.

Lemma 4.10.15 Suppose δ is an ordinal, f is a permutation on A, q is a one-to-one function from A onto a pure set a such that the restriction of q to V_δ is a one-to-one function on V_δ, and h is the following function on A: $h=q*f$. Then the following three statements are true:

1. h is a one-to-one function from A onto α.
2. For each x in V, $h(x)=q[f(x)]$.
3. The restriction of h to V_δ is a one-to-one function on V_δ.

Proof. 1. Since f is a one-to-one function from A onto itself and q is a one-to-one function from A onto α, it immediately follows that h is a one-to-one function from A onto α.

2. Suppose it were not the case that for each x in V, $h(x)=q[f(x)]$. A contradiction

will be shown. Let γ and a be such that γ is an ordinal, a is an entity that has rank γ, $h(a)\neq q[f(a)]$, and for all entities x with rank $<\gamma$, $h(x)=q[f(x)]$. $\gamma\neq 0$, for if $\gamma=0$ then $a\in A$, and thus

$$h(a) = (q * f)(a) = q[f(a)],$$

which contradicts the choice of a. Thus $\gamma\neq 0$, and therefore a is a set. By Theorem 3.5.6, each element of a has rank $<\gamma$. Therefore by the choices of γ and a, $h(x)=q[f(x)]$ for each x in a, that is,

$$h(a) = \{h(x)\,|\,x \in a\} = \{q[f(x)]\,|\,x \in a\} = q[\{f(x)\,|\,x \in a\}] = q[f(a)],$$

contrary to the choice of a.

3. Suppose a and b are arbitrary elements of V_δ and $a\neq b$. Then by Theorem 3.6.1, $f(a)$ and $f(b)$ are in V_δ and $f(d)\neq f(b)$. Since by hypothesis q is a one-to-one function on V_δ, it follows that $q[f(a)]\neq q[f(b)]$. Thus by Statement 2,

$$h(a) = q[f(a)] \neq q[f(b)] = h(b),$$

which, since a and b are arbitrary elements of V_δ such that $a\neq b$, implies that h is a one-to-one function on V_δ.

Lemma 4.10.16 Suppose δ is an ordinal, q and t are one-to-one functions from A onto a pure set a such that the restriction of q to V_δ and the restriction of t to V_δ are one-to-one functions on V_δ. Suppose h is the following function on A: $h=q^{-1}$ $*$ t. Then the following three statements are true:

1. h is a permutation on A.
2. Range of the restriction of q to V_δ=range of the restriction of t to V_δ.
3. For each x in $V\delta$, $h(x)=q^{-1}[t(x)]$.

Proof. Statement 1 immediately follows since q and t are one-to-one functions from A onto the same set α.

Statement 2 will be shown by transfinite induction. *(i)* By hypothesis, the range of the restriction of q to A=the range of the restriction of t to $A=\alpha$. Since $A=V_0$, it then follows that the range of the restriction of q to V_0=the range of the restriction of t to V_0. *(ii)* Suppose β is an ordinal and the range of the restriction of q to V_β=the range of the restriction of t to V_β. Let a be an arbitrary element of $V_{\beta+1}$. To show that the range of the restriction of q to $V_{\beta+1}$=the range of the restriction of t to $V_{\beta+1}$ we need only show that there exist x and y in $V_{\beta+1}$ such that $q(a)=t(x)$ and $t(a)=q(y)$. If $a\in V_\beta$, then by hypothesis there are such x and y in V_β and hence in $V_{\beta+1}$. So suppose $a\in V_{\beta+1} - V_\beta$. Then $a\subseteq V_\beta$. By hypothesis, for each u in a there exists a r_u in V_β. such that $q(u)=t(r_u)$. So let $b=\{r_u|u\in a\}$. Then $q(a)=t(b)$. Similarly, c in V_β can be found so that $t(a)=q(c)$. *(iii)* Suppose $\gamma>0$ is a limit ordinal and for all $\mu<\gamma$, the range of the restriction of q to V_μ=the range of the restriction of t to V_μ. Suppose a be an arbitrary element of V_γ. Then we can find $\mu<\gamma$ such that $a\in V_\mu$. Thus by hypothesis, we can find b and c in V_μ such that $q(a)=t(b)$ and $t(a)=q(c)$. Since $V\mu\subseteq V_\gamma$ and a is an arbitrary element of V_γ, it then follows that the range of the restriction of q to V_γ=the range of the restriction of t to V_γ.

Statement 3 will be shown by transfinite induction. *(i)* For each x in A, $h(x)=q^{-1}[t(x)]$. Thus for all x in V_0, $h(x)=q^{-1}[t(x)]$. *(ii)* Suppose β is an ordinal and for all x in V_β, $h(x)=q^{-1}[t(x)]$. Let a be an arbitrary element of $V_{\beta+1}$. We need only show that $h(a)=q^{-1}[t(a)]$. If $a\in V_\beta$, then by hypothesis $h(a)=q^{-1}[t(a)]$. So suppose $a\in V^{\beta+1}—V_\beta$. Then $a\subseteq V_\beta$. Thus by hypothesis for each x in a, $h(x)=q^{-1}[t(x)]$. Therefore,

$$
\begin{aligned}
h(a) &= \{h(x)\,|\,x\in a\} = \{q^{-1}[t(x)]\,|\,x\in a\} \\
&= q^{-1}[\{t(x)\,|\,x\in a\}] = q^{-1}[t(a)]\,.
\end{aligned}
\tag{4.12}
$$

(iii) Suppose $\gamma>0$ is a limit ordinal and for all $\mu<\gamma$, $h(x)=q^{-1}[t(x)]$ for each $x\in V_\mu$. Then since $V_\gamma = \bigcup_{\mu<\gamma} V_\mu$, it immediately follows that $h(x)=q^{-1}[t(x)]$ for each x in V_γ.

4.10.6 The logical equivalence of M and TM

Definition 4.10.2 Assume ZFA and L($\in$, A, $\emptyset$ M). Then a is said to *specify* G if and only if a is a entity and

$$G = \{f\,|\,f\in \Pi \wedge f(a) = a\}\,.$$

a is said to *meaningfully specify* G if and only if a is meaningful and a specifies G. (Note that if a specifies G, then G is an entity.)

Theorem 4.10.2 Assume D′. Suppose a meaningfully specifies G. Then axiom system TM is true.

Proof. Suppose TM were not true. A contradiction will be shown. Since TM is false, let b be an entity such that $IG(b)$ and $\neg M(b)$. By Theorem 3.5.8, let δ be an ordinal such that a and b are in V_δ. By Lemma 4.10.14, let α be a pure set and q_0 be a function from A onto α such that the restriction of q_0 to V_δ is a one-to-one function on V_δ. Let

Q={q|q is a one-to-one function from A onto a and
the restriction of q to $V\delta$ is a one-to-one function on V_δ}.

Then Q is meaningful by Theorem 4.3.1, Lemma 4.10.6, and *MC*’. Let α_0 be $q_0(a)$. Then by Lemma 4.10.13, α_0 is a pure set. Let

$$R = \{q\,|\,q\in Q \text{ and } q(a) = \alpha_0\}\,. \tag{4.13}$$

Then since Q is meaningful, α_0 is a pure set, and since a meaningfully specifies G, a is meaningful, and thus it follows by *MC′* that

$$M(R). \tag{4.14}$$

Note that $R \neq \emptyset$ since $q_0 \in R$. By Lemma 4.10.15,

$$q_0 * f \text{ is in } R \text{ for each } f \text{ in } G. \tag{4.15}$$

Let t be an arbitrary element of R and q be the function on A defined by $q = q_0^{-1}$ $*t$, Then by Lemma 4.10.16 and Equation 4.13 it follows that q is a permutation on A and

$$q(a) = q_0^{-1}[t(a)] = q_0^{-1}(\alpha_0) = a\,.$$

Since a meaningfully specifies G, it follows that q is in G. Thus, since i is an arbitrary element of R and $t=q_0 *q$, it then follows from Equation 4.15 that

$$R = \{h \mid h = q_0 * f \text{ for some } f \in G\}\,. \tag{4.16}$$

Let $\beta=q_0(b)$. By Lemma 4.10.13, β is a pure set, and thus by axiom MP, β is meaningful. By Lemma 4.10.6, V_δ is meaningful. Let

$$b' = (q_0 * f)^{-1}(\beta)\,. \tag{4.17}$$

Then by *MG'*, c is meaningful. Since $b \in V_\delta$ and $q_0 \in R$, it follows that $b \in c$. Let b' be an arbitrary element of c. Then $b' \in V_\delta$ by Equation 4.17. Thus by Equation 4.17 and 4.16, let f in G be such that

$$b' = (q_0 * f)^{-1}(\beta)\,.$$

Then $(q_0 *f)(b')\beta$, and thus by Lemma 4.10.15, $q_0[f(b')]=\beta$. Since $\beta=q_0(b)$, it then follows that

$$f(b') = q_0^{-1}(\beta) = b\,,$$

and thus that $f^{-1}(b)=b'$ Since f is in G and by the choice of b $I_G(b)$ is true, it follows that $b=b'$. Therefore, since b' is an arbitrary element of c, $c=\{b\}$. Since c is meaningful, it then follows from Theorem 4.3.1 that b is meaningful, and this contradicts the choice of b.

Lemma 4.10.17 Assume D. Then there exists a set a such that

$$G = \{f \mid f \in \Pi \wedge \forall x(x \in a \rightarrow M(x) \wedge f(x) = x)\}\,.$$

Proof. By axiom *MC,* for each ordinal α let

$$M_\alpha = \{x \mid x \in V_\alpha \wedge M(x)\}$$

and

$$G_\alpha = \{f \mid f \in \Pi \wedge \forall x(x \in M_\alpha \rightarrow f(x) = x)\}\,.$$

Then by axiom *MC,* for all ordinals α and β, M_α and G_α are meaningful and

$$\text{if } \alpha > \beta, \text{ then } G \subseteq G_\alpha \subseteq G_\beta\,.$$

Since for each ordinal α, $G_\alpha \subseteq \Pi$, we can find an ordinal γ such that for all ordinals $\delta > \gamma$, $G_\delta = G_\gamma$. (Otherwise Π would have arbitrarily large cardinality.) We will show by contradiction that $G = G_\gamma$. For suppose

$$G_\gamma - G \neq \emptyset.$$

Then it follows from the definitions of $M\gamma$ and M that $M - M_\gamma \neq \emptyset$. So let $x \in M - M_\gamma$ and $f \in G_\gamma - G$ be such that $f(x) \neq x$. Let k be the rank of x. Let δ be an ordinal such that $\delta > K$ and $\delta > \gamma$. Then $x \in V_\delta$. Thus, since $x \in M$, $x \in M_\delta$. However, since $\delta > \gamma$, $G_\delta == G_\gamma$, and since $x \in M_\delta$, it follows from the definition of G_δ that $f(x) = x$, contrary to the choice of f.

Theorem 4.10.3 M↔TM.

Proof. By Theorems 4.2.4, 4.2.5, 4.2.7, 4.2.8, and 4.2.11, it follows that TM implies M. Suppose the axiom system M. By Lemma 4.10.17, let a be a set of meaningful entities such that

$$G = \{f \mid f \in \Pi \wedge \forall x (x \in a \rightarrow f(x) = x)\}. \tag{4.18}$$

Since each element of a is meaningful, it follows from axiom *MI* that a is meaningful. By Theorem 3.4.9, let j be a one-to-one function from a onto an ordinal β. Let

$$b = \{(x, j(x)) \mid x \in a\}.$$

We will show b is meaningful. If $b = \emptyset$, then it is meaningful by Theorem 4.3.1. Suppose $(x, j(x))$ is an arbitrary element of b. Then x is meaningful since it is an element of a, and since $j(x)$ is a pure set, it is meaningful by Theorems 4.3.1 and the *proof* of Theorem 4.2.6. Thus by *MG′*, $\{x\} = \{y \mid y = x\}$ and $\{x, j(x)\} = \{y \mid y = x \vee y = j(x)\}$ are meaningful, and therefore

$$(x, j(x)) = \{\{x\}, \{x, j(x)\}\} = \{z \mid z = \{x\} \vee z = \{x, j(x)\}\}$$

is also meaningful by *MC′*. Thus, since $(x, j(x))$ is an arbitrary element of b, it follows that b is meaningful by *MI*. Let

$$H = \{f \mid f \in \Pi \wedge f(b) = b\}. \tag{4.19}$$

Since b is meaningful, it follows from Theorem 4.10.2 that to show TM it is sufficient to show that $H = G$. Since b is meaningful, it immediately follows from Definition 4.2.1 that $H \supseteq G$. Let f be an arbitrary element of H. We need to only show that $f \in G$. Since $j(x)$ is a pure set for each x in a, it follows from Theorem 3.6.2 that $f[j(x)] = j(x)$ for each x in a. Thus by Theorems 4.3.1 and 4.2.6 and Lemma 4.10.6,

$$f[(x, j(x))] = (f(x), f(j(x)) = (f(x), j(x)), \tag{4.20}$$

and by the choices of b and j, x is the only element of y of a such that $(y, j(x))$ is in b, and therefore since by Equation 4.19 $f(b) = b$, it must be the case that $((f(x),$

$j(x))=(x,j(x))$, from which it immediately follows that $f(x)=x$. Since $(x, j(x))$ is an arbitrary element of b, we have shown that $f(x)=x$ for all x in a. Thus by Equation 4.20, $f\in G$.

4.10.7 For Externally Finite a, D″ (a) Implies TM

Theorem 4.10.4 Suppose a is externally finite and axiom system D″ *(a)* is true. Then axiom system TM is true and

$$G = \{f \mid f \in \Pi \wedge \forall x[x \in a \rightarrow f(x) = x]\}.$$

Proof. By Theorem 4.3.5, axiom system D′ is true. Thus by Theorem 4.10.2, to show TM we need only to show that there is a set that meaningfully specifies G. Since a is externally finite, let $a_0,\ldots, ak$ be the distinct elements of a. By Definition 4.3.6, $a_0,\ldots, ak$ are meaningful. Thus by MC',

$$a = \{x \mid x = a_0 \vee \cdots \vee x = a_k\}$$

is a meaningful set. (This is where the external finiteness of a is used.) Let

$$H = \{f \mid f \in \Pi \text{ and } f(x) = x \text{ for all } x \text{ in } a\}. \tag{4.21}$$

Therefore by Equation 4.21 and Theorem 4.10.2, to show the theorem it is sufficient to show $H=G$. Because G is the transformation group of M, it follows from Equation 4.21 that $H \supseteq G$. *It* will be shown by contradiction that $H=G$:

Suppose $H\neq G$, that is, suppose $H\supset G$. Since G is the transformation group of M, that is,

$$G = \{f | f \in \Pi \text{ and } \forall x[M(x) \rightarrow f(x) = x]\},$$

it follows from $H\supset G$ that b and f can be found such that

$$M(b),\ f\in H, \text{ and } f(b)\neq b.$$

In other words,

$$M(b) \text{ and } \neg I_H(b). \tag{4.22}$$

Let M_1 be the subcollection of V such that for all x in V

$$x\in M_1 \text{ iff } I_H(x). \tag{4.23}$$

Then $\mathfrak{V}_1 = \langle V, \in, A, M_1\rangle$ satisfies axiom system TM. Thus axioms MP, MC', and AL are true about $\mathfrak{V}_1$. By Equations 4.21 and 4.23, $M_1(a_0),\ldots, M_1(ak)$. Because by hypothesis $D''(a)$ holds, let $\phi(x, u_0,\ldots, uk, v_1,\ldots, v_m)$ be a formula of $L(\in, A, \emptyset)$ and $p_1,\ldots, p_m$ be pure sets such that

$$\phi(b) \wedge \forall x[\phi(x, a_0, \ldots, a_k, p_1, \ldots, p_m) \rightarrow x = b].$$

Then

$$\{b\} = \{x | \phi(x, a_0, \ldots, a_k, p_1, \ldots, p_m)\}.$$

Thus by axiom *MC′* applied to $\mathfrak{V}_1$ $M_1(\{b\})$, and thus by axiom *AL* applied to $\mathfrak{V}_1$ $M_1(b)$. Therefore, by Equation 4.23,

$$I_H(b),$$

contradicting Equation 4.22.

4.10.8 Homogeneity

Theorem 4.10.5 Assume D′. Suppose a is a meaningful set. Then the following two statements are true:

1. If a is homogeneous, then a is minimeaningful.
2. If the axiom system M is true and a is nearly homogeneous, then ^{-}a is minimeaningful.

Proof. 1. Suppose a is homogeneous. If $a=\varnothing$ or $a=\{u\}$ for some entity u, it immediately follows that a is minimeaningful. So suppose a has at least two elements. Suppose a were not minimeaningful. A contradiction will be shown. Since a is not minimeaningful, let b be a proper, nonempty, meaningful subset of a. Since $b\neq\varnothing$, let x be an element of b, and since $b\neq a$, let y be an element of a—b. Since by hypothesis a is homogeneous, let g in G be such that $g(x)=y$. Since $y\notin b$, $g(x)\notin b$. However, since b is meaningful, $I_G(b)$, and thus in particular $g(x)\in b$, and this contradicts $g(x)\notin b$.

By Lemma 4.10.8, ^{-}a is meaningful. Statement 2 then follows by a proof almost identical to the one given for Statement 1.

Theorem 4.10.6 (Theorem 4.5.2) Assume TM. Suppose a is a meaningful set. Then the following two statements are true:

1. a is homogeneous if and only if a is minimeaningful.
2. a is nearly homogeneous if and only if ^{-}a is minimeaningful.

Proof. 1. By Theorem 4.10.5, if a is homogeneous, then it is minimeaningful. Suppose a is minimeaningful. To show Statement 1, we need only show that a is homogeneous. Suppose a were not homogeneous. A contradiction will be shown. Since a is not homogeneous, let x and y be elements of a such that for all g in G, $g(x)\neq y$. Let b be the orbit of x. Because, by hypothesis, axiom system TM is true, G is an entity and thus

$$b=\{z|z=g(x) \text{ for some } g \text{ in } G\}.$$

is a set by the axiom of Comprehension. Then $y\notin b$. Therefore, b is a nonempty, proper subset of a, and by Lemma 4.10.12 b is minimeaningful, and this contradicts the minimeaningfulness of a.

Statement 2 follows by a proof almost identical to the one given for Statement 1.

Theorem 4.10.7 (Theorem 4.5.3) The following two statements are true:

1. If D′ is true and *A* is homogeneous, then *A* is minimeaningful.
2. If TM, then: *A* is homogeneous if and only if it is minimeaningful.

Proof. *A* is meaningful by Theorem 4.3.1. The theorem then immediately follows from Theorems 4.10.5 and 4.10.6.

4.10.9 Meaningful Cardinals

Lemma 4.10.18 (Lemma 4.5.9) Assume D′. Then the following three statements are true:

1. Suppose *a* is an arbitrary meaningful set and ι is the identity function on *a*, i.e., $\iota(x)=x$ for all $x\in a$. Then ι is a meaningful one-to-one function.
2. Suppose *f* is a meaningful one-to-one function. Then f^{-1} is a meaningful one-to-one function.
3. Suppose *f* and *g* are meaningful one-to-one functions. Then $f*g$ is a meaningful one-to-one function.

Proof. 1. Since a is a meaningful set, $\iota = \{(x,x)\,|\,x \in a\}$ is meaningful by the axiom of Meaningful Comprehension'.

2. Since *f* is a meaningful function, $f^{-1} = \{(y,x)\,|\,(x,y) \in f\}$ is meaningful by the axiom of Meaningful Comprehension′. Since *f* is one-to-one, f^{-1} is a one-to-one function.

3. Since *f* and *g* are meaningful functions,

$$f * g = \{(x,z)\,|\,\exists y[(x,y) \in g \wedge (y,z) \in f]\}$$

is meaningful by the axiom of Meaningful Comprehension', and since *f* and *g* are one-to-one functions, $f*g$ is also a one-to-one function.

Theorem 4.10.8 (Theorem 4.5.5) Assume D. Then $\equiv_M$ is an equivalence relation on the collection of meaningful sets.

Proof. *(i)* Reflexivity. Suppose *a* is a meaningful set. Then $a \equiv_M a$ since by Lemma 4.10.18 the identity function on *a* is a one-to-one meaningful function onto *a*.

(ii) Symmetry. Suppose $a \equiv_M b$. Then *a* and *b* are sets. By Definition 4.5.6, let *f* be a meaningful one-to-one function from *a* onto *b*. Then by Lemma 4.10.18, f^{-1} is a meaningful one-to-one function from *b* onto *a*, and thus by Definition 4.5.6, $b \equiv_M a$.

(iii) Transitivity. Suppose $a \equiv_M b$ and $b \equiv_M c$. By Definition 4.5.6 let *g* and *f* be one-to-one meaningful functions such that *g* is from *a* onto *b* and *f* is from *b* onto *c*. Then by Lemma 4.10.18, $f*g$ is a one-to-one meaningful function from *a* onto *c*, and thus $a \equiv_M c$.

(iv) Suppose *a* and *b* are such that $a \equiv_M b$. Then by Definition 4.5.6, *a* and *b* are sets. By Definition 4.5.6, let *f* be a meaningful function from *a* onto *b*. Then by Lemma 4.10.5, *a*=domain of *f* and *b*=range of *f* are meaningful.

(i)-*(iv)* show that $\equiv_M$ is an equivalence relation on the collection of meaningful sets.

Theorem 4.10.9 (Theorem 4.5.6) Assume D. Then

$$\forall x \forall y[x \in P \wedge y \in P \rightarrow (x \equiv_M y \leftrightarrow x \equiv y)].$$

Proof. Suppose a and b are arbitrary pure sets. Then by Theorem 4.5.4, $a \equiv_M b \rightarrow a = b$. Thus to show the theorem, we need only show $a = b \rightarrow a \equiv_M b$. Suppose $a = b$. Let f be a one-to-one function from a onto b. To show $a \equiv M b$, we need only show that f is meaningful. Since a and b are pure sets, let a be an ordinal such that $a \in P_\alpha$ and $b \in P_\alpha$. Then by Statement 3 of Theorem 3.5.7, $x \in P_\alpha$ and $y \in P_\alpha$ for each $x \in a$ and each $y \in b$. Let (x,y) be an arbitrary element of f. Then $x \in a$ and $y \in b$ and thus $x \in P_\alpha$ and $y \in P_\alpha$. Therefore, $\{x\} \in P_{\alpha+1}$, $\{x, y\} \in P_{\alpha+1}$, and

$$(x, y) = \{\{x\}, \{x, y\}\} \in P_{\alpha+2}.$$

Thus $f \subseteq P_{\alpha+2}$, and therefore $f \in P_{\alpha+3}$. Thus f is a pure set. By the axiom of Meaningful Pure Sets, f is meaningful.

Theorem 4.10.10 (Theorem 4.5.7) Assume M. Then

$$\exists x(x \in P \wedge x \equiv_M A) \rightarrow \forall y M(y).$$

Proof. Suppose $a \in P$ and $a \equiv_M A$. Let f be a meaningful one-to-one function from a onto A, w be an arbitrary element of A, and b in a be such that $f(b)=w$. Since a is a pure set, each element of a is a pure set by Statement 4 of Theorem 3.5.7. Thus in particular, b is a pure set. Thus by the axiom of Meaningful Pure Sets, b is meaningful. Therefore by the axiom of Meaningful Comprehension, $\{z|(b, z) \in f\}$ is meaningful. But $\{z|(b, z) \in f\} = \{w\}$ since f is one-to-one. Thus $\{w\}$ is meaningful. Therefore by the axiom of Atomic Legacy, w is meaningful. Since w is an arbitrary element of A, we have shown that each element of A is meaningful.

Suppose $\forall y M(y)$ is false. A contradiction will be shown. Let c be a non-meaningful entity of smallest rank, and let a be the rank of c. Since each element of A is meaningful, $\alpha > 0$. Thus c is a set. $c \neq \emptyset$ since $\emptyset$ is meaningful and c is not. Since by Theorem 3.5.6 each element of c has rank $< \alpha$, each element of c is meaningful by the choice of α. But then c itself is meaningful by the axiom of Meaningful Inheritability, and this contradicts the choice of c.

Theorem 4.10.11 (Theorem 4.5.9) Assume D. Suppose a and b are meaningful sets, $c \in \bar{a}$, and $d \in \bar{b}$. Then

$$\bar{a} \leq \bar{b} \text{ iff } \bar{c} \leq \bar{d}.$$

Proof. Let f and g be meaningful one-to-one functions from a onto c and b. onto d respectively.

Suppose $\bar{a} \leq \bar{b}$. Let h be a meaningful one-to-one function from a into b. Then $g * h * f^{-1}$ is a meaningful one-to-one function from c into d by Lemma 4.10.18. Thus $\bar{c} \leq \bar{d}$.

Now suppose $\overline{c} < \overline{d}$. Let k be a meaningful one-to-one function from c into d. Then by Lemmas 4.10.5 and 4.10.18, $g^{-1}*k*f$ is a meaningful one-to-one function from a into b. Thus $\overline{a} \geq \overline{b}$

Theorem 4.10.12 (Theorem 4.5.10) Assume D. Suppose a, b and c are meaningful sets, $\overline{a} \leq \overline{b}$ and $\overline{b} \leq \overline{c}$. Then $\overline{a} \leq \overline{c}$.

Proof. Let f be a meaningful one-to-one function from a into b and g be a meaningful one-to-one function from b into c. Let

$$h = \{(x, z) \,|\, x \in a \wedge z \in c \wedge \exists y[(x, y) \in f \wedge (y, z) \in g]\}.$$

Then h is a one-to-one function from a into c and h is meaningful by the axiom of Meaningful Comprehension. Thus $\overline{a} \leq \overline{c}$.

Theorem 4.10.13 (Theorem 4.5.11) Assume D. Suppose a and b are meaningful sets and $\overline{a} < \overline{b}$ and $\overline{b} \leq \overline{a}$. Then $\overline{a} = \overline{b}$

Proof. Let f be a meaningful one-to-one function from a into b and g be a meaningful one-to-one function from b into a. Let

$$c = \text{range of } f \text{ and } d = \text{range of } g.$$

Define H on the collection of ordinals as follows:

(i) $H(0)=a—d$,
(ii) $H(\alpha+1)$=range of the restriction of $g * f$ to $H(\alpha)$, and
(iii) $H(\beta)=\bigcup_{\gamma<\beta}H(\gamma)$ for limit ordinals $\beta>0$.

Then by the Recursion Theorem, there is a formula $\Phi(x,y, u_1, u_2, u_3, u_4)$ of L($\in$, A, Ø) such that for all entities x,y,

$$H(x)=y \text{ iff } \Phi(x,y, a, d, f, g).$$

Let h be the restriction of H to the set of finite ordinals. Then by the axiom of Replacement of ZFA, h is a set. Since

$$h=\{(x,y)|\Phi(x,y,a,d,f,g)\wedge x\in w\}$$

and a, d, f, and g are meaningful by hypothesis and Lemma 4.10.5 and w is meaningful by the axiom of Pure Sets, it follows by the axiom of Meaningful Comprehension that h is meaningful.

Now, $h(0)\subseteq a$ and $g*f$ is a function from a into a. From this it easily follows by induction $h(n)\subseteq a$ for each $n\in w$. Therefore, $f*h(n)\subseteq b$ for each $n\in w$. Let

$$\begin{aligned} F=\{(x,y)|x &\text{ is an element of a and either}\\ &(i)\ y=f(x) \text{ and for some } n\in w,\ x\in h(n), \text{ or}\\ &(ii)\ y=g^{-1}(x) \text{ and for all } n\in w,\ x \notin h(n)\}. \end{aligned} \tag{4.24}$$

Then F is a set, and by the axiom of Meaningful Comprehension, F is meaningful. It is immediate from Equation 4.24 that F is a function from a into b. We will show that F is onto b and is one-to-one.

F is onto *b*. Suppose y is an arbitrary element of *b*. There are two cases to consider:

Case 1. There exists $n \in w$ such that y is in the range of $f^*h(n)$. Let n be such an element of w. Let $x \in h(n)$ be such that $y=f(x)$. Then since $h(n) \subseteq a$, $x \in a$. Thus since by Equation 4.24, $F(x)=f(x)$, we have shown that y is in the range of F.

Case 2. For each $n \in w$, y is not in the range of $f^*\, h(n)$. We will first show that $g(y) \notin h(n)$ for each $n \in w$. For suppose not, i.e., suppose $n \in w$ is such that $g(y) \in h(n)$. Since $h(0)=a—d$ and since $g(y) \in d$, it follows that $n \neq 0$. Therefore let $m \in w$ be such that $n=m+1$. But then

$$h(n)=h(m+1)=\text{range of the restriction of } g^*f \text{ to } h(m)$$

and

$$g(y) \in h(n)=h(m+1).$$

Since by hypothesis g is a one-to-one function, it follows that y is in the range of $f^*h(m)$, but this contradicts the hypothesis of Case 2. Therefore, we can conclude that for each $n \in w$, $g(y) \notin h(n)$. Now since $y \in b$, it follows that $g(y) \in a$. Therefore, by Equation 4.24,

$$F[g(y)]=g^{-1}[g(y)]=y,$$

and we have shown that y is in the range of F.

Thus for both Case 1 and Case 2 we have shown that y is in the range of F, and since y is an arbitrary element of b, we have shown that F is onto b.

F is one-to-one. Suppose x and y are arbitrary elements of a such that $F(x)=F(y)$. We will first show that

$$x \in h(m) \text{ for some } m \in w \text{ iff } y \in h(n) \text{ for some } n \in w. \tag{4.25}$$

For suppose that Equation 4.25 is false. There are two cases to consider:

Case (i). There exists m in w such that $x \in h(m)$ and for all $n \in w$, $y \notin h(n)$. Let m be such an element of w. Then in this case, the hypothesis $F(x)=F(y)$ implies that $f(x)=g^{-1}(y)$ and thus that $y \in g^*f(x)$. Thus since $x \in h(m)$, it follows that

$$y \in \text{ range of the restriction of } g^*f \text{ to } h(m)=h(m+1)$$

which contradicts that part of the hypothesis of Case *(i)* that says that $y \notin h(n)$ for all $n \in w$.

Case (ii). $y \in h(n)$ for some $n \in w$ and for all $m \in w$, $x \notin h(m)$. This case leads to a contradiction in a manner similar to Case *(i)*. Thus we have shown Equation 4.25 by contradiction. Now, from Equation 4.25 and the hypothesis $F(x)=F(y)$, it follows from Equation 4.24 that either $f(x)=f(y)$ or $g^{-1}(x)=g^{-1}(y)$. Since both f and g are one-to-one functions, it then follows that $x=y$. Since x and y were chosen as arbitrary elements such that $F(x)=F(y)$, we have shown that F is a one-to-one function.

In summary, we have shown that F is a meaningful, one-to-one function from a onto b. Therefore $\overline{a} = \overline{b}$.

Theorem 4.10.14 (Theorem 4.5.12) Assume M. Suppose a is a homomeaningful set. Then

$$\exists x(x \in P \wedge a \equiv_M x).$$

Proof. By Theorem 3.4.9, let b be an ordinal such that a=b. Since b is an ordinal, $b \in P$ by Theorem 3.5.7. Since a=b, let f be a one-to-one function from a onto b. We need only show that f is meaningful. Let (u, v) be an arbitrary element of f. Then u is meaningful since $u \in a$ and a is homomeaningful. Since b is a pure set and $v \in b$, v is a pure set by Theorem 3.5.7. Therefore, since M implies axiom *MP*, v is meaningful. Thus by appropriate uses of the axiom of Meaningful Comprehension, $\{u\}$, $\{u,v\}$, and $(u,v)=\{\{u\},\{u,v\}\}$ are meaningful. Thus, since (u,v) is an arbitrary element of f, we have shown that each element of f is meaningful. Therefore by the axiom of Meaningful Inheritability, f is meaningful.

4.10.10 Generalized Transformational Meaningfulness

Theorem 4.10.15 GTM′→D′.

Proof. Assume GTM′ Let Δ be a transformational family for *M*. We need to only show axioms *MP*, *AL* and *MC′*.

Since Δ is nonempty and generates *M* by invariance, axiom *MP* follows from Theorem 4.2.3.

To show *AL*, suppose b is an atom and $M(\{b\})$. Then by Definitions 4.4.1 and 4.4.2, let E in A be such that $I_E(\{b\})$. Let f be an arbitrary element of E. Then $f(\{b\})=\{b\}$. But $f(\{b\})=\{f(b)\}$, that is, $\{f(b)\}=\{b\}$, and therefore $f(b)=b$. Since f is an arbitrary element of E, it then follows that $I_E(b)$. Since $E \in \Delta$ $M(b)$. Thus axiom *AL* is true.

To show *MC′*, suppose $\varphi(x, u_1,\ldots, u_n)$ is a formula of L(∈, *A*, Ø),

$$a, a_1,\ldots, a_n$$

are entities, $M(a_1),\ldots, M(a_n)$, and

$$a=\{z|\varphi(x,a_1,\ldots,a_n)\}.$$

Since $M(a_1),\ldots, M(a_n)$, by Definition 4.4.1 let $H(1),\ldots, H(n)$ be elements of Δ such that $I_{H(i)}({}^a i)$ for i=1,…, n. It follows from Condition *(ii)* of Definition 4.4.2 that H in Δ can be found such that $H \subseteq H(1) \cap \ldots \cap H(n)$, that is, that $H \subseteq H(i)$ for i=1,…, n. Define the predicate M' on the collection of entities as follows: for all entities x,

$$M'(x) \text{ iff } I_H(x).$$

Then $\langle V, \in, \varnothing, M' \rangle$ is a model of the axiom system TM. Since $M'(a_1),\ldots, M'(a_n)$ and $\varphi(x, u_1,\ldots, u_n)$ is a formula of L(∈, *A*, Ø) and

$$a=\{x|\varphi(x,a_1,\ldots,an)\}$$

and $\langle V,\in,A,\varnothing,M'\rangle$ is a model of TM, it follows by Theorem 4.2.7 that $M'(a)$. Thus $I_H(a)$. Since $H\in\Delta$, $M(a)$. Thus axiom MC' is true.

Theorem 4.10.16 (Theorem 4.4.1) GTM′↔D′.

Proof. Suppose GTM′. Then D′ follows by Theorem 4.10.15.

Suppose D′. For each external finite set of meaningful entities *a,* let

$$G_a=\{f|f\in\Pi \text{ and } f(x)=x \text{ for all } x \text{ in } a\}. \tag{4.26}$$

Then for each such *a*, G_a is an entity that is a group under function composition. Let the collection Δ be defined as follows: H is in Δ if and only if there is an external finite set a of meaningful entities such that $H=G_a$. (Δ is not necessarily a set since its definition uses *M* and "externally finite.") It needs only to be shown that Δ is a transformational family of *M.*

To show that Δ generates *M* by invariance, suppose *b* is an arbitrary entity. First suppose *M(b)*. Then $G_{\{b\}}$ is in Δ, and by Equation 4.26,

$$I_{G\{b\}}(\{b\})$$

is true. From this it easily follows that

$$I_{G\{b\}}(\{b\}).$$

Second, suppose K is in Δ and $I_K(b)$ is true. Let a be an externally finite set of meaningful entities such that $K=G_a$. Let M' be such that $\langle V,\in,A,\varnothing,M'\rangle$ is a model of D″(a). Since $M(x)$ is true for each element x in a, it follows that $M'\subseteq M$. Since a is externally finite, it follows by Theorem 4.10.4 that $\langle V,\in,A,\varnothing,M'\rangle$ satisfies axiom system TM with transformation group G_a. Therefore, since $K=G_a$ and $I_K(b)$ is true, it follows that b is in M'. Thus, since $M'\subseteq M$, $M(b)$ is true. The just completed establishes that Δ generates *M* by invariance. Now suppose *F* and *H* are in Δ. Let *c* and *d* be externally finite sets such that

$$c=\{x|f(x)=x \text{ for all } f \text{ in } F\}$$

and

$$d=\{x|f(x)=x \text{ for all } f \text{ in } H\}.$$

Then $c\cup d$ is an externally finite set of meaningful entities. Let

$$E=\{f|\ f(x)=x \text{ for all } x \text{ in } c\cup d\}.$$

Then E is in Δ and $E\subseteq F$ and $E\subseteq H$. This completes showing Δ is a transformational family of *M.*

Theorem 4.10.17 GTM* → GTM → GTM′.

Proof. Immediate from Definitions 4.4.3, 4.4.4, and 4.4.5.

Theorem 4.10.18 (Theorem 4.4.2) The following two statements are true:

1. GTM→$\bar{\mathsf{D}}$.
2. Assume *EFI*. Then GTM↔$\bar{\mathsf{D}}$.

Proof. 1. Suppose GTM. Then by Theorems 4.10.17 and 4.10.16, axiom system D′ is true. Since GTM is true, let Δ be an entity such that Δ is a transformational family of *M*. Then by Definitions 4.4.2 and 4.4.1,

$$\forall x[M(x) \leftrightarrow \exists H(H \in \Delta \wedge I_H(x))], \tag{4.27}$$

and thus axiom *DM* is true.

2. Assume *EFI*. By Statement 1, GTM → $\bar{\mathsf{D}}$. Suppose $\overline{\mathsf{D}}$. By *DM,* let $\Phi(x, u_1,\ldots, un)$ be a formula of L(∈, *A,* Ø) and $a_1,\ldots, a_n$ be entities such that

$$\forall x[M(x) \leftrightarrow \Phi(x, a_1, \ldots, a_n)].$$

Let

Δ={*H*|*H*⊆*Π*,*H is a* group under function composition,
x[*x* is a finite set, $\forall y(y \in x \rightarrow \Phi(y, ai, \ldots, a_n))$, and
∈*H*↔*f*∈*Π* and *f(y)*=*y* for all *y* in *x*)]}.

Then, by the Axiom of Comprehension of ZFA, Δ is an entity. It needs to only be shown that Δ is a transformational family of *M*.

To show Δ generates *M* by invariance, suppose *b* is an arbitrary entity. First suppose *M(b)*. Then by Theorem 4.2.9, *M({b})*. Let

$$H = \{f \mid f \in \Pi \wedge f(b) = b\} = \{f|f \in \Pi \wedge \forall y(y \in \{b\} \rightarrow f(y) = y)\}.$$

Then *H* is in Δ and $I_H(b)$ is true. Second suppose *K* is in Δ and $I_K(b)$ is true. We will show *M(b)*. By the definition of Δ, let, for each *F* in Δ, s_F be a finite set of meaningful entities such that

$$F = \{f|f \in \Pi \text{ and } f(y) = y \text{ for all } y \text{ in } s_F\}.$$

By the proof of Theorem 4.10.16, for each *F* in Δ such that s_F is *externally finite* and for each entity *c,*

$$I_F(c) \rightarrow M(c). \tag{4.28}$$

Let $\psi(y)$ be the following: *y* is a finite ordinal and for each *E* in *Δ,* if s_E has cardinality ≤ *y,* then

$$\forall x[I_E(x) \rightarrow \Phi(x, a_1, \ldots, a_n)].$$

Since $\psi(y)$ is formulated in terms of L(∈, *A,* Ø) and entities, axiom *EFI* can be applied to it. By Equation 4.28, $\psi(\alpha)$ is true for each externally finite ordinal α. Let β be the cardinality of s_k. Since K is an element of Δ, β is finite (but not necessarily externally finite). Thus by axiom *EFI*, $\psi(\beta)$ is true. Thus, since $I_K(b)$ is true, it follows that $\Phi(b, a_1,\ldots, a_n)$ is true, and thus that $M(b)$ is true. This completes showing A generates M by invariance.

Now suppose F and H are arbitrary elements of Δ. Let d and e be finite sets of meaningful entities such that

$$d=\{x|f(x)=x \text{ for all } x \text{ in } F\} \text{ and } e=\{x|f(x)=x \text{ for all } x \text{ in } H\}.$$

Then $d\cup e$ is a finite set of meaningful entities. Let

$$K = \{f\,|\,f(x) = x \text{ for all } x \text{ in } d \cup e\}.$$

Then K is in Δ and $K\subseteq F$ and $K\subseteq H$. This completes showing Δ is a transformational family of M.

Theorem 4.10.19 (Theorem 4.4.3) The following three statements are true:

1. GTM*→D*.
2. Assume *EFI*. Then GTM*↔D*.
3. Assume *EFI**. Then GTM*↔D* D.

Proof. 1. Suppose GTM*. Then by Theorem 4.10.16 axiom system D′ is true. Since GTM* is true, let Δ be a meaningful entity such that Δ is a transformational family of *M*. Then by Definitions 4.4.2 and 4.4.1,

$$\forall x[M(x) \leftrightarrow \exists H(H \in \Delta \wedge I_H(x))],$$

and thus axiom *DM** is true. Therefore axiom system D* is true.

2. Assume *EFI*. By Statement 1, we need only show D* → GTM*. Suppose D*. Let $\Phi(x,u_1,\ldots,u_n)$ be a formula of L(∈, *A*, Ø) and $a_1,\ldots,a_n$ be meaningful entities such that

$$\forall x[M(x) \leftrightarrow \Phi(x, a_1, \ldots, a_n)].$$

Let

$\Delta=\{H|H\subseteq\Pi$, H is a group under function composition and
$\exists x[x$ is a finite set $\wedge\forall y(y \in x \to \Phi(y, a_1, \ldots, a_n)) \wedge$
$\forall f(f \in H \leftrightarrow f \in \Pi \text{ and } f(y) = y \text{ for all } y \text{ in } x)]\}$.

Then, since Π and $a_1,\ldots, a_n$ are meaningful, it follows from axiom *MC′* that Δ is a meaningful entity. By the proof of Theorem 4.10.18 (which uses principle *EFI*), Δ is a transformational family of *M*. Thus by Definition 4.4.5, axiom system GTM* is true.

3. Assume *EFI**. Since *EFI** → EFI and D*→D, by Statement 2 we need only show that D→GTM*. Suppose D. Let

$$\Delta=\{H|H\subseteq\Pi, H \text{ is a group under function composition, and } \exists x[x \text{ is a finite set } \wedge\forall y(y\in x\rightarrow M(y)\wedge \in f(f\leftrightarrow f\in\Pi \text{ and } f(y)=y \text{ for all } y \text{ in } x)]\}.$$

Then, since Π is meaningful, it follows from axiom *MC* that Δ is a meaningful entity. Using *EFI** in place of *EFI* in a simple modification of the proof of Theorem 4.10.18 then yields Δ is a transformational family of *M*. Thus by Definition 4.4.5, axiom system GTM* is true.

Theorem 4.10.20 Assume *EFI*. Suppose a is an entity and $D''(a)$. Then the following two statements are true:

1. Axiom system $\bar{D}$ is true.
2. If a is meaningful, then axiom system D* is true.

Proof. 1. Since a is an entity and $D''(a)$ is true, a is a set and each element of a is meaningful and (by Theorem 4.3.5) axiom system D′ is true. Let B be the set of finite subsets of a, and for each b in B let

$$G(b)=\{f|f\in\Pi \text{ and } f(x)=x \text{ for all } x \text{ in } b\}.$$

Then, by the axiom of Comprehension of ZFA, $G(b)$ is an entity for each b in a. Let

$$\Delta=\{G(b)\ |b\in B\}.$$

Then Δ is an entity by the axiom of Comprehension. By Theorem 4.10.18 we need only show that GTM is true. By the end of the proof of Theorem 4.10.17, for each F and H in Δ, there exists a K in Δ such that $K\subseteq F\cap H$. Thus by Definitions 4.4.2 and 4.4.4, we need only show Δ is a transformational family, that is,

$$\forall x[M(x)\leftrightarrow\exists b(b\in B\wedge I_{G(b)}(x)].$$

To do this, let c be an arbitrary entity.

Suppose $M(c)$. Let

$$\Phi(x,u_1,\ldots,u_n)$$

be a formula of L($\in$, A, $\varnothing$) and $a_1,\ldots,a_n$ be entities such that for each $1\leq i\leq n$, either a_i is in a or a_i is a pure set and

$$\{c\}=\{x|\Phi(x,a1,\ldots,a_n)\}. \tag{4.29}$$

Let $d=\{a_1,\ldots,a_n\}$ and $b=d\cap a$. Then $b\in B$. Let M' be the subcollection of V such that for each entity x, x is in M' if and only if $I_{G(b)}(x)$. Then $\langle V,\in,A,\varnothing,M'\rangle$ is a model of TM with transformation group $G(b)$. By the choices of b and $G(b)$, it immediately follows that $M'(a_1),\ldots, M'(a_n)$ are true. Since TM$\rightarrow$*MC′* (Theorem

4.2.7), it follows from Equation 4.29 that $M'(\{c\})$. Therefore $I_{G(b)}(\{^c\})$, from which it immediately follows $I_{G(b)}(c)$.

Suppose $\exists b[b\in B\wedge I_{G(b)}(c)]$. Let d be an element of B such that $I_{G(d)}(c)$. We will show by contradiction that $I_{G(e)}(c)$ for some externally finite e in B. For suppose $\neg I_{G(e)}(c)$ for all externally finite e in B. Then using axiom *EFI* it is easy to show that $\neg I_{G(b)}(c)$ for all b in B, and thus in particular, $\neg I_{G(d)}(c)$, contrary to the choice of d. Thus let e in B be externally finite and such that $I_{G(e)}(c)$. Let M' be defined on the collection of entities as follows: for all entities x, $M'(x)$ if and only if $I_{G(e)}(x)$. Then $M'(c)$ is true. Let $\langle V, \in, A, \varnothing, M''\rangle$ be a model of axiom system D″(e). Since $e\subseteq a$, $M''\subseteq M$. Since e is externally finite, it follows from Theorem 4.3.6 that $M'=M''$. Thus $M''(c)$ is true. Since $M''\subseteq M$, it then follows that $M(c)$ is true.

The above argument showed that GTM is true. Thus by Theorem 4.10.18, $\overline{\mathsf{D}}$ is true. This completes the proof of Statement 1.

2. Suppose $M(a)$. By Theorem 4.3.5, axiom system D′ is true. From this and the meaningfulness of a, it easily follows that $M(\Delta)$, where Δ is as defined in the proof of Statement 1. $M(\Delta)$ and the proof of Statement 1 then yield that axiom system GTM* is true. D* then follows by Theorem 4.10.19.

Theorem 4.10.21 Assume *EFI*. Then the following two statements are true:

1. $\overline{\mathsf{D}}\leftrightarrow\exists a[a$ is a set and D″$(a)]$.
2. D*$\leftrightarrow\exists a[a$ is a meaningful set and D″$(a)]$.

Proof. 1. Suppose $\exists a[a$ is a set and D″$(a)]$. Then a set a can be found so that axiom system D″(a) is true. Then $\overline{\mathsf{D}}$ follows by Theorem 4.10.20.

Suppose $\langle V, \in, A, \varnothing, M\rangle$ satisfies $\overline{\mathsf{D}}$. By Theorem 4.10.18, $\langle V, \in, A, \varnothing, M\rangle$ satisfies axiom system GTM. Therefore let Δ' be an entity that is a transformational family for M. For each H in Δ', let M_H be such that $\forall x[M_H(x)\leftrightarrow I_H(x)]$. Then $\mathfrak{V}_H = \langle V, \in, A, \varnothing, M_H\rangle$ satisfies axiom system TM with transformational group G_H. (Note that $G_H\supseteq H$.) Because for each H in Δ',

$$G_H = \{f|f \in \Pi \text{ and } \forall x[I_H(x) \rightarrow f(x) = x]\},$$

it follows from the axiom of Comprehension that G_H is an entity for each H in Δ' and

$$\Delta=\{G_H|H\in \Delta'\}$$

is an entity. Let G_H be an arbitrary element of Δ. Then by axiom system TM applied to $\mathfrak{V}$ and Lemma 4.10.17 there exists a set a_H such that

$$G_H=[f|f\in \Pi\wedge\forall x(x\in a_H\rightarrow f(x)=x)\}. \tag{4.30}$$

By TM→M (Theorem 4.2.12) and Equation 4.30, it then follows that a_H meaningfully specifies (with respect to M_H) G_H (Definition 4.10.2). In particular, a_H is meaningful in $\mathfrak{V}_H$. Let α_H be the rank of a_H. Then α_H is an ordinal.

Let $\Theta(y, z)$ be the following formula of L($\in$, A, $\emptyset$):

z is an ordinal $\wedge$ $\exists x[x \in V_z \wedge \forall f(f \in \Pi \rightarrow [f(x) = x \leftrightarrow f \in Y])]$. Note that $\Theta(G_H, \alpha_H)$ is true. Since G_H is an arbitrary element of Δ, it then follows that

$$\forall K[K \in \Delta \rightarrow \exists \alpha \Theta(K, \alpha)].$$

Thus for each K in Δ, let $\beta(K)$ be the least ordinal such that $\Theta(K, \beta(K))$ is true. Since Δ is an entity, the following set a exists by the axiom of Comprehension:

$$a = \{x | \exists K[K \in \Delta \wedge x \in V_{\beta(K)} \wedge \forall f(f \in \Pi \wedge f(x) = x \leftrightarrow f \in K)]\}. \quad (4.31)$$

By the above and the definition of a the following is true:

> For each G_H in Δ there exists an element b_H in a that meaningfully specifies (in terms of) M_H the transformation group G_H of $\langle V, \in, A, \emptyset, M_H \rangle$.

Since, by the way that the set a was defined, each element x of a satisfies the condition $I_K(x)$ for some K in Δ, and therefore it follows from $\langle V, \in, A, \emptyset, M \rangle$ satisfying axiom system GTM that $M(x)$ is true for each element x of a. Let M' be such that $\langle V, \in, A, \emptyset, M' \rangle$ satisfies axiom system D"(a). Since $M(x)$ holds for each x in a, it easily follows from Definitions 4.3.6 and 4.3.5 that $M' \subseteq M$. Thus to conclude the proof, it is only necessary to show $M \subseteq M'$. Suppose b is an entity such that $M(b)$. It needs to only be shown that $M'(b)$. Since $\langle V, \in, A, \emptyset, M \rangle$ satisfies axiom system GTM with transformational family Δ, let K be an element of Δ such that $I_K(b)$. Then $M_K(b)$. By Equations 4.31 and the comment following it, let b_k in a be such that

$$K = \{f | f \in \Pi \wedge f(b_K) = b_K\},$$

that is, be such that $\{b_K\}$ meaningfully specifies M_K (within $\langle V, \in, A, \emptyset, M_K \rangle$). By Theorem 4.10.4, $\langle V, \in, A, \emptyset, M_K \rangle$ satisfies axiom system D"$(\{b_K\})$. Since $\{b_K\} \subseteq a$, it then follows that $M_K \subseteq M'$, and thus that $M(b)$.

2. $\exists a$[a is a meaningful set and D"(a)] implies axiom system D* by Theorem 4.10.20.

Suppose D*. The proof of $\exists a$[a is a meaningful set and D"(a)] follows from an almost identical argument as given in the proof of Statement 1. The exception is that since the more powerful axiom system D*, is being assumed, we can conclude that Δ is meaningful. From this and Equation 4.31 and axiom *DM**, it immediately follows that a is meaningful.

4.10.11 Independence Results

Theorem 4.10.22 D* does not imply axiom *MI*.

Proof. Assume ZFA and that A is denumerable. Let

$$B = \{a | a \text{ is a finite subset of } A\},$$

and for each a in B, let

$$H_a = \{f \mid f \in \Pi \text{ and } f(x) = x \text{ for all } x \text{ in } a\},$$

and let

$$\Delta = \{H_a \mid a \in B\}.$$

Then Δ is an entity by the axiom of Comprehension. Define the subcollection M of V as follows: for all x in V,

$M(x)$ if and only if there exists H in Δ such that $I_H(x)$.

Let $\mathfrak{V}_1 = \langle V, \in, A, \varnothing, M\rangle$. Since for each a and b in B, $a \cup b$ is in B and $H_{a\cup b} \subseteq H_a \cup H_b$, it is easy to verify that in the model $\mathfrak{V}_1$ Δ is a transformational family for M. Since $\varnothing \in B$, $H_\varnothing = \Pi$ is in Δ. It is easy to verify that $I_\Pi(\Delta)$. Thus $M(\Delta)$. Therefore axiom system GTM* is true. Thus by Theorem 4.10.19, axiom D* is true about $\mathfrak{V}_1$ To show the theorem, it is sufficient to show that axiom *MI* is false about $\mathfrak{V}$ This will be done by contradiction.

Suppose axiom *MI* were true about $\mathfrak{V}_1$ Since A is denumerable, let F be a one-to-one function from A onto w. First, $M(F)$ will be shown. Let (a, b) be an arbitrary element of F. Since $a \in A$, $f(a)=a$ for each f in $H_{\{a\}}$. Since b is a pure set, $f(b)=b$ for each $f \in H_{\{a\}}$. Since axiom D* is true about $\mathfrak{V}_1$, it then follows Lemma 4.10.6 that

$$f[(a, b)]=(f(a), f(b))=(a, b)$$

for each $f \in H_{\{a\}}$. Since $\{a\} \in B$, $H_{\{a\}} \in \Delta$, and thus $M[(a,b)]$. Since (a, b) is an arbitrary element of F, it has been shown that each element of F is meaningful. Thus $M(F)$ by the assumption of *MI* Since F is a meaningful one-to-one function from A onto a pure set, it follows from Theorem 4.10.10 that $\forall x M(x)$ is true about $\mathfrak{V}_1$. Since A is denumerable, let $D \subseteq A$ be such that D and $A—D$ are denumerable. Since $\forall x M(x)$, it follows that in particular that $M(D)$ and $M(A—D)$. Therefore, let u and v be elements of B such that for each f in H_u and each g in H_v, $f(D)=D$ and $g(A—D)=A—D$. Let $s=u \cup v$. Then $s \in B$ and $H_S \subseteq H u \cap H_v$ and for all $h \in H_s$, $h(D)=D$ and $h(A—D)=A—D$. Since s is finite and A and $A—D$ are infinite, let d in D and e in $A—D$ be such that $d \notin s$ and $e \notin s$. Let B be the element of Π such that $k(d)=e$, $k(e)=d$, and $k(z)=z$ for all $z \in A—\{d, e\}$. Then $K \in H_s$. But $k(D) \neq D$ since $k(d) \notin D$, and this contradicts $f(D)=D$ for all $f \in H_s$.

Theorem 4.10.23 $\overline{\text{D}}$ does not imply either *DM** or *MC*.

Proof. Assume ZFA. Suppose A is denumerable. Let B be a nonempty subset of A such that B is infinite and $A—B \neq \varnothing$. Let

$\Gamma = \{H \mid$ there exists a nonempty finite subset S of B

$$\text{such that } H = \{f \mid f \in \Pi \wedge \forall x[x \in S \rightarrow f(x) = x]\}\}..$$

Then by *MC'*, Γ is an entity. Define the subcollection M of entities as follows:

for each entity *x*, *x* is in *M* if and only if for some *H* in Γ, $I_H(x)$. Then it is easy to verify that Γ is a transformational family of *M*. Thus by Definition 4.4.4, $\mathfrak{V} = \langle V, \in, A, \varnothing, M\rangle$ satisfies axiom system GTM. Therefore by Theorem 4.10.18 $\mathfrak{V}$ satisfies axiom system $\overline{\mathsf{D}}$. $\neg[DM^* \vee MC]$ will be shown by contradiction. Since $\overline{\mathsf{D}}$ and *DM** implies *MC* (Theorem 4.2.11), it need only be shown that *MC* leads to a contradiction. So suppose axiom *MC* is true about $\mathfrak{V}$. By using Γ and the definition of *M*, it easily follows that

$$B = \{x \mid x \in A \wedge M(x)\}\,.$$

Thus by axiom *MC*, *M(B)*. So let *H* in Γ be such that $I_H(B)$. By the definition of Γ let *S* be a nonempty finite subset of *B* such that

$$H = \{f \mid f \in \Pi \text{ and } f(x) = x \text{ for all } x \in S\}\,.$$

Let *b* and *c* be entities such that $b \in B—S$ and $c \in A—B$. Let *h* be the following permutation on *A*:

$$h(b)=c,\ h(c)=b, \text{ and for } x \neq b \text{ and } x \neq c,\ h(x)=x.$$

Then $h \in H$ since $h(x)=x$ for all $x \in S$. However, $h(B) \neq B$, since for $b \in B$, $h(b) \notin B$. Thus $\neg I_H(B)$, contrary to the choice of *H*.

It will next be shown that D′ and *MI* do not imply either axiom *DM* or axiom *MC*. The method of proof uses the Löwenheim-Skolem theorem of mathematical logic. Its basic idea is to produce a model $\mathfrak{W}$ of ZFA and L($\in$, *A*, Ø, *M*) for which there does not exist a transformation group within $\mathfrak{W}$ but for which there exists outside of $\mathfrak{W}$ a group of permutations that, for meaningfulness purposes, acts very much like a transformation group for *M*.

Theorem 4.10.24 D′ and *MI* do not imply either axiom *DM* or axiom *MC*.

Proof. Let

$$\mathfrak{T} = \langle T, R_1, \ldots, R_i, c_1, \ldots, c_j, \ldots\rangle$$

be a structure, where $R_1, \ldots, R_i$ are countably many finitary relations on *T* and $c_1, \ldots, c_j$ are countably many elements of *T*. ($c_1, \ldots, c_j$ are called the *individual constants* of $\mathfrak{T}$). Then

$$\mathfrak{T}' = \langle T', R'_1, \ldots, R'_i, \ldots, c_1, \ldots, c_j, \ldots\rangle$$

is said to be a *countable submodel* of $\mathfrak{T}$ if and only if $\mathfrak{T}'$ is a structure, *T′* is a countable subset of *T*, and $R'_1, \ldots, R'i\ldots$, are countable subrelations of $R_1, \ldots, R_i, \ldots$

Note that by this definition $\mathfrak{T}$ and $\mathfrak{T}'$ have the same individual constants.

Let **Ł($R_1, \ldots, Ri, \ldots, c_1, \ldots, cj, \ldots$)** be an appropriate first-order language for $\mathfrak{T}$, and let Γ be the collection of true sentences about $\mathfrak{T}$ in this language. Then it follows from the Löwenheim-Skolem theorem of mathematical logic that there exists a countable submodel of $\mathfrak{T}$ for which each sentence of Γ is true.

Let $\mathfrak{V} = \langle V, \in, A, \varnothing \rangle$ be a model of ZFA and the sentence **A is infinite**. Then A is an infinite subcollection of V. By the Löwenheim-Skolem theorem, let $\mathfrak{V}' = \langle V', \in', A', \varnothing \rangle$ be a countable submodel of $\mathfrak{V}$ such that ZFA and the sentence **A is infinite** are true about $\mathfrak{V}'$ Then it easily follows that V' and A' are denumerable collections.

Each "set" a in $\mathfrak{V}'$ will be identified with a subcollection $\hat{a}$ of V' in the following manner: e is in $\hat{a}$ if an only if "$e \in' a$" is a true statement about $\mathfrak{V}'$. Since V' is a denumerable collection and A' has uncountably many subcollections, let β be a subcollection of A' such that $\beta \neq \hat{a}$ for all $a \in' p'(A')$, where p' is the power set operator in the model $\mathfrak{V}$.

Note that by the way β was chosen, there is no set a of $\mathfrak{V}'$ such that $\beta = \hat{a}$.

Let B be the subcollection of V such that b is an element of B if and only if $b \in \square'$ p'(A') and β is a subcollection of $\hat{\mathfrak{b}}$.

For each b in B, let (within $\mathfrak{V}'$)

$$G_b = \{f | f \text{ is a permutation on } A' \text{ and } \forall x[x \in' b \rightarrow f(x) = x]\}.$$

Then for each b in B, G_b is in V'.

For b in B, define the subcollection M_b of V' as follows: u is in M_b if and only if u is in V' and "$\forall f[f \in' G_b \rightarrow f(u) = u]$" is a true statement about $\mathfrak{V}'$. It then follows that for each b in B, the structure

$$\mathfrak{W}_b = \langle V', \in', A', \varnothing, M_b \rangle$$

satisfies the axiom system TM.

Define the subcollection M of V' as follows: u is in M if and only if u is in M_b for each b in B. Consider the structure

$$\mathfrak{W} = \langle V', \in', A', \varnothing, M \rangle.$$

It will we shown that $\mathfrak{W}$ satisfies axiom system D′ and axiom *MI*, but not either axiom *DM* or axiom *MC*.

By the choice of $\mathfrak{V}'$, $\langle V', \in', A', \varnothing \rangle$ satisfies axiom system ZFA. Suppose within $\mathfrak{W}$ that u is an arbitrary pure set. Since $\mathfrak{W}_b$ satisfies axiom system TM for b in B, it follows that $\mathfrak{W}_b$ satisfies axiom *MP*. Therefore, for each b in B, $M_b(u)$. Thus, by the definition of M, $M(u)$ is true. Since u is an arbitrary pure set within $\mathfrak{W}$, it then follows that the axiom *MP* is true about $\mathfrak{W}$. Now suppose x is an arbitrary atom within $\mathfrak{W}$ such that $M(\{x\})$. Then for each b in B, $M_b(\{x\})$ is true about $\mathfrak{W}_b$. Since $\mathfrak{W}_b$ satisfies TM and therefore axiom *AL*, it then follows that $M_b(x)$ is true for each b in B, and from this it follows that $M(x)$ is true. Thus, axiom *AL* is also true about the structure $\mathfrak{W}$. It will now be shown that the axiom *MC*′ is true about $\mathfrak{W}$. Suppose $\Phi(y, z_1, \ldots, z_n)$ is an arbitrary formula of L($\in$, A, $\varnothing$), and $e_1, \ldots, e_n$ are elements of V' such that $M(e_1), \ldots, M(e_n)$, and s is an element of V' such that within $\mathfrak{W}$

$$s = \{y | \Phi(y, e_1, \ldots, e_n)\}.$$

Since $M(e_1),\ldots, M(e_n)$, it follows that for each b in B, $M_b(e_i)$ for $i= 1,\ldots, n$, and since $\mathfrak{W}_b$ satisfies TM, it then follows that $\mathfrak{W}_b$ satisfies axiom MC'. Thus $M_b(s)$ is true for each b in B. Therefore $M(s)$ is true, and it has been shown that axiom MC' is true about $\mathfrak{W}$. By combining the above results, it follows that axiom system D′ is true about $\mathfrak{W}$. Now to show that axiom MI is also true about $\mathfrak{W}$, suppose that within $\mathfrak{W}$ z is an arbitrary set in V' such that for each $y \in' z$, $M(y)$ is true. Then for each b in B and each $y \in' z$, $M_b(y)$. Since $\mathfrak{W}_b$ satisfies TM and therefore MI, it follows that $M_b(z)$ for each b in B. Thus $M(z)$. Thus $\mathfrak{W}$ satisfies axiom MI.

It will now be shown $\mathfrak{W}$ that does not satisfy either axiom DM or axiom MC. Suppose $\mathfrak{W}$ satisfied either axiom DM or axiom MC. A contradiction will be shown. Since either DM or MC is valid, within $\mathfrak{W}$ let

$$c = \{x \mid x \in' A' \wedge M(x)\}.$$

Then within $\mathfrak{W}$, c is a subset of A'. Since β is a subcollection of $\hat{b}$ for each b in B, it follows that for each x in β, $M_b(x)$ is true for each b in B. Thus for each x in β, $M(x)$ is true. From this and the definition of c it follows that β is a subcollection of $\hat{c}$. By the way β was chosen, it cannot be the case that β is $\hat{c}$. So let d be an element of $\hat{c}$—β Since within $\mathfrak{W}$

$$k = \{x \mid x \neq d \wedge x \in' c\}$$

is an element of V', it follows from the choice of β that it cannot be the case that β is $\hat{k}$. Since d was not an element of β, it follows that β is a proper subcollection of $\hat{k}$. So let e be an element of $\hat{k}$—β. Within $\mathfrak{W}$, let

$$q = \{x \mid x \neq e \wedge x \in' k\}.$$

Then, since q is in V', it follows from the choice of β that q is not β. Further, by the way q was defined, it follows that β is a subcollection of q and q is a subcollection of A'. Thus q’ is in B. Within $\mathfrak{W}_q$, let f be the following permutation on A'

$$f(x) = x \text{ for } x \neq d \text{ and } x \neq e, \text{ and } f(d) = e \text{ and } f(e) = d.$$

Then within $\mathfrak{W}_q$, $f \in' G_q$. From the definition of M_q it then follows that d is not in M_q. Thus from the definition of M, it then follows that d is not in M. However, by the choice of d, d is in $\hat{c}$, and by the definition of c, all elements of $\hat{c}$ are in M. This is a contradiction.

5. Representational Theory of Measurement

5.1 Introduction

5.1.1 Overview

Measurement, which consists of the assignment of numbers to objects, is fundamental to science. This chapter presents a synopsis of the dominant theory of measurement in the literature—*the representational theory (of measurement)*—with special emphasis on its meaningfulness concept, which is based on invariance. The representational theory holds that measurement consists of structure preserving mappings of an empirical or qualitative structure into a numerically based one.

Section 5.2 describes two variants of the representational theory, one for which the structure preserving mappings are homomorphisms, and the other isomorphisms. The representational theory has generated much criticism in the literature, and some of the more prominent of these are discussed in Section 5.3, particularly criticisms of the representational theory as a *general* theory of measurement.

Section 5.4 presents brief descriptions many of the most important measurement structures encountered in the representational theory and the kinds of measurement scales they produce.

Chapter 2 considered a variety of intuitive uses of the meaningfulness concept. Many theorists have employed the representational theory's original meaningfulness concept (described in Section 5.2) as the theoretical justification for these and other intuitive uses. However, over time it became apparent that there were a number of serious difficulties with the original concept, and the representationalists developed other, related concepts of meaningfulness for justifying the intuitive uses. This development is described in Section 5.5.

Section 5.6 employs the axiomatic theories of meaningfulness of Chapter 4 to develop the concept of "meaningful scale." This new concept of "scale" is then related to the representational theory's concept of "scale," and it is argued that the representational theory can be interpreted as a form of "meaningful measurement."

Sections 5.7 to 5.10 present applications of the representational theory, primarily to issues discussed in earlier chapters. Section 5.7 revisits Luce's "possible psychophysical laws" (Section 2.5) and shows that in many circumstances the natural and normal requirement that the law be meaningful further limits the possible quantitative form of the law. Section 5.8 presents a measurement-theoretic version of Stevens' method of magnitude estimation (Subsection 2.8.1). Here meaningfulness is put to a novel use as a method for describing a "minimal cognitive theory." Section 5.9 provides a measurementtheoretic analysis of a highly revered psychological law, discovered in the nineteenth century by the physiologist Ernst Weber. It is argued that a much used concept resulting from Weber's Law—the "Weber constant"—is not meaningful in terms of the psychological primitives. Section 5.10 presents a measurement-theoretic description and analysis of the physical technique of dimensional analysis (Section 1.5). This section provides *(i)* a discussion of the key concepts of dimensional analysis, *(ii)* a qualitative treatment of dimensional analysis in terms of physically observable relationships, *(iii)* generalizations of dimensional analysis to nonphysical situations, *(iv)* a discussion of dimensional analysis as a kind of meaningfulness theory, and *(v)* a discussion of issues involving derived physical measurement, dimensionless constants, and alternative methods of measuring physical dimensions.

Section 5.11 is an evaluation of the strengths and weaknesses of the representational theories in light of the results and discussions of the prior sections. Finally, Section 5.12 provides additional proofs for results stated in prior sections.

5.1.2 Preliminaries

The following basic concepts and results are used throughout the rest of the book:

Ordering Relations

Definition 5.1.1 Let X be a set and $\succsim$ be a binary relation on X. Then $\succsim$ is said to be:

(i) transitive if and only if for all x, y, and z in X, if $x \succsim y$ and $y \succsim z$, then $x \succsim z$;
(ii) *connected* if and only if for all x and y in X, either $x \succsim y$ or $y \succsim x$; and
(iii) *antisymmetric* if and only if for all x and y in X, if $x \succsim y$ and $y \succsim x$, then $x = y$.

Definition 5.1.2 $\succsim$ is said to be a *weak ordering* on X if and only if X is a nonempty set and $\succsim$ is a transitive and connected binary relation on X.

Suppose $\succsim$ is a weak ordering on X and B is a subset of X. Then the following three definitions hold:

1. An element u of X is said to be an $\precsim$-*upper* bound of B if and only if $u \precsim b$ for each $b \in B$.
2. B is said to be $\precsim$-*bounded* if and only if an upper bound for B exists.
3. An element u of X is said to be a $\precsim$ *-least upper bound* of B, if and only if u is a $\precsim$-upper bound of B and for all $\precsim$-upper bounds v of B, $v \precsim u$.

$\precsim$ is said to be a *total ordering* on X if and only if $\precsim$ is a weak ordering on X that is antisymmetric.

Suppose $\precsim$ is a weak ordering on X. Then by definition $\sim$ is the binary relation defined on X by: for all x and y in X, $x \sim y$ if and only if $x \precsim y$ and $y \precsim x$. Then it is easy to show that $\sim$ is an equivalence relation on X. Let $X/\sim$ be the set of $\sim$-equivalence classes of X. Define the binary relation $\succeq$ on $X/\sim$ as follows: For all B and C in $X/\sim$,

$$B \succeq C \text{ iff } b \precsim c \text{ for some } b \in B \text{ and some } c \in C.$$

Then it is easy to verify that $\succeq$ is a total ordering on $X/\sim$.$\succeq$ b is called the *induced total ordering (of* $\precsim$*)* on $X/\sim$.

Structural Homogeneity and Uniqueness

Definition 5.1.3 $\langle X, R_j \rangle_{j \in J}$ is said to be a structure if and only if for each $j \in J$, R_j is either an element of X, or a relation on X, or a relation of relations and/or elements of X, or a relation of relations of relations of X, or etc. For brevity, the Rj will often be referred to as *relations,* even though some of them may be elements of X or "higher-order" relations based on X. Unless explicitly stated otherwise, *n-ary relations on X* will, as before, refer to subsets of the Cartesian product of X with itself n times, with, by convention, 0*-ary relations on X* being elements of X. The phrase *"based on"* will often be used when referring to the more general kinds of relations. In many contexts, the relations R_j based on X will be referred to as *entities,* particularly when axiom system ZFA is assumed and X is identified with the set of atoms A.

$\langle X, \succeq, R_j \rangle_{j \in J}$ is said to be a *totally ordered structure* if and only if $\langle X, \succeq, R_j \rangle_{j \in J}$ is a structure and $\succeq$ is a total ordering on X.

Let $\mathfrak{X} = \langle X, R_j \rangle_{j \in J}$ and $\mathfrak{Y} = \langle X, S_k \rangle_{k \in K}$ be structures. Then f is said to be an *isomorphism* of $\mathfrak{X}$ onto $\mathfrak{Y}$ if and only if f is a one-to-one function from X onto Y, J=K, and for all $j \in J$ and all entities $a_1, \ldots, a_{m(j)}$,

$$R_j[a_1, \ldots, a_{m(j)}] \text{ iff } S_j[f(a_1), \ldots, f(a_{m(j)})].$$

g is said to be a *homomorphism* of $\mathfrak{X}$ *into* $\mathfrak{Y}$ if and only if g is a function from X into Y, J=K, and for all $j \in J$ and all entities $a_1, \ldots, a_{m(j)}$,

$$R_j[a_1, \ldots, a_{m(j)}] \text{ iff } S_j[g(a_1), \ldots, g(a_{m(j)})].$$

(Note the use of "iff" in the above definition of "homomorphism.")

Let $\mathfrak{X} = \langle X, R_j \rangle_{j \in J}$ be a structure. Then, by definition, the domain X of $\mathfrak{X}$ and the relations R_j, $j \in J$, are the *primitives* of $\mathfrak{X}$ By definition, an *automorphism* of $\mathfrak{X}$ is an isomorphism of $\mathfrak{X}$ onto itself. By assuming axiom system ZFA and identifying X with the set of atoms A of ZFA, it is easy to show that the automorphisms of $\mathfrak{X}$ are exactly those permutations on X that leave invariant J, its elements, and the primitive relations of $\mathfrak{X}$ It is also easy to show that the automorphisms of $\mathfrak{X}$ form a group under function composition.

Note that by Definition 5.1.3 the domain of a structure $\mathfrak{X}$ is considered a primitive relation of $\mathfrak{X}$.

Definition 5.1.4 Let $\mathfrak{X} = \langle X, R_j \rangle_{j \in J}$ be a structure. By definition, n-ary relations on X for each nonnegative integer n are called *first-order relations* on X. Other relations on X are called *higher-order relations (based)* on X. Note that by this usage, first-order relations are not higher order and visa versa.

Assume all the primitives of $\mathfrak{X}$ are all first-order relations. Then a first-order predicate calculus **L** can be used to describe $\mathfrak{X}$ Relations that are definable in terms of the primitives of $\mathfrak{X}$ through **L** are said to be *first-order definable relations* of $\mathfrak{X}$ All such relations are first-order relations on $\mathfrak{X}$. In general, the converse is not true.[1] Sentences about $\mathfrak{X}$ formulated in **L** are called *first-order statements (about* $\mathfrak{X}$*)*. Sentences about $\mathfrak{X}$ formulated in some higher-order language, which may include a first-order language as a sublanguage, are called *higher-order statements (about* $\mathfrak{X}$*)* if and only if they are not first-order statements.

Definition 5.1.5 Suppose $\langle X, \succsim, R_j \rangle_{j \in J}$ is a totally ordered structure, and A is a subset of automorphisms of $\mathfrak{X}$. Then A is said to be:

- *0-point homogeneous* if and only if there exist x and y in X such that for all α in A, $\alpha(x) \neq y$.
- *1-point homogeneous* if and only if for all x and y in X there exists α in A such that $\alpha(x)=y$.
- For integers $m>1$, *m-point homogeneous* if and only if for all $x_1, \ldots, x_m$ and $y_1, \ldots, y_m$ in X, if $x_1 \succ, \ldots, \succ x_m$ and $y_1 \succ, \ldots, \succ y_m$, then there exists α in A such that for $i=1, \ldots, m$, $\alpha(x_i)=y_i$.
- *∞-point homogeneous* if and only A is m-point homogeneous for each non-negative integer m.
- *0-point unique* if and only if A is trivial, that is, A has the identity as its only element.
- *1-point unique* if and only if for all a and β in A, if $\alpha(x)=\beta(x)$ for some x in X, then $\alpha=\beta$.
- For positive integers n, *n-point unique* if and only if for all a and β in A, if a and β agree at n distinct points of X, then $\alpha=\beta$.

[1] If, for example, X is denumerable and J is finite, then the cardinality of the set of first-order relations on X is $2^{\aleph_0}$, whereas the cardinality of the definable first-order relations of $\mathfrak{X}$ is No.

- *finitely-point unique* if and only if *A* is *n*-point unique for some nonnegative integer *n.*
- *∞-point unique* if and only if *A* is not finitely-point unique.

Definition 5.1.6 Let $\mathfrak{X}$ be a totally ordered structure and A be the set of automorphisms of $\mathfrak{X}$. Then for nonnegative integers *m*, $\mathfrak{X}$ is said to be *m-point homogeneous* if and only if *A* is *m-point* homogenous. Analogous definitions hold for $\mathfrak{X}$ being *∞-point homogeneous, n-point unique, finitely-point unique,* and so forth. $\mathfrak{X}$ (respectively, *A*) will often be called homogeneous if and only if $\mathfrak{X}$ (respectively, *A*) is 1-point homogeneous.

Note that if $\mathfrak{X}$ is infinite, then *m*-point homogeneity, $m>1$, and ∞-point homogeneity for $\mathfrak{X}$ (respectively *A*) imply $\mathfrak{X}$ (respectively *A*) is homogeneous.

Definition 5.1.7 Let $\mathfrak{X}$ be a structure. Then α is said to be a *translation* of $\mathfrak{X}$ if and only if either α is the identity automorphism of $\mathfrak{X}$ or α is an automorphism of $\mathfrak{X}$ such that for all *x* in the domain of $\mathfrak{X}$ $\alpha(x) \neq x$.

The following concepts and theorems about results of Chapter 4 are used throughout this and later chapters. The following straightforward theorem is particularly useful.

Theorem 5.1.1 Let $\mathfrak{X} = \langle A, R_j \rangle_{j \in J}$ be a structure and *H* be the group of automorphisms of $\mathfrak{X}$. Assume

$$\forall x[I_H(x) \leftrightarrow M(x)].$$

Then *G* is an entity and *G*=*H.*

Proof. It immediately follows from the hypothesis of the theorem that *M(Rj)* for each *j* in *J*. It then follows from the definition of *G* (Definition 4.2.1) that each element of *G* is an automorphism of $\mathfrak{X}$, and thus that $G \subseteq H$. Therefore, to show the theorem it needs only to be shown that $H \subseteq G$. Let *h* be an arbitrary element of *H.* Then by hypothesis, for each entity *x*, if *M(x)* then $I_H(x)$. Thus by the definition of *G*, $h \in G$. Because *h* is an arbitrary element of *H*, it then follows that $H \subseteq G$.

Convention 5.1.1 Theorem 5.1.1 will be used throughout this and later chapters, often without explicit reference. This is especially the case when a structure $\mathfrak{X}$ is used to specify meaningful entities as those entities that are invariant under the automorphisms of $\mathfrak{X}$. In such cases the group of automorphisms of $\mathfrak{X}$ often will be immediately identified with the transformation group *G* of *M.*

As in Chapter 4 definability plays an important role in the theory presented in this chapter. The following definition formalizes one of the major definability concepts employed in Chapter 4.

Definition 5.1.8 Let $\mathfrak{X} = \langle X, R_j \rangle_{j \in J}$ be a structure and *R* be a relation (possibly 0-ary, possibly "higher-order"). Then *R* is said to be *set-theoretically definable in terms of the primitives* of $\mathfrak{X}$ if and only if *(i)* axiom system ZFA is assumed, *(ii)* X is the set of atoms A of ZFA, and *(iii)* there exist a formula $\Psi(x, u_1, \ldots, u_m, v_1, \ldots, v_1)$ of $\pounds(\in, A, \emptyset)$, primitives $R_{j(1)}, \ldots, R_{j(m)}$ of $\mathfrak{X}$, and pure sets $b_1, \ldots, b_n$ such that

$$\Psi(R, R_{j(1)}, \ldots, R_{j(m)}, b_1, \ldots, b_n)$$

and

$$\forall x[\Psi(x, R_{j(1)}, \ldots, R_{j(m)}, b_1, \ldots, b_n) \leftrightarrow x = R].$$

Theorem 5.1.2 Suppose $\mathfrak{X} = \langle A, R_j \rangle_{j \in J}$ is a structure and R is an entity. Then the following two statements are true:

1. If R is set-theoretically definable in terms of the primitives of $\mathfrak{X}$, then R is invariant under the automorphisms of $\mathfrak{X}$.
2. If J is externally finite, then R is set-theoretically definable in terms of the primitives of $\mathfrak{X}$ if and only if it is invariant under the automorphisms of $\mathfrak{X}$.

Proof. Since the meaningfulness predicate M has not been specified, we may, without loss of generality, specify it. To show Statement 1, we specify M by assuming axiom system TM with transformation group the set of automorphisms of $\mathfrak{X}$. Then Statement 1 follows from Definition 5.1.8 and Theorem 4.2.10

Suppose J is externally finite. To show Statement 1, we specify M by assuming axiom system $\mathrm{D}''(\{R_j \mid j \in J\})$. Then Statement 2 follows by *(i)* and *(ii)* below.

(i) Suppose R is set-theoretically definable in terms of the primitives of $\mathfrak{X}$. Then by Theorem 4.3.6 and Definition 4.2.3, R is invariant under the automorphisms of $\mathfrak{X}$.

(ii) Suppose R is invariant under the automorphisms of $\mathfrak{X}$. Then, because by Theorem 4.3.6 axiom system TM is true, $M(R)$. Because by hypothesis axiom system $\mathrm{D}''(\{R_j \mid j \in J\})$ is true, it follows from Definitions 4.3.6 and 5.1.8 that R is set-theoretically definable in terms of the primitives of $\mathfrak{X}$.

In terms of the systems of meaningfulness of Chapter 4 the content of Statement 2 of Theorem 5.1.2 states that for externally finite a, $\mathrm{D}''(a) \leftrightarrow$TM. Because this latter result is referenced later, it is formalized as the following theorem.

Theorem 5.1.3 Assume a is externally finite (Definition 4.3.7). Then axiom system $\mathrm{D}''(a)$ holds if and only if axiom system TM holds.

Proof. See proof of Theorem 5.1.2.

5.2 Representational Theories of Measurement

Measurement in its broadest sense consists of assigning numerically based entities to qualitative objects. The representational theories of measurement presented in this chapter are one approach to the numerical assignment process. This approach dates back at least to the foundational work on physical measurement by Helmholtz (1887) and Hölder (1901). It was also used by Hilbert (1899) to assign numerical coordinates to points of a qualitative Euclidean geometry. Much later abstract versions of it for general measurement theory were given by Scott and Suppes (1958), Suppes and Zinnes (1963), and Pfanzagl (1959, 1968). The version of Scott and Suppes was later adopted as the theoretical basis for measurement in the seminal work on the subject, *Foundations of Measurement, Vols. I, II, and III* (Krantz, et al. 1971; Suppes, et al. 1990; Luce, et al. 1991). Over the last three decades, this abstract version has dominated the theoretical measurement literature. For lack of better terms, I call this approach to measurement the *received representational theory* or *homomorphism theory.* It is contrasted with a closely related theory that I call the *isomorphism theory.*

The formal part of the received representational theory consists of five steps or parts:

1. A qualitative situation is specified by a structure $\mathfrak{X}$ that has finitely many first-order primitives. In terms of these primitives, *measurement axioms* are stated that capture certain important qualitative characteristics of $\mathfrak{X}$.
2. A numerically based structure $\mathfrak{N}$, called the *numerical structure,* is selected, and for each structure $\mathfrak{Y}$ that satisfies the measurement axioms, the set $S_{\mathfrak{Y}}$ of homomorphisms of $\mathfrak{Y}$ into $\mathfrak{N}$ is considered. In the representational theory, such $S_{\mathfrak{Y}}$ are called *scales* and their elements are called *representations.*
3. *(Existence Theorem)* It is shown that $S_{\mathfrak{Y}}$ is nonempty for each $\mathfrak{Y}$ that satisfies the measurement axioms.
4. *(Uniqueness Theorem)* For $\mathfrak{Y}$ that satisfy the measurement axioms, an explicit statement is given about how the elements of $S_{\mathfrak{Y}}$ are related to one another. In practice, this statement usually consists of specifying the scale groups (Definition 2.3.7) for the $S_{\mathfrak{Y}}$.
5. Stevens' concept of meaningfulness is used for judging the qualitative significance of quantitative statements; that is, a m-ary relation S on the domain of the numerically based structure $\mathfrak{N}$ is said to be *meaningful for* $S_{\mathfrak{Y}}$ if and only if for all $x_1, \ldots, x_m$ in the domain of $\mathfrak{Y}$ and all φ and ψ in $S_{\mathfrak{Y}}$,

$$S[\varphi(x_1), \ldots, \varphi(x_m)] \text{ iff } S[\psi(x_1), \ldots, \psi(x_m)] .$$

On the more informal side of the received representational theory, there is a strong tendency towards positivism. This is primarily reflected in the above by restricting the primitive relations to be first-order and finite in number. It is more strongly reflected by the fact that the received representational theory strives to formulate the measurement axioms in a manner such that the ones that carry the key structural information can be closely matched by experiments or empirical observations. In addition, it usually gives methods for constructing or finding homomorphisms that are applicable to many kinds of data sets.

Also on the informal side, there is a strong tendency of many representational theorists to consider the received representational approach as a *general* approach to measurement, to the exclusion of other approaches.

The emphasis of the received representational theory has been more towards the production of numerical representations than to considerations of meaningfulness: Scott and Suppes (1958) remark:

> A primary aim of measurement is to provide a means of convenient computation. Practical control or prediction of empirical phenomena requires that unified, widely applicable methods of analyzing the important relationships between the phenomena be developed. Imbedding the discovered relations in various numerical relational systems is the most important such unifying method that has yet been found. *(pp. 116–117)*

The other representational approach—the *isomorphism theory*—is the theory of measurement that is implicit in Narens (1981a,b; 1985). It is very close to the received representational theory, with the primary formal difference between them being that isomorphisms, instead of homomorphisms, are taken as representations. The main reason for the change is that certain difficulties with meaningfulness part of the received representational theory disappear when the isomorphism theory is adopted. However, for most important measurement cases, the two representational approaches are reconcilable.

Informally, the isomorphism approach of Narens (1981a,b; 1985) is less concerned with empirical and practical matters of measurement than the received theory. Its main foci have been meaningfulness issues and theoretical descriptions of the kinds of measurement structures. Because of this, it also considers structures where the primitives may be infinite in number and higher-order. The isomorphism approach is also more flexible in allowing higher-order axioms, and it is much less concerned with providing methods for constructing representations.

Throughout the book, when the representational approach to measurement is applied, it is most likely the isomorphism theory.

5.3 Criticisms of the Representational Approach

5.3.1 Adams' Criticisms

The representational theories of measurement are not universally accepted as adequate theories of measurement. The main complaints against them tend to be very similar to those expressed in Adams (1966), Niederée (1989, 1992a), and Michell (1990). In this subsection, a brief summary of these concerns is presented. We begin with Adams (1966).

Adams' first concern is that the received representational theory, which purports to be a general theory of measurement, puts too much emphasis on the assignment of numbers:

> It seems to me that in characterizing measurement as the assignment of numbers to objects according to rule, the proponents of the representational theory have fastened on something which is undoubtedly of great importance on modern science, but which is not by any means an *essential* feature of measurement. What is important is that the real numbers provide a very sophisticated and convenient conceptual framework which can be employed in *describing* the results of making measurements: but, what can be conveniently described with numbers can be less conveniently described in other ways, and these alternative descriptions no less 'give the measure' of a thing than do the numerical descriptions.... Note, too, that the ancient Greeks did not have our concepts of rational, much less real numbers, yet it seems absurd to say that they could not measure because they did not assign numbers to objects. In sum, I would say that the employment of numbers in describing the results of measurement is not essentially different from their employment in other numerical descriptions, and that this employment is neither a necessary nor a sufficient condition for making or describing measurements. (pp. 129–130)

His second concern is also about the generality of the representational approach:

> Even granting the assumption that measurement necessarily involves assigning numbers, it seems to me to be far from true that in making these assignments it is always the case that mathematical operations and relations are made to correspond to or represent empirical relations and operations.... The situation is worse with most of the widely used measures in the behavioral sciences, like I.Q.s and aptitude test scores. It may be claimed, of course, that these are not really measurements at all, but to justify this, some argument would have to be given, unless the theory of representational measurement is not to degenerate into a mere definition

> (I.Q.s are not measurements *because* they do not establish numerical representations of empirical operations and relations). (pg. 130)

His third concern is that the qualitative axioms used by the representational theories for the establishment of representations are often empirically false, and at best should only be considered as idealizations. He concludes,

> However, it is absurd to hold that these axioms state conditions which must be satisfied in order that measurement be possible, or that it be justified, since measurement clearly is possible and justified even though some of the measurement axioms are false. (pg. 131)

In our terminology, his fourth and fifth concerns are about the necessity that scales need to be specified in terms of a representation and a set of permissible transformations (e.g., a scale group) and about the narrowness of the representational approach to meaningfulness:

> But, one example should show that even the invariance criterion for statements cannot in general be maintained as a condition for scientific significance in an intuitive sense. Among statements which are not invariant are, in fact, all statements reporting the results of particular measurements—e.g., "Ernest Adams weighs 190 lbs." If such statements were excluded on the grounds that they are not 'meaningful', all *data* would be banished from science.

His final concern is that it is assumed implicitly by the representational theories of measurement that "the objects measured do not *change* in relevant observable relations to one another over time."

Comments

1. It should be noted that many of Adams' concerns are about the received representational theory, whose intended scope includes the construction of numerical assignments from empirical data, and which claims to capture through its concepts the essence of actual, practical measurement. The isomorphism approach, as I have described it in the previous section, is not directly concerned with such matters. In particular, I do not consider the isomorphism theory to be a *general* theory of measurement. Instead, I see it as having important uses in the scientific process, including *(i)* making clear the role of various qualitative concepts in specific mathematical models, *(ii)* specifying the possible kinds of mathematical models for various classes of qualitative situations, and *(iii)* eliminating being relevant to the topic under consideration certain kinds of quantitative sentences and formulae.

2. I agree with Adams that the necessity of numbers for a general concept of measurement has not been justified in the received representational theory.

However, his examples are not convincing, because they are about completely uninteresting cases of measurement. It seems reasonable to expect that if measurement is to have *interesting* uses, then minimally it should provide a more convenient means of formulating relevant concepts and reaching relevant conclusions than what would be available by purely qualitative means; that is, measured qualities should be easily manipulable in ways that would be difficult or impossible to do qualitatively. If such "manipulations" had structure, then they would form (i.e., be isomorphic to) some sort of mathematical system, although not necessarily a standard system of "numbers." Thus one should expect *interesting* cases of measurement to consist of mappings of qualitative situations into mathematical ones, and this is a modest extension of the classical representational view. A different approach to this topic is pursued in Chapter 7, where the key qualitative properties of numbers that allow for their flexible use in science are isolated, and measurement is generalized to non-numerically based representing structures that have domains whose elements satisfy those key qualitative properties.

3. I agree with Adams that as a *general* approach to measurement, the received representational theory lacked justification. In Section 5.6, justifications of the representational theories based on meaningfulness considerations are presented.

5.3.2 Niederée's Criticisms

Like Adams, Niederée (1989, 1992a) criticizes the use of numbers in the representational approach: He first notes that the criterion that a structure be numerically representable into structures based on the real numbers puts very little restriction on the structure except for a cardinality restriction, and even that restriction has not been justified. He goes on to note that this is too liberal because

> ...it does not involve any concept of measurement whatsoever; and measurement theorists would in fact hardly be prepared to accept [it] as a *sufficient* criterion for a structure to be called representable in terms of fundamental measurement. What seems to be lacking here is an analysis of what it should mean for 'numbers' assigned to objects to express (ideal) *values of measurement*....
>
> This indicates that for a theory of measurement to be both general and intellectually satisfactory, it should start out, conceptually, with a suitable general explication of the notion of a 'value of measurement', rather than with the prefabricated and allegedly all-purpose concept of the reals. Some of the resulting 'numerical domains' will then turn out to be closely related to the reals—which will eventually endow us with a deeper understanding of the role of the reals from a measurement-theoretic point

> of view—whereas others might be of a different nature (think, e.g., of *p*-adic numbers). At any rate, such an approach should be both more general and more specific than an approach based on an a priori reference to the reals as in the present measurement-theoretic set-up.
>
> A key to a corresponding refinement of that set-up—which is also of interest in its own right—lies in the explicit incorporation of an appropriate (abstract/ideal) notion of an elementary *measurement procedure*.... A suggestion of such an approach is already discernible in Stevens' idea of an "assignment of numbers according to rules" (although his account is impaired, among other things, by a confusion of *real number* and *numeral*). In many situations iso/homomorphic representations turn out to be related to 'measurement procedures', but as such the Scott-Suppes paradigm does not provide us with an explication of that notion, and in general such representations may exist, in a set-theoretic sense, without there being a corresponding measurement procedure. *(Niederée, 1992)*

Comments

1. I agree with Niederée's criticisms of the received representational theory. Although it has no *formal* concept of measurement process, the received theory uses repeatedly *informal* ones based on establishing scales constructively through use of "standard sequences" for infinite structures and through solutions to sets of inequalities for finite structures. Niederée (1987, 1992a) presents a general concept of "measurement process" that easily encompasses these two kinds of measurement processes.

2. Because the isomorphism approach does not claim to capture a concept of "elementary measurement procedure," Niederée's criticisms are about issues orthogonal to the isomorphism approach.

5.3.3 Michell's Criticisms

In his book about psychological measurement, Michell (1990) presents a number of criticisms of the representational approach. Although his book is specifically concerned with psychological measurement, his analyses, critiques, and criticisms of the representational theory are general and extend to other sciences.

Michell takes as a given that the appropriateness of a theory of measurement in a scientific application depends on the scientist's philosophy of science, with different philosophies of science implying different theories of measurement. Many of his criticisms about the representational approach involve perceived incompatibilities between a philosophy he assumes to be inherent in the received representational theory with a philosophy he assumes to be inherent in the conduction of empirical science. In his view, much of the incompatibility results from the representational approach to numbers:

> Representationalism, as a theory of measurement, is motivated by a quite different concern. It is motivated by the belief that numbers are not part of the furniture of the universe. Its aim is to explain how, given this premise, numbers find their way into the practice of measurement. Whatever the philosophical motivations for this view about numbers (and the motives are many and varied), it is mistaken. There is no place in science for entities thought to be outside of space and time or for "convenient" fictions. When we speculate in science, our speculations are about real things located in the same spatiotemporal universe as ourselves.... an attempt will be made to explain just where within the universe the numbers found in measurement are located. *(pg. 24)*

Comments

1. In the above quotation, Michell does not distinguish between the *domain* of scientific inquiry and the *process* of scientific inquiry. Of course, for most scientific applications, numbers as platonic, intuitionistic, or formalist entities should not be considered as elements of the scientific domain. (Neither should logic!) However, exclusion from the scientific domain is not sufficient reason for exclusion from the process of scientific inquiry.

2. In many important measurement situations, numbers can be given qualitative or empirical interpretations (e.g., see Subsection 5.10.8 and Chapter 7). Thus for some scientific situations, there is the possibility of replacing the use of numbers by their qualitative or empirical interpretations, thereby "reducing" the metaphysical content inherent in the scientific inquiry. It is of philosophical interest when such a reduction can take place. However, if the reduction adds no new scientific insights, it is of little scientific importance.

Stevens' theory of measurement (Section 2.2) is nonrepresentational and somewhat vague. Its vagueness tends to mask the role he intended for his meaningfulness concept. Michell (1990) suggests the following wedding of measurement through isomorphisms and Stevens' meaningfulness concept:

> ... This, in a nutshell, is the solution to Stevens' problem of permissible statistics with the framework of the representational theory of measurement. Measurement can only validly lead to those empirical conclusions already entailed by the empirical premises. These premises will be propositions about the empirical entities and the empirical relations that they enter into. The empirical relations concerned are those involved in the empirical relational system on which the scale is based. The relations within the numerical relational system will have been chosen because they possess the same formal properties as these empirical relations. Thus for each

> valid empirical argument there will be a correspondingly valid numerical argument involving just those numerical relations and no others. Thus, if all numerical inferences from measurements are restricted to valid argument forms involving just those numerical relations representing the relevant empirical relations and to no others, then all inferences to the appropriate empirical interpretations of the numerical conclusions obtained will be valid, and no valid inferences of that kind will be excluded. The use of argument forms involving any other numerical relations will then be unnecessary to those whose interest is validity, for the use of such numerical relations is liable to lead to invalidity. *pg. 45*

Using the above idea that the role of meaningfulness is to limit the empirical consequences obtained through measurement to those of the empirical measurement axioms, Michell sees the potential for havoc as a result of the representational theory's use of platonic numbers as values of measurements:

> … What, according to representationalism, is the literal meaning of such "metaphorical" statements as 10?
>
> 10. The product moment correlation coefficient between height and weight amongst school children in Sydney is .65.
>
> A product moment correlation is a numerical relationship and its coefficient is a real number. Yet what is held to stand in this numerical relationship is height and weight, both of which are empirical (nonnumerical) properties. It may be tempting for a representationalist to suggest that 10 is really a statement about numbers (i.e., the measurements of height and weight), but it is not. Statement 10 may be based on measurements, but is asserts the existence of a relationship between the attributes measured. That some relationship exists between the children's height and weight in the empirical domain, given that 10 is true, cannot be denied. And it will be in virtue of this relationships's existence that 10 is true. Yet, it is a relationship that is remarkably difficult to describe without recourse to numerical terms such as "mean," "variance," "covariance," and "ratio," also it is one that scientists never refer to in anything other than numerical terms.
>
> However, the solution given to Stevens' problem of permissible statistics required that any valid inferences from measurements to empirical facts must be implications already entailed by the empirical data itself. It now appears that even in cases of moderated complexity we know of no way to make the inference other than via the use of numbers, because our nonnumerical empirical language is too poor. As far as most scientific results and generalizations of a quantitative kind are concerned, the language used to describe them does not penetrate the representation to the reality represented. The representationalist is forced to believe in the existence of relationships that he is in practice unable to describe other

than indirectly or metaphorically, through the medium of the language of mathematics. Because this is held to be the language of "abstract" (i.e., nonspatio-temporal) entities, the curious picture that the representationalist paints of the quantitative scientist is of one made to use the language of the nonempirical in order to discourse on the empirical.

The representationalist is led to this puzzling destination through his conviction that numbers are external to the empirical facts involved in measurement. How has such a conviction gained currency in the face of scientific practice, which treats numbers otherwise? The answer is that those who have developed this theory stand within a philosophical tradition that takes it for granted that numbers are not empirical kinds of things. From such a point of view the application of mathematics to reality is a deep problem in need of explanation, and the representational theory was devised to serve that purpose. On the other hand, if numbers are accepted in the same way as quantitative scientists implicitly treat them (i.e., as empirical) then the fact of their use is hardly a problem at all. Then the representational theory becomes a liability, for it pries apart the numerical concepts used in science from the empirical reality studied. *pp. 48–49*

Comments

1. In terms of the concepts developed in this book, the above approach of Michell (1990) makes more sense to me as an approach to qualitativeness than empiricalness:

In a qualitative analysis of a scientific domain one has a qualitative structure, qualitative axioms, and some classical form of logic for making derivations, generally a fragment of higher-order logic. In the carrying out of qualitative derivations and definitions of qualitative concepts, it may be useful to go outside the qualitative domain and use exogenous concepts and derivations, for example, quantitative mathematical concepts and derivations. It is natural to ask when and how such exogenous methods have endogenous counterparts. The answer to this question is useful because then one can use exogenous arguments for qualitative derivations. However, it is clear from other passages of Michell (1990) that he considers his analysis to be about *empiricalness* rather than *qualitativeness.*

The key difference between qualitativeness and empiricalness is that qualitativeness is based on *truth,* whereas empiricalness is based on *observation, verification, and refutability.* I find Michell's approach to be one of several reasonable approaches to qualitativeness. However, because of the different natures of truth and those of observation, verification, and refutability, this does not extend to empiricalness:

First, unlike qualitativeness, which has well-defined formal theories of inference attached with it (i.e., fragments of higher-order logic), there are no corresponding *general* formal theories of inference for empiricalness. Indeed, it is the failure of philosophy to produce such general formal theories, despite many noble attempts by leading philosophers, that have led many—including me—to question whether a general, workable, formal concept of "empirical entailment" is possible. Because the concept of "empirical entailment" is central in Michell's approach, this in itself cast considerable doubt on the feasibility of his approach, except for severely constrained, isolated parts of science.

Second, because empiricalness is based on observation, verification, and refutability and not on truth, a formal theory of "empirical entailment" is not needed in the proper conduction of science:

Scientists generally consider empirical evaluation as the strongest test for empirical propositions. One makes a name for oneself in science by either finding interesting propositions to be empirically evaluated or devising means for evaluating empirically interesting propositions. The propositions can be obtained in many ways: generalization of empirical facts, logical deduction with respect to a theory, reading them in a reverie off of a slate written upon by the goddess Namakkal, and so forth. What is of ultimate importance is whether or not they pass the empirical evaluation, not how they were obtained. Meaningfulness—as representationalists use the concept—can play an important role in the obtaining of interesting propositions. For example, the scientist may use known scientific facts and a mathematical theory to derive a set of candidate theoretical relationships that he or she believes might underlie an important scientific phenomenon. The scientist then applies the representational meaningfulness concept *to eliminate the meaningless ones.* The rationale for doing this is that if the mathematical theory is adequate for describing the relationship underlying the phenomenon in question, then none of the meaningless candidate relations can be the correct relationship. (See the quotation of Narens, 1987, at the end of Section 5.5 for a rationale for this.) That leaves the scientist with the meaningful candidates for serious consideration. If the scientist is lucky, there will only be one. The scientist's problem, and it is a standard very common one in science, is to find empirical ways of evaluating the meaningful candidates. In contrast, Michell's methods of measurement and meaningfulness make it very difficult for scientists to find "interesting" propositions. Although Michell's process, if implementable, will lead to propositions that are empirical and valid, its stringent meaningfulness and deductive requirements are almost a guarantee that these propositions will turn out to be of the "dull and uninteresting" variety and will be easily derivable by other means from the known facts.

2. As mentioned previously, in many cases in science the use of numbers can be given qualitative or empirical interpretations. There are also many cases in science when no such interpretations are possible, cosmology and quantum physics being prominent places where numerous examples abound. Because of the latter, I find highly questionable Michell's assertion that quantitative scientists implicitly treat numbers as empirical entities. Even in those subareas

of science where numbers can be so interpreted, I see no reason to demand that such interpretations be part of the scientific process.

3. Formal distinctions between qualitativeness and empiricalness are made in Chapter 7, which also provides a theory of qualitative numbers.

5.3.4 Criticisms of the Representational Meaningfulness Concept

Over time the representational theory expanded its original meaningfulness concept to include the meaningfulness of qualitative relations. Various versions of such expansions are given in Section 5.5. Both the original concept and the expansions have received much criticism. These generally take the form of producing a situation where there is some relation or statement that should be meaningful by the critics' interpretation of the term "meaningful," but which is meaningless according to the representational theory.

One interpretation of "meaningful" that has been applied is "having meaning," where "meaning," is interpreted in absolute terms. Of course, because the meaningful relations in the representational theory change with the choice of primitives, the representational meaningfulness concepts cannot support an absolute interpretation, and thus are inadequate for situations where an absolute concept is desired. Perhaps relatedly, "meaningful," has been interpreted by some (including some representationalists) as "empirical." As an interpretation of the representational concept, this is indeed strange, because, obviously, even for cases involving empirical primitives and powerful automorphism groups, the relations left invariant need not be empirical, with some in fact only be able to be shown to exist through applications of the axiom of Choice.

A frequent criticism of the representational theory is that important objects are often taken to be meaningless. For example, for the usual extensive measurement of length considered by the representational theory, the meter representation as well as the meter stick in Paris are meaningless. Some critics believe that because these play an important role in the scientific use of length, they should be meaningful for considerations involving length. Of course, the representational theory can accommodate this by including the meter stick as a primitive. However, if this were done, the resulting representational scale would become an absolute scale, and thus would have every relation be meaningful. This would render useless the customary applications of "meaningfulness" by the representational theory. I interpret such "counterexamples" involving absolute scales as illustrating the need for broadening the representational concept of meaningfulness. I do not consider them to be a devastating blow to representationalism, because they do not invalidate the productive use of the representational meaningfulness concept in situations involving other scale types, for example the usual extensive measurement of length. Except for Dzhafarov (1995), I do not know of any "critic" who has attempted a *workable and useful* alternative concept for cases involving absolute scales.

5.4 Continuous Measurement Structures

5.4.1 Introduction

The theory of measurement has produced descriptions and analyses of a vast number of measurement situations. This section discusses the most important of these that are based on continua (Definition 5.4.1 below). Many of the measurement structures considered in this section are used elsewhere in the book for purposes of illustration, exposition, and theory development.

Throughout this section, as well for most of the book, the statement of existence and uniqueness theorems about measurement structures are formulated in terms of the isomorphism version of the representational theory.

5.4.2 Continua

Totally ordered structures that are isomorphic to an open interval of real numbers are of considerable importance to mathematics. They are called "continua," and the following qualitative axiomatization in terms of ordering properties was given by Cantor (1895).

Definition 5.4.1 $\langle X, \succsim \rangle$ is said to be a *continuum* if and only if the following five statements are true:

1. *Total ordering:* $\succsim$ is a total ordering on X.
2. *Unboundedness:* $\langle X, \succsim \rangle$ has no greatest or least element.
3. *Density:* For all x and z in X, if $x \succ z$, then there exists y in X such that $x \succ y$ and $y \succ z$.
4. *Dedekind completeness:* Each $\succsim$-bounded nonempty subset of X has a $\succsim$-least upper bound.
5. *Denumerable density:* There exists a denumerable subset Y of X such that for each x and z in X, if $x \succ z$ then there exists y in Y such that $x \succsim y$ and $y \succsim z$.

Cantor (1895) showed the following theorem.

Theorem 5.4.1 *(Existence Theorem)* $\langle X, \succsim \rangle$ is a continuum if and only if it is isomorphic to $\langle \mathbb{R}^+, \geq \rangle$

Proof. Cantor (1895). (A proof is also given in Theorem 2.2.2 of Narens, 1985.)

The following uniqueness theorem is an easy consequence of Theorem 5.4.8:

Theorem 5.4.2 *(Uniqueness Theorem)* Suppose $\mathfrak{X} = \langle X, \succsim \rangle$ is a continuum and $\mathcal{S}$ is a set of isomorphisms of $\mathfrak{X}$ onto $\langle \mathbb{R}^+, \geq \rangle$. Then for all φ and ψ in $\mathcal{S}$ there exists a strictly increasing function f from $\mathbb{R}^+$ onto $\mathbb{R}^+$ such that for all x in X,

$$\psi(x) = f[\varphi(x)] .$$

In terms of Stevens' classification (Section 2.2), Theorems 5.4.1 and 5.4.2 say that $\mathfrak{X}$ can be measured by an ordinal scale of representations.

It is easy to show that the axioms for a continuum are necessary for $\langle X, \succsim \rangle$ to have a representation onto $\langle \mathbb{R}^+, \geq \rangle$.

The primitives for a continuum are first-order, and the axioms of Total Ordering, Unboundedness, and Density are formulable in first-order predicate logic. It is well-known in logic that the first-order theory of these three axioms are *complete* in the sense that for each first-order statement Θ formulated in terms of the primitives, X and $\succsim$, either Θ or $\neg\, \Theta$ is a logical consequence of Total Ordering, Unboundedness, and Density.

The axioms of Dedekind Completeness and Denumerable Density are not formulable in terms of first-order predicate logic: Dedekind completeness uses the higher-order quantifier "for all subsets" and Denumerable Density uses the higher-order quantifier "there exists a denumerable subset." In the presence of the first-order axioms, Denumerable Density guarantees the existence of representations of $\mathfrak{X}$ into $\langle \mathbb{R}^+, \geq \rangle$, and the additional assumption of Dedekind Completeness guarantees that all such representations are onto $\mathbb{R}^+$.

Because the theory T of axioms of Total Ordering, Unboundedness, and Density is complete, the expanded theory consisting of T and the axioms of Dedekind Completeness and Denumerable Density yield the same first-order consequences as T. If one were to assume—as is generally the case in the literature but not in this book—that empirical propositions are first-order, it then follows that if T were assumed to be an empirical theory then the additional assumption of the non-empirical measurement axioms of Dedekind Completeness and Denumerable Density would add no new empirical consequences.

5.4.3 Continuous Threshold Structures

Definition 5.4.2 $\langle X, \geq_\star, T \rangle$ is said to be a *continuous threshold structure* if and only if the following three statements are true:

1. $X, \geq_\star$ is a continuum.
2. T is a function from X onto X such that for all x and y in X,

$$x >_\star y \text{ iff } T(x) >_\star T(y)\,.$$

3. For each x in X, $T(x) >^* x$.

Let $\mathfrak{X} = \langle X, \geq_\star, T \rangle$ be a continuous threshold structure. Then T is called the *threshold function* of $\mathfrak{X}$.

Continuous threshold structures naturally appear in many scientific applications, particularly in psychology where T is viewed as a discrimination threshold such that for stimuli x and y, "y is discriminatively more intense than x" if and only if $y>T(x)$. Continuous threshold structures may be viewed as a continuous version of a "finite semiorder," a much studied measurement structure introduced by Luce (1956).

Definition 5.4.3 By definition, the *canonical, numerical, continuous threshold structure* is the structure $\langle \mathbb{R}, \geq, S \rangle$ where $\geq$ is the usual ordering on the reals, and S is the function $\mathbb{R}$ on defined by

$$S(x) = x + 1 .$$

Theorem 5.4.3 *(Representation Theorem)* Suppose $\mathfrak{X} = \langle X, \geq_{\star}, T \rangle$ is a continuous threshold structure. Then $\mathfrak{X}$ is isomorphic to the canonical, numerical continuous threshold structure.

Proof. Theorem 3.4 of Narens (1994).

It is easy to show that all the axioms for a continuous threshold structure are necessary for the existence of a representation onto the canonical, numerical continuous threshold structure.

Theorem 5.4.4 Let $\mathfrak{X} = \langle X, \geq_{\star}, T \rangle$ be a continuous threshold structure. Then $\mathfrak{X}$ is homogeneous.

Proof. By Theorem 5.4.3, let φ be an isomorphism of $\mathfrak{X}$ onto the canonical, numerical threshold structure, $\mathfrak{S} = \langle \mathbb{R}^{+}, \geq, S \rangle$, where for each real u, $S(u)=u+1$. Let x and y be arbitrary elements of X and $r=\varphi(y)—\varphi(x)$. Let α be the function on $\mathbb{R}$ such that for all $u \in \mathbb{R}^{+}$, $\alpha(u) = u + r$. Then α is an automorphism of $\mathfrak{S}$ and $\alpha[\varphi(x)] = \varphi(y)$. Let $\beta = \varphi^{-1}(\alpha)$. Then, by isomorphism, β is an automorphism $\mathfrak{X}$ of and $\beta(x) = y$.

The uniqueness theorem for continuous threshold structures is somewhat complicated. We begin its discussion by first providing a description of the automorphism group of a continuous threshold structure.

Convention 5.4.1 Throughout the rest of this subsection, let A be the set of strictly increasing functions from the half-open interval (0,1] of the reals onto itself. Also by definition, for each α in A and each r in $\mathbb{R}$, let α_r be the function on $\mathbb{R}$ such that for all x in $\mathbb{R}$ and all m in $\mathbb{I}$,

$$\text{if } x \in (m, m+1], \text{ then } \alpha_r(x) = m + r + \alpha(x - m) .$$

Also, throughout the rest of this subsection, let

$$\mathcal{H} = \{\alpha_r \,|\, \alpha \in \mathcal{A} \text{ and } r \in \mathbb{R}\} .$$

Theorem 5.4.5 $\mathcal{H}$ is the set of automorphisms of the canonical, numerical continuous threshold structure.

Proof. Theorem 4.1 of Narens (1994).

Note that it is an immediate consequence of Theorems 5.4.4, 5.4.3, and 5.4.5 that continuous threshold structures are 1-point homogeneous and ∞-point unique. Thus the scale type of continuous threshold structures is not among those covered by Stevens' classification.

Theorem 5.4.6 *(Uniqueness Theorem)* Suppose $\mathfrak{X}$ is a continuous threshold structure and φ and ψ are isomorphisms of $\mathfrak{X}$ onto the canonical, numerical structure $\mathfrak{S}$ Then the following two statements are true:

1. There exists an automorphism γ of $\mathfrak{S}$ such that $\psi = \gamma * \varphi$
2. For each automorphism δ of $\mathfrak{S}$ $\delta * \varphi$ is an isomorphism of $\mathfrak{X}$ onto $\mathfrak{S}$.

Proof. Theorem 4.2 of Narens (1994).

The following definition is useful later in the book.

Definition 5.4.4 Let ι be the identity function on $\mathbb{R}$. By definition for each r in $\mathbb{R}$, let ι_r be the function on $\mathbb{R}$ such that for x in $\mathbb{R}$ and all integers m,

$$\text{if } x \in (m, m+1] \text{ then } \iota_r(x) = m + r + \iota(x - m)\,.$$

Then it follows from Definition 5.4.3 that ι_r is an automorphism of the canonical, numerical continuous threshold structure $\mathfrak{S}$ for each r in $\mathbb{R}$.. It also easily follows that for each x and r in $\mathbb{R}$,

$$\iota_r(x) = r + x\,.$$

β is said to be a *pure translation* of $\mathfrak{S}$ if and only if $\beta = \iota_r$ for some r in $\mathbb{R}$..

5.4.4 Continuous Extensive Structures

A qualitative theory for fundamental physical qualities like length, mass, time, and so forth, was provided by Helmholtz (1887). The idea behind this theory was that each fundamental physical quality had a natural, observable, qualitative ordering and a combination operation (called "concatenation") with certain specific, observable, algebraic properties. Helmholtz gave axioms about this ordered algebraic situation and showed that all ordered algebraic structures that satisfied the axiomatization were isomorphic to one another, and in particular isomorphic to the structure of positive reals with $\geq$ as its ordering relation and $+$ as its concatenation operation. It was also shown that the set of isomorphisms from any of these qualitative structures into $\langle \mathbb{R}^+, \geq, + \rangle$ was a ratio scale. To my knowledge, these results of Helmholtz constitute the first explicit example of the representational theory of measurement. (Hölder, 1901, presented a deeper mathematical analysis of this kind of measurement.)

The qualitative structures used by Helmholtz for measuring fundamental physical qualities are called "continuous extensive structures." Their formal qualitative properties are essentially those of the positive elements of a continuous ordered group:

Definition 5.4.5 $\mathfrak{X} = \langle X, \succsim, \oplus \rangle$ is said to be a *continuous extensive structure* if and only if the following seven axioms are true:

1. *Total Ordering:* $\succsim$ is a total ordering on X (Definition 5.1.2).
2. *Density:* For all x and z in X, if $x \succ z$ then for some y in X, $x \succ y \succ z$.
3. *Associativity:* $\oplus$ is a binary operation that is *associative*; that is,

$$(x \oplus y) \oplus z = x \oplus (y \oplus z)$$

for all x, y, and z in X.

4. *Monotonicity:* For all x, y, and z in X,

$$x \succsim y \text{ iff } x \oplus z \succsim y \oplus z \text{ iff } z \oplus x \succsim z \oplus y\,.$$

5. *Solvability:* For all x and y in X, if $x \succ y$, then for some z in X, $x = y \oplus z$.
6. *Positivity:* $x \oplus y \succ x$ and $x \oplus y \succ y$, for all x and y in X.
7. *Dedekind Completeness:* Each nonempty, $\succsim$–bounded above subset of X has a $\succsim$–least upper bound.

In essence Helmholtz (1887) showed the following Theorem.

Theorem 5.4.7 Suppose $\mathfrak{X} = \langle X, \succsim, \oplus \rangle$ is a continuous extensive structure. Then the set S of representations of $\mathfrak{X}$ into $\mathfrak{N} = \langle \mathbb{R}^+, \geq, + \rangle$ is a ratio scale, and each element of S is an isomorphism onto $\mathfrak{N}$.

Proofs of generalizations of Theorem 5.4.7 are in Chapter 3 of Krantz et al. (1971) and in Section 9 of Chapter 2 of Narens (1985). To obtain a proof of Theorem 5.4.7 from these generalizations, one merely uses solvability and continuous ordering to show that X satisfies the "Archimedean axiom," and uses simple consequences of the axioms to show that all representations in S are onto $\mathbb{R}^+$.

Suppose $\mathfrak{X} = \langle X, \succsim, \oplus \rangle$ is a continuous extensive structure. Observe that it follows from Theorem 5.4.7 that $\langle X, \succsim \rangle$ is a continuum (Definition 5.4.1), because it is isomorphic to the continuum $\langle \mathbb{R}^+, \geq \rangle$. Thus, in particular, it follows that the axioms $\mathfrak{X}$ for being a continuous extensive structure imply that $\langle X, \succsim \rangle$ satisfies denumerable density (Definition 5.4.1).

The following quotation from Narens (1985) illustrates how the concept of "continuous extensive structure" can be used to describe contexts appropriate for the measurement of fundamental physical variables.

> …An example of this sort of measurement is the measurement of mass by use of a equal arm pan balance. Physical objects a and b are said to be "equivalent in mass," $a \sim b$, if and only if when placed in opposite pans a balances b. It is assumed that $\sim$ is an equivalence relation. Let X be the set of $\sim$–equivalence classes of physical objects. The binary relation $\succsim$ can be defined on X as follows: $\alpha \succsim \beta$ if and only if there exist physical objects x in α and y in β such that if x and y are placed in opposite pans, either they balance or the pan with x becomes lower than the one with y. $\oplus$ is defined on X as follows: $\alpha \oplus \beta = \gamma$ if and only if there exist x in α, y in β, and z in γ

> such that when x and y are placed in the same pan and z in the opposite, the result balances. It is assumed that $\langle X, \succeq, \oplus \rangle$ satisfies the abovementioned properties [of a continuous extensive structure]. Another example is the measurement of length. Here R is the set of measuring rods, which ideally look like line segments. Rods a and b are said to be equivalent in length, $a \sim b$, if and only if a and b can be laid side by side with endpoints exactly corresponding. It is assumed that $\sim$ is an equivalence relation. Let X be the set of $\sim$—equivalence classes of elements of R. Then $\succeq$ can be defined on X as follows: $\alpha \succeq \beta$ if and only if there exist x in α and y in β such that either $x \sim y$ or when x and y are placed side by side with left endpoints exactly corresponding, then the right endpoint of x extends beyond the right endpoint of y. $\oplus$ is defined on X as follows: $\alpha \oplus \beta = \gamma$ if and only if there exist x in α, y in β, and z in γ such that when x and y are placed on an oriented line with the right endpoint of x touching the left endpoint of y (i.e., "x is abutted to y"), they form a rod w such that $w \sim z$. In theoretical classical physics, it is assumed that $\langle X, \succeq, \oplus \rangle$ satisfies the properties mentioned above [i.e., those of a continuous extensive structure]. *(pp. 1–2)*

He then describes how to construct isomorphisms of $\langle X, \succeq, \oplus \rangle$ onto $\langle \mathbb{R}^+, \geq, + \rangle$. He concludes with the following comment about this process of measurement:

> In practice, mass and length are no longer measured as described above. Today, they are measured using rather sophisticated instruments, and the justifications for the correctness of the resulting measurements rest heavily on physical theory, which in turn assumes a theory of measurement for mass and length. Extensive measurement, as described above, is useful as a theory of measurement for *theoretical* physics: it justifies how numbers are assigned to ideal physical entities. It is important in physical theory to make explicit the measurement process, for the rules by which numbers are assigned to physical entities necessarily condition the mathematical form of physical laws. *(pg. 3)*

Continuous extensive structures are also used as a basis for the measurement of non-physical variables. However, in the non-physical sciences—particularly in the social and psychological sciences—it is usually very diffi- cult to find primitive associative concatenation operations and orderings that are observable and useful for the measurement of basic qualities. The next subsection will provide general techniques for establishing ratio scales for qualitative structures that do not have primitive associative concatenation operations.

All the axioms for continuous extensive structures are necessary for representations onto $\langle \mathbb{R}^+, \geq, + \rangle$. All axioms are first-order except for the axiom of Dedekind Completeness. If Dedekind Completeness is dropped, then the theory T of the remaining axioms is not first-order complete: Theorem 8.3 and the discussion following it of Chapter 2 of Narens (1985) shows that neither the commutativity of $\oplus$ ($x \oplus y = y \oplus x$ for all x and y in X) nor its negation is a consequence of T. There are several conclusions to draw from this result:

1. The commutativity of $\oplus$ is a consequence of the axioms for a continuous extensive structure because $+$ is commutative and by Theorem 5.4.7, $\langle X, \succeq, \oplus \rangle$ is isomorphic to $\langle \mathbb{R}^+, \geq, + \rangle$.
2. Because the commutativity of $\oplus$ is not a consequence of T, the derivation of the commutativity of $\oplus$ from the axioms of a continuous extensive structure must use Dedekind Completeness.
3. If $\oplus$ is an empirically observable operation, then under the assumption of T, the negation of Dedekind Completeness may be empirically verifiable by verifying a failure of the commutativity of $\oplus$; that is, under the assumption of the first-order axioms T, the higher-order axiom of Dedekind Completeness is potentially falsifiable.

The commutativity of $\oplus$ does not appear among the axioms for measuring $\mathfrak{X}$ because measurement theorists, for reasons of parsimony and elegance, have followed the mathematical practice of eliminating logically redundant axioms from their axiomatizations. I believe that for the purposes of the theory of measurement, pursuing this generally worthy goal of elegant, parsimonious axiomatizations sometimes leads to suboptimal axiomatizations. The above axiomatization of continuous extensive structures is an example of this: When possible, the higher-order axioms should only be used to guarantee representations into the numerical structure; they should not have empirical (or more generally, first-order) consequences that are not derivable from the first-order axioms.

Another consequence of axioms for a continuous extensive structures is *divisibility:* Inductively define nx for n in $\mathbb{I}^+$ as follows: $1x=x$, and $(n+1)x=(nx)\oplus x$. Then for each n in $\mathbb{I}^+$, *n-divisibility (of* $\oplus$*)* is said to hold if and only if for all x in X there exists y in X such that $ny=x$. *Divisibility (of* $\oplus$*)* is said to hold if and only if n-divisibility holds for each n in $\mathbb{I}^+$. It is not difficult to show that continuous extensive structures are divisible.

Note that for each n in $\mathbb{I}^+$, n-divisibility can be formulated as a first-order statement. Thus divisibility can be formulated as an axiom schemata of first-order statements.

Let T' be the axiom system composed of T, the commutativity of $\oplus$, and the n-divisibility of $\oplus$ for each n in $\mathbb{I}^+$. Then axioms for T' are the axioms for the restriction to positive elements of a much investigated mathematical structure—a densely ordered, divisible, abelian group. It is well-known in logic that the theory of first-order axioms for densely ordered, divisible, abelian groups is complete, and it is not difficult to modify proofs of this result so that they apply to the axiom system T'. Thus, the addition of Dedekind Completeness to T'

cannot produce new first-order consequences. Because of these considerations, I believe that *for measurement purposes*, the axiom system composed of T' and Dedekind Completeness is superior to the more parsimonious one given at the beginning of this subsection consisting of T' and Dedekind completeness.

5.4.5 Continuous Scalar Structures

This subsection presents part of the very general approach of Narens (1981a) for describing qualitative situations that have ratio scale representations.

We first start with some remarks about axiomatizing continua: Let $\mathfrak{X} = \langle X, \succsim, \oplus \rangle$ be a continuous extensive structure. Then it immediately follows by Definition 5.4.5 that $\langle X, \succsim \rangle$ satisfies total ordering, density, and Dedekind completeness. As remarked after Theorem 5.4.7, by using the other axioms for continuous extensive structures, it can be shown that $\langle X, \succsim \rangle$ is a continuum, and therefore satisfies denumerable density (Definition 5.4.1).

Let $\succsim$ be a total ordering on *X*. In terms of traditional logic, Dedekind completeness and denumerable density are "higher-order" properties of $\langle X, \succsim \rangle$; that is, they are not formulated in terms of (first-order) relations on *X* and first-order logic. By an easy application of the Löwenheim-Skolem Theorem of mathematical logic, it follows that for infinite *X* there exist no logically equivalent first-order formulations of Dedekind completeness and denumerable density, even allowing for infinitely many sentences and infinitely many relations.

Let $\succsim$ be a total ordering on *X*. Then it is not difficult to give formulas $\Phi(x,y)$ and $\Psi(x,y)$ of L($\in$, *A*, Ø) such that $\Phi(X, \succsim)$ says that " $\langle X, \succsim \rangle$ is Dedekind complete," and $\Psi(X, \succsim)$ says that "$\langle X, \succsim \rangle$ has a denumerable dense subset." Thus if *X* and $\succsim$ are meaningful and axiom *MC'* (Definition 4.2.7) is assumed, then the assertions that $\langle X, \succsim \rangle$ is Dedekind complete and denumerably dense can be made in a meaningful way. It is easy to see that the other axioms of a continuum (Definition 5.4.1) are also formulable in a meaningful way under the assumption of *MC′* and the meaningfulness of *X* and $\succsim$.

It is interesting to note, however, that for the structure $\langle X, \succsim \rangle$ above, "denumerable density" is in many respects a more complicated and abstract concept than "Dedekind completeness": Both use the power set of *X,* but in addition "denumerable density" uses the concept of subset of From the ω. point of view of a qualitativeness concept that will be developed in Chapter 7, "the power set of X" is at a very low level of abstraction and can be used to specify additional qualitative entities, whereas the concept "ω" is at an infinite level of abstraction and cannot be properly used for the specification of additional qualitative entities. Because it is sometimes demanded that axioms about certain kinds of qualitative structures be made in qualitative ways, it is important for some theories of qualitativeness that "denumerable density" gets implied by other conditions that are expressible in qualitative ways. Fortunately, this is the case for the concept of "continuous scalar structure," which may be viewed as the ultimate generalization of "continuous extensive structure."

Definition 5.4.6 $\mathfrak{X} = \langle X, \succsim, R_j \rangle_{j \in J}$ is said to be a *continuous scalar structure* if and only if $\mathfrak{X}$ is a totally ordered structure, X has at least two elements, $\langle X, \succsim \rangle$ is dense and Dedekind complete, and $\mathfrak{X}$ is 1-point homogeneous and 1-point unique (Definition 5.1.3).

Theorem 5.4.8 Suppose $\mathfrak{X} = \langle X, \succsim, R_j \rangle_{j \in J}$ is a continuous scalar structure. Then $\langle X, \succsim \rangle$ is a continuum.

Proof. Theorem 5.12.1.

Theorem 5.4.9 Suppose $\mathfrak{X} = \langle X, \succsim, R_j \rangle_{j \in J}$ is a continuous scalar structure. Then there exists a numerical structure $\mathfrak{N} = \langle \mathbb{R}, \geq, S_j \rangle_{j \in J}$ such that the set S of isomorphisms of $\mathfrak{X}$ onto $\mathfrak{N}$ is a ratio scale.

Proof. Theorem 2.6 of Narens (1981a).

Let $\mathfrak{X} = \langle X, \succsim, R_j \rangle_{j \in J}$ be a continuous scalar structure. In many scientific applications, one would want the primitives of $\mathfrak{X}$ to be idealizations of empirical processes or observations, and one would want the assertion of $\mathfrak{X}$ being a continuous scalar structure to be the conclusion of empirically based assumptions about $\mathfrak{X}$ rather than a direct hypothesis about $\mathfrak{X}$. Although the conditions of 1-point homogeneity and 1-point uniqueness of $\mathfrak{X}$ are abstract and not directly empirically based, they are sometimes derivable from empirically based conditions about $\mathfrak{X}$, given, of course, certain standard theoretical assumptions about $\mathfrak{X}$, for example, the Dedekind completeness of $\langle X, \succsim \rangle$. For example, it easily follows from Theorem 5.4.7 that continuous extensive structures are scalar structures, and thus the 1-point homogeneity and 1-point uniqueness of continuous extensive structures are derivable from empirically based assumptions about their primitives, namely, monotonicity, associativity, positivity, and solvability of the primitive operations.

As a practical matter, it is usually easy to find empirically based conditions about the primitives that imply 1-point uniqueness. The problem is consistently finding elegant, minimal conditions. 1-point homogeneity is a different matter: While often a natural and plausible assumption about idealizations, 1-point homogeneity is nevertheless a relatively abstract concept. Below, a method is discussed for reducing 1-point homogeneity to empirically based conditions. It involves finding empirically based ways of saying that certain functions defined from the primitives are in fact automorphisms, and showing that the existence of these together with 1-point uniqueness imply 1-point homogeneity.

First, consider the case of a continuous extensive structure $\mathfrak{E} = \langle X, \succsim, \oplus \rangle$ By definition, let $C_2(x) = x \oplus x$. By simple algebraic techniques, it can be shown that $\oplus$ is commutative, i.e., that $x \oplus y = y \oplus x$ for all x and y in X. From this and the associativity of $\oplus$, it easily follows that

$$C_2(x \oplus y) = C_2(x) \oplus C_2(y)$$

for all x and y in X. From the monotonicity of $\oplus$ it easily follows that

$$x \succeq y \text{ iff } C_2(x) \succeq C_2(y)$$

for all x and y in X. From this and assumptions about $\mathfrak{E}$, it can be shown that C_2 is a one-to-one function from X onto X. Thus, by the previous two equations, C_2 is an automorphism of $\mathfrak{E}$. Similarly, it can be shown that C_3, defined by

$$C_3(x) = (x \oplus x) \oplus x\,,$$

is an automorphism of $\mathfrak{E}$. Similarly, C_4, C_5,...can be defined in the obvious ways and shown to be automorphisms of $\mathfrak{X}$. By the commutativity and associativity of $\oplus$ and the definitions of C_k and C_n, it can be shown through simple algebraic techniques that $C_k * C_n = C_n * C_k$ for all positive integers k and n, that is, that the automorphisms C_k and C_n commute.

Next, consider the structure $\mathfrak{X} = \langle X, \succeq, R_j \rangle_{j \in J}$ where J is a finite set. Suppose $\mathfrak{X}$ is 1-point unique, $\oplus$ is set-theoretically defined in terms of the primitives of $\mathfrak{X}$ (Definition 5.1.8) and $\mathfrak{E} = \langle X, \succeq, \oplus \rangle$ is a continuous extensive structure. Then it can be shown (see Theorem 5.4.10 below) that the following two statements are equivalent:

1. $\mathfrak{X}$ is a continuous scalar structure.
2. C_n (as defined above for the structure $\mathfrak{E}$) is an automorphism of $\mathfrak{X}$ for each positive integer n.

Observe that the C_n are defined in terms of $\oplus$, which is defined in terms of the primitives of $\mathfrak{X}$ and pure sets. Also observe that because $\mathfrak{E}$ is a continuous extensive structure, it follows that C_n is a one-to-one function from X onto X and for all x and y in X,

$$x \succeq y \text{ iff } C_n(x) \succeq C_n(y)\,.$$

Thus C_n being an automorphism of $\mathfrak{X}$ is expressible in the following way: For all j in J and all $x_1, \ldots, x_{m_j}$ in X,

$$R_j(x_1, \ldots, x_{m_j}) \text{ iff } R_j[C_n(x_1), \ldots, C_n(x_{m_j})]\,. \tag{5.1}$$

Thus, since C_n is definable in terms of $\oplus$, which in turn is set-theoretically definable in terms of the primitives of $\mathfrak{X}$ (Definition 5.1.8), the statement "C_n is an automorphism of $\mathfrak{X}$ can be formulated through a formula of $\mathcal{L}(\in, A, \oplus)$ with the primitives of $\mathfrak{X}$ and pure sets.

Theorem 5.4.10 Suppose $\mathfrak{X} = \langle X, \succeq, R_j \rangle_{j \in J}$, J is a nonempty finite set, and $\langle X, \succeq \rangle$ is a continuum. Then the following two statements are equivalent:

1. $\mathfrak{X}$ is a scalar structure.
2. There is an operation $\oplus$ such that *(i)* $\oplus$ is set-theoretically definable in terms of the primitives of $\mathfrak{X}$, *(ii)* $\mathfrak{C} = \langle X, \succsim, \oplus \rangle$ is a continuous extensive structure, *(iii)* for each n in $\mathbb{I}^+$, the function C_n, defined inductively by $C_1(x)=x$ and if $n>1$ then $C_n(x)=C_{n-1}(x)\oplus x$ for all x in X, is an automorphism of $\mathfrak{X}$, and *(iv)* the primitives of $\mathfrak{X}$ are set-theoretically definable in terms of the primitives of $\mathfrak{C}$.

Proof. Theorem 5.12.4.

Thus to summarize, let $\mathfrak{X} = \langle X, \succsim, R_j \rangle_{j \in J}$ be a structure. To establish $\mathfrak{X}$ is a continuous scalar structure one needs to establish that it is 1-point unique and 1-point homogeneous. The former is usually done through structural properties of $\mathfrak{X}$ that are expressible in terms of the primitives of $\mathfrak{X}$ and presents no special problem. The 1-point homogeneity of $\mathfrak{X}$, however, on the surface appears to present considerable difficulties, because it is formulated as an abstract concept, and as such it is difficult to justify it in an empiricallike manner, even in highly "idealized" situations. Theorem 5.4.10, however, shows that for cases with finitely many primitives, it can always be stated in terms of the primitives of $\mathfrak{X}$ by "defining" an appropriate operation in terms of the primitives and pure sets. Of course, to completely eliminate the "problems" one would want this "definition" to be done in an empirical-like way, and the above characterization does not guarantee that this can be done in general. Note that if the primitives are empirical-like and it is done in an empirical-like way, then under any reasonable concept of "empirical-like," the functions C_n are also empirical-like, and thus so are the conditions given in Equation 5.1.

For the case where J above is infinite, a similar result follows, except "set-theoretically defined" becomes "defined in terms of primitives of $\mathfrak{X}$ and pure sets through a formula of L($\in$, *A*, Ø) *and through applications of axiom MI* (Definition 4.2.5)."

Cohen and Narens (1979) presented a generalization of continuous extensive structures that provides much more flexibility than continuous extensive structures in asserting in empirical-like ways the 1-point homogeneity of structures in which they are defined. These structures, which have operations that need be neither associative nor commutative, are called "continuous, solvable PCSs." Because PCSs are restricted algebraically far less than continuous extensive structures, they have empirical-like definitions in a much greater variety of empirical settings than continuous extensive structures.

The following three theorems show some of the relationships between continuous solvable PCSs and continuous scalar structures:

Definition 5.4.7 $\mathfrak{X} = \langle X, \succsim, \oplus \rangle$ is said to be a *continuous, solvable* PCS if and only if $\mathfrak{X}$ satisfies all the conditions for a continuous extensive structure (Definition 5.4.5) except possibly for associativity.

$\mathfrak{X} = \langle X, \succsim, \oplus \rangle$ is said to be a *unit* PCS if and only if $\mathfrak{X}$ is a continuous, solvable PCS and $\mathfrak{X}$ is 1-point homogeneous.

Theorem 5.4.11 Suppose $\mathfrak{X} = \langle X, \succsim, \oplus \rangle$ is a unit PCS. Then $\mathfrak{X} = \langle X, \succsim, \oplus \rangle$ is a continuous scalar structure.

Proof. Cohen and Narens (1979) show that continuous, solvable PCS's are 1-point unique. The Theorem then follows from Definition 5.4.7.

Definition 5.4.8 Let X be a nonempty set and $\oplus$ be a binary operation on X. Then the *n-copy operators* (*determined by* $\oplus$), C_n, are defined inductively for positive integers n as follows: for all positive integers $n>1$ and all x in X,

$$C_1(x) = x \text{ and } C_n(x) = [C_{n-1}(x)] \oplus x\,.$$

Theorem 5.4.12 Let $\mathfrak{X} = \langle X, \succsim, R_j \rangle_{j \in J}$ be a continuous scalar structure, and $\oplus$ be a binary operation on X that is set-theoretically definable from the primitives of $\mathfrak{X}$ (Definition 5.1.8), $n \in \mathbb{I}^+$, and C_n be the n-copy operator determined by $\oplus$. Suppose C_n is $\succsim$-strictly increasing and is onto X. Then C_n is an automorphism of $\mathfrak{X}$.

Proof. Theorem 5.12.2.

Theorem 5.4.13 Suppose $\mathfrak{X} = \langle X, \succsim, R_j \rangle_{j \in J}$, J is a nonempty finite set, and $\langle X, \succsim \rangle$ is a continuum. Then the following two statements are equivalent:

1. $\mathfrak{X}$ is a scalar structure.
2. There is an operation $\oplus$ such that *(i)* $\oplus$ is set-theoretically definable in terms of the primitives of $\mathfrak{X}$, *(ii)* $\mathfrak{Y} = \langle X, \succsim, \oplus \rangle$ is a continuous PCS, *(iii)* for each n in $\mathbb{I}^+$, the n-copy operator determined by $\oplus$ is an automorphism of $\mathfrak{X}$, and *(iv)* the primitives of $\mathfrak{X}$ are set-theoretically definable in terms of the primitives of $\mathfrak{Y}$.

Proof. Theorem 5.12.3.

5.4.6 Continuous Additive Conjoint Structures

Many applications in science are concerned with ordered situations that can be factored into subsituations. In such applications the major consideration is usually a description of the interaction of the subsituations. Two examples of this are *(i)* the ordering of physical energy of objects in terms of their qualitative mass and qualitative velocity with the interaction of these qualitative variables, when appropriately measured, being interpretable quantitatively as multiplication, and *(ii)* the ordering of subjective loudness of ordered pairs of qualitative sounds *(x, y)*, where x is placed in the subject's left ear and y in the subject's right ear, with the interaction of these qualitative variables when appropriately measured being interpretable quantitatively as addition. This subsection presents a qualitative theory of such multiplicative and additive interactions is presented.

Convention 5.4.2 By convention, ordered pairs *(x, y)* will often be written in a more condensed form as *xy,* and similarly for ordered *n*-tuples.

Definition 5.4.9 $\mathfrak{C} = \langle X \times Y, \succsim \rangle$ is said to be a *solvable conjoint structure* if and only if the following four statements are true:

1. *Weak ordering:* $\succsim$ is a weak ordering on $X \times Y$.
2. *(Conjoint) independence: (i)* for all *a* and *b* in *X,* if for some *z* in *Y,* $az \succsim bz$, then for all *y* in *Y,* $ay \succsim by$; and *(ii)* for all *c* and *d* in *Y,* if for some *w* in *X,* $wc \succsim wd$, then for all *x* in *X,* $xc \succsim xd$.
3. *Component antisymmetry: (i)* for all *a* and *b* in *X,* if for some *z* in *Y,* $az \sim bz$, then $a=b$; and *(ii)* for all *c* and *d* in *Y,* if for some *w* in *X,* $wc \sim wd$, then $c=d$.
4. *(Unrestricted) solvability:* for all *a* and *b* in *X* and all *c* and *d* in *Y,* there exist *x* in *X* and *y* in *Y* such that $ac \sim xd$ and $ac \sim by$.

Definition 5.4.10 Let $\langle X \times Y, \succsim \rangle$ be a solvable conjoint structure. Define $\succeq X$ and $\succeq Y$ on *X* and *Y* respectively as follows: for all *a* and *b* in *X* and all *c* and *d* in *Y,*

$$a \succeq X\, b \text{ iff } ay \succsim by \text{ for some } y \in Y,$$

and

$$c \succeq y\, d \text{ iff } xc \succsim xd \text{ for some } x \in X.$$

Then by using independence and component antisymmetry, it is easy to verify that $\succeq X$ and $\succeq Y$ are total orderings on *X* and *Y* respectively. $\succeq X$ and $\succeq Y$ are called respectively the $\succsim$*induced (component) (total) ordering* on *X* and the $\succsim$*-induced (component) (total) ordering* on *Y*. By definition, $\langle X \times Y, \succsim \rangle$ is said to be *continuous* if and only if $\langle X, \succeq_X \rangle$ and $\langle Y, \succeq_Y \rangle$ are continua.

Definition 5.4.11 Suppose $\langle X \times Y, \succsim \rangle$ is a continuous, solvable conjoint structure. Then *(φ, ψ)* is said to be an *additive representation* for $\langle X \times Y, \succsim \rangle$ if and only if φ and ψ are functions respectively from *X* and *Y* onto $\mathbb{R}^+$ and for all *xp* and *yq* in $X \times Y$,

$$xp \succsim yq \text{ iff } \varphi(x) + \psi(p) \geq \varphi(y) + \psi(q)\,.$$

Definition 5.4.12 Suppose $\langle X \times Y, \succsim \rangle$ is a solvable conjoint structure. Then the *Thomsen condition* is said to hold if and only if for all *x, y,* and *z* in *X* and all *p, q,* and *r* in *Y,* if

$$xp \sim yq \text{ and } yr \sim zp,$$

then

$$xr \sim zq.$$

Suppose $\langle X \times Y, \succsim\rangle$ is a continuous, solvable conjoint structure. Then it is easy to verify that the Thomsen condition is a necessary condition for the existence of an additive representation of $\langle X \times Y, \succsim\rangle$.

Definition 5.4.13 $\langle X \times Y, \succsim\rangle$ is said to be an *additive conjoint structure* if and only if $\langle X \times Y, \succsim\rangle$ is a continuous, solvable conjoint structure and the Thomsen condition (Definition 5.4.12) holds.

The following theorem can be arrived at in many ways, most notably through methods of Debreu (1961), Luce and Tukey (1964), and Holman (1971):

Theorem 5.4.14 Suppose $\langle X \times Y, \succsim\rangle$ is a continuous, solvable conjoint structure. Then the following three statements are true:

1. *(Existence Theorem)* There exists an additive representation (Definition 5.4.12) for $\langle X \times Y, \succsim\rangle$.
2. *(Uniqueness Theorem, first part)* If (φ, ψ) is and additive representation for $\langle X \times Y, \succsim\rangle$, then for each r in $\mathbb{R}^+$ and each s and t in $\mathbb{R}$, $(r\varphi+s, r\psi+t)$ is an additive representation for $\langle X \times Y, \succsim\rangle$..
3. *(Uniqueness Theorem, second part)* If (φ, ψ) and (φ', ψ') are additive representations for $\langle X \times Y, \succsim\rangle$, then for some r in $\mathbb{R}^+$ and some s and t in $\mathbb{R}$,

$$\varphi' = r\varphi + s \text{ and } \psi' = r\psi + t\,.$$

Let $\langle X \times Y, \succsim\rangle$ be an additive conjoint structure, and let S be the set of additive representations of $\langle X \times Y, \succsim\rangle$. Let

$$\mathcal{S}_X = \{\varphi \mid \exists\psi[(\varphi, \psi) \in \mathcal{S}]\} \text{ and } \mathcal{S}_Y = \{\psi \mid \exists\varphi[(\varphi, \psi) \in \mathcal{S}]\}\,.$$

Then by Theorem 5.4.14, S_X and S_Y are respectively interval scales (Definition 2.3.1) on X and Y.

Theorem 5.4.14 shows that the X-component and Y-component variables of an additive conjoint structure, when appropriately measured, interact additively. Thus by use of an exponential transformation, the theorem also shows that the component variables, when appropriately measured, interact multiplicatively.

Theorem 5.4.14 can be used to establish representation and uniqueness results about a variety of other measurement structures. Two examples are briefly mentioned:

A *continuous difference structure* is a structure of the form $\langle (X \times X) \times (X \times X), \succsim\rangle$, where $\succsim$ is a binary relation on $X{\times}X$ and the following two statements are true:

1. There exists a function φ from X onto $\mathbb{R}$ such that for all xp and yq in $X{\times}X$,

$$xp \succsim yq \text{ iff } \varphi(x) - \varphi(p) \geq \varphi(y) - \varphi(q)\,. \tag{5.2}$$

2. The set of functions φ that satisfy Equation 5.2 for all *xp* and *yq* in $X \times X$ forms an interval scale.

Continuous difference structures have a simple qualitative axiomatization. For details see Chapter 4 of Krantz, et al. (1971).

A *continuous, intensive bisymmetric structure* is a structure of the form $\langle X, \succsim, \oplus \rangle$, where $\succsim$ is a total ordering on *X* and $\oplus$ is a binary operation on *X*, and is such that there exists positive real numbers *r* and *s*, $r+s=1$, such that the following two statements are true:

1. There exists a function φ from *X* onto $\mathbb{R}$ such that for for all *x* and *y* in *X*,

$$x \succsim y \text{ iff } \varphi(x) \geq \varphi(y)\,, \tag{5.3}$$

and

2.

$$\varphi(x \oplus y) = r\varphi(x) + s\varphi(y)\,. \tag{5.4}$$

The set of functions φ that satisfy Equations 5.3 and 5.4 forms an interval scale (Definition 2.3.1).

Continuous, intensive bisymmetric structures have a simple qualitative axiomatization. For details see Chapter 6 of Krantz, et al. (1971).

5.4.7 Continuous Linear Structures

In this subsection, a qualitative characterization is provided for structures that have interval scales of isomorphisms onto $\mathbb{R}$.

Definition 5.4.14 Let $\mathfrak{X} = \langle X, \succsim, R_j \rangle_{j \in J}$ be a structure. Then $\mathfrak{X}$ is said to be a *continuous linear structure* if and only if *(i)* $\langle X, \succsim \rangle$ is a continuum and *(ii)* $\mathfrak{X}$ is 2-point homogeneous and 2-point unique (Definition 5.1.5).

Theorem 5.4.15 Suppose $\mathfrak{X} = \langle X, \succsim, R_j \rangle_{j \in J}$ is a continuous linear structure. Then there exists a numerical structure $\mathfrak{N} = \langle \mathbb{R}, \geq, S_j \rangle_{j \in J}$ such that the following two statements are true:

1. The set of isomorphisms of $\mathfrak{X}$ onto $\mathfrak{N}$ forms an interval scale.
2. For each *r* in $\mathbb{R}^+$ and each *s* in $\mathbb{R}$, the function α on $\mathbb{R}$ defined by

$$\alpha(u) = ru + s$$

is an automorphism of $\mathfrak{N}$, and all automorphisms $\mathfrak{N}$ of have this form.

Proof. Narens (1981b) or Section 5 of Chapter 2 of Narens (1985).

Let $\mathfrak{X} = \langle X, \succsim, R_j \rangle_{j \in J}$ be a continuous linear structure and $\mathfrak{N} = \langle \mathbb{R}, \geq, S_j \rangle_{j \in J}$ be a numerical structure such that there is an interval scale of isomorphisms of $\mathfrak{X}$ onto $\mathfrak{N}$. Then it easily follows that $\langle X, \succsim \rangle$ is a continuum and that $\mathfrak{X}$ is 2-point homogeneous and 2-point unique.

5.4.8 Homogeneous Measurement Structures

Definition 5.4.15 $\mathfrak{X} = \langle X, \succsim, R_j \rangle_{j \in J}$ is said to be a *continuous structure* if and only if $\mathfrak{X}$ is a structure and $\langle X, \succsim \rangle$ is a continuum.

Narens (1981a,b) began the systematic investigation and classification of homogeneous structures that might be of use to measurement theory. Besides showing that scalar structures were ratio scalable (Theorem 5.4.9) and linear structures were interval scalable (Theorem 5.4.15), he showed that m-point homogeneous m-point unique structures were impossible for integers m>2 (Narens, 1981b).

Later Alper (1985), using the method of proof of Narens (1981b), showed the following: If $\mathfrak{X}$ is a continuous structure that is m-point homogeneous and m+1-point unique for some positive integer m, then m=1 and there exists a scale S of isomorphisms of $\mathfrak{X}$ onto a numerical structure such that S is a subscale of an interval scale. Given this result, it is easy to show S must have the following form: for each φ in S,

$$S = \{r\varphi + s \mid r \in K \text{ and } s \in \mathbb{R}\},$$

where K is a nontrivial subgroup of the multiplicative group of the positive reals. Narens (1981b) method of proof breaks down at a critical stage for m-point homogeneity and n-point uniqueness, where n−m>1, and thus a different argument was apparently needed for this case. This was supplied by Alper (1987), where it was shown that if m>2 and $\mathfrak{X}$ is a continuous structure that is m-point homogeneous, then $\mathfrak{X}$ is ∞-point unique. Alper's (1987) method of proof is subtle. (Currently the best presentation of Alper's results is Chapter 20 of Luce, et al., 1990.)

The following summarizes the above results:

Theorem 5.4.16 Suppose $\mathfrak{X}$ is a homogeneous, finitely unique continuous structure. Then one of the following three cases obtains:

1. $\mathfrak{X}$ is a scalar structure.
2. $\mathfrak{X}$ is a linear structure.
3. $\mathfrak{X}$ is neither 2-point homogeneous nor 1-point unique and there exists a subscale S of an interval scale of isomorphisms of $\mathfrak{X}$ onto a numerical structure that has the following form: for each φ in S,

   $$S = \{r\varphi + s \mid r \in K \text{ and } s \in \mathbb{R}\},$$

   where K is a nontrivial subgroup of the multiplicative group of the positive reals.

Continuous extensive structures and continuous PCS's are subclasses of a more general class of measurement structures called "continuous concatenation structures":

Definition 5.4.16 $\mathfrak{X} = \langle X, \succsim, \oplus \rangle$ is said to be a *continuous concatenation structure* if and only the following two statements are true:

1. $\langle X, \succsim \rangle$ is a continuum.
2. $\oplus$ is a binary operation on X such that for all x, y, and z in X,

$$x \succsim y \text{ iff } x \oplus z \succsim y \oplus z \text{ iff } z \oplus x \succsim z \oplus y\,.$$

Definition 5.4.17 Let $\mathfrak{X} = \langle X, \succsim, \oplus \rangle$ be a continuous concatenation structure. Then the following three definitions hold:

1. $\mathfrak{X}$ is said to be *weakly positive* if and only if $x \oplus x \succ x$ for all x in X.
2. $\mathfrak{X}$ is said to be *weakly negative* if and only if $x \oplus x \prec x$ for all x in X.
3. $\mathfrak{X}$ is said to be *idempotent* if and only if $x \oplus x = x$ for all x in X.

Theorem 5.4.17 Suppose $\mathfrak{X}$ is a continuous concatenation structure and $\mathfrak{X}$ is homogeneous. Then the following three statements are true:

1. $\mathfrak{X}$ is either weakly positive or weakly negative or idempotent.
2. $\mathfrak{X}$ is 2-point unique.
3. If $\mathfrak{X}$ is not 1-point unique, then it is idempotent.

Proof. Statements 1 and 3 are shown in Theorem 2.2 of Luce and Narens (1985). (See also Theorem 9 of Chapter 20 of Luce, et al, 1990, for a proof.) Statement 2 is an unpublished result of Michael Cohen.

The following definition characterizes an important class of numerical representing structures for homogeneous concatenation structures.

Definition 5.4.18 $\langle \mathbb{R}^+, \geq, \oplus \rangle$ is said to be a *real unit structure* if and only if there exists a function f from $\mathbb{R}^+$ into $\mathbb{R}^+$ such that the following three conditions are satisfied for all r and s in $\mathbb{R}^+$:

(1) If $r > s$, then $f(r) > f(s)$.

(2) If $r > s$, then

$$\frac{f(r)}{r} < \frac{f(s)}{s}\,.$$

(3) $\oplus$ and f are related by the following equation:

$$r \oplus s = s \cdot f\left(\frac{r}{s}\right)\,.$$

Note that $\langle \mathbb{R}^+, \geq, + \rangle$ is a real unit structure with

$$r + s = s \cdot (1 + \frac{r}{s})\,,$$

that is, with $f(t) = 1 + t$, where f is as in Definition 5.4.18.

Let $\mathfrak{R}$ be a real unit structure. Then it immediately follows from Definition 5.4.18 that $\mathfrak{R}$ is a continuous concatenation structure. It is easy to verify through

Definition 5.4.18 that for each r in $\mathbb{R}^+$, multiplication by r is an automorphisms of $\mathfrak{N}$. Thus $\mathfrak{N}$ is also homogeneous. The following shows that real unit structures characterize all homogeneous continuous concatenation structures.

Theorem 5.4.18 Let $\mathfrak{X}$ be a homogeneous continuous concatenation structure. Then the following two statements are true:

1. *(Existence Theorem)* There exists a real unit structure that is isomorphic to $\mathfrak{X}$.
2. *(Uniqueness Theorem)* By Statement 1, let $\mathfrak{N}$ be a real unit structure that is isomorphic to $\mathfrak{X}$ and S be the set of isomorphisms of $\mathfrak{X}$ onto $\mathfrak{N}$. Then if $\mathfrak{X}$ is 1-point unique, then S is a ratio scale. If S is 2-point homogeneous, then S is a log-interval scale (Definition 2.3.1). And if $\mathfrak{X}$ is neither 2-point homogeneous nor 1-point unique, then S is a subscale of a log-interval scale that is neither a ratio scale nor a log-interval scale.

Proof. Luce and Narens (1985).

2-point homogeneous, 2-point unique continuous concatenation structures are highly structured, as the following theorem shows:

Theorem 5.4.19 Let $\mathfrak{X} = \langle X, \succsim, \oplus \rangle$ be a continuous concatenation structure that is 2-point homogeneous and 2-point unique. Then the following two statements are true:

1. $\mathfrak{X}$ is isomorphic to $\langle \mathbb{R}^+, \succsim, \oplus' \rangle$ where $\oplus'$ is such that there exists $0 < r < 1$ and $0 < s < .5$ such that for all a and b in $\mathbb{R}^+$,

$$a \oplus b = ra + (1 - r)b + s\,|a - b| \,. \tag{5.5}$$

2. The set of isomorphisms of $\mathfrak{X}$ onto $\mathfrak{N}$ is an interval scale.

Proof. Theorem 3.12 of Luce and Narens (1985).

An important special case of Theorem 5.4.19 arises when $r=\frac{1}{2}$ and $s=0$ in Equation 5.5. This special case, which is important to psychological measurement theory, is characterized in the following definition and theorem.

Definition 5.4.19 $\mathfrak{X} = \langle X, \succsim, \oplus \rangle$ is said to be a *continuous bisection structure* if and only if $\mathfrak{X}$ is a continuous concatenation structure and the following four statements are true for all x, y, and z in X,

(1) If $x \succ y$, then $x \succ x \oplus y \succ y$.
(2) $x \oplus y = y \oplus x$.
(3) $(x \oplus y) \oplus (z \oplus w) = (x \oplus z) \oplus (y \oplus w)$.
(4) If $x \succ y$, then there exist u and v in X such that

$$x = u \oplus y \text{ and } y = x \oplus v\,.$$

Theorem 5.4.20 Suppose $\mathfrak{X} = \langle X, \succsim, \oplus \rangle$ is a continuous bisection structure. Then $\mathfrak{X}$ is isomorphic to $\langle \mathbb{R}, \geq, \oplus' \rangle$, where $\oplus'$ is such that for all u and v in $\mathbb{R}$,

$$u \oplus' v = \frac{u}{2} + \frac{v}{2}.$$

Proof. Follows from Theorem 10 of Chapter 6 of Krantz, et al. (1971).

5.5 Representational Concepts of Meaningfulness

Although twenty years old, the following quotation of Narens is still an apt description of the measurement literature on "meaningfulness":

> ...meaningfulness is a term used to describe relations and concepts that are relevant to the underlying measurement situation. 'Relevant' here is used loosely since no one has really described what properties a satisfactory meaningfulness concept should have. What has happened is that various researchers have invoked various meaningfulness concepts for particular measurement contexts. In this paper, I will take 'meaningfulness' to refer to those properties and concepts that are consistent with the particular underlying measurement situation, where by 'consistent' I mean that the relevant qualitative properties and concepts are incorporable into the qualitative structure without changing the underlying measurement situation. For example, one might consider the relation T on Y to be meaningful for the structure $\langle Y, P, Q \rangle$ if and only if $\langle Y, P, Q \rangle$ and $\langle Y, P, Q, T \rangle$ satisfy the same measurement processes, where 'measurement processes' for one kind of meaningfulness may be taken as the set of $\mathfrak{N}$ -representations for a particular [numerical structure] $\mathfrak{N}$, and for another kind of meaningfulness as the set of automorphisms of $\langle Y, P, Q \rangle$, etc. *(Narens, 1981a, pg. 3)*

This section presents a brief description of the most important measurement-theoretic concepts of meaningfulness of the representational literature. The development follows Narens (1981a, 1985). However, changes in the definitions and terminology have been made so that the text conforms more closely with the terminology and concepts used elsewhere in this book. Also the quotes from Narens, 1981a, have sometimes been slightly altered to conform with terminology used here.

The measurement-theoretic literature has generally concerned itself with meaningfulness of first-order relations that are either on the domain of qualitative objects or on the set of real numbers. The reasons in the literature for restricting the meaningfulness concept to first-order relations is far from clear. If forced to hazard a reason, I would say that the theorists involved believed meaningfulness to be an approach to *empirical* relationships, which by their philosophies of science were necessarily first-order relationships. (Unlike most of the literature, the theory of empiricalness presented in Section 7.11 is not restricted to first-order relationships and has no necessary connection to meaningfulness.)

Definition 5.5.1 Let S be a scale from a set of qualitative objects X into $\mathbb{R}$. (Thus S is a nonempty functions from X into $\mathbb{R}$, with no additional restrictions put on S.) Let T be a n-ary relation on $\mathbb{R}$, where $n>0$. Then T is said to be *quantitatively S-invariant* if and only if for each $x_1, \ldots, x_n$ in X and each φ and ψ in S,

$$T[\varphi(x_1), \ldots, \varphi(x_n)] \text{ iff } T[\psi(x_1), \ldots, \psi(x_n)]\,.$$

Quantitative S-invariance is what Stevens called "meaningfulness." Suppes and Zinnes (1963) used quantitative S-invariance with S being a scale of homomorphisms into a numerical structure as the meaningfulness concept for their extension of the Scott-Suppes (1958) theory of measurement. (The Scott-Suppes theory had no meaningfulness concept.) Roberts (1985, 1990) and others have applied the Suppes and Zinnes approach to meaningfulness to many situations in the behavioral sciences, a few of which have been discussed in Chapter 2.

A fundamental problem encountered with taking quantitative S-invariance as a theory of meaningfulness is that the quantitative S-invariance of quantitative relations depend on the scale S used to measure the qualitative structure. In terms of the representational theories of measurement, this translates into a problem about the selection of the numerical representing structure.

The measurement literature has paid scant attention to what properties a "good" numerical representing structure should have. There seems to be implied consensus in the literature, however, that any "good" representing structure for effectively analyzing a qualitative structure $\mathfrak{X}$ should be just as effective as any other "good" representing structure for analyzing $\mathfrak{X}$, or to put it another way, the final results of the analysis should be independent of the particular "good" representing structure chosen.

Most of the formal results of Narens (1981a) are based on the idea that measurement and meaningfulness should only depend on structural properties of the qualitative structure $\mathfrak{X}$; that is, the final conclusions of the analysis should be the same for every qualitative structure isomorphic to $\mathfrak{X}$. The following is a consequence of this idea for quantitative S-invariance:

Suppose S is a scale of homomorphisms of $\mathfrak{X}$ into $\mathfrak{N}$ and f is a isomorphism of the structure $\mathfrak{Y}$ with domain Y onto $\mathfrak{X}$. Then it easily follows that

$$\mathcal{S}' = \{\varphi[(f(y))] \,|\, y \in Y \text{ and } \varphi \in \mathcal{S}\}$$

is a scale of homomorphisms of $\mathfrak{Y}$ into $\mathfrak{N}$. Furthermore, suppose T is a n-ary relation on the domain of $\mathfrak{N}$. Then for all φ and ψ in S and all $x_1, \ldots, x_n$ in the domain of $\mathfrak{X}$,

$$T[\varphi(x_1),\ldots,\varphi(x_n)] \leftrightarrow T[\psi(x_1),\ldots,\psi(x_n)],$$

if and only if for $y_1 = f^{-1}(x_1),\ldots,y_n = f^{-1}(x_n)$,

$$T[\varphi(f(y_1)),\ldots,\varphi(f(y_n))] \leftrightarrow T[\psi(f(y_1)),\ldots,\psi(f(y_n))],$$

that is, if and only if for all φ' and ψ' in S' and all $z_1, \ldots, z_n$ in the domain of $\mathfrak{Y}$,

$$T[\varphi'(z_1),\ldots,\varphi'(z_n)] \leftrightarrow T[\psi'(z_1),\ldots,\psi'(z_n)].$$

Thus T is quantitatively S-invariant if and only if it is quantitative S'-invariant.

Narens (1981a) implicitly assumes the following as a necessary condition for a "good" representing structure:

Definition 5.5.2 Let S be a scale of homomorphisms of $\mathfrak{X}$ into $\mathfrak{N}$. Then $\mathfrak{N}$ is said to be *S-compatible* with $\mathfrak{X}$ if and only if for all representations φ and ψ in S there exists an automorphism α of $\mathfrak{N}$ such that all x in the domain of $\mathfrak{X}$,

$$\psi(x) = \alpha[\varphi(x)]. \tag{5.6}$$

The following theorem is an immediate consequence of Definition 5.5.2:

Theorem 5.5.1 Suppose S is a scale of representations $\mathfrak{X}$ into $\mathfrak{N}$, $\mathfrak{N}$ is S-compatible with $\mathfrak{X}$. Then the following two statements are true:

1. S is a regular scale (Definition 2.3.7).
2. For all n-ary relations T on the domain of $\mathfrak{N}$, if T is invariant under the set of automorphisms of $\mathfrak{N}$, then T is quantitatively S-invariant.

Suppose S is a scale of isomorphisms of $\mathfrak{X}$ into $\mathfrak{N}$, $\mathfrak{N}$ is S-compatible with $\mathfrak{X}$, and φ is and element of S, and X is the domain of $\mathfrak{X}$. Then $\varphi(\mathfrak{X})$ is a substructure of $\mathfrak{N}$, and all automorphisms of $\varphi(\mathfrak{X})$ are φ-images of automorphisms of $\mathfrak{X}$. Let β be an automorphism of $\mathfrak{X}$. Then $\psi=\varphi^*\beta$ is in S and is onto $\varphi(X)$. From this it follows that $\varphi(\beta)$ is an automorphism of $\varphi(\mathfrak{X})$. By compatibility, let g be an automorphism of $\mathfrak{N}$ such that for each x in X,

$$g[\varphi(x)] = \psi(x) = \varphi[\beta(x)].$$

Then g is an extension of $\varphi(\beta)$. Thus *each automorphism of* $\varphi(\mathfrak{X})$ *extends to an automorphism of* $\mathfrak{N}$.

Because $\varphi(\mathfrak{X})$ isomorphically imbeds $\mathfrak{X}$ into $\mathfrak{N}$, $\mathfrak{N}$ may be viewed as an extension of $\mathfrak{X}$, which we will call $\overline{\mathfrak{X}}$. When this is done, compatibility is equivalent to the following statement: *each automorphism of* $\mathfrak{X}$ *extends to an automorphism of* $\overline{\mathfrak{X}}$.

Narens (1981a) makes the following comments about this situation for totally ordered $\mathfrak{X}$:

The best case for measurement is where the various measurement considerations lead to a unique (up to isomorphism) Dedekind completion $[\overline{\mathfrak{X}}]$ of $\mathfrak{X}$; the other case, having many non-isomorphic Dedekind completions, greatly compounds—and perhaps makes insolvable— the problem of finding the appropriate numerical representing structure. One natural condition for limiting the set of Dedekind completions of $\mathfrak{X}$ is to interpret the compatibility condition [Definition 5.5.1] solely in terms of the structure $\mathfrak{X}$, i.e., to require that Dedekind completions $\overline{\mathfrak{X}}$ of $\mathfrak{X}$ satisfy the following condition: each automorphism α of $\mathfrak{X}$ extends to an automorphism $\overline{\alpha}$ of $\overline{\mathfrak{X}}$. Unfortunately, this condition by itself is in general not enough to insure the existence of a unique Dedekind completion. *(pg. 14)*

Definition 5.5.3 Let S be a scale from X into $\mathbb{R}$ and R be a n-ary relation on X, where $n>0$. Then R is said to be *qualitatively S-invariant* if and only if there exists a n-ary relation T on $\mathbb{R}$ such that for all $x_1, \ldots, x_n$ in X and all φ in S,

$$R(x_1, \ldots, x_n) \text{ iff } T[\varphi(x_1), \ldots, \varphi(x_n)].$$

Pfanzagl (1968) introduced qualitative S-invariance, with S being a scale of homomorphisms into a numerical structure, as a meaningfulness concept.

Note that the relation T in Definition 5.5.3, if it exists, has the property that for all $x_1, \ldots, x_n$ and for all φ and ψ in S,

$$T[\varphi(x_1), \ldots, \varphi(x_n)] \text{ iff } T[\psi(x_1), \ldots, \psi(x_n)],$$

that is, the numerical relation T is S-invariant.

Let $\mathfrak{X}$ be a qualitative structure. A weakness of the homomorphism approach to the representational theory of measurement is that it allows for the existence of too many different kinds of scales S for $\mathfrak{X}$. A consequence of this is that the qualitative S-invariance of relations on the domain are not in general entirely determined by $\mathfrak{X}$ but depend also on S.

Example 5.5.1 For each k in $\mathbb{R}^+$ let $\oplus_k$ be the following 3-ary relation on $\mathbb{R}^+$: For all x, y, and z in $\mathbb{R}^+$, $x \oplus_k y=z$ if and only if either

$$x,\ y, \text{ and } z \le k \text{ and } \frac{x+y}{1+\frac{xy}{k^2}} = z$$

or

$$x,\ y, \text{ and } z > k \text{ and } \frac{x+y}{1+\frac{xy}{k^2}} = z\,.$$

For each k, r, and s in $\mathbb{R}^+$ let $\varphi_{k,r,s}$ be the following function on $\mathbb{R}^+$: For all x and y in $\mathbb{R}^+$, $\varphi_{k,r,s}(x)=y$ if and only if either

$$x < k \text{ and } \varphi_{k,r,s}(x) = \tanh\left[r \cdot \tanh^{-1}\left(\frac{x}{k}\right)\right]$$

or

$$x > k \text{ and } \varphi_{k,r,s}(x) = \tanh\left[s \cdot \tanh^{-1}\left(\frac{x}{k}\right)\right]$$

or

$$x = k \text{ and } \varphi_{k,r,s}(x) = k\,.$$

Let

$$\oplus = \bigcup_{k \in \mathbb{R}^+} \oplus_k\,.$$

And let

$$\mathfrak{X} = \mathfrak{N} = \langle \mathbb{R}^+ - \{1\}, \geq, \oplus_1 \rangle$$

and

$$\mathfrak{N}' = \langle \mathbb{R}^+, \geq, \oplus \rangle\,.$$

Then, through use of the identity,

$$\tanh(x+y) = \frac{\tanh(x) + \tanh(y)}{1 + \tanh(x)\tanh(y)}\,,$$

and its consequence that tanh is an isomorphism of $\langle \mathbb{R}^+, \geq, + \rangle$ onto $\langle R, \geq, \oplus_1 \rangle$ where $R = \{x \mid 0 < x < 1\}$, it is not difficult to show the following:

1. All homomorphisms of $\mathfrak{X}$ into $\mathfrak{N}$ are isomorphisms of $\mathfrak{X}$ onto $\mathfrak{N}$.
2. $\{\varphi_{1,r,s} \mid r \in \mathbb{R}^+ \text{ and } s \in \mathbb{R}^+\}$ is the set of isomorphisms of $\mathfrak{X}$ onto $\mathfrak{N}$.
3. For each *k, r,* and *s* in $\mathbb{R}^+$, $\varphi_{k,r,s}$ is a homomorphism of $\mathfrak{X}$ into $\mathfrak{N}'$.
4. $\mathfrak{N}'$ is compatible with $\mathfrak{X}$.

Let *S* be the set of homomorphisms of $\mathfrak{X}$ onto $\mathfrak{N}$, *T* be the set of homomorphisms of $\mathfrak{X}$ into $\mathfrak{N}'$, $R = \{x \mid 0 < x < 1\}$. Then it is easy to verify that *R* is qualitatively *S*-invariant. However, *R* is not qualitatively *T*-invariant: For suppose *R* were qualitatively *T*-invariant. A contradiction will be shown. By Definition 5.5.3, let *U* be a subset of $\mathbb{R}^+$ such that for each *x* in $\mathbb{R}^+$ and each φ in *T,*

$$R(x) \text{ iff } U[\varphi(x)]\,.$$

But this is impossible, since $R(\frac{2}{3})$, $\neg R(2)$, and

$$U(\varphi_{1,1,1}(\tfrac{2}{3})) = U(\tfrac{2}{3}) = U(\varphi_{3,1,1}(2)).$$

Because R is simply defined in terms of the primitives of $\mathfrak{X}$ by

$$R(x) \text{ iff } x \oplus_1 x > x\,,$$

one sees that qualitative *T*-invariance can have unpleasant consequences when the measurement scale is defined in terms of homomorphisms—even with $\mathfrak{N}'$ -compatibility holding.

The weakness of the homomorphism approach exhibited in Example 5.5.1 does not extend to the isomorphism approach:

Theorem 5.5.2 Let $\mathfrak{X}$ be a qualitative structure, $\mathfrak{N}$ and $\mathfrak{N}'$ be numerical structures, and R be a n-ary relation on the domain of $\mathfrak{X}$. Suppose S and T are the sets of isomorphisms onto respectively $\mathfrak{N}$ and $\mathfrak{N}'$. Then R is qualitative S-invariant if and only if it is qualitatively T-invariant.

Proof. Left to reader.

Because of difficulties like the one exhibited in Example 5.5.1, proponents of the homomorphism approach (especially, Luce, et al., 1990) have developed an alternative approach to qualitative S-invariance based on endomorphisms.

Definition 5.5.4 Let $\mathfrak{X} = \langle X, R_j \rangle_{j \in J}$ be a qualitative structure and R be a n-ary relation on X. Then

(1) R is said to be *endomorphism invariant* if and only if for all $x_1, \ldots, x_n$ in X and all endomorphisms α of $\mathfrak{X}$,

$$R(x_1, \ldots, x_n) \text{ iff } R[\alpha(x_1), \ldots, \alpha(x_n)];$$

(2) R is said to be *automorphism invariant* if and only if for all $x_1, \ldots, x_n$ in X and all automorphisms α of $\mathfrak{X}$,

$$R(x_1, \ldots, x_n) \text{ iff } R[\alpha(x_1), \ldots, \alpha(x_n)].$$

The following theorem and two examples of Narens (1981a) show the logical relationships of qualitative S-invariance, endomorphism invariance, and automorphism invariance of n-ary relations on X:

Theorem 5.5.3 Let $\mathfrak{X} = \langle X, R_1, \ldots, R_j, \ldots, \rangle_{j \in J}$ be a structure, R be a n-ary relation on X, and S be a scale of homomorphisms of $\mathfrak{X}$ into the numerical structure $\mathfrak{N}$. Then the following two statements are true:

1. If R is qualitatively S-invariant, then R is endomorphism invariant.
2. If R is endomorphism invariant, then R is automorphism invariant.

Proof. The proof is straightforward and easy and can be found in Theorem 3.1 of Narens (1981a).

The following two examples show that the converses of Statements 1 and 2 of Theorem 5.5.3 need not hold:

Example 5.5.2 Let X={1,2,3}. Then $\mathfrak{Y} = \langle X, \geq \rangle$ has the identity, ι, as its only endomorphism. Therefore, every relation on X is endomorphism invariant. Let $R(z)$ hold if and only if z=2. Then R is endomorphism invariant. Let $\mathfrak{N} = \langle \mathbb{R}^+, \geq \rangle$. Let

$$S = \{\varphi \mid \varphi \text{ is a homomorphism of } \mathfrak{Y} \text{ into } \mathfrak{N}\}.$$

Then the identity function ι on $\mathfrak{Y}$ is in S. It will be shown by contradiction that R is not qualitatively S-invariant:

Suppose R were qualitatively S-invariant. Let T be the 1-ary relation on $\mathbb{R}^+$ such that for all x in X and all φ in S,

$$R(x) \text{ iff } T[\varphi(x)].$$

Let ψ be an element of S such that

$$\psi(1) = 1, \quad \psi(2) = 3, \quad \psi(3) = 4\,.$$

Then because $\iota \in S$,

$$R(3) \text{ iff } T(3). \tag{5.7}$$

Because ψ is in S,

$$R(2) \text{ iff } T(3). \tag{5.8}$$

It then follows from Equations 5.7 and 5.8 that

$$R(3) \text{ iff } R(2),$$

which is impossible.

The next example shows that automorphism invariance and endomorphism invariance need not coincide—even for structures with rich automorphism groups:

Example 5.5.3 Let π be a positive transcendental number and

$$Y = \{1, \pi, \pi^2, \pi^3, \ldots\}.$$

Let X be in the intersection of all sets Z such that

(1) $Y \subseteq Z$,
(2) $rz \in Z$ for each positive rational r and each $z \in Z$, and
(3) $y+z \in Z$ for each y and z in Z.

Let $\mathfrak{X} = \langle X, \succsim, + \rangle$. For each $r \in \mathbb{R}^+$, define the function α_r from Z into $\mathbb{R}^+$ as follows: For each $z \in Z$, $\alpha_r(z) = rz$. Then it easily follows that for each positive rational r, α_r is an automorphism of $\mathfrak{X}$. It also easily follows that the automorphism group of $\mathfrak{X}$ is 'dense' in the following sense: if x, y, and z; are elements of X such that $x > y$, then for some automorphism β of $\mathfrak{X}$, $x > \beta(z) > y$. α_π is an endomorphism of $\mathfrak{X}$. However, α_π is not an automorphism of $\mathfrak{X}$, because there is no element x of X such that $\alpha_\pi(x)=1$.

Let $\mathbb{Q}^+$ be the set of positive rationals. $\mathbb{Q}^+$ will be shown to be automorphism invariant by showing all automorphisms of $\mathfrak{X}$ are multiplications by elements of $\mathbb{Q}^+$.

It follows from the uniqueness theorem of extensive measurement (e.g., Theorem 3 of Chapter 3 of Krantz, et al., 1971, Theorem 8.2 of Chapter 2 of Narens, 1985) that each automorphism of $\mathfrak{X}$ is a multiplication by a positive constant. Suppose α is an automorphism of $\mathfrak{X}$. Let k be the positive real number such that $\alpha(x)=kx$. Then $\alpha(1)=k$, and because α is an automorphism, k must be in X. It then follows from the definition of X that k must be a polynomial $P(\pi)$ in π (possibly of degree zero) with positive rational coefficients. Since a is an automorphism of $\mathfrak{X}$, α^{-1} is also an automorphism of $\mathfrak{X}$, and

$$\alpha^{-1}(1) = \frac{1}{k} = q(\pi),$$

where $q(\pi)$ is a polynomial in π with positive rational coefficients. However, because π is transcendental and $p(\pi)q(\pi)=1$, p and q must be polynomials of degree 0, that is, k must be a positive rational. Therefore $\alpha(\mathbb{Q}^+) = \mathbb{Q}^+$, that is, $\mathbb{Q}^+$ is automorphism invariant. $\mathbb{Q}^+$ is not endomorphism invariant, because $1 \in \mathbb{R}^+$ and $\alpha_\pi(1) = \pi \notin \mathbb{Q}^+$.

For many important measurement situations, qualitative S-invariance, endomorphism invariance, and automorphism invariance coincide for first-order qualitative relations:

Theorem 5.5.4 Let $\mathfrak{X} = \langle X, R_j \rangle_{j \in J}$ be a structure, S the set of homomorphisms of $\mathfrak{X}$ into some numerical structure, and R be a n-ary relation on X. Suppose the following three conditions are satisfied:

(1) *Existence:* $S \neq \emptyset$.
(2) *1-point S-uniqueness:* For all φ and ψ in S, if for some x in X, $\varphi(x)=\psi(x)$, then $\varphi=\psi$.
(3) $\mathfrak{X}$*-homogeneity:* For each x and y in X there exists an automorphism α of $\mathfrak{X}$ such that $\alpha(x)=y$.

Then R is qualitatively S-invariant if and only if it is endomorphism invariant if and only if it is automorphism invariant.

Proof. Theorem 3.2 of Narens (1981a).

Definition 5.5.5 Let $\mathfrak{X} = \langle X, R_j \rangle_{j \in J}$ be a structure and R be a set of finitary relations on X.

Then R is said to be *automorphism invariant* if and only if for each $R \in R$ and each automorphism a of $\mathfrak{X}$, $\alpha(R) \in R$.

R is said to be *endomorphism invariant* if and only if for each $R \in R$ and each endomorphism of $\mathfrak{X}$, $\alpha(R) \in R$.

Narens (1981a) shows the following theorem:

Theorem 5.5.5 Let $\mathfrak{X} = \langle X, R_j \rangle_{j \in J}$ be a structure, A the set of automorphisms of $\mathfrak{X}$, and ε the set of endomorphisms of $\mathfrak{X}$. Then the following two statements are true:

1. A and ε are automorphism invariant.
2. If ε has an element that is not onto X, then A and ε are not endomorphism invariant.

Proof. 1. It is easy to show that the sets of automorphisms and endomorphisms of $\mathfrak{X}$ are automorphism invariant.

If an endomorphism γ of $\mathfrak{X}$ is not onto X, then neither A nor ε are endomorphism invariant, because $\gamma(\iota)$, where ι is the identity function on X, is neither an endomorphism nor an automorphism of $\mathfrak{X}$.

Using the current terminology, Narens (1981a) makes the following comment about Theorem 5.5.5:

> [Theorem 5.5.5] indicates that endomorphism invariance is probably not the correct way of formulating the invariance condition for relations on X [for meaningfulness considerations]. However, a variety of families of transformations can agree with endomorphism meaningfulness on (the first-order) relations on X yet disagree on higher-order relations. *(pg. 39)*

Narens (1981a) also suggests that meaningfulness should be reflexive in the sense that it should be "meaningful." Meaningfulness based on automorphism invariance has this property, that is, the concept of "automorphism invariance" for a structure $\mathfrak{X}$ is invariant under the automorphisms of $\mathfrak{X}$; endomorphism invariance does not have this property. A straightforward extension of this idea to the formal meaningfulness axioms Chapter 4 suggests that axiom *DM** should be assumed. A more subtle variant is that *if* axiom *DM* is assumed, *then* axiom *DM** should be assumed. While reflexivity appears to be a desirable property for meaningfulness concepts, I see no argument for its necessity.

Narens (1981a) also investigated meaningfulness as invariance under sets of partial endomorphisms (i.e., structure preserving functions from a *subset* of the domain of the structure *into* the domain of the structure) and partial automorphisms (i.e., partial endomorphisms that are one-to-one functions). Narens (1981a) showed that invariance under appropriate sets partial endomorphisms and partial automorphisms had better meaningfulness theories than invariance under endomorphisms, particularly invariance under certain naturally generated "pseudo-groups" of partial automorphisms.

Narens (1981a) made the following comment about basing meaningfulness on the above invariance concepts:

> In this section [of Narens, 1981a], a number of concepts of qualitative meaningfulness have been presented, and the problem remains of deciding which, if any, is the "correct" concept. It is my view that there is no single correct concept of meaningfulness. I believe that in the final analysis the choice of the "correct" invariance concept for a structure will not be determined solely by the structure, but in general will depend upon features of the intended measurement application. What we have today is a handful of successful applications of the various meaningfulness concepts; what

is still lacking is a general theory of invariance and inference based upon invariance. The meaningfulness concepts presented above are attempts to abstract the common core of this handful of successful applications, and are not based upon any detailed philosophical analysis, and thus their usefulness and generality are somewhat in doubt. Hopefully in the future someone will find a more direct and comprehensive approach to this important problem.

[For purposes of basing theories of meaningfulness on] the invariance concepts considered in this section, automorphism invariance has the greatest applicability, mainly because the most important structures that appear in measurement have an abundance of automorphisms. Endomorphism and qualitative *S*-invariance, when they do not coincide with automorphism invariance, thus far have had far fewer applications. I also believe that these two latter concepts have inherent difficulties [as bases for theories of meaningfulness], which arise from the fact that representations of the qualitative structure are only required to be *into* (rather than *onto*) the [numerical] representing structure. Interesting enough, it is this "into" property of representations that make [meaningfulness based on endomorphisms and qualitative *S*-invariance of qualitative relations] natural concepts for measurement.... However, to my knowledge, the practice of using "into" representations for the general measurement case has never been philosophically justified. The situations where "into" representations have been useful are rather special and are characterized by conditions similar to compatibility [Definition 5.5.2], or as I prefer to see it, characterized by the representing structure being isomorphic to an extension of the qualitative structure where the automorphisms (or in some situations certain key partial automorphisms) of the qualitative structure extend to automorphisms of the extension. It is my belief that measurement of general structures should be based either upon representations that are isomorphisms onto the representing structure, or, if the situation demands it, upon an appropriate set of partial isomorphisms. (A *partial isomorphism* is an isomorphism of a restriction of the qualitative structure into the representing structure.) This should be done in a way so that the corresponding qualitative concept is either automorphism invariance or [invariance under some appropriate pseudo-group of partial automorphisms]. *(Narens, 1981a, pp. 45–47)*

It should be noted that although the representational theory has characterized meaningfulness as various kinds invariance, its *applications* of the meaningfulness concept (e.g., those considered in Chapter 2 and later in this chapter) generally use invariance only as a necessary condition for meaningfulness. The following quote of Narens (1987) link such representational uses of invariance to the definitional concepts of meaningfulness of Chapter 4:

Suppose in a particular setting we are interested in finding the functional relationship of the qualitative variables, *x, y,* and *z*. We believe that the primitive relations (which are known) completely characterize the current situation. Furthermore, our understanding (or insight) about the situation tells us that *x* must be a function of *y* and *z*. (This is the typical case for an application of dimensional analysis in physics.) This unknown function—which we will call "the desired function"—must be determined by the primitives and the qualitative variables *x, y,* and *z*. Therefore, it should somehow be "definable" from these relations and variables. Even though the exact nature of the definability condition is not known, (it can be argued that) it must be weaker than the enormously powerful methods of definability encompassed by the conjunction of the axioms *MC′*, *AL, DM**, *MP,* and *MI*. Thus by [Theorem 4.2.10] we know that any function relating the variable *x* to the variables *y* and *z* that is *not* invariant under the automorphisms of the primitives cannot be the desired function. In many situations, this knowledge *of knowing that functions not invariant under the automorphisms of the primitives cannot be the desired function* can be used to effectively find or narrow down the possibilities for the desired function.

It appears likely to the author that the near future will bring better theories of meaningfulness that will more precisely specify the nature of definability properties of the meaningfulness predicate, and that this additional knowledge will likely prove useful in strengthening the techniques of dimensional analysis of physics and other meaningfulness methods of drawing inferences about qualitative relationships. *(pp. 70–71)*

Narens (2001) expanded on this theme as follows:

Scientific inquiry is a complicated issue with many overlapping parts. I believe meaningfulness belongs primarily to the theoretical part of scientific inquiry. Because of the overlap of the theoretical part of a science with its experimental and applied parts, meaningfulness often has important ramifications in the experimental and applied parts.

Meaningfulness is essentially a theoretical position about scientific content and its role in (theoretical) inference. For example, consider the case where by extra-scientific means (e.g., intuition, experience, etc.) a scientist is led to believe that a function $z=F(x, y)$ that he needs to describe from a subset of $A \times A$ into A is completely determined by the observable, first-order relations $R_1, \ldots, R_n$ on A. Then it is reasonable for the scientist to proceed under the hypothesis that F belongs to the scientific content of $\mathfrak{X} = \langle A, R_1, \ldots, R_n \rangle$, which for this discussion may be taken as the set of meaningful entities determined by $[\mathrm{D}''(R_1, \ldots, R_n)]$. Thus the scientist assumes F has a scientific definition in terms of $\mathfrak{X}$ and its primitives [i.e., assumes F is meaningful]. By [Theorem 4.3.6], F is invariant under the automorphisms of $\mathfrak{X}$. Suppose the scientist knows enough properties about $\mathfrak{X}$ and has the mathematical skill to determine the automorphism group G of $\mathfrak{X}$. Then methods of analyses involving automorphisms may be employed to provide information helpful in characterizing F. There are several methods in the literature for accomplishing this.

Note that in the above process, scientific definability [axiom *MC'*] is used *to justify* F belonging to the appropriate topic, invariance is used *as a mathematical technique* to find helpful information for characterizing F, and that these two uses are connected *by a theorem* of mathematical logic. Also note that the scientist's belief that F belonged to the topic generated by $\mathfrak{X}$ is extra-scientific. Therefore, the deductions based on information obtained through the above process should be either checked by experiment or be derived from accepted scientific theory and facts; i.e., they should be treated as scientific hypotheses that need corroboration. Thus, for the purposes of science, the above process is a method of generating hypotheses and not facts: If the scientist's extra-scientific beliefs are correct, then the generated hypotheses will be facts; however, the scientist has no *scientific guarantee* that his beliefs are correct. *(Narens, 2001.)*

5.6 Meaningful Scales

The traditional form of the representational theory of measurement proceeds as follows: (1) A qualitative structure $\mathfrak{X}$ is selected to capture the domain of interest; (2) a numerical structure $\mathfrak{N}$ is selected to measure $\mathfrak{X}$ in terms of the scale S of homomorphisms of $\mathfrak{X}$ into $\mathfrak{X}$; and (3) meaningfulness is identified with a form of invariance associated with S, for example, quantitative S-invariance (Definition 5.5.1; Suppes and Zinnes, 1963), qualitative S-invariance (Definition 5.5.3; Pfangagl, 1968), or endomorphism invariance (Definition 5.5.4; Luce, et. al., 1990). In this section the process is inverted: *(i)* An axiomatic theory of meaningfulness is assumed; *(ii)* in terms of this theory of meaningfulness, the concept of a *meaningful scale* is formulated; and *(iii)* for each meaningful scale S a structure $\mathfrak{N}$ with domain a pure set and a qualitative structure $\mathfrak{X}$ with meaningful primitives and domain the set of atoms, A, are constructed such that S is a set of homomorphisms of $\mathfrak{X}$ into $\mathfrak{N}$.

Definition 5.6.1 Assume axiom system ZFA. Then the following three definitions hold:

1. A *pure structure* is a structure (Definition 5.1.3) that has a pure set as its domain.

2. A *numerical structure* is a structure that has a subset of the set-theoretic reals (see Section 4.1.3 of Chapter 4) as its domain.
3. A *numerically based structure* is a structure that has a set of elements based on the set-theoretic reals, for example, a Cartesian product of set-theoretic reals) as its domain.

Note that by Definition 5.6.1 numerical structures and numerically based structures have domains of pure sets.

Definition 5.6.2 Assume D′. A *meaningful scale* is a nonempty set of functions *S* such that *M(S)* and the range of each element of *S* is a pure set.

The following theorem is an immediate consequence of axiom system D′.

Theorem 5.6.1 Assume axiom system D′. Suppose $\mathfrak{X} = \langle A, R_j \rangle_{j \in J}$ is a structure, *J* is a pure set, $\mathfrak{X}$ and the primitives of $\mathfrak{X}$ are meaningful, $\mathfrak{N} = \langle N, S_j \rangle_{j \in J}$ is a pure structure, *S* is either the set of homomorphisms of $\mathfrak{X}$ into $\mathfrak{N}$ or the set of isomorphisms of $\mathfrak{X}$ onto $\mathfrak{N}$, and $S \neq \emptyset$. Then *S* is a meaningful scale.

The following example shows that different meaningful scales can exist for the same qualitative structure.

Example 5.6.1 Assume axiom system D′. Let $\mathfrak{X} = \langle X, \succsim \rangle$ be a continuum, $\mathfrak{N} = \langle \mathbb{R}^+, \geq \rangle$, *S* be the set of isomorphisms of $\mathfrak{X}$ onto $\mathfrak{N}$ and *T* be the set of homomorphisms of $\mathfrak{X}$ into $\mathfrak{N}$. Assume that $\mathfrak{X}$ and the primitives of $\mathfrak{X}$ are meaningful. Then $S \neq T$, and by axiom system D′, *S* and *T* are meaningful scales.

Theorem 5.6.2 Assume $\mathfrak{Y} = \langle A, T_j, \rangle_{j \in J}$, $M(T_j)$ for each $j \in J$,

$$a = \{T_j \mid j \in J\},$$

axiom system D”*(a)*, and *S* is a meaningful scale. Let

$$N = \bigcup_{\varphi \in S} \varphi(A),$$

and for each $j \in J$, let

$$S_j = \bigcup_{\varphi \in S} \varphi(T_j),$$

and let

$$\mathfrak{N} = \langle N, S_j, \rangle_{j \in J}.$$

For each $j \in J$ define R_j on *A* as follows:

For each $a_1, \ldots, a_{m(j)}$ of *A*, $R_j(a_1, \ldots, a_{m(j)})$ if and only if there exists φ in *S* such that $S_j(\varphi(a_1), \ldots, \varphi(a_{m(j)})$.

Let

$$\mathfrak{X} = \langle A, R_j \rangle_{j \in J}.$$

Then S is a set of homomorphisms of $\mathfrak{X}$ into $\mathfrak{N}$.

Proof. Since by hypothesis, the range of each element of S is a pure set, the union of these ranges, N, is a pure set. Because for each $j \in J$, S_j is a pure set, it follows by the hypothesis D″*(a)* and the hypothesis S is a meaningful scale that R_j is meaningful for each j in J. Let j be an arbitrary element of J. Without loss of generality, suppose R_j is a $m(j)$-relation. Let $a_1, \ldots, a_{m(j)}$ be $m(j)$ arbitrary elements of A, and ψ be an arbitrary element of S. Then by the definition of R_j given above,

$$R_j(a_1, \ldots, a_{m(j)}) \text{ iff } S_j(\psi(a_1), \ldots, \psi(a_{m(j)})),$$

establishing that ψ is a homomorphism of $\mathfrak{X}$ into $\mathfrak{N}$.

Recall that by Definition 4.5.3, a set was said to be minimeaningful if and only if it was meaningful and no proper subset of it was meaningful. The following two theorems show that the minimeaningfulness of domains of structures and scales produce highly desirable measurement situations.

Theorem 5.6.3 Assume J is a set, $\mathfrak{X} = \langle A, R_j \rangle_{j \in J}$, $M(R_j)$ for each $j \in J$, $a = \{R_j \mid j \in J\}$, and axiom system D″*(a)*. Then the following two statements are true:

1. If $\mathfrak{X}$ is homogeneous (Definition 4.5.5), then A is minimeaningful (Definition 4.5.3).
2. Suppose a is (externally) finite. Then $\mathfrak{X}$ is homogeneous if and only if A is minimeaningful.

Proof. Since a is an entity and axiom system D″*(a)* is true, it follows that the transformation group G for M is an entity, and thus Statement 1 is true by Theorem 4.5.1. If a is (externally) finite, then by Theorem 4.3.6 axiom system TM is true, and thus by Theorem 4.5.3 Statement 2 is true.

Theorem 5.6.4 Assume axiom system D′. Suppose $\mathfrak{X} = \langle A, R_j \rangle_{j \in J}$, J is a pure set, the primitives of $\mathfrak{X}$ are meaningful, and S is a meaningful scale that is minimeaningful. Then the following two statements are true:

1. For all f and g in S, range(f)=range(g).
2. If each element of S is a one-to-one function, then there exists a pure structure $\mathfrak{N}$ such that *(i)* S is a subset of isomorphisms of $\mathfrak{X}$ onto $\mathfrak{N}$, and *(ii)* $\mathfrak{N}$ is compatible with $\mathfrak{X}$, that is, for each φ and ψ in S, there exists an automorphism γ of $\mathfrak{N}$ such that $\psi = \gamma * \varphi$.

Proof. 1. Let φ be an element of S. Let

$$\mathcal{W} = \{\psi \mid \psi \in S \text{ and } \psi(A) = \varphi(A)\}.$$

Because $\varphi(A)$ is a pure set and S is meaningful, W is meaningful by axiom system D′. Since S is minimeaningful and $\emptyset \subseteq W \subseteq S$, it follows that $S=W$.

2*(i)*. Suppose each element of S is a one-to-one function. Let φ be an arbitrary element of S. Because S is a meaningful scale, the range of φ is a pure set N. For each $j \in J$, let $S_j = \varphi(R_j)$, and let $\mathfrak{N} = \langle N, S_j \rangle_{j \in J}$. Then, because φ is a one-to-one function, φ is an isomorphism of $\mathfrak{X}$ onto the pure structure $\mathfrak{N}$. Let

$$\mathcal{T} = \{\psi \mid \psi \in \mathcal{S} \text{ and } \psi \text{ is an isomorphism onto } \mathfrak{N}\} .$$

Then, because $\mathfrak{N}$ is a pure structure and J, the primitives of $\mathfrak{X}$, and S are meaningful, it follows from axiom D′ that T is meaningful. Because φ is in T, $T \neq \emptyset$. Thus, since $T \subseteq S$ and S is minimeaningful, $T=S$. Therefore each element of S is an isomorphism onto $\mathfrak{N}$.

2*(ii)*. Suppose each element of S is a one-to-one function. Let φ and ψ be arbitrary elements of S. Then it is easy to verify that $\gamma = \psi * \varphi^{-1}$ is an automorphism of $\mathfrak{N}$ and that $\gamma * \varphi = \psi$.

5.7 Possible Psychophysical Laws Revisited

Luce's 1959 theory of "possible psychophysical laws" was discussed in Section 2.5. In that theory Luce concluded that psychophysical laws were impossible in several measurement situations, for example, in situations where the independent variable is measured on an interval scale and the dependent variable on a ratio scale. In this section, it is argued that Luce (1959) misidentified the possible psychophysical laws for those cases where the independent and dependent have different scale types. The argument for this position is based on meaningfulness considerations not available to Luce in 1959.

For concreteness, we consider the case where the dependent variable is a ratio scale and the independent variable is an interval scale. Similar conclusions for the other cases follow by similar arguments. In order to discuss qualitative as well as quantitative issues, we suppose that the independent variable results from measurement of the qualitative structure $\mathfrak{X}$ by the ratio scale S onto $\mathbb{R}^+$ the dependent variable results from measurement of the qualitative structure $\mathfrak{Y}$ by interval scale T onto $\mathbb{R}$, and the domains X of $\mathfrak{X}$ and Y of $\mathfrak{Y}$ are disjoint. Let f be a function between the measurements of $\mathfrak{X}$ and $\mathfrak{N}$ that is a "law" relating $\mathfrak{X}$ and $\mathfrak{N}$ in sense of Luce (1959), that is, let $\varphi \in S$, $\psi \in T$, and f be a continuous function from $\mathbb{R}^+$ onto $\mathbb{R}$, and let the following condition be satisfied:

For each $r \in \mathbb{R}^+$ there exists ψ' in T such that for each x in X,

$$\psi[f(r\varphi(x))] = \psi'[f(\varphi(x))] .$$

Under the above conditions, Luce (1959; see Section 2.5) showed that f has the following form: There exist nonzero a and b in $\mathbb{R}$ such that for all x 'in X,

$$f(\varphi(x)) = a \log f(\varphi(x)) + b\,. \tag{5.9}$$

A consequence of this is that f is a one-to-one function. Under any reasonable concept of "law" it appears to be eminently reasonable that if a one-to-one function is a "law", then its inverse should also be a "law". However, by Luce's 1959 theory this need not be the case, and in fact by Luce's theory it is not the case for many important situations, including the "law" represented by the function f above. The reason for this is that the "inverse" of the above "law" would be a "law" with the independent variable an interval scale and the dependent variable a ratio scale—one of the impossible cases for a "law" of Luce (1959). This failure about inverses suggests that there is something amiss about Luce's method of obtaining possible psychophysical laws. One way to investigate the nature of this problem is through meaningfulness considerations.

Luce (1959) attempted to extend the methods of physical dimensional analysis to case of psychophysical laws involving a single independent variable, particularly the extension of the concept of dimensional invariance. It turns out that dimensional invariance is just invariance under a special group of transformations, which when viewed qualitatively is automorphism invariance (Luce, 1978; see Subsection 5.10.3), that is, possible psychophysical laws, when viewed qualitatively, should be automorphism invariant.

Let us assume, as is implicit in Luce (1959), that the "law" f above corresponds to a qualitative one-to-one function F from X onto Y. Then the underlying "lawful" qualitative situation is described by the structure

$$\mathfrak{Q} = \langle X \cup Y, F, X, \succsim_X, R_k, Y, \succsim_Y, S_j \rangle_{k \in K,\ j \in J}\ ,$$

where

$$\mathfrak{X} = \langle X, \succsim_X, R_k \rangle_{k \in K} \ \text{ and } \ \mathfrak{Y} = \langle Y, \succsim_Y, S_j \rangle_{j \in J}$$

are the totally ordered qualitative structures used above to generate S and T as scales of isomorphisms of $\mathfrak{X}$ and $\mathfrak{Y}$ respectively into the respective numerical structures

$$\mathfrak{N}_1 = \langle \mathbb{R}^+, \geq, R_k^1 \rangle_{k \in K} \ \text{ and } \ \mathfrak{N}_2 = \langle \mathbb{R}, \geq, S_j^1 \rangle_{j \in J}\ .$$

Let H be the automorphism group of $\mathfrak{Q}$.

Theorem 5.7.1 H has the following three properties:

(1) All elements of H are of the form $\alpha \cup \beta$, where a is an automorphism of $\mathfrak{X}$ and β is an automorphism of $\mathfrak{Y}$ that is a translation (Definition 5.1.7).
(2) For each automorphism α of $\mathfrak{X}$ there is exactly one translation β of $\mathfrak{Y}$ such that $\alpha \cup \beta$ is in H.

(3) For each translation β of $\mathfrak{Y}$ there exists an automorphism α of $\mathfrak{X}$ such that $\alpha \cup \beta$ is in H.

Proof. Theorem 5.12.5

Let

$$\mathfrak{N} = \langle \mathbb{R}, f, \mathbb{R}^{+}, \geq, R_{k}^{1}, \mathbb{R}, \geq, S_{j}^{1} \rangle_{k \in K,\, j \in J}$$

and

$\mathcal{T}' = \{\psi \mid \psi \in \mathcal{T}$ and ψ is a translation representation (Definition 2.3.10)$\}$.

Then by the representational theory, $\mathfrak{Q}$ is properly measured by the set $\mathcal{U}$ of homomorphisms of $\mathfrak{Q}$ into $\mathfrak{N}$. By using property (1) of Theorem 5.7.1 it easily follows that

$$\mathcal{U} \subseteq \{\varphi \cup \psi \mid \varphi \in \mathcal{S} \text{ and } \psi \in \mathcal{T}'\} ,$$

and that for each φ in S there exists ψ in T' such that for each x in X and each y in Y,

$$y = F(x) \text{ iff } \psi(y) = f[\varphi(x)] .$$

From (1), (2), and (3) of Theorem 5.7.1, it is easy to see that *as part of the structure* $\mathfrak{Q}$ that $\mathfrak{X}$ should be measured by ratio scale $\mathcal{S}$ and $\mathfrak{Y}$ should be measured by the translation scale T' (Definition 2.3.10). Thus in formulating the functional equation for the psychophysical law based on f, Luce (1959) and many subsequent researchers failed to realize that *the inclusion of f into the measurement situation altered the way in which $\mathfrak{Y}$ should be measured.* Put in meaningfulness terms, *the meaningfulness of the law f was ignored.* With simple consideration about what group of transformations should be chosen to make f meaningful, it becomes apparent that $\mathfrak{Y}$ should not be measured by the full interval scale $\mathcal{T}$ for the purpose of "measuring" the (meaningful) law f.

The above analysis is based on viewing the law linking the independent and dependent variables as a single function. Falmagne and Narens (1983) view laws as families of functions linking independent and dependent variables, with the "lawfulness" of a family consisting of various kinds of invariances, including the kind inherent in Luce's Principle (see Section 2.6). By this view it is possible to have a law relating a ratio scalable independent variable with an interval scalable dependent variable with the functional relationship being a family of functions. However, in many theoretical applications of Luce's Principle it is inherent in the situation that a single function and not a family of functions is the correct underlying functional relationship. Examples of this arise in psychophysics, where frequently the independent variable is the physical intensity of the stimulus and the dependent variable is the subjective intensity of the stimulus. Often in such psychophysical situations psychological reasoning tells us that psychological intensity is related to physical intensity through a single function.

A deeper analysis of possible psychophysical laws is presented in Subsection 6.4.2.

5.8 Magnitude Estimation Revisited

5.8.1 Introduction

Magnitude estimation is a controversial form of measurement developed by the psychologist S.S.Stevens that is widely used in the behavioral and social sciences. A psychophysical application of it was discussed in Section 2.2. This section presents, for this slippery subject, a measurement-theoretic foundation of Narens (1996). Another formulation, perhaps more in line with Stevens' original ideas, is given in Section 7.7.

Let X be a set of stimuli to be presented to a subject in a psychological experiment. *Stevens' method of magnitude estimation* essentially proceeds by having the subject produce a function φ_t from X into $\mathbb{R}^+$ as follows: An element t—called the *modulus*—is selected from X. The subject is told to consider the number 1 as representing his or her subjective intensity of t, and keeping this consideration in mind to give his or her numerical estimation of subjective intensity value of another stimulus x in X. The experimenter uses these verbal estimates of the subject to construct the function φ_t by assigning the number corresponding to the subject's numerical estimate of x as the value of $\varphi_t(x)$.

I find Stevens writings to be somewhat vague as to what is being accomplished by magnitude estimation. I believe the following two assumptions, which I refer to as *Stevens's Assumptions,* are inherent in his ideas about his method of magnitude estimation:

1. The function φ_t is an element of a ratio scale $\mathcal{S}$ that adequately measures the subject's subjective intensity of stimuli in X.
2. Each element x in $\mathfrak{X}$ can be used as a modulus and the resulting representation φ_x is in the ratio scale $\mathcal{S}$ that is, there exists r in $\mathbb{R}^+$ such that $\varphi x = r\varphi t$.

Let $D = \{\varphi_t \mid t \in X\}$ Then D is the complete data set that is generated by conducting all possible magnitude estimations of stimuli in X with all possible moduli from $\mathfrak{X}$. D can be recoded as the set E of 3-tuples of the form (x, p, t), where

$$(x, \mathbf{p}, t) \in E \text{ iff } \varphi_t(x) = \mathbf{p}.$$

(The reason for the bold "$\mathbf{p}$" in the above equation is explained later.)

Definition 5.8.1 Let X and E be as above. Then E is said to have the *multiplicative property* if and only if for all x and t in X and all p in $\mathbb{R}^+$, if $(x, \mathbf{p}, t) \in E$, $(y, \mathbf{q}, x) \in E$ and $(y, \mathbf{r}, t) \in E$ then $\mathbf{r} = \mathbf{p} \cdot \mathbf{q}$.

The multiplicative property puts powerful constraints on the subject's magnitude estimation behavior. Both theoretically and experimentally one would want would want theories of magnitude estimation that have weaker constraints on E.

Theorem 5.8.1 Let X and E be as above. Suppose Stevens' Assumptions. Then E has the multiplicative property.

Proof. Suppose $(x, \mathbf{p}, t) \in E$, $(y, \mathbf{q}, x) \in E$, and $(y, \mathbf{r}, t) \in E$. Then $\varphi_t(x) = p$, $\varphi_x(y) = q$, and $\varphi_t(y) = r$. By Stevens' Assumptions, let u in $\mathbb{R}^+$ be such that

$$\varphi_x = u\varphi_t \, .$$

Then $1 = \varphi_x(x) = u\varphi_t(x) = u \cdot p$, that is,

$$u = \frac{1}{p} \, .$$

Therefore,

$$q = \varphi_x(y) = \frac{1}{p} \cdot \varphi_t(y) = \frac{1}{p} \cdot r \, ,$$

that is, $r = p \cdot q$.

Throughout the rest of this section the following convention is observed.

Convention 5.8.1 X is a nonempty set and $\succsim$ is a binary relation on X. Elements of X are possible stimuli to be presented to a subject for judgment, and $\succsim$ is an experimenter determined, intensity ordering on X. An example of $\langle X, \succsim \rangle$ is where X is a set of physical lights and $\succsim$ is a total ordering of X in terms of physical energy.

E is a nonempty set of ordered triples of the form $(x, \mathbf{p}, t)$, where x and t are elements of X and $\mathbf{p}$ is a numeral denoting the positive real number p. (In most of the axiomatic theories considered in this section, p is a positive integer.) E is interpreted as the behavior of the subject in a magnitude estimation paradigm. For most kinds of magnitude estimation tasks, the experimental results can be coded as such a set E.

In triples $(x, \mathbf{p}, t)$ in E, $\mathbf{p}$ put is in bold typeface because it represents a *numeral* and not a *number*. In this section, numbers are assumed to be *highly abstract scientific objects,* and it is not assumed that subjects understand or use such scientific objects in their calculations or responses, nor is it assumed that subjects have or use a philosophically sound correspondence between (scientific) numbers and numerals.

The axiomatic theories of magnitude estimation presented generally do not depend on the details of the instructions given to the subject. Instead, they generally depend only on the set E and an explicit theory of the relationship of E to psychological processes. This restriction does not prevent different interpretations of the axiomatic theories according to the kinds of instructions given to the subject. For example, the same axiomatic theory may apply to data collected from the following three instructions:

(1) Find a stimulus in X which appears to be p times greater in intensity than the stimulus t.
(2) Pick the number p which best describes the stimulus x as being p times as more intense than the stimulus t.
(3) Find the stimulus which in your subjective valuation is $p+$ the valuation of the stimulus t.

For (1) and (2) one may want to represent stimuli numerically so that the numerical interpretation of (x, p, t) is that the numerical value of x is p times the numerical value of t, whereas in (3) one may want to interpret (x, p, t) so that the numerical value of x equals p plus the numerical value of t.

5.8.2 Ratio Magnitude Estimation

In *ratio magnitude estimation,* the subject is asked to give subjective estimation of ratios. As previously discussed, this can be done in several ways. For concreteness, unless otherwise stated, it is assumed that the subject has been instructed to "Find a stimulus in X that appears to be p times greater in intensity than the stimulus t."

The axioms for ratio magnitude estimation considered in this section are divided into the following three kinds:

Behavioral assumptions. These consist of axioms about *(i)* the behavior of stimuli, *(ii)* the behavior of the subject, and *(iii)* the relationships between these two behaviors. (Traditionally, axioms about *(i)* have been called "physical.") The primitives that appear in these axioms are observable to the experimenter. (In Section 7.7, it will be important to separate out the above subforms *(i), (ii),* and *(iii)* of behavioral axioms. However, for the purposes of this section there is no need to do this.)

Inner psychological assumptions. These are axioms about the mental activity of the subject, often involving subjective experience. Because these axioms are formulated in terms of relationships that are not observable to the experimenter, they should be considered theoretical.

Psychobehavioral assumptions These are axioms that link behavioral objects and relationships with inner psychological ones. Such linkages are necessarily theoretical and are non-observable to the experimenter.

5.8.3 Behavioral Axiomatization

Behavioral Assumptions

Axiom 5.8.1 $\langle X, \succeq \rangle$ is a continuum.

In Axiom 5.8.1, $\succeq$ is intended to be a total ordering of the stimulus set selected by the experimenter.

The assumption that $\langle X, \succeq \rangle$ is a continuum can be weakened so that the results presented will generalize. However, to achieve such generalizations more complicated axiomatic systems would be needed.

Axiom 5.8.2 The following five statements are true:

1. $E \subseteq \{(x, p, t) \mid x \in X,\ p \in \mathbb{I}^+, \text{ and } t \in X\}$.
2. For all *(x, p, t)* in *E*, $x \succeq t$.
3. For all *t* in *X*, *(t*, 1, *t)* is in *E*.
4. For all *x* and *t* in *X* and all *p* in $\mathbb{I}^+$ there exist exactly one *z* in *X* and exactly one *s* in *X* such that

$$(z, p, t) \in E \text{ and } (x, p, s) \in E.$$

5. For all *x, y, t,* and *s* in *X*, if *(x, p, t)* $\in$ *E*, and *(y, p, s)* $\in$ *E*, then

$$x \succeq y \text{ iff } t \succeq s.$$

Statements 1 to 5 of Axiom 5.8.2 are straightforward. Subsection 5.8.7 considers a more general situation in which the numerals need not correspond to integers.

Axiom 5.8.3 The following three statements are true:

1. For all *(x, p, t)* and *(y, q, t)* in *E*,

$$x \succ y \text{ iff } p > q.$$

2. For all *x* and *t* in *X*, if $x \succ t$, then there exist *y* in *X* and *p* in $\mathbb{I}^+$ such that $y \succ x$ and $(y, p, t) \in E$.
3. For all *x* and *t* in *X*, if $x \succ t$, then there exist *y* and *z* in *X* and *p* in $\mathbb{I}^+$ such that

$$(y, p+1, t) \in E\,,\ (y, p, z) \in E\,,\ \text{and } x \succ z \succ t.$$

Axiom 5.8.3 describes natural conditions for a ratio magnitude estimation paradigm. Statement 1 provides the linkage of the usual ordering on numbers, and consequently the usual ordering on numerals to the experimenter determined ordering on stimuli. Statement 2 is an “Archimedean axiom” which guarantees that no element of *X* is “infinitely large” in terms of magnitude estimation with respect to another element of *X*. Statement 3 is also an “Archimedean axiom” that essentially says that no two distinct elements are “infinitesimally close” in terms of magnitude estimation.

Axiom 5.8.4 For all *p* and *q* in $\mathbb{I}^+$ and all *x, y, z, t,* and *w* in *X*, if *(x, p, t)* $\in$ ***E*, *(z, q, x)*** $\in$ ***E*, *(y, q, t)*** $\in$ *E*, and ***(w, p**, y)* $\in$ *E*, then *z*=*w*.

Let ***q•p*** stand for first estimating *p* times a stimulus *t* and then *q* times that estimated stimulus. Then Axiom 5.8.4 says that ***q•p*=*p•q*.**

Definition 5.8.2 Assume the behavioral assumptions Axioms 5.8.1 to 5.8.4. For each *p* in $\mathbb{I}^+$ define the binary relation $\overline{p}$ on *X* as follows: For all *x* and *t* in *X*,

$$x = \overline{p}(t) \text{ iff } (x, p, t) \in E.$$

$\overline{p}$ is the *behavioral interpretation* of $\boldsymbol{p}$.

Because $\overline{p}$ is defined entirely in terms of behavioral concepts, it is a behavioral concept.

It easily follows from Axioms 5.8.1 to 5.8.4 that for each p in $\mathbb{I}^+$, $\overline{p}$ is a function on X.

Definition 5.8.3 Let

$$\mathfrak{B} = \langle X, \succeq, \overline{1}, \ldots, \overline{p}, \ldots \rangle_{p \in \mathbb{I}^+} .$$

By definition, $\mathfrak{B}$ is called the *behavioral structure (associated with E).*

Convention 5.8.2 Throughout the rest of this section, $\mathfrak{B}$ will denote the behavioral structure associated with E.

Note that each primitive of $\mathfrak{B}$ is a behavioral concept.

Definition 5.8.4 φ is said to be a *multiplicative representing function* for $\mathfrak{B}$ if and only if φ is a function from X into $\mathbb{R}^+$ such that for each $p \in \mathbb{I}^+$ $\varphi(\overline{p})$ is a function that is multiplication by a positive real.

A scale S on X is said to be a *multiplicative scale* for $\mathfrak{B}$ if and only if each element of S is a multiplicative representing function for $\mathfrak{B}$.

Theorem 5.8.2 Assume Axioms 5.8.1 to 5.8.4. Then the following two statements are true:

1. There exists a numerical structure $\mathfrak{N}$ such that the set of isomorphisms of $\mathfrak{B}$ onto $\mathfrak{N}$ is both a ratio scale and a multiplicative scale for $\mathfrak{B}$.
2. Suppose S is a ratio scale of isomorphisms of $\mathfrak{B}$ onto a numerical structure. Then S is a multiplicative scale for $\mathfrak{B}$.

Proof. Narens (1996).

5.8.4 Cognitive Axiomatization

Narens (1996) makes the following comments about the behavioral axiomatization:

> Some researchers are content to deal with only behavioral issues. For these, there is no need to go beyond behavioral primitives and behavioral assumptions.
>
> Others are interested in the interplay between cognition and behavior. For this case one needs to include additional pyschobehavioral and psychological primitives and assumptions. With the addition of these primitives one has the ability to formulate clearly conditions for measurements of the behavioral structure $\mathfrak{B}$ to translate into measurements of a cognitive structure based on sensations.

He then continues with the following analysis of what he believes is minimally needed for an adequate cognitive theory of magnitude estimation:

> For magnitude estimation, the obvious cognitive question is "How is the subject producing his or her responses in the magnitude estimation paradigm?"...[A] *minimal theory* (Axioms 5.8.5 to 5.8.8 below) is presented for answering this. ("Minimal" is meant to convey here the author's belief that any plausible cognitive theory at the same level of idealization as the minimal theory and designed to answer the question will imply the minimal theory.)
>
> The minimal theory is based on the idea that the responses of the subject correspond to inner psychological functions that are computed by the subject from an "inner psychological measurement structure." The exact form of the "computation" and the specific primitives that make up the "inner psychological measurement structure" are not given; only their most general features are specified. (This is what gives the "minimalness" to the theory.)
>
> Magnitude estimation is usually not the primary goal of empirical studies: It is generally used as an instrument to investigate a substantive domain of interest. In such situations, the choice between behavioral and cognitive scales will depend on the particular objectives of the research.
>
> Although the minimal theory together with Axioms 5.8.1 to 5.8.4 force a strong relationship between the behavioral and cognitive scales, they do not force these scales to have identical measurement properties: By Theorem 5.8.2, X is measured behaviorally by a ratio scale of isomorphisms of $\mathfrak{B}$ [onto a numerical structure]; whereas by Theorem 5.8.3 below, the inner psychological measurement structure upon which the magnitude estimations depend is measured by a scale that is a subscale of a ratio scale. Theorem 5.8.3 shows that a necessary and sufficient condition for the this inner psychological measurement structure to be measurable by a ratio scale is that it be homogeneous. With the additional assumption of the homogeneity, the behavioral and cognitive scales resulting from magnitude estimation are, for practical purposes, identical (Theorem 5.8.4 below).

Psychobehavioral Assumptions

Axiom 5.8.5 Ψ is a function from X into the set of the subject's sensations.

Technically, the use of "sensation" to describe mental impressions of stimuli in X may not be appropriate for some magnitude estimation situations, as for example in the magnitude estimation of the seriousness of crime. For such cases, other concepts can be substituted for "sensation" without affecting the theory or results presented here.

Axiom 5.8.6 For all x and y in X,

$$x \succ y \text{ iff } \Psi(x) \succ_{\psi} \Psi(y)\,,$$

where $\succsim_{\psi}$ is the inner psychological intensity ordering on the set of sensations $\Psi(X)$ described next in Axiom 5.8.7.

Inner Psychological Assumptions

Axiom 5.8.7 The subject has an inner psychological structure

$$\mathfrak{J} = \langle \Psi(X), \succsim_{\psi}, R_1, \ldots, R_j, \ldots \rangle_{j \in J}$$

for "measuring" the intensity of sensations in $\Psi(X)$.

It follows immediately from Axioms 5.8.1 and 5.8.6 that $\succsim_{\psi}$ is a total ordering and Ψ is a one-to-one function.

In the inner psychological structure $\mathfrak{J}$, the primitives $\Psi(X)$, $\succsim_{\psi}$ and R_j, $j \in J$, are considered to be inner psychologically meaningful. Thus in particular, the primitive $\succsim_{\Psi}$ is an inner psychological meaningful ordering of subjective intensities.

Except for the domain $\Psi(X)$ and the primitive $\succsim_{\psi}$, which is linked to the behavioral ordering $\succsim$ through the psychobehavioral Axiom 5.8.6, other *individual* primitives of $\mathfrak{J}$ will not be explicitly mentioned in the axioms of this subsection. However, the *structure* $\mathfrak{J}$ of primitives will play an important role in various assumptions throughout the section, for example, *(i)* in hypotheses which in later theorems assert $\mathfrak{J}$ is homogeneous, and *(ii)* in Axiom 5.8.8 below.

Definition 5.8.5 For each p in $\mathbb{I}^{+}$, define $\hat{p}$ from $\Psi(X)$ into $\Psi(X)$ as follows: For each x and t in X,

$$\hat{p}[\Psi(t)] = \Psi(x) \text{ iff } (x, p, t) \in E\,.$$

It follows from the previous axioms that $\hat{p}$ is a function.

Because the expression "$x = \hat{p}(t)$" is a natural inner psychological correlate of the expression $(x, p, t) \in E$, the function $\hat{P}$ is taken to be an inner psychological interpretation of the numeral p.

The next axiom says that the function $\hat{p}$ is inner psychologically meaningful. Narens (1996) gives the following intuitive justification for this:

> Intuitively, this is how meaningfulness enters in the present context: Subjective magnitude is captured by the inner psychological structure $\mathfrak{J}$. We do not know much about $\mathfrak{J}$ except that its domain consists of sensations of stimuli, and we believe that $\mathfrak{J}$ has a primitive ordering of sensations corresponding to "subjective intensity." We assume that the subject's magnitude estimations involve the structure $\mathfrak{J}$—that is, the subject somehow performs a calculation or evaluation involving $\mathfrak{J}$ to produce his or her responses to trials in the magnitude estimation experiment. We

> assume that the subject does this in a way that gives a constant meaning to each numeral ***p;*** i.e., *it is assumed that the interpretation that the subject gives to the numeral p is calculated or defined in terms of the primitives of* $\mathfrak{I}$. Of course, something needs to be said about the subjective methods of calculation or definition of the numeral ***p***. They are inner psychological, and it is natural, therefore, to suspect that they would have special properties reflecting that they are products of mental activity. Nevertheless, without knowing the details of these properties, it is reasonable to believe they can be captured formally in terms of the extremely powerful logical languages (which among other things contain the equivalents of all known mathematics), and therefore [by Theorem 4.2.10] that *these inner psychological methods of calculation or definition are invariant under the automorphisms of* $\mathfrak{I}$. These intuitive considerations are summarized in [Axiom 5.8.8].

Axiom 5.8.8 For each p in $\mathbb{I}^+$, $\hat{p}$ is set-theoretically definable (Definition 5.1.8) in terms of primitives of $\mathfrak{I}$.

Note that in Axiom 5.8.8, $\hat{p}$ may have many definitions in terms of formulae of set theory, the primitives of $\mathfrak{I}$, and pure sets. Many of these definitions are unreasonable as *inner psychological* definitions of $\hat{p}$ Axiom 5.8.8 asserts that only at least one of them is appropriately inner psychological.

Consider the axiom system consisting of Axioms 5.8.1 to 5.8.8. In this system, most of the mathematical structure about magnitude estimation is contained in the behavioral axioms 5.8.1 to 5.8.4, often as testable hypotheses. The mathematical content in the remaining axioms is very minimal and appear to me, via the intuitive argument previously cited, to be necessary for theories of ratio magnitude estimation of mental phenomena.

Definition 5.8.6 φ is said to be a *multiplicative representing function* for $\mathfrak{I}$ if and only if φ is a function from $\Psi(X)$ into $\mathbb{R}^+$ such that for each $p \in \mathbb{I}^+$, $\varphi(\hat{p})$ is a function that is multiplication by a positive real.

A scale S on $\Psi(X)$ is said to be a *multiplicative scale* for $\mathfrak{I}$ if and only if each element of S is a multiplicative representing function for $\mathfrak{I}$.

The following theorem is a consequence of Axioms 5.8.1 to 5.8.8:

Theorem 5.8.3 Assume Axioms 5.8.1 to 5.8.8. Then there exists a numerical $\mathfrak{N}$ structure with $\mathbb{R}^+$ domain such that the following three statements are true:

1. The set S of isomorphisms of the inner psychological structure $\mathfrak{I}$ onto $\mathfrak{N}$ is a subscale of a ratio scale.
2. S (as defined in Statement 1) is a multiplicative scale for $\mathfrak{I}$.
3. If $\mathfrak{I}$ is homogeneous, then the following two statements are true:

(i) $\mathcal{S}$ (as defined in Statement 1) is a ratio scale.

(ii) Let t be an arbitrary element of X, and by *(i)* let φ be the unique element of $\mathcal{S}$ such that $\varphi(\Psi(t)) = 1$ and by Statement 2, for each p in $\mathbb{I}^+$, let c_p be the positive real such that multiplication by c_p is $\varphi(\hat{p})$. Then for all p in $\mathbb{I}^+$ and x in X,

$$(x, p, t) \in E \text{ iff } \varphi(\Psi(x)) = c_p \,.$$

Proof. Narens (1996).

Assume Axioms 5.8.1 to 5.8.8. By Theorem 5.8.2, let $\mathcal{S}$ be a ratio scale of isomorphisms of $\mathfrak{B}$ onto a numerical structure. Then $\mathcal{S}$ measures the *data* in E behaviorally. Although $\mathcal{S}$ only measures in terms of the subject's observable behavior, it can be used to define a closely related scale $\mathcal{S}'$ called the *derived sensory scale* $\mathcal{S}'$ that measures intensity of sensations of stimuli of X: Let

$$\mathcal{S}' = \{\varphi' \,|\, \varphi \in \mathcal{S} \text{ and for each } x \text{ in } X,\ \varphi'(\Psi(x)) = \varphi(x)\}\,.$$

Although $\mathcal{S}'$ is a scale on sensations that is consistent with the subject's behavior, it may be inappropriate for measuring subjective intensity. One reason is that the qualitative structure for subjective intensity—the structure $\mathfrak{J}$ in Axiom 5.8.7—may not be homogeneous, and $\mathcal{S}'$ like $\mathcal{S}$, is a ratio scale and therefore is homogeneous. Thus at least the homogeneity of $\mathfrak{J}$ is needed to make $\mathcal{S}'$ coordinate with a ratio scale for measuring the intensity of the subject's sensations. The following theorem shows that with the additional assumption of homogeneity a strong relationship obtains between behaviorally based scales and inner psychologically based ones.

Theorem 5.8.4 Assume Axioms 5.8.1 to 5.8.8 and that $\mathfrak{J}$ is homogeneous. Then *(i)* for each scale $\mathcal{S}$ of isomorphisms of $\mathfrak{B}$ onto a numerical structure, its derived sensory scale $\mathcal{S}'$ (discussed above) is a scale of isomorphisms of $\mathfrak{J}$ onto a numerical structure, and *(ii)* for each scale $\mathcal{T}$ of isomorphisms of $\mathfrak{J}$ onto a numerical structure, there exists a scale $\mathcal{U}$ of isomorphisms of $\mathfrak{B}$ onto a numerical structure such that $\mathcal{U}' = \mathcal{T}$ and $\mathcal{U}'$ is the derived sensory scale for $\mathcal{U}$.

Proof. Narens (1996).

The principal difference between measuring the behavioral structure $\mathfrak{B}$ through scales based Axioms 5.8.1 to 5.8.4 and measuring the inner psychological structure $\mathfrak{J}$ through scales based on Axioms 5.8.1 to 5.8.8 is the choice of primitives: In $\mathfrak{B}$ the primitives are

$$X, \succsim, \overline{1}, \ldots, \overline{p}, \ldots\,,$$

whereas in $\mathfrak{J}$ the primitives are

$$\Psi(X),\ \succsim_\Psi,\ R_1, \ldots, R_j, \ldots\,.$$

Thus the inner psychological structure $\mathfrak{B}'$ that is coordinate to $\mathfrak{B}$ has the form

$$\mathfrak{B}' = \langle \Psi(X), \succsim_{\psi}, \hat{1}, \ldots, \hat{p}, \ldots \rangle_{p \in \mathbb{I}^+} .$$

It easily follows from Axioms 5.8.7 and 5.8.6 that $\mathfrak{B}$ and $\mathfrak{B}'$ are isomorphic. However, it should be noted that the primitives of $\mathfrak{B}'$ are not the inner psychological primitives for subjective sensitivity (i.e., are not the primitives of $\mathfrak{J}$). They are only definable from the primitives of $\mathfrak{J}$ (Axiom 5.8.8). Because of this, $\mathfrak{B}'$ (and therefore $\mathfrak{B}$) may have a different scale type than $\mathfrak{J}$ and thereby be inappropriate for measuring *subjective* intensity.

5.8.5 Additive Scales

Definition 5.8.7 φ is said to be an *additive representing function* for $\mathfrak{J}$ if and only if φ is a function from $\Psi(X)$ into $\mathbb{R}$ such that for each $p \in \mathbb{I}^+$, $\varphi(\hat{p})$ is a function that is an addition by a nonnegative real.

A scale S on $\Psi(X)$ is said to be an *additive scale* for $\mathfrak{J}$ if and only if each element of S is an additive representing function for $\mathfrak{J}$.

Theorem 5.8.5 Assume Axioms 5.8.1 to 5.8.8. Then the following two statements are true:

1. There exists an additive scale of isomorphisms of $\mathfrak{J}$ onto a numerical structure.
2. If $\mathfrak{J}$ is homogeneous, then there exists a translation scale (Definition 2.3.10) of isomorphisms of $\mathfrak{J}$ onto a numerical structure that is an additive scale for $\mathfrak{J}$.

Proof. Transform the structure $\mathfrak{N}$ in Theorem 5.8.3 by the function $r \rightarrow \log(r)$.

It is interesting to inquire what happens when a subject engages in *different kinds* of magnitude estimation tasks on the *same* set of stimuli. For example, suppose for one task the subject is instructed to estimate ratios and in another instructed to estimate differences, and suppose Axioms 5.8.1 to 5.8.8 hold for both tasks. Then by Theorem 5.8.2 the subject's data can be measured *separately* by multiplicative scales of the behavioral structures associated with the data sets from the two tasks. If in addition it is assumed that the *same homogeneous* inner psychological structure is used by the subject to "compute" his or her responses in both tasks, then it is a consequence of Theorem 5.8.6 below that a scale for X exists that is *simultaneously* a multiplicative scale of the data collected in the first task and of the data collected in the second.

Although the representational theory of measurement does not justify the selection of one scale of isomorphisms used to measure a structure over another, it can still make *relative* distinctions between various scales of isomorphisms of *different* structures with the *same* domain. For example, in the above situation with two magnitude estimation tasks, suppose it turned out to be the case that a multiplicative scale of isomorphisms of the the behavioral structure associated with the first task is also an additive scale of isomorphisms of the behavioral structure associated with the second task. (Such examples occur in the magnitude estimation literature. See the quotation of Torgerson below.) For such examples it would then follow by the previous discussion and Axioms 5.8.1 to 5.8.8 that relative distinctions between the multiplicative and additive scale types can be made only when either the subject has *different* inner psychological measurement structures for each task or the subject has a *nonhomogeneous* inner psychological measurement structure for both tasks.

The following theorem is the technical result used in reaching the above conclusions:

Theorem 5.8.6 Suppose for Task 1 *(i)* the subject has been instructed to "Find a stimulus in X which appears to be p times greater in intensity than the stimulus t," *(ii)* E is the subject's responses to this task, *(iii)* Axioms 5.8.1 to 5.8.8 hold and the inner psychological structure $\mathfrak{J}$ is homogeneous, and *(iv)* $\mathcal{S}$ is a multiplicative scale of isomorphisms of $\mathfrak{J}$. (The existence of $\mathcal{S}$ follows from *(iii)* and Theorem 5.8.3.)

Also suppose in Task 2 that different instructions are given to the subject, for example, "Find the stimulus which in your subjective valuation is q+ the valuation of the stimulus t," and as a result of these instructions the subject produces the partial data set H where elements of H have the form (x, g, t), where q is a fixed positive integer, t ranges over the elements of X, and (1) for each t in X there exists exactly one x in X such that $(x, q, t) \in H$, (2) for all (x, g, t) in H, $x \succsim t$, and (3) for all x, y, t, and v in X, if (x, q, t) and (y, g, v) are in H and $t > v$, then $x > y$. Let $\hat{q}$ be the following function on $\Psi(X)$: For all x and t in X,

$$\hat{q}(\Psi(t)) = \Psi(x) \text{ iff } (x, q, t) \in H .$$

Assume $\hat{q}$ is meaningful with respect to $\mathfrak{J}$. Then there exists a positive real r such that for all φ in $\mathcal{S}$, $\varphi(\hat{q})$ = the function that is multiplication by r.

Proof. Narens (1996).

Theorem 5.8.6 provides a theoretical basis for the following empirical findings discussed in Torgerson (1961):

> The situation turns out to be much the same in the quantitative judgment domain. Again, we have both distance methods, where the subject is instructed to judge subjective differences between stimuli, and ratio methods, where the subject is instructed to judge subjective ratios. Equisection and equal appearing intervals are examples of distance methods. Fractionation and magnitude estimation are examples of ratio methods.
>
> In both classes of methods, the subject is suppose to tell us directly what the differences and ratios are. We thus have the possibility of

settling things once and for all. Judgments of differences take care of the requirements of the addition commutative group. Judgments of ratios take care of the multiplication commutative group. All we need to show is that the two scales combine in the manner required by the number system. This amounts to showing that scales based on direct judgments of subjective differences are linearly related to those based on subjective ratios.

Unfortunately, they are not. While both procedures are subject to internal consistency checks, and both often fit their own data, the two scales are not linearly related. But when we plot the logarithm of the ratio scale against the difference scale spaced off in arithmetic units, we usually do get something very close to a straight line. Thus, according to the subject's own judgments, stimuli separated by equal subjective intervals are also separated by approximately equal subjective ratios.

This result suggests that the subject perceives or appreciates but a single quantitative relation between a pair of stimuli. This relation to begin with is neither a quantitative ratio or difference, of course—ratios and differences belong only to the formal number system. It appears that the subject simply interprets this single relation in whatever way the experimenter requires. When the experimenter tells him to equate differences or to rate on an equal interval scale, he interprets the relation as a distance. When he is told to assign numbers according to subjective ratios, he interprets the same relation as a ratio. Experiments on context and anchoring show that he is also able to compromise between the two. *(pp. 202–203)*

5.8.6 Numeral Multiplicative Scales

Definition 5.8.8 φ is said to be a *numeral* multiplicative representing function for $\mathfrak{I}$ (respectively, $\mathfrak{B}$) if and only if it is a multiplicative representing function for $\mathfrak{I}$ (respectively, $\mathfrak{B}$) and for each $p \in \mathbb{I}^+$, $\varphi(\hat{p})$ is the function that is multiplication by p.

A scale $\mathcal{S}$ on $\Psi(X)$ is said to be a *numeral* multiplicative scale for $\mathfrak{I}$ (respectively, $\mathfrak{B}$), if and only if it is a multiplicative scale for $\mathfrak{I}$ (respectively, $\mathfrak{B}$) such that each of its elements is a numeral multiplicative representing function for $\mathfrak{I}$ (respectively, $\mathfrak{B}$).

The following theorem is an immediate consequence of Definition 5.8.8.

Theorem 5.8.7 Suppose φ is a numeral multiplicative representing function for $\mathfrak{I}$ (respectively, $\mathfrak{B}$) and $r \in \mathbb{R}^+$. Then $r\varphi$ is a numeral multiplicative representing function for $\mathfrak{I}$ (respectively, $\mathfrak{B}$).

It is easy to show, by using Theorem 5.8.7, that numerical multiplicative scales are the variety of scales that follow from Stevens' method of magnitude estimation together with Stevens' Assumptions.

The following *behavioral* axiom is important for establishing the existence of numeral multiplicative scales.

Axiom 5.8.9 For all *p, g,* and *r* $\mathbb{I}^+$ in and all *t, x* and *z* in *X,* if $(x, p, t) \in E$, $(z, q, x) \in E$, and $r=qp$, then $(z, r, t) \in E$.

Note that Axiom 5.8.9 implies Axiom 5.8.4.

Also note that Axiom 5.8.9 is the multiplicative property, which at the beginning of this section was shown to be a consequence of Stevens' Assumptions.

Axiom 5.8.9 is a very stringent condition that most behavioral scientists would expect to fail in behavioral experiments designed to test it.

Theorem 5.8.8 Assume Axioms 5.8.1 to 5.8.4. Then the following two statements are logically equivalent:

1. Axiom 5.8.9.
2. There exists a numerical representing structure

$$\mathfrak{N} = \langle \mathbb{R}^+, \geq, T_1, \ldots, T_i, \ldots \rangle_{i \in \mathbb{I}^+}$$

such that the set $\mathcal{S}$ of isomorphisms of the behavioral structure $\mathfrak{B}$ onto $\mathfrak{N}$ is a ratio scale and is a numeral multiplicative scale for $\mathfrak{B}$ (Definition 5.8.8).

Proof. Narens (1996).

The following is the "cognitive version" of Theorem 5.8.8:

Theorem 5.8.9 Assume Axioms 5.8.1 to 5.8.8. Then the following two statements are logically equivalent:

1. Axiom 5.8.9.
2. There exists a numerical representing structure

$$\mathfrak{N} = \langle \mathbb{R}^+, \geq, S_1, \ldots, S_j, \ldots \rangle_{j \in J}$$

such that the set $\mathcal{S}$ of isomorphisms of the inner psychological structure $\mathfrak{I}$ onto $\mathfrak{N}$ is a numeral multiplicative scale for $\mathfrak{I}$ (Definition 5.8.8).

Proof. Narens (1996).

Assume Axioms 5.8.1 to 5.8.9. Then it is easy to show that the scale $\mathcal{S}$ in Statement 2 of Theorem 5.8.9 is a ratio scale if and only if $\mathfrak{I}$ is homogeneous.

5.8.7 Magnitude Estimation with Generalized Numerals

The previous results about ratio magnitude estimation generalize to other sets of numerals.

Let $\mathbb{N}$be a nonempty set, and, with a mild abuse of notation, let $\geq$ be a binary relation on $\mathbb{N}$. Elements $\mathbf{p}$, $p \in \mathbb{N}$ are called *generalized numerals*.

The following are the two most important cases of generalized numerals:

(1) $\mathbb{I}^+ \subseteq \mathbb{N} \subseteq \{r \mid r \in \mathbb{R}^+ \text{ and } r \geq 1\}$, and
(2) $\langle C, \succeq \rangle$ is a physical continuum, $e \in C$, and $\mathbb{N} = \{a \mid a \in C \text{ and } a \succeq e\}$.

(1) generalizes ratio magnitude estimation to situations with non-integral numerals, and (2) applies to situations with generalized numerals based on physical stimuli can be used, for example, as in experiments where *(x, p, t)* stands for the pressure *p* that results when the subject squeezes a ball to display how much he or she believe that a crime *x* is more serious than a crime *t*.

The axioms for generalized magnitude estimation are the same as Axioms 5.8.1 to 5.8.8 with the following exceptions:

1. $\mathbb{N}$ is substituted throughout for $\mathbb{I}^+$.
2. $\geq$ is assumed to be a total ordering on $\mathbb{N}$ with a least element *e* and no greatest element.
3. Statement 3 of Axiom 5.8.2, which states,

   For all *t* in *X*, *(t,* 1, *t)* is in *E*,

   is replaced by,

   For all t in *X*, *(t, e, t)* is in *E*.
4. Statement 3 of Axiom 5.8.3, which states,

   For all *x* and *t* in *X*, if $x \succ t$, then there exist *y* and *z* in *X* and *p* in $\mathbb{I}^+$ such that

   $$(y, \mathbf{p}+\mathbf{1}, t) \in E\ ,\ (y, \mathbf{p}, z) \in E\ ,\ \text{and } x \succ z \succ t\,,$$

   is replaced by,

   For all *x* and *t* in *X*, if $x \succ t$, then there exist *y* and *z* in *X* and *p* and *q* in $\mathbb{N}$ such that

   $$q > p\ ,\ (y, q, t) \in E\ ,\ (y, p, z) \in E\ ,\ \text{and } x \succ z \succ t\,.$$

Narens (1996) makes the following comment about the above axiomatization:

> It is an easy (but somewhat tedious) matter to verify that all the above consequences of the behavioral axiomatization consisting of Axioms 5.8.1 to 5.8.4 and the cognitive axiomatization consisting of Axioms 5.8.1 to 5.8.8, when appropriately reformulated using the above conventions, are consequences of the corresponding axiom systems made up from the above axioms for generalized magnitude estimation.

5.9 Weber's Law

The physiologist Ernst Weber (1795–1878) conducted experiments that showed that all of the senses, whose physical stimuli could be measured precisely on a one-dimensional physical scale, obeyed a uniform law. This law, which is now called Weber's Law, has figured prominently in the history of psychology, and has been at the root of a number of long-lasting debates. This section presents a measurement-theoretic treatment of Weber's Law together with some meaningfulness issues generated by it. The presentation closely follows Narens (1994).

5.9.1 Weber Representations

Definition 5.9.1 Let $\mathfrak{X} = \langle X, \succsim_\star, T\rangle$ be a continuous threshold structure (Definition 5.4.2.) Then φ is said to be a *Weber representation* of $\mathfrak{X}$ if and only if for all x and y in X,

(i) $x \succsim_\star y$ iff $\varphi(x) \geq \varphi(y)$, and
(ii) $\varphi[T(x)] = k \cdot \varphi(x)$ where $k > 1$.

The real number k in *(ii)* is called the *modified Weber constant,* and *(ii)* is called the *modified Weber formula*. These are related to the (usual) *Weber constant c* and *Weber formula* by the following:

$$k = 1 + c \text{ and } \frac{\varphi[T(x)] - \varphi(x)}{\varphi(x)} = c\,.$$

Convention 5.9.1 Throughout the rest of this subsection, let

$$\mathfrak{N} = \langle \mathbb{R}, \geq, S\rangle\,,$$

where S is the function $x+1$ on $\mathbb{R}$ and for each $k>1$ let

$$\mathfrak{N}_k = \langle \mathbb{R}^+, \geq, S_k\rangle\,,$$

where S_k is defined as follows: For all x in $\mathbb{R}^+$,

$$S_k(x) = k \cdot x\,.$$

In this subsection, $\mathfrak{N}$ and $\mathfrak{N}_k$ roles will be numerical representing structures for measurements of continuous threshold structures.

Theorem 5.9.1 For each $k>1$, let φ_k be the function from $\mathbb{R}$ onto $\mathbb{R}^+$ that is defined by

$$\varphi_k(x) = k^x\,.$$

Then the following two statements are true for each $k>1$:

1. φ_k is an isomorphism of $\mathfrak{N}$ onto $\mathfrak{N}_k$.

2. Let β be a pure translation of $\mathfrak{N}$ (Definition 5.4.4), that is, let r in $\mathbb{R}$ be such that $\beta(x)=x+r$. Then the image γ of β under the isomorphism φ_k is such that for all y in $\mathbb{R}^+$, $\gamma(y) = k^r \cdot y$.

Proof. Theorem 5.2 of Narens (1994).

Theorem 5.9.2 Suppose $\mathfrak{X}$ is a continuous threshold structure and $k>1$. Then $\mathfrak{X}$ has a Weber representation with modified Weber constant k.
Proof. Theorem 5.3 of Narens (1994).

Theorem 5.9.3 Let $\mathfrak{Y} = \langle X, \succsim_\star, R_j\rangle_{j\in J}$ be a scalar structure, and let $\mathfrak{X} = \langle X, \succsim_\star, T\rangle$ be a continuous threshold structure. By Theorem 5.4.9, let $\mathcal{S}$ be a ratio scale of isomorphisms of $\mathfrak{Y}$ onto $\mathfrak{D} = \langle \mathbb{R}^+, \geq, D_j\rangle_{j\in J}$. Then φ in $\mathcal{S}$ is a Weber representation for $\mathfrak{X}$ if and only if T is an automorphism of $\mathfrak{Y}$.
Proof. Theorem 5.4 of Narens (1994).

As a special case of Theorem 5.9.3, consider the situation where $\mathfrak{E} = \langle X, \succsim_\star, \oplus\rangle$ is a continuous extensive structure, $\mathcal{S}$ is a ratio scale of isomorphisms of $\mathfrak{E}$ onto $\langle \mathbb{R}^+, \geq, +\rangle$ and $\mathfrak{X} = \langle X, \succsim_\star, T\rangle$ is a continuous threshold structure. Then it easily follows that $\mathfrak{E}$ is homogeneous and 1-point unique. To make the example more concrete, first assume $\mathfrak{E}$ is a physical structure. Then assume that the ordering $\succsim_\star$ can also be determined by purely psychological means, that is, can be determined without reference to physical measurement. Finally assume that the threshold function T is determined by purely psychological means. Let $\varphi \in S$. Then by Theorem 5.9.3, the following two statements are logically equivalent:

1. For all x and y in X, $T(x \oplus y) = T(x) \oplus T(y)$.
2. There exists $k>1$ such that for all x in X, $\varphi[T(x)] = k \cdot \varphi(x)$.

Note that Statement 1 corresponds to possible experimental observations. Therefore, Statement 2 can be tested experimentally via the experimental testing of Statement 1; that is, there is a simple way of testing Statement 2 without resorting to "fitting curves to $\varphi[T(x)]$."

Theorem 5.9.2 implies that each continuous threshold structure has a modified Weber representation. Householder and Young (1940) mistakenly confused a very similar conclusion with Weber's Law. The following quotation of Narens (1994) explains the difference between the two concepts:

> Let $\mathfrak{X} = \langle X, \succsim_\star, T\rangle$ be a continuous threshold structure. To obtain a Weber representation of $\mathfrak{X}$ one finds an isomorphism of $\mathfrak{X}$ onto $\mathfrak{N}_k$ for some positive real k. There is obviously nothing "lawful" about this. To obtain a *Weber's Law* representation of $\mathfrak{X}$, one *first* obtains a function φ of a ratio scale of isomorphisms of another structure $\mathfrak{Y}$ which does not has T as a primitive relation, *and then* verify that φ is a Weber representation for $\mathfrak{X}$. What is "lawful" about the later is that φ is simultaneously an isomorphism of a ratio scale for $\mathfrak{Y}$ and a Weber representation for $\mathfrak{X}$. Qualitatively, this

> law reduces to saying that T is an automorphism of $\mathfrak{Y}$—an experimentally testable condition if T and the primitive relations of $\mathfrak{Y}$ are empirically determinable. *(pg. 318)*

5.9.2 Weber's Law and Meaningfulness

Definition 5.9.2 Let $\mathfrak{X} = \langle X, \succsim_\star, T\rangle$ be a continuous threshold structure. An automorphism β of $\mathfrak{X}$ is said to be $\mathfrak{X}$-*automorphism invariant* if and only if for all x in X and all automorphisms γ of $\mathfrak{X}$,

$$\gamma[\beta(x)] = \beta[\gamma(x)]\,,$$

that is, if and only if

$$\gamma * \beta = \beta * \gamma\,,$$

that is, if and only if

$$\gamma * \beta * \gamma^{-1} = \beta\,.$$

Theorem 5.9.4 Suppose $0<r<1$ and β is the pure translation of $\mathfrak{N}$ defined by

$$\beta(x) = x + r\,.$$

Then β is not $\mathfrak{N}$-automorphism invariant.

Proof. Theorem 6.1 of Narens (1994)

Let $\mathfrak{X} = \langle X, \succsim_\star, T\rangle$ be a continuous threshold structure and $\mathcal{S}$ be a scale of isomorphisms of $\mathfrak{X}$ onto $\mathfrak{N}_k$. Then it can be shown using Theorem 5.9.2 that for all φ and ψ in $\mathcal{S}$ and all x in X,

$$\varphi[T(x)] = k \cdot \varphi(x) \text{ and } \psi[T(x)] = k \cdot \psi(x)\,, \tag{5.10}$$

that is, all isomorphisms in $\mathcal{S}$ yield the same modified Weber constant k. Narens (1994) comments following about this:

> ...Some measurement theorists might want to use this result to say that "k is meaningful." I think this would be a error: This *by itself* is not enough to conclude that "k is meaningful;" it is only enough to conclude that the sentence
>
> $$\varphi[T(x)] = k \cdot \varphi(x)$$
>
> is a meaningful assertion. To properly conclude "k is meaningful," additional observations like the following are needed:
>
> > Multiplication by the constant k is an automorphism of $\mathfrak{N}_k$ and it is $\mathfrak{N}_k$-automorphism invariant since it is the threshold function S_k. Through the isomorphisms of $\mathcal{S}$, it has an interpretation in $\mathfrak{X}$ as the threshold function T.
>
> *(pg. 320)*

He then discusses the differences in meaningfulness between the modified Weber constant k and the Weber constant c:

Observations similar to the above do not hold for the Weber constant c: Consider the Weber formula

$$\varphi[T(x)] - \varphi(x) = c \cdot \varphi(x)\,, \tag{5.11}$$

where $\varphi \in$S and c=k—1. [Assume 0<c<1.] By Theorem 5.9.4 and the isomorphism of $\mathfrak{N}$ and $\mathfrak{N}_k$, multiplication by c is not $\mathfrak{N}_k$-automorphism invariant. By [Theorem 4.2.10], this means that the automorphism of $\mathfrak{X}$ that corresponds to multiplication by c *via* φ^{-1} is not definable in terms of the primitives X, $\succsim_\star$, and T, no matter how powerful a logical language is used.

Note that the statement in Equation 5.11 is meaningful in the sense that if ψ is any element of $\mathcal{S}$, then

$$\psi[T(x)] - \psi(x) = c \cdot \psi(x)\,.$$

The fact that this statement is "meaningful" does not mean that every part of it—for example, the constant c—has a proper interpretation in$\mathfrak{X}$.

Let $\mathfrak{E} = \langle X, \succsim_\star, \oplus\rangle$ be a continuous extensive structure and $\mathfrak{X} = \langle X, \succsim_\star, T\rangle$ be a continuous threshold structure [and T be an automorphism of $\mathfrak{E}$]. For this discussion, X will be considered as a set of physical objects as well as a set of psychological stimuli, $\succsim_\star$ will be considered as a physical relation on physical objects as well as a psychological relation on psychological stimuli, $\oplus$ will be considered as a physical operation on physical objects, and T will be considered as a psychological function on psychological stimuli. Thus $\mathfrak{E}$ characterizes a physical situation and $\mathfrak{X}$ characterizes a psychological situation. Let $\mathcal{S}$ be a ratio scale of isomorphisms of $\mathfrak{E}$ onto $\langle \mathbb{R}^+, \geq, +\rangle$. As discussed above, the modified Weber constant that results from measurement by $\mathcal{S}$ [always] has an interpretation in the psychological structure $\mathfrak{X}$, whereas the Weber constant [usually] has no such interpretation [e.g., when 0<c<1]. By [Theorem 4.3.6], both constants [always] have interpretations in the physical structure $\mathfrak{E}$. [Here "interpretability" means definable from the primitives of $\mathfrak{E}$ and pure sets through L($\in$, A, Ø).] *(pp. 320–321)*

5.10 Dimensional Analysis

5.10.1 Overview

Traditionally dimensional analysis has been a set of procedures used by physicists and engineers to discover solutions to some highly complex problems.

Some aspects of this subject were discussed in Section 1.5, where an elementary illustrative example was presented.

Most treatments of dimensional analysis are concerned with *applications* of the technique and do not attempt to provide serious *foundations* for the subject. Notable exceptions to this are Causey (1969), Chapter 10 of Krantz et al. (1971), Chapter 22 of Luce et al. (1991), and Dzhafarov (1995). The focus of this section is entirely on foundations; for substantive applications, the reader is referred to Sedov (1943, 1956).

There are four principal ideas that are used throughout dimensional analysis: the vector space-like structure of physical units, the ratio scalability of individual physical units, dimensional invariance, and the so-called "Π-Theorem." Because the fourth turns out to be a straightforward mathematical consequence of the first three, it is not discussed here, because it contributes nothing new to the foundations of the subject.

Subsection 5.10.2 provides brief descriptions of the vector space-like structure of physical units and dimensional analysis. These descriptions rest on the following three ideas: *(i)* there are certain fundamental physical qualities whose measurements form ratio scales; *(ii)* all other physical qualities have ratio scale measurements that are proportional to products of powers of fundamental ones; and *(iii)* physically significant functional relationships between physical units satisfy a mathematically specifiable condition called "dimensional invariance," which is defined in Subsection 5.10.2.

Subsection 5.10.3 provides a basis for a theory due to Narens and Luce (1976) for the ratio scalability of non-fundamental—that is, "derived"—physical qualities and a rationale for their representations being products of powers of representations of fundamental qualities. It also very briefly discusses a result of Luce (1978) which characterizes a qualitative correlate of "dimensional invariance."

Subsection 5.10.3 also provides a qualitative foundation for dimensional analysis in terms of particular observable primitive physical relations and operations. However, because the resulting quantitative theory of dimensional analysis is not formulated in terms of quantitative correlates of the primitive qualitative relations and operations, there is the possibility of generalizing dimensional analysis to situations whose qualitative formulations are very different than in physics. The feasibility of such a program is demonstrated in Subsections 5.10.4 and 5.10.5, where very general and highly abstract generalizations of the physical case are given. In particular, Subsection 5.10.5 approaches the foundations of dimensional analysis from a different point of view than has been pursued in the foundations of measurement, a point of view that strongly stresses the idea that the fundamental qualities can be measured independently of one another.

The emphasis of this section is to show, by the example of dimensional analysis, that the kinds of meaningfulness considerations of the previous chapter can be grounded in a qualitative theory so that issues about the interpretability of quantitative and qualitative statements within a scientific topic can be given a precise meaning.

Many of the concepts and results discussed in the section are outlined and/or sketched; proofs are usually omitted. Relevant descriptions and references, however, are provided so that the proofs can be constructed or found elsewhere.

5.10.2 Dimensional Analysis: Quantitative Theory

In physics the quantities of interest are products of powers of certain fundamental quantities. For example, energy, md^2t^{-2}, is the product of the fundamental quantity mass, m, and the square of the fundamental quantity distance, d, and the negative square of the fundamental quantity time, t. Such products are called "derived (physical) units," and correspondingly the fundamental quantities are often called "fundamental (physical) units." Although in practice, only relatively few physical units are employed, it is useful to consider all possible products of physical units so that a richer mathematical structure results. This richer mathematical structure strongly resembles a vector space, except that the "vector quantities"—that is, the units—multiply instead of add, and the "scalars"—that is, the positive reals—act on the units through the raising to powers instead of through multiplication. In this formulation, the fundamental units are also chosen so they act like a basis for this "multiplicative vector space," that is, they are "independent" of one another in the sense that no fundamental unit is proportional to a product of powers of others. In principle, any set of such "independent" units that "span" the space will do, but in fact the "basis" has historically been chosen in ways that the fundamental units are easy and practical to measure.

The physical units have another important property: each can be looked at as a measurement of a physical quality by a ratio scale. Put in more precise terms: *(i)* the proper representations of each fundamental quality form a ratio scale; *(ii)* fundamental qualities are represented independently of one another, for example, the representation chosen for mass—say the kilogram representation—is independent of the representation chosen for length—say the mile representation; and *(iii)* the proper representations of each nonfundamental quality are products of powers of the representations of fundamental qualities from which it is derived, and the set of these representations form a ratio scale. Some, for example, Krantz et al., 1971, have questioned the ratio scalability of derived physical qualities. This is a subtle measurement issue. It is argued in Subsection 5.10.6 that ratio scales are the proper scales for derived physical qualities.

The remaining important concept for the foundations of dimensional analysis is "dimensional invariance" (Definition 5.10.1 below). The proper application of this concept in dimensional analysis is often a complicated matter, because dimensional invariance requires that the relevant physical variables and constants that determine a lawful physical phenomenon must be completely specified. As a practical matter, this means that the researcher must have considerable insight into the phenomenon: he or she must know exactly the physical variables and constants that determine it, and know that the phenomenon is governed by a physically lawful functional relationship between one of the variables and the rest of the variables and constants. In dimensional analysis, "physically lawful" is taken to be that the functional relationship is "(quantitatively) dimensionally invariant." However, the reasons for taking "dimensionally invariant" as an adequate mathematical formulation of "physically lawful" is very opaque in the physical literature.

The complete specification of the relevant physical variables and constants is often tricky. For example, Bridgman writes the following about a defense by Lord Rayleigh of a particular application of dimensional analysis:

> This reply of Lord Rayleigh is, I think likely to leave us cold. Of course we do not question the ability of Lord Rayleigh to obtain the correct result by the use of dimensional analysis, but must we have the experience and physical intuition of Lord Rayleigh to obtain the correct result also? Might not perhaps a little examination of the logic of the method of dimensional analysis enable us to tell whether temperature and heat are "really" independent units or not, and what the proper way of choosing our fundamental units is?
>
> Besides the prime question of the proper number of units to chose in writing our dimensional formulas, this problem of heat transfer raises many others also of a physical nature. For instance, why are we justified in neglecting the density, or the viscosity, or the compressibility, or the thermal expansion of the liquid, or the absolute temperature? We will probably find ourselves able to justify the neglect of all these quantities, but the justification will involve real argument and a considerable physical experience with physical systems of the kind which we have been considering. The problem cannot be solved by the philosopher in his armchair, but the knowledge involved was gathered only by someone at some time soiling his hands with direct contact. *(Bridgman, 1931 pp. 11–12)*

There are several ways of formulating the concept of "dimensional invariance." The following is one often found in the literature. It assumes that physical (quantitative) variables and constants are measured in terms of physical units.

Definition 5.10.1 Let $x_1,\ldots,x_i,\ldots,x_{n-1}$ be physical variables or constants and let x_n be a physical variable. Then a function f of the form

$$x_n = f(x_1,\ldots,x_i,\ldots,x_{n-1})$$

is said to be *(quantitatively) dimensionally* invariant if and only if f is such that whenever the fundamental units undergo ratio transformations, transforming the quantities $x_1,\ldots, x_i,\ldots, x_n$ into $x'_1,\ldots,x'_i,\ldots,x'_n$ respectively, then

$$x'_n = f(x'_1, \ldots, x'_i, \ldots, x'_n) .$$

Note that Definition 5.10.1 is for the case where the law under consideration can be formulated in terms of a *single* function. Often laws can only be formulated in terms of a family of functions. I believe that the version of dimensional analysis provided here can be expanded to include such cases. (Dzhafarov, 1995, presents a quantitative theory for such families of functions.)

In Subsections 5.10.3, 5.10.4, and 5.10.5 foundations for physical dimensional analysis are given by providing precise, qualitative descriptions of its three main concepts—fundamental physical units, the multiplicative vector space of physical quantities, and dimensional invariance. A qualitative description of how to obtain a complete set of physical variables and constants that determine a phenomenon for an application of dimensional analysis will not be attempted, because I agree with the above quote from Bridgman that this is a matter of intuition and experience and not one of logic, and thus should not be considered as part of the formal foundations of the subject.

5.10.3 Distributive Triples with Associative Operations

Various kinds of cartesian products involving continuous extensive structures will be employed to obtain the qualitative equivalent of the multiplicative vector space of physical units. The basic idea is that the cartesian product and its components correspond to physical qualities. In natural ways, qualitative orderings and operations are put on the physical qualities so that they become continuous extensive structures. Additional qualitative conditions in terms of the operations and orderings can be stated so that the representations of the quality corresponding to the cartesian product are products of powers of the representations of the component qualities, which are themselves the ratio scale representations of the continuous extensive structures defined on the component qualities.

The just described qualitative approach to the structure of physical units was first undertaken by Causey (1969) and later improved upon by Krantz et al. (1971). (See Chapter 10 of Krantz et al. for an excellent and complete exposition of the foundations of dimensional analysis up to 1971.) Later Narens (1976) discovered, through the use of distributive operations discussed below, a simpler and more elegant way of obtaining a critical mathematical result that allows for a better qualitative characterization. Narens (1976) was concerned with qualitative descriptions of expected utility theory, and no connection was made between the mathematical results of that paper and dimensional analysis. The connection, however, was made in Narens and Luce (1976), which systematically investigated distributivity in a context appropriate to dimensional analysis—distributive triples discussed below—and concluded that it provided an appropriate foundation for dimensional analysis and gave rise to representing derived physical units as products of powers of fundamental units. Ramsay (1976) independently realized that distributive operations were a major driving force in the qualitative theory of dimensional analysis. He did not, however, provide numerical representation theorems. Luce (1978), using the results about distributive triples of Narens and Luce (1976), carried out a systematic construction of the qualitative correlate of the multiplicative vector space of physical quantities and showed that dimensional invariance corresponded qualitatively to a form of automorphism invariance. Luce noted that, because of this correspondence, dimensional invariance was formally a variant of the representational concept of meaningfulness.

Definition 5.10.2 Let $\langle X \times Y, \succsim \rangle$ be a solvable conjoint structure (Definition 5.4.9) and $\oplus$ be a binary operation. Then the following three definitions hold:

1. $\oplus$ is said to be X-*distributive* if and only if $\oplus$ is an operation on X and for all a, b, c, and d in X and all y and z in Y:

$$\text{if } ay \sim cz \text{ and } by \sim dz, \text{ then } (a \oplus b)y \sim (c \oplus d)z\,.$$

2. $\oplus$ is said to be Y-distributive if and only if $\oplus$ is an operation on Y and for all e, f, g, and h in Y and all w and x in X:

$$\text{if } we \sim xg \text{ and } wf \sim xh, \text{ then } w(e \oplus f) \sim x(g \oplus h)\,.$$

3. $\oplus$ is said to be *X×Y-distributive* if and only if $\oplus$ is an operation on $X \times Y$ and for all a, b, and c in X and all y and z in Y,

$$(ay) \oplus (by) \sim cy \text{ iff } (az) \oplus (bz) \sim cz\,.$$

"Distributive triples" are situations where two of the three dimensions, X, Y, and $X \times Y$, of a solvable conjoint structure $\langle X \times Y, \succsim \rangle$ have distributive operations on them. For the development in this section, it is assumed that these two distributive operations give rise to continuous extensive structures on their respective dimensions. By Theorem 5.4.7, it then follows that the two dimensions with these operations can be given ratio scale representations. This subsection shows that a particular product of powers of ratio scale representations of these dimensions induce ratio scale representations of the third dimension.

Definition 5.10.3 Let $\mathfrak{T} = \langle X \times Y, \succsim, \oplus_X, \oplus_Y \rangle$. Then $\mathfrak{T}$ is said to be an *X, Y-distributive triple* if and only if $\langle X \times Y, \succsim \rangle$ is a solvable conjoint structure and $\oplus_X$ and $\oplus_Y$ are respectively X-distributive and Y-distributive operations. $\mathfrak{T}$ is said to be a *continuous, extensive, X, Y-distributive triple* if and only if it is an X, Y-distributive triple and $\langle X, \succsim_X, \oplus_X \rangle$ and $\langle Y, \succsim_Y, \oplus_Y \rangle$ are continuous extensive structures.

Definition 5.10.4 Let $\mathfrak{T} = \langle X \times Y, \succsim, \oplus_X, \oplus_{X\times Y}\rangle$. Then $\mathfrak{T}$ is said to be an *X, X×Y-distributive triple* if and only if $\langle X \times Y, \succsim\rangle$ is a solvable conjoint structure and $\oplus_X$ and $\oplus_{X\times Y}$ are respectively *X*-distributive and *X×Y*-distributive operations. $\mathfrak{T}$ is said to be a *continuous, extensive, X, X×Y– distributive triple* if and only if it is an *X, X×Y*-distributive triple and $\langle X, \succsim_X, \oplus_X\rangle$ and $\langle X \times Y/\sim, \succsim/\sim, \oplus/\sim\rangle$ are continuous extensive structures, where $\succsim/\sim$ is the induced total ordering (Definition 5.1.2) on $X \times Y/\sim$ and $\oplus/\sim$ is the binary operation defined on $X \times Y/\sim$ such that for all *B, C,* and *D* in $X \times Y/\sim$,

> *B* $\oplus/\sim$ *C=D* if and only if there exist *xy* in *B, uv* in *C,* and *zw* in *D* such that $xy \oplus uv \sim zw$.

Definition 5.10.5 Let $\mathfrak{T} = \langle X\times Y, \succsim, \oplus_Y, \oplus_{X\times Y}\rangle$. The definition of $\mathfrak{T}$ being a *continuous, extensive, Y, X×Y-distributive triple* is given in the obvious manner similar to Definition 5.10.4

The following theorem is the core result of the mathematical theory of continuous, extensive distributive triples:

Theorem 5.10.1 Let $\mathfrak{T} = \langle X \times Y, \succsim, \oplus_X, \oplus_Y\rangle$ be a continuous, extensive *X, Y*-distributive triple, $\mathfrak{R} = \langle \mathbb{R}^+, \geq, +\rangle$ $\mathcal{S}_X$ and $\mathcal{S}_Y$ be respectively sets of representations of $\langle X, \succsim_X, \oplus_X\rangle$ onto $\mathfrak{R}$ and of $\langle Y, \succsim_Y, \oplus_Y\rangle$ onto $\mathfrak{R}$. Then

(i) $\mathcal{S}_X$ and $\mathcal{S}_Y$ are ratio scales, and
(ii) there exists exactly one positive real number *r* such that for all *xy* and *uv* in *X×Y,* and all φ in $\mathcal{S}_X$ and all ψ in $\mathcal{S}_Y$,

$$xy \succsim uv \text{ iff } \varphi(x)\cdot\psi(y)^r \geq \varphi(u)\cdot\psi(v)^r\,. \tag{5.12}$$

Proof. Theorem 3.5.5 of Narens, 1985. (See also Section 5 of Narens and Luce, 1976.)

The exponent *r* in Equation 5.12 can be qualitatively specified to any positive numerical value. To see this for the case of *r*=2, consider the following sequence of valid equations (some of which follow from Equation 5.12):

$$4\varphi(x) = [((\varphi(x) + \varphi(x)) + \varphi(x)) + \varphi(x)] = \varphi[((x \oplus_X x) \oplus_X x) \oplus_X x]\,,$$

$$2\psi(y) = \psi(y) + \psi(y) = \psi(y \oplus_Y y)\,,$$

and

$$\begin{aligned} r = 2 \text{ iff } & [4\varphi(x)]\psi^r(y) = \varphi(x)[4\psi^r(y)] \\ & = \varphi(x)[2\psi(y)]^r = \varphi(x)[\psi(y) \oplus_Y \psi(y)]^r \\ \text{iff } & [\varphi(((x \oplus_X x) \oplus_X x) \oplus_X x)]\psi^r(y) = \varphi(x)[\psi(y \oplus_Y y)]^r \\ \text{iff } & [((x \oplus_X x) \oplus_X x) \oplus_X x]y \sim x[y \oplus_Y y]\,. \end{aligned}$$

Thus the qualitative equation,

$$\forall x \forall y ([((x \oplus_X x) \oplus_X x) \oplus_X x]y \sim x[y \oplus_Y y])\,,$$

characterizes r=2 in Equation 5.10.1. Other positive numerical values of *r* can similarly be characterized. (See pp. 223–224 of Narens, 1985, or Section 5 of Narens and Luce, 1976.)

Theorem 5.10.1 can be used to obtain the following result about *Y*, *X*×*Y*-distributive triples:

Theorem 5.10.2 Let $\mathfrak{T} = \langle X \times Y, \succsim, \oplus_Y, \oplus_{X\times Y}\rangle$ be a continuous, extensive *Y*, *X*×*Y*-distributive triple, $\mathfrak{N} = \langle \mathbb{R}^+, \geq, +\rangle$,$\mathcal{S}_Y$ and $\mathcal{S}_{X\times Y}$ be respective sets of representations of $\langle Y, \succsim_Y, \oplus_Y\rangle$ onto $\mathfrak{N}$ and of $\langle X \times Y, \succsim /\!\sim, \oplus_{X\times Y}/\!\sim\rangle$ onto $\mathfrak{N}$. Then $\mathcal{S}_Y$ and $\mathcal{S}_{X\times Y}$ are ratio scales, and there exists exactly one positive real number *r* such that for all *xy* and *uv* in *X*×*Y*, and all φ in $\mathcal{S}_{X\times Y}$ and all ψ in $\mathcal{S}_Y$,

$$xy \succsim uv \text{ iff } \varphi(xy) \cdot \psi(y)^{-r} \geq \varphi(uv) \cdot \psi(v)^{-r}\,.$$

Proof. Section 5 of Chapter 3 of Narens (1985) or Section 5 of Narens and Luce (1976).

A result analogous to Theorem 5.10.2 follows similarly for continuous, extensive *X*, *X*×*Y*-triples.

The following outlines how distributive triples are employed to formulate a qualitative basis for physical dimensional analysis:

Continuous extensive distributive triples have three physical variables: two with continuous extensive structures on them, and the other variable. The process starts with the case where the variables with continuous extensive structures are fundamental physical variables and the other variable is non-fundamental-that is, is a "derived" physical variable. By Theorems 5.10.1 and 5.10.2, ratio scale measurements of the fundamental variables produce measurements of the derived variable which are products of powers of the measurements of the fundamental variables. As easily follows by results of Narens and Luce (1976), a continuous extensive structure $\mathfrak{Z}$ can be defined on the derived variable in terms of the primitives of the distributive triple, and an appropriate product of powers of the ratio scale representations of the fundamental variables will form a ratio scale $\mathcal{U}$ for the "derived continuous extensive structure" on the derived variable. It is shown that each element of $\mathcal{U}$ is of the form η^r, where η is an isomorphism of $\mathfrak{Z}$ onto $\langle \mathbb{R}^+, \geq, +\rangle$ and *r* is a positive real. By using this derived, continuous extensive structure as part of a distributive triple involving other similarly derived continuous extensive structures and/or fundamental continuous extensive structures, and by using Theorems 5.10.1 and 5.10.2 above, more complicated products of powers of measurements of fundamental extensive structures involving three or more fundamental qualities can be constructed. By this method the multiplicative vector space of physical quantities discussed at the beginning of Subsection 5.10.2 is given a qualitative description in terms of continuous extensive structures of qualitative physical qualities.

The construction described above is given in more detail in Luce (1978). That paper uses it to show that when the qualitative fundamental qualities are measured on ratio scales, the dimensional invariance of quantitative relations exactly correspond (through this measurement process) to invariance of relations on the fundamental qualities under the group of automorphisms of the structure of physical qualities. In this case, the latter takes the following form: the *n-ary* relation R is invariant in the quantitative manner described above if and only if whenever $R(a_1,\ldots, a_n)$ is true for elements $a_1,\ldots, a_n$ of fundamental qualities $Q_1,\ldots, Q_n$, then $R[\alpha_1(a_1),\ldots, \alpha_n(a_n)]$ is true for all automorphisms $\alpha_1,\ldots, \alpha_n$ respectively of the continuous extensive fundamental structures corresponding to $Q_1,\ldots, Q_n$. In this way Luce showed the convention of dimensional analysis of restricting the attention to only dimensionally invariant functions is a variant of the Erlanger Program's perspective about meaningfulness. In the terms of Chapter 4 this conclusion may be restated as follows: Dimensional analysis, through its use of dimensional invariance, assumes the axiom of Transformational Meaningfulness *(TM)* as an important part of its method.

The next two subsections consider extending dimensional analysis to much more general settings. Most of the ideas in the two subsections follow from Narens (1981a).

5.10.4 Generalized Distributive Triples

Definition 5.10.6 $\mathfrak{C} = \langle X \times Y, \succsim\rangle$ is said to be a *continuous conjoint structure* if and only if $\mathfrak{C}$ is a solvable conjoint structure and $\langle X, \succsim_X\rangle$ where $\succsim_X$ is the $\succsim$-induced ordering on X, is a continuum (Definition 5.4.1). It easily follows from the properties of a solvable conjoint structure that if $\mathfrak{C}$ is continuous, then $\langle Y, \succsim_Y\rangle$ is also a continuum and each nonempty $\succsim$-bounded subset of $X\times Y$ has a $\succsim$-least upper bound in $X\times Y$.

Let $\mathfrak{C} = \langle X \times Y, \succsim\rangle$ be a continuous conjoint structure. In Subsection 5.10.3, an approach to dimensional analysis was outlined based on the structure $\mathfrak{C}$ together with associative operations on two of the three sets, X, Y, and $X \times Y/\sim$. In that development, these sets and operations (with the appropriate orderings) became continuous extensive structures with the operations distributing in $\mathfrak{C}$—that is, formed continuous, extensive, distributive triples. In this subsection, generalizations of this approach due to Narens (1981a) and Luce (1987) are discussed. In addition, other qualitative methods are presented that can also be used as the basis for a qualitative treatment of dimensional analysis.

Narens (1981a) saw that generalizing distributive triples essentially rested upon the following idea, which for convenience will be stated for the X and Y components of a continuous conjoint structure $\mathfrak{C}$: There exists ratio scales $\mathcal{S}$ and $\mathcal{T}$ on X and Y respectively such that for all φ in $\mathcal{S}$ and all Ψ in $\mathcal{T}$, there exist functions τ from Y onto $\mathbb{R}^+$and σ from X onto $\mathbb{R}^+$ so that for all xy and uv in $X \times Y$,

$$xy \succsim uv \text{ iff } \varphi(x)\tau(y) \geq \varphi(u)\tau(v)\,, \tag{5.13}$$

and

$$xy \succsim uv \text{ iff } \sigma(x)\psi(y) \geq \sigma(u)\psi(v)\,. \tag{5.14}$$

From Equations 5.13 and 5.14 and properties of continuous conjoint structures, it is not difficult to show that τ and σ must have the following forms:

$$\tau = r\psi^t \text{ and } \sigma = s\varphi^{\frac{1}{t}}\,,$$

for some positive r, s, and t. Thus the problem for Narens (1981a) became one of describing when $\mathcal{S}$ and $\mathcal{T}$ exist so that Equations 5.13 and 5.14 obtain. He provided three separate approaches to this problem. These, and a fourth approach by Luce (1978) are discussed next.

Definition 5.10.7 Let $\mathfrak{C} = \langle X \times Y, \succsim\rangle$ be a continuous conjoint structure.

Let a and u be arbitrary elements of X and b be an arbitrary element of Y. Let v in Y be such that

$$av \sim bu\,.$$

Then, by definition, β_{abu} is the function on X such that for each x in X,

$$\beta_{abu}(x) = z \text{ iff } zv \sim bu\,.$$

It then easily follows from the properties of a continuous conjoint structure (Definition 5.4.9) that β_{abu} is a function from X onto X.

By definition, β is said to be a right translation of $\mathfrak{C}$ if and only if $\beta = \beta_{abu}$ for some a and u in X and some b in Y.

It what follows, the right translations of $\mathfrak{C}$ will generally be automorphisms of structures based on X. Because $\mathfrak{C}$ satisfies unrestricted solvability, it follows from the next theorem that the automorphism groups of these structures are 1-point homogeneous:

Theorem 5.10.3 Let $\mathfrak{C} = \langle X \times Y, \succsim\rangle$ be a continuous conjoint structure. Then for all x and y in X, there exists a right translation β of $\mathfrak{C}$ such that $\beta(x)=y$.

Proof. Immediate from the definitions of "right translation" (Definition 5.10.7) and "unrestricted solvability" (Definition 5.4.9).

One consequence of the next theorem is that in many important settings, Theorem 5.10.3 above is useful for establishing the homogeneity of structures based on X:

Theorem 5.10.4 Suppose $\mathfrak{C} = \langle X \times Y, \succsim\rangle$ is a continuous conjoint structure, and $\mathfrak{X} = \langle X, \succeq_X, R_j\rangle_{j \in J}$ is 1-point unique, where $\succeq_X$ is the $\succsim$ ordering on X. Then the following two statements are logically equivalent:

1. Each right translation of $\mathfrak{C}$ is an automorphism of $\mathfrak{X}$.
2. $\mathfrak{X}$ is a scalar structure and there is a numerical structure $\mathfrak{N} = \langle \mathbb{R}^+, \geq, S_j \rangle_{j \in J}$ such that the set $\mathcal{S}$ of isomorphisms of $\mathfrak{X}$ onto $\mathfrak{N}$ is a ratio scale, and for each φ in $\mathcal{S}$ there exists a function γ from Y onto $\mathbb{R}^+$ such that for all xy and uv in $X \times Y$,

$$xy \succsim uv \text{ iff } \varphi(x)\gamma(y) \geq \varphi(u)\gamma(v)\,.$$

Proof. Section 4 of Narens (1981a).

Note by the equivalence of Statements 1 and 2 in Theorem 5.10.4 and by Theorem 5.10.3 and Definition 5.10.6 that the 1-point homogeneity of $\mathfrak{X}$ canbe easily stated in terms of the primitives, $X \times Y$, $\succsim$, X, $\succsim_X$, and R_j, $j \in J$.

"Left translations of $\mathfrak{C}$ can be analogously defined, and a theorem similar to Theorem 5.10.4 can be shown. As mentioned earlier, these two theorems are sufficient to establish the product of powers representation.

A different, but related, approach of Narens (1981a) is to generalize directly the results about distributive triples of Subsection 5.10.3. Narens (1981a) shows the following two theorems:

Theorem 5.10.5 Suppose $\mathfrak{C} = \langle X \times Y, \succsim \rangle$ is a continuous conjoint structure and $\mathfrak{X} = \langle X, \succsim_X, R_j \rangle_{j \in J}$, where $\succsim_X$ is the $\succsim$-induced ordering on X. Also suppose $\oplus$ is set theoretically definable in terms of the primitives of $\mathfrak{X}$ (Definition 5.1.8) and each right translation of $\mathfrak{C}$ is an automorphism of $\mathfrak{X}$. Then $\oplus$ is $\mathfrak{X}$ distributive in $\mathfrak{C}$.

Proof. Section 4 of Narens (1981a) and Theorem 5.1.2.

Theorem 5.10.6 Suppose $\mathfrak{C} = \langle X \times Y, \succsim \rangle$ is a continuous conjoint structure and $\mathfrak{X} = \langle X, \succsim_X, R_j \rangle_{j \in J}$, where $\succsim_X$ is the $\succsim$-induced ordering on X. Also suppose $\oplus$ is set theoretically definable in terms of the primitives of $\mathfrak{X}$ (Definition 5.1.8). Then the following two statements are equivalent:

1. $\langle X, \succsim_X, \oplus \rangle$ is 1-point unique and $\oplus$ is $\mathfrak{X}$-distributive in $\mathfrak{C}$.
2. $\mathfrak{X}$ is a scalar structure and there is a numerical structure

$$\mathfrak{N} = \langle \mathbb{R}^+, \geq, S_j \rangle_{j \in J}$$

such that the set $\mathcal{S}$ of isomorphisms of $\mathfrak{X}$ onto $\mathfrak{N}$ is a ratio scale, and for each φ in $\mathcal{S}$ there exists a function γ from Y onto $\mathbb{R}^+$ such that for all xy and uv *in* $X \times Y$,

$$xy \succsim uv \text{ iff } \varphi(x)\gamma(y) \geq \varphi(u)\gamma(v)\,.$$

Proof. Section 4 of Narens (1981) and Theorem 5.1.2.

The third approach of Narens (1981a) is to use a concept called "component invariance":

Definition 5.10.8 Let $\mathfrak{C} = \langle X \times Y, \succsim \rangle$ be a continuous conjoint structure. Then $\mathfrak{C}$ is said to satisfy *X-component invariance relative to* $\mathfrak{H}$ if and only if $\mathfrak{H}$ is a subgroup of automorphisms of $\langle X, \succeq_X \rangle$ such that the following two conditions hold: *(i)* $\mathfrak{H}$ is *homogeneous;* that is, for all x and y in X, there exists κ in $\mathfrak{H}$ such that $\kappa(x) = y$ and *(ii)* for all xy and uv in $X \times Y$ and all θ in $\mathfrak{H}$,

$$xy \succsim uv \text{ iff } \theta(x)y \succsim \theta(u)v\,.$$

An analogous definition of *Y-component invariance relative to* $\mathfrak{H}$ holds.

Theorem 5.10.7 Suppose $\mathfrak{C} = \langle X \times Y, \succsim \rangle$ is a continuous conjoint structure, and suppose $\mathfrak{X} = \langle X, \succeq_X, R_j \rangle_{j \in J}$, where $\succeq_X$ is the $\succsim$-induced ordering on X, is a continuous scalar structure with automorphism group $\mathfrak{H}$. Then the following two statements are true:

1. $\mathfrak{C}$ satisfies X-component invariance relative to $\mathfrak{H}$.
2. There is a numerical structure $\mathfrak{N} = \langle \mathbb{R}^+, \geq, S_j \rangle_{j \in J}$ such that the set $\mathcal{S}$ of isomorphisms of $\mathfrak{X}$ onto $\mathfrak{N}$ is a ratio scale, and for each φ in $\mathcal{S}$ there exists a function γ from Y onto $\mathbb{R}^+$ such that for all xy and uv in $X \times Y$,

$$xy \succsim uv \text{ iff } \varphi(x)\gamma(y) \geq \varphi(u)\gamma(v)\,.$$

Proof. Section 4 of Narens (1981a).

A fourth approach, which is due to Luce (1987), is to generalize the concept of distributive operations so that it applies to relations:

Definition 5.10.9 Suppose $\mathfrak{C} = \langle X \times Y, \succsim \rangle$ is a solvable conjoint structure and R is a k-ary relation on X. Then R is said to be *X-distributive in* $\mathfrak{C}$ if and only if for all $a_1, \ldots, a_k$ and $b_1, \ldots, b_k$ in X, if $R(a_1, \ldots, a_k)$ and there exist u and v in Y such that $a_i u \sim b_i v$ for $i=1, \ldots, k$, then $R(b_1, \ldots, b_k)$.

Theorem 5.10.8 Suppose $\mathfrak{C} = \langle X \times Y, \succsim \rangle$ is a solvable conjoint structure and $\mathfrak{X} = \langle X, \succeq, R_j \rangle_{j \in J}$ is a continuous scalar structure and for each $j \in J$, R_j is X-distributive in $\mathfrak{C}$ Let $\mathcal{S}$ be an arbitrary ratio scale of $\mathfrak{X}$ onto $\mathbb{R}^+$ Then for each φ in $\mathcal{S}$ there exists a function ψ from Y onto $\mathbb{R}^+$ such that for all xy and zw *in* $X \times Y$,

$$xy \succsim zw \text{ iff } \varphi(x)\psi(y) \geq \varphi(z)\psi(w)\,.$$

Proof. Theorem 6.2 of Luce(1987).

Equations 5.13 and 5.14 will follow under the following additional assumption to the hypotheses of Theorem 5.10.8: $\mathfrak{Y} = \langle Y, \succeq_Y, S_k \rangle_{k \in K}$ is a scalar structure and each S_k, $k \in K$ is Y-distributive in $\mathfrak{C}$.

Any of the four approaches presented above can be used to provide a qualitative foundation for a generalization of dimensional analysis by employing methods similar to those described in Subsection 5.10.3 for constructing the space of physical quantities and showing the correspondence between dimensional invariance and invariance under automorphisms of the appropriate qualitative structure. The next section presents a different, but related, approach that applies component invariance simultaneously to several variables.

5.10.5 Qualitative Dimensional Structures

From the point of view of the representational theory of measurement, it is desirable to found dimensional analysis and its generalizations on qualitative structures. For the received representational theory, the primitives of the qualitative structures should also be observable. The theory of distributive operations discussed in the previous subsection was developed by Narens and Luce as a means for achieving this latter objective of the received representational theory. Dimensional analysis and its generalizations can also be founded on component invariance. Component invariance is a much more abstract concept than distributivity and in most applications is not directly observable. Nevertheless, it appears to me that from many perspectives component invariance is a superior qualitative concept for founding dimensional analysis and its generalizations. Reasons for this will become apparent in the next chapter. They have to do with distinguishing "meaningfulness" from "lawfulness" and the insight that dimensional analysis is concerned with the description of "lawful" relationships. Because of these considerations, component invariance instead of distributivity is chosen as the means for generalizing dimensional analysis in this subsection.

Definition 5.10.10 Let *J(i), i*=1,…, *n,* be a sequence of nonempty sets of indices and

$$\mathfrak{X}_i = \langle X_i, \succsim_i, R_{ji} \rangle_{ji \in J(i)}$$

be totally ordered structures for i=1,…, n. Without loss of generality, suppose $J_i \cap J_k = \varnothing$ for $1 \leq i \leq k \leq n$. Let

$$J = \bigcup_{i=1}^{n} J(i)\,.$$

Then the structure

$$\mathfrak{X} =< X, X_i, \succsim_i, R_{ji} \rangle_{ji \in J(i),\ i=1,\ldots,n}$$

is said to be a *dimensional structure* if and only if the following five statements are true:

1. J is finite.
2. $X = X_1 \cup \cdots \cup X_n$ and $X_i \cap X_k = \varnothing$ for $1 \leq i < k \leq n$.
3. For $i = 1, \ldots, n$, $\mathfrak{X}_i$ is 1-point unique.
4. For $i = 1, \ldots, n$, $\langle X_i, \succsim_i \rangle$ is dense and Dedekind complete.

5. For i=1,…, n and for each x and y in X_i, there exists an automorphism α of $\mathfrak{X}$ such that (1) $\alpha(x)=y$ and (2) for each $k \neq i$, $k \in \{1, ..., n\}$, and each z in X_k, $\alpha(z)=z$.

In Definition 5.10.10, J is taken to be finite for convenience. This is so that the theorems can be stated using "set-theoretical definability in terms of the primitives of $\mathfrak{X}$" instead of "invariance under the automorphisms of $\mathfrak{X}$" or a mixture of the two concepts. The extensions of results to cases with an infinity of primitives will be obvious.

Note that it follows from Statements 3 and 4 and Part (1) of Statement 5 of Definition 5.10.10 that the $\mathfrak{X}_i$ are scalar structures. Also note that Part (2) of Statement 5 is a generalized form of component invariance (Definition 5.10.8).

Definition 5.10.11 Let $\mathfrak{X} = \langle X, X_i, \succeq_i, R_{ji}\rangle_{ji \in J(i),\ i=1,\dots,n}$ be a dimensional structure. Then "fundamental" and "derived" dimensional qualities are defined as follows: $\mathfrak{Y}$ is said to be a *fundamental (dimensional) quality of* $\mathfrak{X}$ if and only if $\mathfrak{Y} = \mathfrak{X}_i$ for some i in {1,…, n}, where

$$\mathfrak{X}_i = \langle X_i, \succeq_i, R_{ji}\rangle_{j \in J(i)} .$$

The domains X_i *of* $\mathfrak{X}_i$ are called *fundamental dimensions*. For purposes of exposition, $\mathfrak{X}_i$ and its domain are sometimes harmlessly interchanged, for example, X_i will be referred to as a "fundamental quality."

$\mathfrak{C} = \langle Z, \succsim, \succeq\rangle$ is said to be a *derived dimensional quality of* $\mathfrak{X}$ if and only if $Z=D_1\times\ldots\times D_k$, for some fundamental dimensions $D_1\ldots, D_k$ of $\mathfrak{X}$ where k is a positive integer ≥2, and the following four axioms are true:

1. $\succsim$ is a weak ordering on Z and $\succeq$ is the $\succsim$-induced total ordering on $Z/\sim$ (Definition 5.1.2).
2. *(Conjoint) independence:* For all $1 \leq m \leq k$ and all u_m and v_m in D_m, if for some $z_1\ldots, z_k$ in Z,

$$z_1 \cdots z_{m-1} u_m z_{m+1} \cdots z_k \succsim z_1 \cdots z_{m-1} v_m z_{m+1} \cdots z_k ,$$

then for all $w_1\ldots w_k$ in Z,

$$w_1 \cdots w_{m-1} u_m w_{m+1} \cdots w_k \succsim w_1 \cdots w_{m-1} v_m w_{m+1} \cdots w_k .$$

3. *Fundamentally based component ordering:* For each $1 \leq m \leq k$, let $\succeq'_m$ be the binary relation defined on X_m as follows: for all u and v in D_m,

$$u \succeq'_m v \text{ iff } z_1 \cdots z_{m-1} u z_{m+1} \cdots z_k \succsim z_1 \cdots z_{m-1} v z_{m+1} \cdots z_k ,$$

for some $z_1\ldots z_k$ in Z. Then for each $1 \leq m \leq k$, $\succeq'_m$ is the relation $\succeq_i$ or its converse $\preceq_i$, where i is such that D_m is the fundamental quality X_i.

4. *Set-theoretic definability:* $\succsim$ is set-theoretically definable in terms of the primitives of $\mathfrak{X}$ (Definition 5.1.8).

The two orderings, $\precsim$ on Z and $\preceq$ on $Z/\sim$ in Definition 5.10.11 are somewhat redundant, especially since $\preceq$ is simply and directly definable in terms of $\precsim$. The redundancy is included for convenience to make it easy to formulate the needed measurements of derived qualities in terms of isomorphisms. Sometimes in talking about qualitative correlates of quantitative (derived) units, these two orderings are harmlessly confused, for example, in some instances, the weak ordering $\precsim$ is called its "qualitative correlate" and in others the total ordering $\preceq$ is designated its "qualitative correlate." This minor confusion, which makes exposition and notation easier, should not pose any trouble to the reader.

In Definition 5.10.11, the fact that the component orderings of a derived dimensional quality can be converses of the primitive orderings of fundamental qualities allows for the possibility of representations with negative exponents, for example, $v=dt^{-1}$, where v is the quantitative correlate of the derived quality of velocity $V^\star$ and d and t are the respective quantitative correlates of the fundamental qualities of distance D and time T. In this case the qualitative correlate to the usual ordering of $V^\star$ is a weak ordering $\precsim$ on $D\times T$.

Definition 5.10.12 $\mathfrak{W} = \langle X, X_i, \preceq_i, R_{ji}, T_k, \precsim_k, \preceq_k\rangle_{ji\in J(i),\ i=1,\dots,n,\ k\in K}$ is said to be a *structure of dimensional qualities* if and only if

$$\mathfrak{X} = \langle X, X_i, \preceq_i, R_{ji}\rangle_{ji\in J(i),\ i=1,\dots,n}$$

is a dimensional structure and

$$\mathfrak{T} = \{\langle T_k, \precsim_k, \preceq_k\rangle \mid k \in K\}$$

is the set of derived dimensional qualities for $\mathfrak{X}$.

Note that in Definition 5.10.12, the set K is in general infinite.

The "numerical" representing structures for structures of dimensional qualities will look very much like the multiplicative vector space of physical qualities described in Section 5.10.5. However, for various reasons they will not exactly be multiplicative vector spaces—even for the physical case. The main reason for this is a technical one: I want the representations of a structure of dimensional qualities to be isomorphisms, and because the fundamental qualities are disjoint, their isomorphic images must also be disjoint. In Chapter 6, additional theoretically based reasons for keeping the images disjoint are given. For the physical case this means that distinct continuous extensive structures of physical qualities cannot be mapped onto the single numerical structure $\langle \mathbb{R}^+, \geq, +\rangle$. Because of this, multiple "copies" of $\langle \mathbb{R}^+, \geq, +\rangle$ are used as representing structures so that different continuous extensive structures of fundamental qualities can be mapped onto different copies of $\langle \mathbb{R}^+, \geq, +\rangle$. Structures based on copies of numerical structures will often be referred to as "numerical-like." Implicit in this idea is that numerical functions and concepts have numerical-like correlates, for example, each copy of $\langle \mathbb{R}^+, \geq, +\rangle$ will have its "squaring function", which by convention will be denoted the numerical squaring function $s(x)=x^2$. That is, each copy is seen as part of a richer numerical-like context, and all relevant aspects of this richer numerical-like context will, by convention, be denoted the corresponding aspect in the numerical context.

The measurement literature, as well as physics, do not employ multiple copies of $\langle \mathbb{R}^+, \geq, + \rangle$; instead, they identify all the copies in the obvious way with $\langle \mathbb{R}^+, \geq, + \rangle$. This identification substantially simplifies notation, and thus makes presentation and calculation much simpler. However, in dealing with meaningfulness matters, it is very important to have a clear understanding of just where qualitatively the quantitative concepts are coming from, and from this point of view the identification used in the literature may suggest inappropriate interpretations, especially for identities involving higher-order quantitative concepts.

As an example, physics measures both length and mass in the same real number system. To avoid ambiguity, the measurements of length and mass of an object are kept distinct through the use of units. However, this kind of distinction disappears for dimensionless numbers. Thus for the case where the ratio of a's length to b's length is 2 and the ratio of a's mass to b's mass is 2, there is no equivalent physical convention to describe whether a particular instance of the pure number 2 resulted from length measurement or mass measurement or some other kind of measurement. Particularly disturbing is the fact that the dimensionless 2 resulting from length measurement=the dimensionless 2 that results from mass measurement, because, as will be argued in Subsection 5.10.8, the qualitative interpretation of the dimensionless 2 from length measurement is an automorphism α of the qualitative length structure, and the qualitative interpretation of the dimensionless 2 arising from mass measurement is an automorphism β of the qualitative mass structure, and not only is $\alpha \neq \beta$, but in general the identification of α and β is not valid; and in the special circumstances where it is valid to make such an identification, the identification depends critically on relations that are not part of the topics determined by the qualitative structures for length and mass measurement. The above scheme of measuring different physical dimensions onto different isomorphic copies of the reals do not have these and related kinds conceptual difficulties.

Thus, I find for *foundational purposes* the more notationally clumsy approach of multiple copies of the reals philosophically sounder and conceptually clearer than using a single real number system and the current conventions of physics. Of course, for *doing* physics, I prefer measuring into a single real number system and using the current conventions of physics.

Theorem 5.10.9 Suppose

$$\mathfrak{W} = \langle X, X_i, \succeq_i, R_{ji}, T_k, \precsim_k, \succeq_k \rangle_{j \in J(i),\ i=1,\ldots,n,\ k \in K}$$

is a structure of dimensional qualities (Definition 5.10.12). Then there exists a numerical-like structure

$$\mathfrak{N} = \langle N, N_i, \geq_i, S_{ji}, U_k, \succsim'_k, \succeq'_k \rangle_{ji \in J(i),\ i=1,\ldots,n,\ k \in K}\,,$$

such that the following five statements are true:

1. $\mathfrak{M}$ and $\mathfrak{N}$ are isomorphic.
2. For i=1,…, n, the numerical-like structure $\langle N_i, \geq_i \rangle$ is isomorphic to (i.e., is a "copy of") $\langle \mathbb{R}^+, \geq \rangle$.
3. For i=1,…, n, the set $\mathcal{S}_i$ of isomorphisms of $\langle X_i, \succeq_i, R_{ji} \rangle_{ji \in J(i)}$ onto $\langle N_i, \geq, S_{ji} \rangle_{ji \in J(i)}$ is a ratio scale.
4. Let k be an arbitrary element of K, and let $k(1)$,…, $k(p)$ be such that

$$T_k = X_{k(1)} \times \cdots \times X_{k(p)}\,,$$

and let $\varphi_{k(m)}$ be arbitrary representations in scales $\mathcal{S}_{k(m)}$ (as defined in Statement 3) for m=1,…, p. Then there exist $k(p)$ non-zero real numbers $r(k(1))$,…, $r(k(p))$ such that for all $(a_{k(1)},\ldots, a_{k(p)})$, $(b_{k(1)},\ldots, b_{k(p)})$ in $T_{k(p)}$,

$$\begin{aligned}
&(a_1, \ldots, a_{k(p)}) \succsim_k (b_1, \ldots, b_{k(p)}) &(5.15)\\
&\text{ iff } ((\varphi_{k(1)}(a_1))^{r(k(1))}, \ldots, (\varphi_{k(p)}(a_{k(p)}))^{r(k(p))})\\
&\qquad \succsim'_k ((\varphi_{k(1)}(b_1))^{r(k(1))}, \ldots, (\varphi_{k(p)}(b_{k(p)}))^{r(k(p))})\\
&\text{ iff } \varphi_{k(1)}(a_1))^{r(k(1))} \cdots \varphi_{k(p)}(a_{k(p)}))^{r(k(p))}\\
&\qquad \geq \varphi_{k(1)}(b_1))^{r(k(1))} \cdots (\varphi_{k(p)}(b_{k(p)}))^{r(k(p))} &(5.16)\\
&\text{ iff } (a_1/\sim, \ldots, a_{k(p)}/\sim) \succeq_k (b_1/\sim, \ldots, b_{k(p)}/\sim)\,.
\end{aligned}$$

Note that in Equation 5.15, $\succsim'_k$ is an ordering over ordered k-tuples of reallike numbers, whereas in Equation 5.16, $\geq$ is the usual numerical-like ordering over numerical-like products of real-like numbers. (In Equation 5.16, the ordering $\geq$ may be taken as the usual greater than or equal ordering on the reals and the products of real-like numbers may be taken as the products of real numbers that result when when each real number is substituted for its copy.)

5. With the understanding that the N_i are fundamental quantities, a function f is quantitatively dimensionally invariant (Definition 5.10.1) if and only if its image under an isomorphism of $\mathfrak{N}$ onto $\mathfrak{M}$ (which exists by Statement 1) is set-theoretically definable in terms of the primitives of $\mathfrak{M}$.

Proof. Statement 3 can be shown by using Theorem 5.10.6. Then Statements 1, 2, and 4 are easy consequences of Theorem 5.10.7 extended to cases possibly involving more than two variables. (The proof is not be given here.) Statement 5 easily follows by noting that the quantitative dimensional invariance of f when translated into $\mathfrak{M}$ by an isomorphism of $\mathfrak{N}$ onto $\mathfrak{M}$ says that the isomorphic image of f is invariant under the automorphisms of $\mathfrak{M}$, which by Theorem 5.1.2 yields that it is set-theoretically definable in terms of the primitives of $\mathfrak{M}$.

The following definition is useful in discussions:

Definition 5.10.13 Let

$$\mathfrak{W} = \langle X, X_i, \succeq_i, R_{ji}, T_k, \precsim_k, \succeq_k \rangle_{ji \in J(i),\ i=1,\ldots,n,\ k \in K}$$

be a structure of dimensional qualities Then a numerically based structure

$$\mathfrak{N} = \langle N, N_i, \geq_i, S_j, U_k, \precsim'_k, \succeq'_k \rangle_{ji \in J(i),\ i=1,\ldots,n,\ k \in K}$$

that satisfies Statements 1 to 5 of Theorem 5.10.9 is called a *scalar-based numerical-like representing structure* for $\mathfrak{W}$.

Definition 5.10.14 Let $\mathfrak{W}$ be a structure of dimensional qualities (as in Definition 5.10.11). Then it is easy to show that there exist many distinct scalar-based numerical-like representing structures $\mathfrak{N}$ for $\mathfrak{W}$. For the physical case, the following are selected as canonical representing structures: The physical case assumes that the fundamental qualities of $\mathfrak{W}$ are the fundamental physical qualities, and that the scalar structures

$$\mathfrak{X}_i = \langle X_i, \succeq_i, R_j \rangle_{j \in J(i)}$$

are continuous extensive structures. $\mathfrak{N}$ is chosen as a canonical scalar-based numerical-like structure for $\mathfrak{W}$ such that under the isomorphisms of $\mathfrak{W}$ onto $\mathfrak{N}$, each $\mathfrak{X}_i$ is isomorphically mapped onto a numerical-like copy of $\langle \mathbb{R}, \geq, + \rangle$. It can be shown that such a $\mathfrak{N}$ exists and is unique up to substitutions of appropriate copies of $\langle \mathbb{R}, \geq, + \rangle$. Such $\mathfrak{N}$ an is called an additive numerical-like *representing structure for (the structure of physical qualities)* $\mathfrak{W}$.

5.10.6 Alternative Physical Measurements

Vector Space of Physical Units

In dimensional analysis, the physical units form a vector space with *product* of units being *vector addition* and the *raising of units to powers* being *scalar multiplication.* Examples of this are the product of the unit of mass *m* with the unit of velocity *υ* producing the unit of momentum *mv,* and the raising of the unit of length *l* to the power 2 producing the unit of area l^2. The physical units have additional operators *that are not part of their forming a vector space,* namely multiplications of units by positive reals. Such operators are called *similarity transformations*. (Under the similarity transformation of multiplication by 100, the length unit "centimeter" becomes the length unit "meter".) In the literature, similarity transformations are sometimes referred to as "multiplication by scalars," a convention that invites confusion in the vector structure of physical units, because it clashes with the structure's vector concept of scalars, which are real powers.

Definition 5.10.15 B is said to be a *similarity basis* for the physical units if and only if the following three conditions are satisfied:

(1) Each element of B is a physical unit.
(2) For each physical unit u there exist a positive integer k, elements $u_1, \ldots, u_k$ of B, and positive real numbers s, $r(1), \ldots, r(k)$ such that

$$u = s u_1^{r(1)} \cdot u_k^{s(k)} .$$

(3) For all elements v of B there do not exist a positive integer k, k elements $u_1, \ldots, uk$ of B different from v, and $k+1$ positive real numbers s, $r(1), \ldots, r(k)$ such that

$$v = s u_1^{r(1)} \cdot u_k^{s(k)} .$$

A core principle of physical dimensional analysis is that from the point of physical theory any similarity basis of the vector space of physical units is just as good as any other similarity basis. However, some care should be taken in employing this principle: In the axiomatization of dimensional analysis given in this section, only a degenerate form of the 3-dimensional geometry of physical space was utilized, namely the extensive measurement of length. Sometimes in dimensional analysis, a unit of angle is added as a separate dimension to the structure of dimensional units; but even with this addition, a highly degenerate form of 3-dimensional geometry of physical space still results. As a consequence, one has to be careful that the law that one is seeking in an application of dimensional analysis does not depend on the geometry of physical space beyond the relevant, degenerate form inherent in the dimensional analysis utilized. This is another example of the extra kind of knowledge required in applications of dimensional analysis that Bridgman alluded to in the quotation presented in Subsection 5.10.2.

A special case of the above principle occurs when a basis element is replaced by a power of it. As will be seen later in this subsection, this is qualitatively equivalent to replacing the extensive structure corresponding to the basis unit with one that containing a different but "physically equivalent" associative operation. The possibility of such replacements have been acknowledge by several researchers. However, in my reading of the literature, the justifications for the use of such replacements are at best obscure.

The following example due to Ellis (1966) with a commentary from Krantz, et al. (1971) gives an example of an alternative extensive structure for measuring physical length. The measurement of the length through additive representations of this alternative extensive structure results in length units that are the power 2 of the usual physical length units:

> As Ellis (1966) pointed out, at least one other totally different interpretation of concatenation also satisfies the axioms [for a continuous extensive

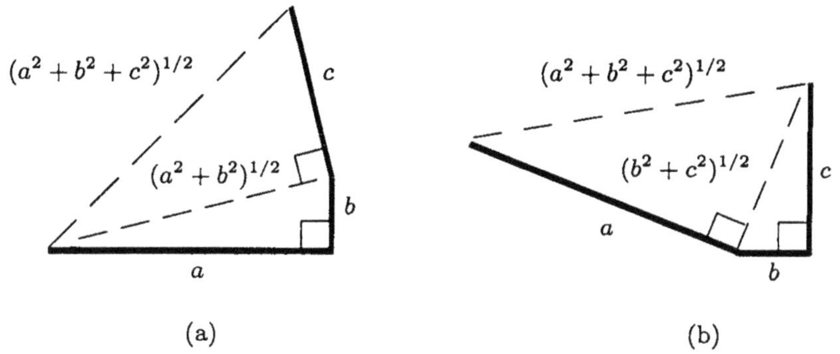

Fig. 5.1. Orthogonal concatenation for length measurement illustrating left term (a) and right term (b) of the associative property

structure on lengths] and so leads to an additive representation; this measure of length is not linearly related to the usual one. Campbell (pp. 290–294 of 1957 edition) discussed other examples of a similar nature.

To present Ellis' interpretation we begin with a collection of rods. Let [the concatenation] $a * b$ be the hypotenuse of the right triangle whose sides are a and b. The comparison relation $\succsim$ is determined by placing two rods side by side, with one end coinciding, and observing which one extends at the other end. Using properties of right triangles it is easy to verify that [the axioms of a continuous extensive structure] are satisfied. The only property that might present a slight difficulty is associativity. It is explained in Figure 5.1 where the lines are labeled by their lengths in the usual measure. Since [the axioms of a continuous extensive structure are] satisfied, [Helmholtz' result obtains], hence there is a measure ψ that is order preserving and additive over this new concatenation. Since the usual measure φ is also order preserving, ψ and φ must be monotonically related, and by the properties of triangles it is easy to see that ψ is proportional to φ^2. To most people, the new interpretation seems much more artificial than the original one. In spite of this strong feeling, neither Ellis nor the authors know of any argument for favoring the first interpretation except familiarity, convention, and, perhaps convenience. We are used to length being measured along straight lines, not along the hypotenuses of right triangles, but no empirical reasons appear to force that choice. Indeed, we could easily reconstruct the whole of physics in terms of ψ by replacing all occurrences of φ by $\psi^{\frac{1}{2}}$. This would make some equations appear slightly more complicated; others would be simpler. In fact, when φ^2 happens to be the more convenient measure, it is common to assign it a name and to treat it as the fundamental measure. Examples are the moment of inertia and the variance of a random variable. In the present case, if a and b are rods, the squares with side a and with side b can be concatenated by forming the square on the hypotenuse; φ^2 will be an additive (area) measure for such concatenation of squares. *(Krantz et al, 1971, pp. 87–88)*

Let X be the set of lengths and $\oplus$ be the operation of concatenating lengths by abutting rods, and let $\oplus_2$ be the operation of concatenating lengths by the "right triangle method" previously described. Let $\gtrsim_\star$ be the total ordering of lengths described above. Then by the above discussion, both $\mathfrak{X} = \langle X, \gtrsim_\star, \oplus\rangle$ and $\mathfrak{X}_2 = \langle X, \gtrsim_\star, \oplus_2\rangle$ are continuous extensive structures. Let φ be a representation from $\mathfrak{X}$ onto $\langle \mathbb{R}^+, \geq, +\rangle$. Then from the above it is easy to show that φ^2 is a representation from $\mathfrak{X}_2$ onto $\langle \mathbb{R}^+, \geq, +\rangle$ and that for all x,y in X,

$$x \oplus_2 y = (\varphi^2)^{-1}[\varphi^2(x) + \varphi^2(y)]\,.$$

I agree with Ellis and Krantz et al. that *for the purposes of doing dimensional analysis* in classical physics, $\oplus_2$ is just as good operation for measuring length as $\oplus$. However, it should be noted that the above example uses the Euclidean structure of space in a critical way in obtaining its results, and therefore its methods do not extend to other physical dimensions. This issue and some of its implications are discussed more fully in the next chapter, particularly in Example 6.3.1 and the discussion following it.

Physically Equivalent Continuous Extensive Structures

Convention 5.10.1 Throughout the remainder of this subsection the following conventions and notation are observed:

1. Language $\mathsf{L}(\in, A, \varnothing, M)$ and system ZFA are assumed.
2. $\mathfrak{X} = \langle A, \succeq, \oplus\rangle$ is a continuous extensive structure.
3. $\mathfrak{N} = \langle \mathbb{R}^+, \geq, +\rangle$.
4. φ is an isomorphism of $\mathfrak{X}$ onto $\mathfrak{N}$. (φ exists by Theorem 5.4.7.)
5. For each $r \in \mathbb{R}^+$, $\mathfrak{N}_r = \langle \mathbb{R}^+, \geq, +_r\rangle$, where $+_r$ is as defined on $\mathbb{R}^+$ by

$$s +_r t = [s^r + t^r]^{\frac{1}{r}}\,.$$

6. For each $r \in \mathbb{R}^+$, $X_r = \langle A, \succeq, \oplus_r\rangle$, where $\oplus_r = \varphi^{-1}(+_r)$.

Note that the function f from $\mathbb{R}^+$ onto $\mathbb{R}^+$ such that $f(x) = x^r$ for each x in $\mathbb{R}^+$ is an isomorphism of $\mathfrak{N}$ onto $\mathfrak{N}$. Thus the following theorem is an easy consequence of Convention 5.10.1:

Theorem 5.10.10 Assume Convention 5.10.1. Then for each r in $\mathbb{R}^+$, $\mathfrak{N}_r$ and $\mathfrak{X}_r$ are continuous extensive structures.

Theorem 5.10.11 Assume Convention 5.10.1 and axiom system $\mathsf{D}''(\succeq, \oplus)$ (Definition 4.3.6). Then the following two statements are equivalent:

1. $\oplus'$ is meaningful and $\mathfrak{E} = \langle A, \succeq, \oplus' \rangle$ is a continuous extensive structure.
2. For some r in $\mathbb{R}^+$, $\oplus' = \oplus_r$.

Proof. Theorem 5.12.6.

In obtaining Theorem 5.10.11 the axiom of Meaningful Pure Sets has to be used in an essential way. This is not surprising, because there are $2^{\aleph_0}$ operations $\oplus_r$ and the language $\mathsf{L}(\in, A, \varnothing, M)$ has only $\aleph_0$ many formulas. (The end of Appendix of Narens and Mausfeld, 1992, shows that for rational *r*, the meaningfulness of $\oplus_r$ can be establish without the use of the axiom of Meaningful Pure Sets.)

Definition 5.10.16 Assume Convention 5.10.1 and axiom system $\mathsf{D}''(\succeq, \oplus)$, and interpret *M* as the predicate of physical meaningfulness for concepts and relations based on A and pure sets. Then structures $\mathfrak{X}$ and $\mathfrak{Y} = \langle A, \succeq, \oplus' \rangle$ are said to be *physically equivalent* if and only $\oplus'$ is meaningful and $\mathfrak{X}$ and $\mathfrak{Y}$ are isomorphic.

Theorem 5.10.12 Assume Convention 5.10.1 and $\mathsf{D}''(\succeq, \oplus)$ Suppose

$$\langle A, \succeq, \oplus' \rangle$$

is physically equivalent to $\mathfrak{X}$. Then the following two statements are true:

1. $\oplus'$ is set-theoretically definable in terms of *A*, $\succeq$, and $\oplus$.
2. $\oplus$ is set-theoretically definable in terms of *A*, $\succeq$, and $\oplus'$.

Proof. Theorem 5.12.7

Assume Convention 5.10.1 and $\mathsf{D}''(\succeq, \oplus)$ and suppose $\mathfrak{Y} = \langle A, \succeq, \oplus' \rangle$ is physically equivalent to $\mathfrak{X}$. Then, because their primitives are meaningful (and therefore physical by Definition 5.10.16), both $\mathfrak{X}$ and $\mathfrak{Y}$ are physical structures. And because they isomorphic, $\mathfrak{X}$ and $\mathfrak{Y}$ have identical structural properties, that is, they are indistinguishable structurally. Furthermore, with respect to set-theoretic definability, they are by Theorem 5.10.12 interdefinable with each other, that is, $\oplus'$ can be viewed part of the structure $\mathfrak{X}$ through set-theoretic definability and similarly $\oplus$ can be viewed as part of the structure $\mathfrak{Y}$. From these considerations I find it reasonable to conclude that there is no physical reason for taking $\mathfrak{X}$ over $\mathfrak{Y}$ for the purpose of measuring *A*. Tradition and convenience do not count as "physical reasons."[2]

In the definition of "physical equivalence" given in Definition 5.10.16, the ordering relation remains constant across structures while the operation varies. Why this asymmetry? An answer is provided in Section 6.2.

[2] The philosophies of science of some philosophers and scientists would restrict "physical reasons" to only empirical considerations. While I consider this to be a valid philosophical view in regards to how one might want to conceptualize, reason about, and explain physical phenomena, it is not a view I hold. In such an empirical approach to physics, the above concept of physical equivalence would be too broad, because it does not require $\oplus$, $\oplus'$, and "interpretability" to be empirical.

Theorem 5.10.13 Assume Convention 5.10.1 and $D''(\succeq,\oplus)$. Let

$$\mathfrak{Y} = \langle A, \succeq, \oplus' \rangle.$$

Then the following two statements are logically equivalent:

1. $\mathfrak{X}$ and $\mathfrak{Y}$ are physically equivalent.
2. For some positive real *r*, $\oplus' = \oplus_r$.

Proof. Immediate from Definition 5.10.16 and Theorem 5.10.11.

Theorem 5.10.14 Assume Convention 5.10.1 and $D''(\succeq,\oplus)$. Let *h* and *k* be positive reals ≥ 1, $\mathcal{S}_k$ be the scale of isomorphisms of $\mathfrak{X}$ onto $\mathfrak{N}_k$, and $\mathcal{T}_h$ be the scale of isomorphisms of $\mathfrak{X}_h$ onto $\mathfrak{N}$. Then the following two statements are true:

1. $\mathcal{S}_k$ and $\mathcal{T}_h$ are ratio scales.
2. For all $\psi \in \mathcal{S}_1$, $\eta \in \mathcal{S}_k$ and $\theta \in \mathcal{T}_h$ there exist *r* and *s* in $\mathbb{R}^+$ such that

$$\psi = r\eta^{\frac{1}{k}} = s\theta^{\frac{1}{h}}.$$

Proof. Left to reader.

5.10.7 Scale Types of Derived Physical Qualities

In the theory of physical qualities presented in Subsection 5.10.5, both fundamental and derived physical qualities are measured by ratio scales. This is in contrast to the view of Krantz, et al. (1991) that fundamental physical qualities are measured by ratio scales and some derive physical qualities—such as density—are measured by log-interval scales:

> …if the set of possible volumes is denoted by A_1 and the set of possible materials by A_2, then $A_1 \times A_2$ defines a set of objects that vary in mass. Let the qualitative ordering by mass be denoted by $\succsim$, then if the law $m=Vd$ [where *m, V,* and *d* denotes the respective measurements of mass, volume, and density] is true, $\langle A_1 \times A_2, \succsim \rangle$ must exhibit all the necessary conditions of [a continuous additive conjoint structure (Definition 5.4.13)]… Thus, we know that there exist numerical scales ψ_k on A_k such that $\psi = \psi_1\psi_2$ preserves the qualitative ordering by mass, $\succsim$. Since log ψ_k is [a representation from] an interval scale, each $\psi_k$is [a representation from a log-interval scale (Section 2.2.2)], i.e., it is unique up to power transformations of the form $\gamma_k\psi_k^{\alpha}$, where γ_k, $\alpha>0$, for $k=1, 2$. *(Krantz, et al., 1971, pg. 484)*

...The ψ_2 resulting is assigned a name and is, thenceforth, treated as if it were a ratio scale. So, for example, when both the volume and the material are varied, [the exponent can be chosen so that] density is defined to $\frac{m}{V}$ be not, for example, $(\frac{m}{V})^{3.83}$ (which is just as satisfactory a measure of density except for those who would continually have to write the exponent 3.83). Presumably, however, if $(\frac{m}{V})^2$ always appeared in the equations of physics, there would have be a strong tendency to define that to be density. Ellis (1966, pp. 118– 121, 125–126) has also pointed out the highly conventional nature of treating these so-called derived measures as ratio scales, when in fact no experiment determines the exponent. This convention is especially apparent when it is realized that such derived measures arise as part of a conjoint structure and they are not extensively measurable; so they must have the two free constants inherent in any conjoint measure. *(Krantz, et al., 1971, pg. 481)*

To make matters concrete, consider the derived quality of density discussed in the above quotes of Krantz, et al. (1971). According to the theory of derived qualities developed in Subsection 5.10.5, the derived quality of density is not measured through the conjoint structure $\langle A_1 \times A_2, \precsim \rangle$, but through the structure

$$\mathfrak{D} = \langle A_1 \cup A_2, A_1, A_2, A_1 \times A_2, \precsim, \oplus_1, \oplus_{A_1 \times A_2} \rangle ,$$

where

(1) $\mathfrak{T} = \langle A_1 \times A_2, \precsim, \oplus_1, \oplus_{A_1 \times A_2} \rangle$ is a A_1, $A_1 \times A_2$-distributive triple (Definition 5.10.4);
(2) $\mathfrak{X}_1 = \langle A_1, \succsim_1, \oplus_1 \rangle$ is the continuous extensive structure use to measure volume, where $\succsim_1$ is the $\precsim$-induced ordering on A_1; and
(3) $\mathfrak{X}_{12} = \langle A_1 \times A_2/\sim, \precsim/\sim, \oplus_{A_1 \times A_2} \rangle$ is the continuous extensive structure used to measure mass.

The following argument shows that the measurement of density that results from structure $\mathfrak{D}$ is a ratio scale:

Let H be the automorphism group of $\mathfrak{D}$. It is not difficult to show the following two statements:

1. Each element α of H has the form $\beta \cup \gamma$, where β is an automorphism of $\mathfrak{X}_1$ and γ is a one-to-one function from A_2 onto A_2.
2. For each automorphism δ of $\mathfrak{X}_1$ there exists ξ such that $\delta \cup \xi$ is in H.

Let

$$K = \{\gamma \mid \exists \beta [\beta \text{ is an automorphism of } \mathfrak{X}_1 \text{ and } \beta \cup \gamma \in H]\} .$$

Then using Theorem 5.10.2, it is not difficult to show that $\langle K, * \rangle$ is a 1-point homogeneous, 1-point unique group of functions from A^2 onto A^2. Under measurement of $\mathfrak{D}$ by a scale $\mathcal{S}$ of isomorphisms, K corresponds to the scale group of the measurements of A_2 *derived from the scale* $\mathcal{S}$, which, because of the 1-point homogeneity, 1-point uniqueness of *K,* rules out this *derived scale* from being a log-interval scale:

In particular, through the use of Theorem 5.4.9, let $\mathcal{S}$ be a scale of isomorphisms of $\mathfrak{D}$ onto the numerically based structure

$$\mathfrak{N} = \langle \mathbb{R}_1^+ \cup \mathbb{R}_2^+, \mathbb{R}_1^+, \mathbb{R}_2^+, \mathbb{R}_1^+ \times \mathbb{R}_2^+, \succsim', +_1, +_{12} \rangle ,$$

where

(1) $\mathbb{R}_1^+$ and $\mathbb{R}_2^+$ are copies of $\mathbb{R}^+$, for example, $\mathbb{R}_1^+ = \mathbb{R}^+ \times \{1\}$ and $\mathbb{R}_2^+ = \mathbb{R}^+ \times \{2\}$
(2) the orderings induced by $\succsim'$ on A_1 and A_2 are copies of $\geq$; and
(3) $+_1$ and $+_{12}$ are copies of +; and
(4) for each representation φ in $\mathcal{S}$ *(i)* φ restricted to A_1 is an isomorphism of $\mathfrak{X}_1$ onto $\langle \mathbb{R}_1^+, \succsim_1', +_1 \rangle$, where $\succsim_1'$ is the order induced by $\succsim'$ on A_1, *(ii)* $\hat{\varphi}$ is an isomorphism of $\mathfrak{X}_{12}$ onto $\langle \mathbb{R}_1^+ \times \mathbb{R}_2^+/\sim', \succsim'/\sim', +_{12} \rangle$, where $\hat{\varphi}$ is the function on $A_1 \times A_2/\sim$ such that for each $\sim'$-equivalence class U in $A_1 \times A_2/\sim$ and each (x,y) in U, $\hat{\varphi}(U)$ is the $\sim'$-equivalence class of $\mathbb{R}_1^+ \times \mathbb{R}_2^+/\sim'$ to which $(\varphi(x),\varphi(y))$ belongs, and *(iii)* (ψ_1, ψ_2), where ψ_1 is the restriction of φ to A_1 and ψ_2 is the restriction of φ to A_2, is a conjoint isomorphism of $\langle A_1 \times A_2, \succsim \rangle$ onto $\langle \mathbb{R}_1^+ \times \mathbb{R}_2^+, \succsim' \rangle$.

Then $\mathcal{S}_2 = \{\psi_2 \mid \exists\varphi[\varphi \in \mathcal{S}$ and φ restricted to A_2 is $\psi_2]\}$ is *the ratio scale derived from* $\mathcal{S}$ *used to measure* A_2.

5.10.8 Dimensionless Quantities

Physical quantities are usually measured in dimensional units, for example, length is measured in centimeters, meters, and so forth. By tradition, the ratio of physical quantities with the same unit are called *dimensionless quantities* or *dimensionless numbers*. Dimensionless quantities play a central role in dimensional analysis and other parts of physics. (See, for example Section 1.6 of Chapter 1 which deals with Eddington's theory of pure numbers.) The literature identifies dimensionless quantities with real numbers. I believe that the uncritical identification of dimensionless quantities with real numbers can lead to epistemological difficulties. The following example illustrates the potential for such difficulties.

Let *a* and *b* be objects that have the physical dimensions of length and weight. Let Θ be the following statement:

> Θ: The ratio of *a*'s length to *b*'s length is the same dimensionless quantity *r* as the ratio of *a*'s weight to *b*'s weight.

Suppose Θ is true. The problem is to understand qualitatively what is being asserted by the two ratios being the same dimensionless quantity *r*. In order to accomplish this, let us assume that length is described qualitatively by the continuous extensive structure $\mathfrak{X} = \langle X, \succeq, \oplus \rangle$, weight by the continuous extensive structure $\mathfrak{Y} = \langle Y, \succeq', \oplus' \rangle$, a's and b's lengths have been measured by the representation φ of the set of isomorphisms of $\mathfrak{X}$ onto $\mathfrak{R} = \langle \mathbb{R}^+, \geq, + \rangle$, and a's and b's weights have been measured by the representation ψ of the set of isomorphisms of $\mathfrak{Y}$ onto $\mathfrak{R} = \langle \mathbb{R}^+, \geq, + \rangle$. Then

$$\varphi(a) = r \cdot \varphi(b) \text{ and } \psi(a) = r \cdot \psi(b)\,. \tag{5.17}$$

Thus by Equation 5.17, by the well-known fact that the multiplication by positive reals are the automorphisms of $\mathfrak{R}$, and by the isomorphisms φ and ψ, we see that the dimensionless ratio of lengths, r, is interpretable qualitatively as the automorphism $\alpha = \varphi^{-1}(m_r)$ of $\mathfrak{X}$, where m_r is the function on $\mathbb{R}^+$ that is multiplication by r, and β is interpretable qualitatively as the automorphism $\beta = \psi^{-1}(m_r)$ of $\mathfrak{Y}$. Because α and β are defined on different sets, there is no proper way of comparing them *without assuming additional qualitative structure.*

To make the problem more concrete, consider the following two instances of the above situation:

I1: a is composed of a weightless, frictionless spring with a weight attached. b is also composed of a weightless, frictionless, spring with a weight attached. Both a and b are hanging from a ceiling, and the spring part of a has the same physical properties as the spring part of b. In this situation, Hook's law applies, and thus the equality of the ratio of the lengths of a and b with the ratio of their weight is not only an empirical fact but also a necessary consequence of Hook's law. Qualitatively, Hook's law provides additional structure to the situation in the form of a function F from Y onto X. With this function as *an additional primitive relation* and a proper qualitative formulation of Hook's law, automorphisms of $\mathfrak{X}$ are appropriately identilied with automorphisms oi $\mathfrak{R}$, and under this identification $\varphi^{-1}(r)$ is identified with $\psi^{-1}(r)$ via the rul

$$F[\psi^{-1}(r)] = \varphi^{-1}(r)\,.$$

I2: a and b are people. Unlike the case of springs above, there is no obvious law connecting a person's height with his or her weight that can be used to identify automorphisms of $\mathfrak{X}$ with those of $\mathfrak{R}$.

The literature—in contrast to the presentation given in this book—has the vector space of physical units include a dimension of dimensionless quantities (see Chapter 10 of Krantz, et al. 1971 and Section 7 of Chapter 22 of Luce, et al. 1991). Although the literature's formulation is simpler and more in line with the practice of dimensional analysis than this book's, which uses different copies of real numbers for representing the dimensionless quantities of different dimensions instead of a single copy for representing the dimensionless quantities of all dimensions, the former suffers from the fact that there is no acceptable qualitative counterpart to the single dimension of dimensionless quantities, and therefore it is not a suitable candidate for a numerical representing structure for the qualitative structure of dimensional qualities via the representational theory of measurement. Because of this, the representational theory's meaningfulness concepts are not fully applicable to this situation. In the book's formulation, *distinct* automorphism groups of the qualitative structure of physical qualities are the qualitative counterparts of the *distinct* dimensions of the dimensionless quantities. Thus for *foundational purposes* I prefer the slightly more clumsy formulation given in this and the previous section, because it preserves the integrity of the representational theory and because I believe it to be sounder philosophically.

5.10.9 Summary for Dimensional Analysis

Dimensional analysis is a useful tool in physics, and it is natural to ask if it generalizes to other scientific domains. One way to approach this issue is to isolate principles of dimensional analysis that are fundamental and routinely used, provide an abstract measurement-theoretic foundation for them, and see if that formulation applies to phenomena in other scientific domains. Such a program was carried out in Subsections 5.10.1to5.10.3, where the four highly applicable, fundamental and interconnected principles of dimensional analysis—the vector space-like structure of physical units, the ratio scalability of individual physical units, dimensional invariance, and the "Π-Theorem"— were given a rigorous measurement-theoretic foundation in terms of distributive triples with associative operations. Other sciences share with physics the property that complex attributes can be thought of as conjoint structures of simple attributes. But it is extremely rare outside physics for the simple attributes to have basic, observable associative operations on them so that the attributes with these operations form continuous extensive structures—a condition that exists in physics and is necessary for the formation of associative distributive triples. Thus a foundation based on distributive triples with associative operations is unlikely to have much applicability outside of physics.

In Subsection 5.10.4 associative distributive triples were generalized in three separate ways, and each of these can be used to give a measurementtheoretic foundation for the four abovementioned principles of dimensional analysis. Of the three, I believe that component-invariance (Definition 5.10.8) is, at the theoretical level, most informative in non-physical applications, and a variant of it was used in Subsection 5.10.5 to formulate a general idea of a structure of dimensional qualities (Definition 5.10.12). Theorem 5.10.9, which is a representation theorem for structures of dimensional qualities, shows the first three of the four abovementioned principles of dimensional analysis. As stated in Subsection 5.10.1, that the fourth principle—the "Π-Theorem" is a consequence of the first three principles.

5.11 Discussion and Conclusions

Traditionally, in the theory of measurement, meaningfulness has been a derived concept defined in terms of a measurement scale. This has been reversed in Section 5.6, where meaningfulness is taken as the fundamental concept and measurement scale as the derived concept. As a fundamental concept, meaningfulness is identified with the topic under consideration, that is, the relations and concepts that belong to the topic are considered meaningful and those that do not belong to the topic are considered meaningless. In particular, some scales of measurement belong to the topic. Thus associated with a topic are scales that also belong to the topic. These are the "meaningful scales" described in Section 5.6. Theorem 5.6.2 gives, for each meaningful scale of one-to-one functions, a construction of a structure $\mathfrak{X}$ with meaningful primitives and a pure $\mathfrak{N}$ structure such that $\mathcal{S}$ is a subset of isomorphisms of $\mathfrak{X}$ into $\mathfrak{N}$. This result shows from meaningfulness considerations alone that the homomorphism version of the representational theory of scales is a reasonable approach to a theory of measurement scales. (See also Example 5.6.1 and Theorems 5.6.1 and Theorem 5.6.4.)

The representational approach starts with a structure of primitives $\mathfrak{X}$, which is used to capture the qualitative situation under consideration. Thus from the point of view of this book, the fleshing out or formalization of "capturing the qualitative situation" constitutes a theory of meaningfulness. This theory is, of course, prior to the kind of theory of meaningfulness developed in the representational theory. The representational theory then proceeds by assuming axioms $\mathcal{A}$ about $\mathfrak{X}$ and selecting a numerical (or more generally, a numerically based) representing structure $\mathfrak{N}$, and then defines a scale $\mathcal{S}$ for $\mathfrak{X}$ to be a set of representations of $\mathfrak{X}$ into $\mathfrak{N}$. The only formal constraint imposed by the representational theory being that $\mathfrak{N}$ must be selected in such a way that $\mathcal{S} \neq \varnothing$. This is shown through use of the axioms $\mathcal{A}$.

Narens and Luce (1976) employed a different approach for selecting the numerical representing structure. Slightly generalized, their approach was the following: Instead of choosing a numerical representing structure and using the axioms $\mathcal{A}$ to show the existence of a representation of $\mathfrak{X} = \langle X, R, R_j \rangle_{j \in J}$ onto the numerical representing structure, they used $\mathcal{A}$ to show the existence of a one-to-one function φ from X onto a subset N of $\mathbb{R}$, *and then defined* the numerical representation structure $\mathfrak{N}$ to be

$$\mathfrak{N} = \langle N, \varphi(R), \varphi(R_j) \rangle_{j \in J} .$$

Then the scale of isomorphisms $\mathcal{S}$ from $\mathfrak{X}$ onto $\mathfrak{N}$ is nonempty, since by construction φ is an isomorphism and $\varphi \in S$. In Narens' and Luce's approach, the axioms in $\mathcal{A}$ are used to show the existence of a one-to-one function from the domain of $\mathfrak{X}$ onto a subset of the reals. If X is finite, then such a one-to-one function always exist. If X is infinite and the primitive R of $\mathfrak{X}$ is a total ordering on X, then the existence of such a one-to-one function is equivalent to the qualitative assertion that $\langle X, R\rangle$ satisfies denumerable density (Definition 5.4.1). (See Theorem 2.3 of Chapter 2 of Narens, 1985, or Theorem 2 of Chapter 2 of Krantz, et al., 1971, for proofs.)

Niederée (1987, 1992a) employed ideas similar to Narens and Luce (1976), but with two important twists:

The first is that the one-to-one function φ from X onto a subset N of $\mathbb{R}^+$ is required to be produced from $\mathfrak{X}$ through some constructive or algorithmic procedure in terms of the primitives of $\mathfrak{X}$ and one or two elements of X. One role of the axioms $\mathcal{A}$ about $\mathfrak{X}$ is to guarantee that the procedure yields such a one-to-one function, Then like in Narens and Luce (1976), φ is used to define the numerical representing structure $\mathfrak{N}$ for $\mathfrak{X}$. The axioms $\mathcal{A}$ about $\mathfrak{X}$ are used to show that if the one or two elements of X that were used in the construction of φ were replaced by other element(s) of X and the procedure were applied with these new elements to produce a one-to-one function ψ, *then the same numerical structure* $\mathfrak{N}$ *results.*

Consider the situation where $\mathfrak{X} = \langle X, \succsim, \oplus\rangle$ where $\oplus$ is a binary operation on X and $\succsim$ is a total ordering on X. Let a be an element of X. Consider the following measurement procedure:

> For each positive integer n, let a_n be the n-copy operator determined by $\oplus$ (Definition 5.4.8), and let φ be such that for each b in X and each m and k in $\mathbb{I}^+$,
>
> (1) if $b=\alpha_m(a)$, then $\varphi(b)=m$;
>
> (2) if $\alpha_k(b)=\alpha_m(a)$, then
>
> $$\varphi(b) = \frac{m}{k}\text{; and}$$
>
> (3) if $b_1, \ldots, b_i, \ldots, m_1, \ldots, m_i, \ldots$, and $k_1, \ldots, k_i, \ldots$ are such that
>
> $$\alpha_{k_i}(b_i) = \alpha_{m_i}(a) \text{ and } b = \text{l.u.b. } b_i\,,$$
>
> then
>
> $$\varphi(b) = \text{l.u.b. } \frac{m_i}{k_i}\,.$$

The above procedure produces a function φ from a subset of X onto a subset of $\mathbb{R}^+$. For φ to be a function on X and one-to-one, axioms $\mathcal{A}$ about $\mathfrak{X}$ need to be assumed. Suppose that $\mathcal{A}$ are axioms that $\mathfrak{X}$ say is a continuous extensive structure (Definition 5.4.5). Then it can be shown that φ is a one-to-one function on X. Let

$$\mathfrak{N} = \langle \varphi(X), \varphi(\succsim), \varphi(\oplus)\rangle\,.$$

Then it follows $\mathcal{A}$ from that

$$\mathfrak{N} = \langle \mathbb{R}^+, \geq, +\rangle\,.$$

Furthermore it follows from $\mathcal{A}$ that if an element c were used in place of a in the above procedure, then a one-to-one function ψ from X into $\mathbb{R}^+$ would result such that

$$\psi = \frac{\varphi}{\varphi(c)},$$

and consequently,

$$\langle \psi(X), \psi(\succeq), \psi(\oplus) \rangle = \langle \mathbb{R}^+, \geq, + \rangle .$$

Furthermore, it can be shown from Axioms $\mathcal{A}$ that each isomorphism of $\mathfrak{X}$ onto $\mathfrak{N}$ can be realized by applying the above procedure to an appropriately selected element of X. Thus in summary, the above procedure when applied to continuous extensive measurement yields the particular numerical representing structure $\mathfrak{N}$, and constructs, through the appropriate selections of the element a of X, each isomorphism of $\mathfrak{X}$ onto $\mathfrak{N}$ (see Theorem 5.4.7). Thus by Niederée's view, the selection of $\mathfrak{N}$ to represent continuous extensive structures is not conventional or arbitrary, but results from applying a general kind of measurement procedure to continuous extensive structures. Much of Niederée's research consists of making explicit the concept of "measurement procedure" and describing the various varieties of such procedures.

Niederée's second twist consists of using products of the measurement procedure for the numerical structure, for example, using the n-copy operator α_n to stand for the integer n. Although this gives interesting insights into the nature of several measurement procedures common in science, and is an interesting subject in its own right, it is tangential to meaningfulness issues considered in this chapter, and is not pursued further in this discussion. (It is much more pertinent to the considerations of qualitativeness and empiricalness discussed in Chapter 7.)

Implicit in the approach of Luce and Narens (1976) is that the structure $\mathfrak{X}$ captures the underlying qualitative situation, which, using the terminology of this book, is equivalent to saying that $\mathfrak{X}$ specifies the topic of interest—that is, meaningfulness considerations are already implicit at the very first stages of doing measurement. For Niederée's approach—at least as I understand it—this is not necessarily the case: the qualitative structure $\mathfrak{X}$ is used as part of the measurement process of assigning numbers to the elements of X; there is no need for it to specify the qualitative situation of interest.

Because of its constructive aspect, Niederée's approach cannot be applied to as wide range of cases as the representational theory. For example, consider the case of an arbitrary continuous threshold structure $\mathfrak{X} = \langle X, \succeq_\star, T \rangle$ (Section 5.4.2). Through Theorems 5.4.3, 5.4.5, and 5.4.6, it is not difficult to see that there is no constructive method of producing an isomorphism of $\mathfrak{X}$ onto the numerical structure

$$\mathfrak{N} = \langle \mathbb{R}, \geq, S \rangle ,$$

where S is the function defined on $\mathbb{R}^+$ such that for each r in $\mathbb{R}$,

$$S(r) = r + 1 ,$$

and thus in general Niederée's constructive procedures of measurement cannot be carried out for such structures.

In psychology, the most important application of continuous threshold structures has been in psychophysics. There, the domain of X is considered both as a set of physical stimuli and as a set of psychological stimuli, the relation $\succeq_\star$ is considered both as a physical and a psychological ordering relation, and the threshold hold function T is considered as a psychological relation determined by the subject's behavior. Thus the continuous threshold structure $\mathfrak{X} = \langle X, \succeq_\star, T\rangle$ is considered a description of a psychological situation. *As a practical matter, for this particular kind* of continuous threshold structure, an isomorphism of $\mathfrak{X}$ onto a numerical structure $\mathfrak{N}'$ can be obtained through physical measurement: First a physical operation $\oplus$ is selected so that $\mathfrak{X}' = \langle X, \succeq_\star, \oplus\rangle$ is a continuous extensive structure. Then physical measurement is applied to $\mathfrak{X}'$ yielding a scale $\mathcal{S}$ of isomorphisms of $\mathfrak{X}$ onto $\langle \mathbb{R}^+, \geq, +\rangle$. Let φ be an element of $\mathcal{S}$. Then φ is an isomorphism of $\mathfrak{X}$ onto $\mathfrak{N}' = \langle \mathbb{R}^+, \geq, \varphi(T)\rangle$. However, the above construction of φ involves a concept that is exogenous to the psychological structure $\mathfrak{X}'$ —namely the physical operation $\oplus$. Put into a meaningfulness terms: the physical operation $\oplus$ is not part of the topic determined by the structure $\mathfrak{X}$. (See Subsection 6.4.3 for a fuller discussion of this point.)

Such practical methods of measurement that employ concepts exogenous to the topic under consideration run counter to the intention of the representational theory and its meaningfulness concept. For example, suppose we are interested in the topic determined by the qualitative structure $\mathfrak{Y}$. Then, the Narens and Luce (1976) approach to the representational theory requires for the measurement of $\mathfrak{Y}$ sufficient *axioms in terms of the primitives of* $\mathfrak{Y}$ for the existence of one-to-one function from the domain of $\mathfrak{Y}$ onto a subset of $\mathbb{R}$. Giving interesting sufficient axioms about $\mathfrak{Y}$ to achieve this is a major research thrust of the representational theory, and such an axiomatization is often in itself an important scientific and philosophical contribution. But practical approaches, similar to the above, where the construction of oneto-one functions from the domain of $\mathfrak{Y}$ onto a subset of the positive reals rely not only on axioms about $\mathfrak{Y}$ *but also axioms about other structures with the same domain as* $\mathfrak{Y}$, thwart the intention and value of the representational approach, especially when the axioms about $\mathfrak{Y}$ are not sufficient for the existence of a one-to-one function from the domain of $\mathfrak{Y}$ onto a subset of $\mathbb{R}$.

In this chapter, representation and uniqueness results have been established for a variety of continuous measurement structures. For the case of homogeneous, finitely unique, continuous structures there is a complete understanding of the possible scale types, and for the case of homogeneous, continuous concatenation structures, there is a fairly good understanding about the functional form of the primitive operations of such structures. There is far less known about structures that are not easily interpretable in terms of homogeneous, finitely unique, continuous ones.

In the measurement literature, there are several concepts of meaningfulness based on various kinds of invariance. These concepts are highly interrelated and generally agree for homogeneous continuous structures, where they are interpretable as variants of automorphism invariance—that is, as the Erlanger Program's concept of meaningfulness. They are also closely related to the concept of meaningful scale developed in Section 5.6. Except for a few varieties of special situations, the above representational concepts of meaningfulness unfortunately produce, for structures that have the identity as their only automorphism, a trivial concept of meaningfulness in which all entities are meaningful. For many kinds of infinite situations, this is an undesirable consequence, and one that on intuitive grounds appears to be incorrect. A new approach to meaningfulness which is capable of overcoming this kind of difficulty is presented in Chapter 6.

In the literature, one of the most important roles of the representational theory has been to provide axiomatic characterizations of quantitative models and methods. Such characterizations often reveal hidden assumptions and suggest new kinds of empirical tests. This chapter presented such characterizations for the possible psychophysical laws (Section 5.7), magnitude estimation (Section 5.8), Weber's Law (Section 5.9), and dimensional analysis (Section 5.10). Each of these qualitative characterizations provided a more rigorous foundation for the issues under consideration and revealed subtle and important relationships that were obscured in quantitative form.

In conclusion, the results of this chapter show that the axiomatic theories of meaningfulness of Chapter 4 and the representational theory of measurement can be integrated in fruitful ways, with the axiomatic theories of meaningfulness providing a theoretical foundation and justification for the representational theory and its meaningfulness concept. The representational theory, as construed throughout this chapter, should neither be considered as a general theory of measurement nor be considered as a synthesis of ways of measuring that have appeared in the literature. It definitely should not be considered as a description of empirical measurement, especially if the "empirical measurement" includes concepts of error. In my view, the primary goals of the representational theory are *(i)* to produce techniques for providing qualitative, axiomatic characterizations of quantitative situations, *(ii)* to provide a foundational theory for the concept of scale type, and *(iii)* to provide a meaningfulness theory for quantitative models. I believe that representational theory described in this chapter achieves these goals.

5.12 Additional Proofs and Theorems

Theorem 5.12.1 (*Theorem 5.4–8*) Suppose $\mathfrak{X} = \langle X, \succsim, R_j \rangle_{j \in J}$ is a continuous scalar structure. Then $\langle X, \succsim \rangle$ is a continuum.

Proof. By either Theorem 2.6 of Narens (1981a) or Theorem 2.4.3 of Narens (1985) there exists an isomorphism φ of $\mathfrak{X}$ onto a numerical structure of the form $\langle \mathbb{R}^+, \geq, S_j \rangle_{j \in J}$. Then φ is an isomorphism of $\langle X, \succsim \rangle$ onto $\langle \mathbb{R}^+, \geq \rangle$, and since $\langle \mathbb{R}^+, \geq \rangle$ is a continuum, it follows by isomorphism that $\langle X, \succsim \rangle$ is a continuum.

Theorem 5.12.2 (*Theorem 5.4.12*) Let $\mathfrak{X} = \langle X, \succsim, R_j \rangle_{j \in J}$ be a continuous scalar structure, and $\oplus$ be a binary operation on X that is set-theoretically definable from the primitives of $\mathfrak{X}$ (Definition 5.1.8), $n \in \mathbb{I}^+$ and C_n be the n-copy operator determined by $\oplus$. Suppose C_n is $\succsim$-strictly increasing. Then C_n is an automorphism of $\mathfrak{X}$.

Proof. Let $\mathfrak{Y}$ be a structure that is obtained by adding $\oplus$ to $\mathfrak{X}$ as a primitive, for example, let

$$\mathfrak{Y} = \langle X, \succsim, \oplus, R_j \rangle_{j \in J} .$$

Then it follows from Theorem 5.1.2 that $\mathfrak{X}$ and $\mathfrak{Y}$ have the same set of automorphisms, that is, that $\mathfrak{Y}$ is a scalar structure. By Theorem 5.4.9, let $\mathfrak{N}$ be a structure with domain $\mathbb{R}^+$ and $\mathcal{S}$ be such that $\mathcal{S}$ is a ratio scale of isomorphisms of $\mathfrak{Y}$ onto $\mathfrak{N}$. Then the set of multiplications by positive reals is the set of automorphisms of $\mathfrak{N}$.. Let $\varphi \oplus S$. Let $F=\varphi(C_n)$. Then, since C_n is $\succsim$-strictly increasing, it follows by isomorphism that F is strictly increasing. Since $\mathfrak{X}$ is homogeneous, C_n is onto X and therefore by isomorphism, F is onto $\mathbb{R}^+$. It is well-known in analysis that strictly increasing functions from $\mathbb{R}^+$ onto $\mathbb{R}^+$ are continuous. Thus F is continuous. Since C_n is settheoretically definable from the primitives of $\mathfrak{Y}$ (and in fact set-theoretically definable from $\oplus$), it follows from Theorem 5.1.2 that C_n is invariant under the automorphisms of $\mathfrak{Y}$. Thus, by isomorphism, F is invariant under the automorphisms of $\mathfrak{N}$—i.e., F is invariant under multiplications by positive constants. Thus, for all r and u in $\mathbb{R}^+$,

$$F(r \cdot u) = r \cdot F(u) .$$

It is well-known (and is not difficult to show) that the only continuous solutions of this functional equation have the form

$$F(u) = s \cdot u ,$$

where s is a positive real. Thus F is an automorphism of $\mathfrak{N}$, and by the isomorphism φ^{-1}, C_n is an automorphism of $\mathfrak{X}$.

Theorem 5.12.3 *(Theorem 5.4.13)* Suppose $\mathfrak{X} = \langle X, \succsim, R_j \rangle_{j \in J}$, J is a nonempty finite set, and $\langle X, \succsim \rangle$ is a continuum. Then the following two statements are equivalent:

1. $\mathfrak{X}$ is a scalar structure.
2. There is an operation $\oplus$ such that *(i)* $\oplus$ is set-theoretically definable

in terms of the primitives of $\mathfrak{X}$ *(ii)* $\mathfrak{Y} = \langle X, \succsim, \oplus \rangle$ is a continuous PCS, *(iii)* for each n in $\mathbb{I}^+$ the n-copy operator determined by $\oplus$ is an automorphism of $\mathfrak{X}$, and *(iv)* the primitives of $\mathfrak{X}$ are set-theoretically definable in terms of the primitives of $\mathfrak{Y}$.

Proof. 1. Assume $\mathfrak{X}$ is a scalar structure. Then by Theorem 5.4.9, let $\mathfrak{N}$ and $\mathcal{S}$ be such that $\mathfrak{N}$ is a numerical structure with domain $\mathbb{R}^+$ and $\mathcal{S}$ is a ratio scale of isomorphisms of $\mathfrak{X}$ onto $\mathfrak{N}$ Let φ be an element of $\mathcal{S}$ Because φ is an isomorphism of $\mathfrak{X}$ onto $\mathfrak{N}$ and $+$ is invariant under the automorphisms of $\mathfrak{N}$, $\oplus = \varphi^{-1}(+)$ is invariant under the automorphisms of $\mathfrak{X}$. Thus by Theorem 5.1.2, $\oplus$ is set-theoretically definable in terms of the primitives of $\mathfrak{X}$. Because φ is an isomorphism and $\langle \mathbb{R}^+, \geq, + \rangle$ is a continuous extensive structure, it follows that $\mathfrak{Y} = \langle X, \succsim, \oplus \rangle$ is a continuous extensive structure. By observing that for all isomorphisms of ψ and θ of $\mathfrak{Y}$ onto $\langle \mathbb{R}^+, \geq, + \rangle$ $\psi^{-1}\ \theta$ is an automorphism of $\mathfrak{Y}$, it easily follows from Theorem 5.4.9 that $\mathfrak{Y}$ is a scalar structure. Since $\oplus$ is invariant under the automorphisms of $\mathfrak{X}$ it follows that $\mathfrak{Y}$ is invariant under the automorphisms of $\mathfrak{X}$ and from this and the fact that $\mathfrak{X}$ and $\mathfrak{Y}$ are scalar structures, it easily follows that the set of automorphisms of $\mathfrak{X}$= the set of automorphisms of $\mathfrak{Y}$. Thus to show that the n-copy operators determined by $\oplus$ are automorphisms of $\mathfrak{X}$ it need only be shown that they are automorphisms of $\mathfrak{Y}$. This follows by an argument given in the discussion following Theorem 5.4.8. (It also follows from the more general Theorem 3.1 of Cohen and Narens, 1979, which shows that the n-copy operators of a homogeneous continuous PCS, $\mathfrak{Z}$, are automorphisms of $\mathfrak{Z}$.) Because the automorphisms of $\mathfrak{X}$ and $\mathfrak{Y}$ coincide, each primitive of $\mathfrak{X}$ is invariant under the automorphisms of $\mathfrak{Y}$, and thus by Theorem 5.1.2, the primitives of $\mathfrak{X}$ are set-theoretically definable in terms of the primitives of $\mathfrak{Y}$.

2. Suppose *(i)* $\oplus$ is an operation that is set-theoretically definable in terms of the primitives of $\mathfrak{X}$ *(ii)* $\mathfrak{Y} = \langle X, \succsim, \oplus \rangle$ is a continuous PCS, *(iii)* for each n in $\mathbb{I}^+$ the n-copy operator determined by $\oplus$ is an automorphism of $\mathfrak{X}$, and *(iv)* the primitives of $\mathfrak{X}$ are set-theoretically definable in terms of the primitives of $\mathfrak{Y}$. By *(i)* and Theorem 5.1.2,

$$\text{the set of automorphisms of } \mathfrak{X} \subseteq \text{the set of automorphisms of } \mathfrak{Y}. \quad (5.18)$$

Thus by *(iii)* each n-copy operator determined by $\oplus$ is an automorphism of $\mathfrak{Y}$ It then follows from *(ii)* and Theorem 3.1 of Cohen and Narens (1979) that $\mathfrak{Y}$ is homogeneous and 1-point unique. By *(iv)* and Theorem 5.1.2,

$$\text{the set of automorphisms of } \mathfrak{Y} \subseteq \text{the set of automorphisms of } \mathfrak{X} \quad (5.19)$$

Thus by Equations 5.18 and 5.19, the automorphisms and $\mathfrak{X}$ and $\mathfrak{Y}$ coincide, and therefore, because $\mathfrak{Y}$ is homogeneous and 1-point unique, $\mathfrak{X}$ is a scalar structure.

Theorem 5.12.4 *(Theorem 5.4.10)* Suppose $\mathfrak{X} = \langle X, \succsim, R_j \rangle_{j \in J}$, J is a nonempty finite set, and $\langle X, \succsim \rangle$ is a continuum. Then the following two statements are equivalent:

1. $\mathfrak{X}$ is a scalar structure.
2. There is an operation $\oplus$ such that *(i)* $\oplus$ is set-theoretically definable in terms of the primitives of $\mathfrak{X}$, *(ii)* $\mathfrak{Y} = \langle X, \succsim, \oplus \rangle$ is a continuous extensive structure, *(iii)* for each *n* in $\mathbb{I}^+$ the *n*-copy operator determined by $\oplus$ is an automorphism of $\mathfrak{X}$ and *(iv)* the primitives of $\mathfrak{X}$ are set-theoretically definable in terms of the primitives of $\mathfrak{Y}$.

Proof. Statement 1 implies Statement 2 by part 1 of the *proof* of Theorem 5.12.3. (In that part of the proof, $\mathfrak{Y}$ is a continuous extensive structure by construction.) Statement 2 implies Statement 1 by Theorem 5.12.3.

Theorem 5.12.5 *(Theorem 5.7.1)* *H* has the following three properties:

(1) All elements of *H* are of the form $a \cup \beta$, where a is an automorphism of $\mathfrak{X}$ and β is an automorphism of $\mathfrak{Y}$ that is a translation (Definition 5.1.7).
(2) For each automorphism a of $\mathfrak{X}$ there is exactly one translation β of $\mathfrak{Y}$ such that $\alpha \cup \beta$ is in *H*.
(3) For each translation β of $\mathfrak{Y}$ there exists an automorphism α of $\mathfrak{X}$ such that $\alpha \cup \beta$ is in *H*.

Proof. Let

$$B = \{\beta \mid \beta \text{ is an automorphism of } \mathfrak{X}\}$$

and

$$C = \{\delta \mid \delta \text{ is an automorphism of } \mathfrak{Y}\}.$$

Let a be an arbitrary element of *H*. Let α_1 be the restriction of *a* to *X* and α_2 be the restriction of α to *Y*. Then α_1 is an element of *B* and α_2 is an element of *C*. Then by simple verification,

$$H \subseteq \{\beta \cup \delta \mid \beta \in B \text{ and } \delta \in C\}.$$

The quantitative condition,

For each φ in $\mathcal{S}$ ψ in $\mathcal{T}$ and $r \in \mathbb{R}^+$ there exists ψ' in $\mathcal{T}$ such that for each *x* in *X*,

$$\psi[f(r\varphi(x))] = \psi'[f(\varphi(x))],$$

translates into the qualitative condition,

For each automorphism β of $\mathfrak{X}$ there exists an automorphism δ of the $\mathfrak{Y}$ such that for each *x* in *X*,

$$F[\beta(x)] = \delta[F(x)]. \tag{5.20}$$

Let β in B and δ in C be such that for each x in X,

$$F[\beta(x)] = \delta[F(x)].$$

Define γ on $X \cup Y$ by the following: For each z in $X \cup Y$ *(i)* if $z \in X$ then $\gamma(z)=\beta(z)$, and *(ii)* if $z \in Y$ then $\gamma(z)=\delta(z)$. Then by direct verification

$$\gamma \in H \text{ and } \gamma_1 = \beta \text{ and } \gamma_2 = \delta.$$

Thus

$$B = \{\gamma_1 \,|\, \gamma \in H\}.$$

Suppose α and γ are in H and $\alpha_1=\gamma_1$. Then by the above for each x in X,

$$\alpha_2^{-1}[F(\alpha_1(x))] = F(x) \text{ and } \gamma_2^{-1}[F(\gamma_1(x))] = F(x).$$

Since $\alpha_1=\gamma_1$, it follows that for each x in X,

$$\alpha_2^{-1}[F(\alpha_1(x))] = \gamma_2^{-1}[F(\alpha_1(x))],$$

and thus, since α_1 is onto X and F is onto Y, $\alpha_2^{-1} = \gamma_2^{-1}$..

To summarize, the following two result have been shown:

1. The restrictions of each automorphism α of $\mathfrak{Q}$ to X and Y respectively are automorphisms α_1 and ak of respectively $\mathfrak{X}$ and $\mathfrak{Y}$.
2. For each automorphism β of $\mathfrak{X}$ there is α unique automorphism a of $\mathfrak{X}$ such that $\alpha_1=\beta$.

For i=1, 2, let

$$H_i = \{\alpha_i \,|\, \alpha \in H\}.$$

By Statement 2 above,

$$H_1 = B.$$

We will now characterize H_2. Since $\mathcal{T}$ is an interval scale of isomorphisms of $\mathfrak{Y}$, it easily follows that C is a 2-point homogeneous and 2-point unique set of automorphisms. Elements of C are divided into two sets, *(i)* those that have fixed points and are different from the identity, and *(ii)* the identity and those without fixed points. Elements described by *(ii)* are the translations (Definition 2.3.10).

Suppose that δ is an element of H_2 that is not a translation. It will be shown by contradiction that for all α in H that $\alpha_2 \neq \delta$. For suppose that α in H is such that $\alpha_2=\delta$. Since δ is not a translation, let y in Y be such that

$$\alpha_2(y) = y.$$

Since F is onto Y, let x in X be such that

$$F(x) = y.$$

Then

$$F[\alpha_1(x)] = \alpha_2[F(x)] = F(x)\,.$$

Because by hypothesis F is a one-to-one function, it follows that

$$\alpha_1(x) = x\,.$$

Since by hypothesis, S is a ratio scale for $\mathfrak{X}$, it then follows that B is 1-point unique. Thus α_1 is the identity function ι_X on X. The identity function ι on $X \cup Y$ is in H, and

$$\iota_1 = \iota_X = \alpha_1\,.$$

Let ι_Y be the identity function on Y. Then by Statement 2 above,

$$\alpha_2 = \iota_2 = \iota_Y\,.$$

Thus $\delta = \alpha_2 = \iota_Y$, which contradicts the choice of δ as a non-translation.

Thus H_2 is a subset of the set of translations of $\mathfrak{Y}$. Because the set of translations of $\mathfrak{Y}$ is 1-point homogeneous and 1-point unique, in order to show that H_2 is the set of translations of $\mathfrak{Y}$, it is sufficient to show that H_2 is homogeneous. Let y_1 and y_2 be arbitrary elements of Y. Since F is one-to-one and onto Y, let

$$x_1 = F^{-1}(y_1) \text{ and } x_2 = F^{-1}(y_2)\,.$$

Since S is a ratio scale onto $\mathbb{R}^+$ B is homogeneous. Thus let β in B be such that $\beta(x_1) = x_2$. By Equation 5.20 let δ in H_2 be such that

$$F[\beta(x_1)] = \delta[F(x_1]\,.$$

Then

$$y_2 = F(x_2) = F[\beta(x_1)] = \delta[F(x_1)] = \delta(y_1)\,,$$

and it has been shown that H_2 is homogeneous.

Lemma 5.12.1 Let $\mathfrak{Y} = \langle A, \succsim, \oplus' \rangle$ be a continuous extensive structure and suppose G is a set of automorphisms of $\mathfrak{Y}$ such that for all x and y in A there exists γ in G such that $\alpha(x)=y$. Then G is the set of automorphisms of $\mathfrak{Y}$.

Proof. It immediately follows from the hypothesis of the theorem that G is homogeneous. It is an easy consequence of Theorem 5.4.7 that $\mathfrak{Y}$ is 1-point unique. Let β be an arbitrary automorphism of $\mathfrak{Y}$. It needs to only be shown that β is in G. Let a be an element of A. By hypothesis, let α in G be such that $\alpha(a)=\beta(a)$. Then by the 1-point uniqueness of $\mathfrak{Y}$, $\alpha=\beta$, and therefore $\beta \in G$.

Theorem 5.12.6 *(Theorem 5.10.11)* Assume Convention 5.10.1 and axiom system D″($\succsim$,$\oplus$) (Definition 4.3.6). Then the following two statements are equivalent:

1. $\oplus'$ is meaningful and $\mathfrak{C} = \langle A, \succsim, \oplus' \rangle$ is a continuous extensive structure.

2. For some r in $\mathbb{R}^+$, $\oplus' = \oplus_r$.

Proof. Assume Statement 1. By Theorem 5.1.3, axiom system TM is true. Thus $\oplus'$ is invariant under the set G of automorphisms of $\mathfrak{X}$. Therefore $\mathfrak{E}$ is invariant under G. Because G is homogeneous $\mathfrak{E}$ and is 1-point unique, it follows by Lemma 5.12.1 that G is also the automorphism group of $\mathfrak{E}$. Because $\mathfrak{E}$ is a continuous extensive structure, let ψ be an isomorphism of $\mathfrak{E}$ onto $\langle \mathbb{R}^+, \geq, + \rangle$. Let $\odot = \varphi(\oplus')$. Then φ is an isomorphism of $\mathfrak{E}$ onto $\mathfrak{N}' = \langle \mathbb{R}^+, \geq, \odot \rangle$. Because φ is an isomorphism of $\mathfrak{X}$ onto $\mathfrak{N}$, $\varphi(G)$ is the set of automorphisms of $\mathfrak{N}$, which = the set of multiplications by positive reals. Therefore, because G is also the set of automorphisms of $\mathfrak{E}$, by the isomorphism φ, the set of multiplications by positive reals is the set of automorphisms of $\mathfrak{N}'$. Thus by Theorem 2.7 of Narens (1981a) (or Theorem 2.10.3 of Narens, 1985), $\odot = +_r$ for some $r \in \mathbb{R}^+$. Thus $\oplus' = \varphi^{-1}(\oplus) = \oplus_r$ for some positive r, and Statement 2 has been shown.

Assume Statement 2. Then it is easy to verify that $\mathfrak{N}_r$ is a continuous extensive structure and $\mathfrak{N}_r$ is invariant under multiplications by positive reals, that is, is invariant under the automorphisms of $\mathfrak{N}$. Thus by axiom system $D''(\succeq, \oplus)$ and the isomorphism φ^{-1}, $\mathfrak{E}_r = \langle A, \succeq, \oplus_r \rangle$ is a continuous extensive structure and has the same set of automorphisms as $\mathfrak{X}$. Therefore, $\oplus_r$ is invariant under the automorphisms of $\mathfrak{X}$. By hypothesis $D''(\succeq, \oplus)$ is true. Therefore, by Theorem 5.1.3, axiom system TM is true and G is the automorphism group of $\mathfrak{X}$. Thus $\oplus_r$ is meaningful. Therefore, Statement 1 has been shown.

Theorem 5.12.7 *(Theorem 5.10.12)* Assume Convention 5.10.1 and axiom system $D''(\succeq, \oplus)$. Suppose $\langle A, \succeq, \oplus' \rangle$ is physically equivalent to $\mathfrak{X}$. Then the following two statements are true:

1. $\oplus'$ is set-theoretically definable in terms of A, $\succeq$, and $\oplus$.
2. $\oplus$ is set-theoretically definable in terms of A, $\succeq$, and $\oplus'$.

Proof. By Definition 5.10.16, $\oplus'$ is meaningful. Let G be the automorphism group of $\mathfrak{X}$. Because $D''(\succeq, \oplus)$ is true, it follows by Theorem 5.1.3 that axiom system TM is true and G is the transformation group for M. Thus by Theorem 5.1.2, $\oplus'$ is set-theoretically definable in terms of the primitives of and Statement 1 has been shown. Since $\oplus'$ is meaningful, it is invariant under G. Let $\mathfrak{C} = \langle A, \succeq, \oplus' \rangle$. Then, since by Definition 5.10.16 $\mathfrak{C}$ isomorphic to $\mathfrak{X}$, $\mathfrak{C}$ is 1-point unique. Thus by Lemma 5.12.1, G is the set of automorphisms of $\mathfrak{C}$. Since axiom system TM with transformation group G is true and $\oplus$ is meaningful, it follows that $\oplus$ is definable from the primitives of $\mathfrak{C}$, and Statement 2 has been shown.

6. Intrinsicness

6.1 Overview

Chapter 2 considered several examples of the use of the intuitive meaningfulness concept. Historically, the examples stemmed from the seminal articles of Stevens (1946) and Luce (1959), the former being about meaningfulness and scale types and the latter about scale types and lawfulness. Later, Luce (1978), Falmagne and Narens (1983), and Roberts and Rosenbaum (1986) integrated the meaningfulness and lawfulness concepts by showing that many lawful relations can be viewed as meaningful. In this chapter a related, but different, perspective is followed: Meaningfulness is a necessary condition—but not necessarily a sufficient condition—for lawfulness; in other words, generally lawfulness is meaningfulness plus something extra. For invariance concepts of meaningfulness, the "extra" consists of an additional form of invariance; and for definitional concepts of meaningfulness, the "extra" consists of the "lawfulness" part of a law being formulable in terms of meaningful concepts that do not depend on specific choices of primitives. In dimensional analysis in physics, the latter took the special form of having physical laws be independent of the basis of units used to generate the multiplicative vector space of dimensional quantities.

Throughout this chapter lawfulness and examples of various lawful situations are formulated in terms of "intrinsic" concepts. Intrinsic concepts are discussed and formally defined in Sections 6.2 and 6.3. In particular, for the axiomatic systems of meaningfulness which have axiom DM as a consequence, for example, axiom system TM, it will follow that relations that are defined in terms of *M* and concepts about the meaningfulness predicate *M* are intrinsic. Thus it follows that for these kinds of axiom systems, relations and concepts about the transformation group *G* for *M* that are defined in terms of *G* and other intrinsic relations are intrinsic. This latter result is exploited in Section 6.4 to characterize a broad class of laws as well as to explain how lawfulness and meaningfulness differ.

Intrinsicness is a very flexible concept. Theorems in Section 6.2 show that it is interpretable as a meaningfulness concept satisfying axiom system D′. As a meaningfulness concept, it may be useful for situations where the meaningfulness concepts of Chapter 4 fail to give intuitively acceptable results, for example, in cases where *M* is specified in terms of a structure of primitives that has the identity as its only automorphism.

Section 6.4 formulates lawfulness in terms of intrinsicness and applies the formulation to well-known psychophysical examples. In particular, Luce's 1959 theory of the possible psychophysical laws is interpreted as a form of intrinsicness. Section 6.4 also gives intrinsic formulations of Weber's Law (Section 5.9) and the psychophysical power law that often results from the magnitude estimation of physical stimuli (Sections 2.8 and 5.8). Finally, a reformulation by Luce of his 1959 version of the possible psychophysical laws is described, and it is compared and contrasted with the chapter's approach to laws.

Section 6.5 employs intrinsicness as the key concept for a theory of the relationship of psychology and physics in psychophysics. The theory assumes that the qualitative psychophysical situation under consideration can be cleanly split into qualitative psychological and physical structures that share same domain and possibly some relations. The theory then asserts that the psychological and physical relations used in defining a psychophysical relationship are in an asymmetrical relationship to one another. The asymmetry consists of having the psychological relations be meaningful with respect to the psychological structure and the physical relations be intrinsic with respect to the physical structure. The theory is then employed to formulate a principle to test whether a given quantitative psychophysical relationship is equivalent to a meaningful psychological relation, that is, to test whether the psychophysical relationship has a "purely psychological interpretation" in terms of the psychological structure. The principle is then applied to important psychophysical situations, and it is shown that several important psychophysical relationships that one might expect to have purely psychological interpretations, because psychologists tend to implicitly treat them that way in their theorizing, rely in essential ways on concepts outside of the topic determined by the psychological structure.

Sections 6.6 and 6.7 present applications of two concepts of "informational equivalence." Both are conceptualized and formalized in manners very similar to intrinsicness. Section 6.6 generalizes a concept that is used in algebra and analysis to banish *(i)* elements that are infinitely large or infinitesimal and *(ii)* pairs of distinct elements that are infinitesimally close. This concept is called "Archimedeanness" in the literature, and its traditional definitions depend on having an operation or its equivalent among the primitives of the structure and enough algebraic properties to define "equally spaced sequences." Section 6.6 generalizes the concept of Archimedeanness to cases where primitive operations may be lacking. For continuous structures $\mathfrak{X}$, this is accomplished by calling $\mathfrak{X}$ "Archimedean (in the new sense)" if and only if *(i)* it is informationally equivalent to a structure that is Archimedean in the old sense of the term, and *(ii)* for each structure $\mathfrak{Y}$ if $\mathfrak{Y}$ is informationally equivalent to $\mathfrak{X}$ and the notion

$\mathfrak{Y}$ of being Archimedean or non-Archimedean in the old sense of the term is appropriate, then $\mathfrak{Y}$ is Archimedean in the old sense of the term.

Section 6.7 employs a different concept of informational equivalence to investigate the structuring of a population in terms of characteristics of its members. The concept of two sets of properties containing the same informational content is defined. It is argued that many kinds of interpretations commonly given to various kinds of structurings of populations in terms of a particular set of characteristics of their members are unfounded. The argument consists in showing that such interpretations depend critically on the particular way of representing the information contained in data describing which members of population have which properties. Because *(i)* the choice of the particular way of representing the information is generally not justified by the scientific investigator, and (ii) there are generally other equally appropriate ways of the representing the same information that yield interpretations that conflict with the original ones, it follows that the original interpretations are unfounded, and if correct, depend critically on additional unstated assumptions.

Section 6.8 presents concluding remarks, and Section 6.9 provides additional proofs and theorems.

6.2 ε-Intrinsicness

In this section, a new and very flexible kind of definitional invariance called "ε-intrinsicness" is introduced. ε-intrinsicness is applicable in many different kinds of situations. In particular, it appears to be a useful alternative to the concepts of meaningfulness presented in the previous chapters, especially when the latter fail to capture certain interesting topics by having all entities be meaningful.

Thus far, like most books and articles dealing with foundational issues in science, qualitative situations have been identified with structures of the form $\mathfrak{X} = \langle A, R_j \rangle_{j \in J}$. Because in most cases one would want the qualitative *situation* identified with a structure to be more than the set consisting of the structure's primitives, some further theory is needed to link the structure of primitives with the situation it specifies. The axiomatic theories to meaningfulness of Chapter 4 suggest the following approach to this identification problem:

Conceptualize the qualitative situation as being based on a set of qualitative objects, which we will take as the set of atoms A of ZFA. Next assume that the qualitative situation has a meaningfulness concept M associated with it. This meaningfulness concept is meant to describe the topic or subject matter that we are interested in investigating. The qualitative structure $\mathfrak{X} = \langle A, R_j \rangle_{j \in J}$ is used as a basis to capture M.A definition of " $\mathfrak{X}$ capturing M" or giving necessary conditions for " $\mathfrak{X}$ capturing M" will result in a theory of meaningfulness.

The following is an example of a theory of a structure $\mathfrak{X}$ capturing M.

Example 6.2.1 Let $\mathfrak{X} = \langle A, R_j \rangle_{j \in J}$. Assume axiom system $\mathsf{D}''(\{R_j \mid j \in J\})$ (Definition 4.3.6).

In Example 6.2.1, M is captured by saying an entity a is meaningful if and only if it is definable in terms of the primitives of $\mathfrak{X}$ and pure sets through a formula of L(∈, A, Ø). An instance where one might want to use Example 6.2.1 is plane Euclidean geometry. Here one takes A to be the set of points in the Euclidean plane, $J=\{1,\ldots, 6\}$, R_1 to be the set of Euclidean circles on A, R_2 the set of Euclidean lines on A, R_3 the plane Euclidean incidence relation, and R_4, R_5, and R_6, the Euclidean congruence relations for respectively line segments, triangles, and angles. Then (by Theorem 4.3.6) axiom system $\mathsf{D}''(\{R_j \mid j \in J\})$ captures the Erlanger Program's use of "geometrical" for plane Euclidean geometry.

Because capturing M through $\mathsf{D}''(\{R_j \mid j \in J\})$ plays a central role in the developments of this chapter, it is given the following formal definition.

Definition 6.2.1 Let $\mathfrak{X} = \langle X, R_1, \ldots, R_j, \ldots \rangle_{j \in J}$ be a structure. Then $\mathfrak{X}$ is said to *set-theoretically capture M* if and only if $X=A$ and axiom system $\mathsf{D}''(\{R_j \mid j \in J\})$ (Definition 4.3.6) is true.

Intuitively, structures that set-theoretically capture M contain the "same information" about the topic determined by M. This intuitive observation is made precise in the following two definitions and next theorem.

Definition 6.2.2 Assume axiom system ZFA. Let a and b be sets. Then a and b are said to be L(∈, A, Ø)-*equivalent* if and only if

(1) for each e in b there exist a formula $\varphi(v, x_1,\ldots, x_n, y_1,\ldots,y_m)$ of L(∈, A, Ø), elements $a_1,\ldots, a_n$ of a, and pure sets $q_1,\ldots, q_m$ such that

$$\forall v[v = e \text{ iff } \varphi(v, a_1, \ldots, a_n, q_1, \ldots, q_m)];$$

and

(2) for each d in a there exist a formula $\psi(u, x_1,\ldots, x_k, y_1,\ldots,y_s)$ of L(∈, A, Ø) and elements $b_1,\ldots, b_k$ of b and pure sets $p_1,\ldots, p_s$ such that

$$\forall u[u = d \text{ iff } \psi(u, b_1, \ldots, b_k, p_1, \ldots, p_s)].$$

Definition 6.2.3 Assume ZFA. Let $\mathfrak{X} = \langle A, R_j \rangle_{j \in J}$ and $\mathfrak{Y} = \langle A, S_k \rangle_{k \in K}$ be structures. Then and $\mathfrak{Y}$ are said to be L(∈, A, Ø)-*equivalent* if and only if $\{R_j \mid j \in J\}$ and $\{S_k \mid k \in K\}$ are $\mathsf{L}(\in, A, \varnothing)$ are L(∈, A, Ø)-*equivalent.*

Theorem 6.2.1 Let $\mathfrak{X} = \langle A, R_j \rangle_{j \in J}$ and $\mathfrak{Y} = \langle A, S_k \rangle_{k \in K}$ be structures, and suppose $\mathfrak{Y}$ set-theoretically captures M. Then $\mathfrak{Y}$ set-theoretically captures M if and only if $\mathfrak{X}$ and $\mathfrak{Y}$ are L(∈, A, Ø)-*equivalent.*

Proof. Suppose $\mathfrak{Y}$ set-theoretically captures M. Then axiom system $\mathsf{D}''(\{S_k \mid k \in K\})$ is true. Thus for each k in K, $M(S_k)$. Therefore, since axiom system $\mathsf{D}''(\{R_j \mid j \in J\})$ holds, each S_k has a definition through a formula of

L(∈, A, Ø) in terms of the primitives of $\mathfrak{X}$ and pure sets. Similarly, since for each j in J, $M(R_j)$, and axiom system $\mathsf{D}''(\{S_k \mid k \in K\})$ is true, each R_j has a definition through a formula of L(∈, A, Ø) in terms of the primitives of $\mathfrak{Y}$ and pure sets. Therefore by Definition 6.2.2, $\mathfrak{X}$ and $\mathfrak{Y}$ are L(∈, A, Ø)-equivalent.

Suppose $\mathfrak{X}$ and $\mathfrak{Y}$ are L(∈, A, Ø)-equivalent. Then it easily follows from Definitions 6.2.2, 6.2.1, and the fact that $\mathfrak{X}$ set-theoretically captures M that $\mathfrak{Y}$ set-theoretically captures M.

Definition 6.2.4 ε is said to be a *set of equivalent descriptions of M* if and only if the following three conditions hold:

(i) ε is a nonempty set.
(ii) Each element of ε is a structure that set-theoretically captures M (Definition 6.2.1).
(iii) There exists a set J such that for each $\mathfrak{Y}$ in ε there exist R_j, $j \in J$, such that $\mathfrak{X} = \langle A, R_j \rangle_{j \in J}$.

Note by Condition *(iii)* of Definition 6.2.4 that the primitives of each structure in ε are indexed by the same set J.

By Condition *(ii)* of Definition 6.2.4, each structure in ε captures M. Thus the structures in ε specify the same topic. They are informationally equivalent in the following sense:

Lemma 6.2.1 Suppose ε is a set of equivalent descriptions of M. Then for each $\mathfrak{X}$ and $\mathfrak{Y}$ in ε, $\mathfrak{X}$ and $\mathfrak{Y}$ are L(∈,A, Ø)-equivalent (Definition 6.2.4).

Proof. Immediate from Definitions 6.2.4 and Theorem 6.2.1.

Condition *(iii)* of Definition 6.2.4 requires all structures in ε to have the same domain, A, and the same form of indexing of their primitive relations, J. This makes possible the following definition.

Definition 6.2.5 a is said to be *ε-intrinsic* if and only if ε is a set of equivalent descriptions of M, a is an entity, and there exist

$$\mathfrak{X} = \langle A, R_j \rangle_{j \in J}$$

in ε, a formula

$$\varphi(x, x_1, \ldots, x_n, v_1, \ldots, v_m)$$

of L(∈, A, Ø), elements $j(1), \ldots j(n)$ of J, and pure sets $b_1, \ldots, b_m$ such that the following four conditions are satisfied:

(i) $\varphi(a, R_{j(1)}, \ldots, R_{j(n)}, b_1, \ldots, b_m)$.
(ii) For all entities e, if $\varphi(e, R_{j(1)}, \ldots, R_{j(n)}, b_1, \ldots, b_m)$, then $a{=}e$.
(iii) For all structures $\mathfrak{Y} = \langle A, S_j \rangle_{j \in J}$ in ε,

$$\varphi(a, S_{j(1)}, \ldots, S_{j(n)}, b_1, \ldots, b_m)\,.$$

(iv) For all structures $\mathfrak{Y} = \langle A, S_j \rangle_{j \in J}$ in ε and all entities e, if

$$\varphi(e, S_{j(1)}, \ldots, S_{j(n)}, b_1, \ldots, b_m),$$

then a=e.

Let ε be a set of equivalent descriptions of M and a be an entity that is ε-intrinsic. Then by Definition 6.2.4, for each structure $\mathfrak{X}$ in ε, a is settheoretically definable in terms of the primitives of $\mathfrak{X}$. More importantly, a has the *same* set-theoretic definition in terms of the primitives of $\mathfrak{X}$ for each $\mathfrak{X}$ in ε.

The next two theorems demonstrate that ε-intrinsicness generalizes the concepts of meaningfulness of Chapter 4.

Theorem 6.2.2 Let *M* be the collection of ε-intrinsic entities. Then *M*⊆*M* and $\langle V, \in, A, \varnothing, \mathcal{M} \rangle$ satisfies axiom system D′.

Proof. Theorem 6.9.1.

Theorem 6.2.3 Let *M* be the collection of ε-intrinsic entities and $\mathcal{E} = \{\mathfrak{X}\}$ for some $\mathfrak{X}$ Then *M*=*M*.

Proof. Immediate from Definitions 6.2.4, 6.2.1, and 4.3.6.

Meaningfulness becomes a trivial and usually a useless concept when all entities are meaningful. Unfortunately, such trivial meaningfulness concepts sometimes result from the theories of meaningfulness considered in Chapters 4 and 5 when the underlying situation is specified by a structure with the identity as its only automorphism. This has raised serious doubts in the literature about the adequateness and philosophical correctness of those meaningfulness theories for such situations, and thus has raised doubt about their adequateness and philosophical correctness as general theories of meaningfulness. Theorems 6.2.2 to 6.2.3 suggest that this kind of difficulty might be overcome by judiciously selecting ε and employing ε-intrinsicness as the meaningfulness concept.

Suppose $\mathfrak{X}$ set-theoretically captures *M* and $\forall x M(x)$ is true. Let ε be the set of $\mathfrak{Y}$ structures with domain *A* such that $\mathfrak{Y}$ is isomorphic to $\mathfrak{X}$. Let *M* be the collection of ε-intrinsic entities. Then as a meaningfulness predicate for determining interesting topics, *M* generally allows too many entities to be meaningful and *M* too few. ($\mathcal{M}$ generally allows too few elements to be meaningful, because $\forall x M(x)$ holds, and thus each element of $\mathcal{M}$ must have an invariant definition across all structures based on *A* that are isomorphic to $\mathfrak{X}$, which in general produces a topic too narrow to be of interest.) The following example employs intrinsicness to produce a meaningfulness predicate that is between *M* and $\mathcal{M}$.

Example 6.2.2 Let $\mathfrak{X} = \langle A, \succeq, \oplus \rangle$ and φ be an isomorphism of $\mathfrak{X}$ onto $\mathfrak{N} = \langle \mathbb{R}^+, \geq, \oplus' \rangle$ where for all *r* and *s* in $\mathbb{R}^+$,

$$r \oplus' s = r + s + r^2 s^2 .$$

Then by Example 4.2 of Cohen and Narens (1979) $\mathfrak{X}$ is a continuous, solvable PCS (Definition 5.4.7) that has the identity as its only automorphism. Assume axiom system D″($\succsim$, $\oplus$) Then by Theorem 4.3.6, $\forall x M(x)$. Let *F* be the set of strictly increasing functions f from $\mathbb{R}^+$ onto $\mathbb{R}^+$ such that the derivative of *f* is defined and is continuous at *r* for each *r* in $\mathbb{R}^+$. For each *f* in *F*, let

$$\oplus_f = \varphi^{-1} * f(\oplus') \text{ and } \mathfrak{X}_f = \langle A, \succsim, \oplus_f \rangle .$$

Let

$$\mathcal{E} = \{\mathfrak{X}_f \mid f \in F\} \text{ and } \mathcal{M}\text{=the collection of } \varepsilon\text{-intrinsic entities.}$$

Then with $\mathcal{M}$ taken as a meaningfulness predicate, $\langle V, \in, A, \varnothing, \mathcal{M}\rangle$ satisfies axiom system D′ by Theorem 6.2.2. Let

$$H = \{\varphi^{-1}(f) \mid f \in F\} .$$

Then it easily follows that if $\mathcal{M}$ *M(b)*, then *h(b)=b* for all *h* in *H,* that is, invariance under elements of *H* is a necessary condition for the predicate $\mathcal{M}$ to hold. Note that *H* is not a group, because some *f* in *F* do not have inverses that are in *F*.

6.3 Intrinsicness Relative to M, *{Sj}j∈J*

6.3.1 Definition of Intrinsicness Relative to *M,{Sj}j∈J*

Let $\mathfrak{X} = \langle X, \succsim, \oplus\rangle$ and $\mathfrak{X}' = \langle X, \succsim', \oplus'\rangle$ be continuous extensive structures. The concept physical equivalence of $\mathfrak{X}$ and $\mathfrak{X}'$ formally introduced and discussed in Subsection 5.10.6, is very close to the concept of L(∈, *A,* Ø)-equivalence (Definition 6.2.2) of $\mathfrak{X}$ and $\mathfrak{X}'$. However, there are important differences. In particular, physical equivalence requires *(i) X* to be a physical domain, *(ii)* $\mathfrak{X}$ and $\mathfrak{X}'$ to be isomorphic (because they are continuous extensive structures), and *(iii)* $\succsim = \succsim'$. Physical equivalence is a special form of an important kind of intrinsicness that is specified in the following two definitions.

Definition 6.3.1 ε is said to be the *set of isomorphic descriptions of M based on* $\{S_j\}_{j\in J}$ if and only if

(1) ε is a set of equivalent descriptions of *M* (Definition 6.2.4);
(2) each element of ε is a structure of the form

$$\langle A, S_j, R_k\rangle_{j\in J,\ k\in K} ;$$

(3) all elements of ε are isomorphic;
(4) for all structures of the form

$$\mathfrak{Y} = \langle A, S_j, T_h \rangle_{j \in J,\, h \in H}\ ,$$

where H is some set, if $\mathfrak{Y}$ set-theoretically captures M and $\mathfrak{Y}$ is isomorphic to some element of ε, then $\mathfrak{Y}$ is in ε.

Definition 6.3.2 An entity a is said to be *intrinsic relative to* M, $\{S_j\}_{j \in J}$ if and only if a is ε-intrinsic for the set ε of isomorphic descriptions of M based on $\{S_j\}_{j \in j}$.

Convention 6.3.1 If in Definition 6.3.2 $\{S_j\}_{j \in j}=\emptyset$, then "intrinsic relative to M, Ø" is often written as *intrinsic relative to M.* If in Definition 6.3.2 J is a nonempty finite set, say $J=\{1,\ldots,n\}$, then "intrinsic relative to M, $(S_i,\ldots, S_n\}$" is often written as "intrinsic relative to M, $S_1,\ldots, S_n$."

Note that the notion of F being a set of isomorphic descriptions of M (Definition 6.3.1) is much more structured than then the notion of ε being a set of equivalent descriptions of M (Definition 6.2.4).

Also note the following difference between "ε-intrinsicness" and "intrinsicness relative to M, $\{S_j\}_{j \in J}$": For ε-intrinsicness, *(i)* each element of $\in$ is a structure that set-theoretically captures M, and *(ii)* all equivalent descriptions in ε are indexed by the same set. In addition to *(i)* and *(ii)*, intrinsicness relative to M, $\{Sj\}_{j \in j}$ requires *(iii)* each element of ε to be a structure whose primitives begins with A, $\{Sj\}_{j \in J}$, *(iv)* all elements of ε to be isomorphic to one another, and *(v)* ε to be a maximal set of structures satisfying *(i)-(iv)*.

It is easy to show that the following sets are ε-intrinsic for all sets of equivalent descriptions of M (and thus intrinsic relative to M, $\{S_j\}j_{\in}j$ for all $\{S_j\}_{j \in J}$ that are part of an equivalent description of M): A, Ø, all pure sets, the power set of A, the set of all finite subsets of A, the Cartesian product of A with itself. And if axiom TM holds, then it easily follows that the transformation group G for M is also intrinsic relative to M.

Example 6.3.1 Let $\mathfrak{E} = \langle A, R_1, \ldots, R_6 \rangle$ where A is a set of points in the Euclidean plane, R_1 is the set of Euclidean circles on A, R_2 is the set of Euclidean lines on A, $R2$ the planar Euclidean incidence relation on A, and R_4, R_5, and R_6, the Euclidean congruence relations on A for respectively line segments, triangles, and angles. Assume D″$(R_1,\ldots, R_6)$. Let X be the set of equivalence classes of congruent line segments. Define the binary relation $\succeq$ on X as follows: For each x and y in X, $x \succeq y$ if and only if for some a in x and some b in y, $a \supset b$. (Note that in this axiomatization, line segments are subsets of Euclidean points. Because of this, the relation 2, which is intrinsic relative to M, can be employed to define concepts about line segments.) Define $\oplus$ on X as follows: For each x, y, and z in X, $x \oplus y=z$ if and only if there exist a in x, b in y, and c in z such that $c=a \cup b$ and $a \cap b$ consists of a single point. (Note again the concepts $\cup$, $\cap$, and singleton set used in the above definition of $\oplus$ are intrinsic relative to M.) Using facts about Euclidean geometry, it is not difficult to show that X is intrinsic relative to M. Thus $\succeq$ and $\oplus$ are intrinsic relative to M. By employing simple well-known results of Euclidean geometry, it can be shown that $\mathfrak{X} = \langle X, \succeq, \oplus \rangle$ is a continuous extensive structure.

In many circumstances one may view the domain of a physical quality used to measure a fundamental physical dimension (e.g., mass, charge, length, time, etc.) as a set equivalence classes of sets of atoms. In these circumstances, it is also natural and correct to view such domains D as being intrinsic relative to M. Because of this, there is a natural total ordering on D that is intrinsic relative to M: For all x and y in D, $x \succeq_D y$ if and only if for some a in x and b in y, $a \supseteq b$. ($\succeq_D$ is clearly a total ordering; it is intrinsic relative to M because it is defined in terms of the intrinsic concepts D and $\supseteq$.) This, together with the fact that natural "zero elements" can be added to the domains, is why in the definition of "physical equivalence" given in Definition 5.10.16, I had the equivalent structures share a common ordering, $\succeq$, but have unique operations, $\oplus_r$.

6.3.2 Enervation of Inferential Techniques Due to the Overspecification of Primitives

In Subsection 5.10.6 an example was presented in which length was measured through a continuous extensive structure of the form $\mathfrak{X}_2 = \langle X, \succeq, \oplus_2 \rangle$, where $\oplus_2$ corresponded to concatenating rods through "the right triangle method." Krantz, et al. (1971), echoing the views of Ellis (1966) stated, "For most people, the new interpretation [measuring length in terms of isomorphisms of $\mathfrak{X}_2$ onto $\langle \mathbb{R}^+, \geq, + \rangle$ seems much more artificial than the original one [measuring length in terms of isomorphisms of $\mathfrak{X} = \langle X, \succeq, \oplus \rangle$ onto $\langle \mathbb{R}^+, \geq, + \rangle$, where $\oplus$ corresponded to "abutting" rods on a straight line]. In spite of this strong feeling, neither Ellis nor the authors know of any argument for favoring the [abutting] interpretation except familiarity, convention, and, perhaps convenience. We are used to length being measured along straight lines, not along the hypotenuses of right triangles, but no empirical reasons appear to force that choice." *(Krantz, et al. (1971), pp. 87–88.* See Subsection 5.10.6 for a fuller discussion and additional quoted material from Krantz, et al., 1971, on this subject.)

I believe that $\mathfrak{X}$ should be favored over $\mathfrak{X}_2$ for measuring length in many geometrical settings. However, I also believe that in many physical situations where only an impoverished part of Euclidean geometry is needed to characterize the critical features of a physical law under consideration (as is the case in many applications of dimensional analysis), there is no reason to favor $\mathfrak{X}$ over $\mathfrak{X}_2$ "except familiarity, convention, and, perhaps convenience." Thus for me the favoring of $\mathfrak{X}$ over $\mathfrak{X}_2$ in general depends on the situation under consideration.

By imbedding the structures $\mathfrak{X}$ and $\mathfrak{X}_2$ into a Euclidean plane, one sees that the continuous extensive structure $\mathfrak{X}$ measures length of line segments according to the usual Euclidean definition; whereas, the continuous extensive structure $\mathfrak{X}_2$ measures area of squares in terms of the length of their sides according to the usual Euclidean definition. Thus *in terms of the larger structure in which they are jointly imbeddable,* $\mathfrak{X}$ and $\mathfrak{X}_2$ are distinguishable in the sense that they have different planar Euclidean properties. However in many settings, there is no reason to do the imbedding, and for reasons discussed below, such an imbedding may cloud the empirical relationship one is investigating.

Whether in a particular applied situation meaningfulness, ε-intrinsicness, or a form of relative intrinsicness is the appropriate concept to employ is often a subtle matter, involving not only the goals of the application, but also how the underlying situation is to be conceptualized. In many ways this is similar to the problem in dimension analysis of specifying the relevant variables on which to base the dimensional analysis. In dimensional analysis, adding variables irrelevant to the particular application under consideration (but perhaps relevant to a complete physical characterization of the situation) generally reduces the effectiveness of the inferential techniques. (See the quote by Bridgman in Subsection 5.10.2.) Similarly, overspecification of the situation under consideration tends to limit the effeqtiveness of epistemological uses of the meaningfulness and the intrinsicness concepts.

For example, consider the situation of the extensive measurement of length. Let $\mathfrak{X} = \langle \succsim, \oplus \rangle$ be the continuous extensive structure described in Example 6.3.1 for measuring length. Let *m* stand for the equivalence class of line segments that are 1 meter, and let $\mathfrak{X}_m = \langle X, \succsim, \oplus, m \rangle$. Both $\mathfrak{X}$ and $\mathfrak{X}_m$ may appropriately be considered as "physical structures." Both can be used to measure length in appropriate manners through the representational theory. However, in fundamental ways they are different: Mathematically, they exhibit different kinds of invariance; in terms of meaningfulness, they determine different topics. In the characterization of many lawful situations involving length, *m,* although a physically valid concept, is irrelevant. If this irrelevancy is included in inferences involving invariance (e.g., in inferences using techniques similar to dimensional analysis), then in general, the inferences will yield less sharp results than in the case where the irrelevancy was excluded.

Similarly, in many physical applications using dimensional analysis, the "length dimension" may be measured in a way corresponding to measurement by a structure that is physically equivalent to $\mathfrak{X}$. However, in other physical applications, which employ length in an essential way as part of a multidimensional Euclidean space, only $\mathfrak{X}$ rather than a different structure that is physically equivalent to it is appropriate for measuring the "length dimension."

6.3.3 The Relativity of Meaningfulness and Intrinsicness

Throughout this book, meaningfulness and intrinsicness are used as *relative* concepts that vary from application to application. For example, in an application of fluid dynamics a certain concept may be meaningful while in another application of fluid dynamics be non-meaningful, and similarly, in one application an intrinsic approach to length may be called for, while in another application the particular extensive structure $\mathfrak{X}$ described above is needed for capturing the role of length in the application.

Unfortunately, in the literature various authors appear to interpret meaningfulness as an absolute concept—even in cases where they are commenting on articles where it has been explicitly presented otherwise. Perhaps this has to do with the word "meaningfulness" containing the term "meaning," the latter connoting to some readers an absolute property. In any case, the meaningfulness literature—particularly the part involving "meaningful statistics"—has generated a number of controversies, most of which disappear when "meaningfulness" is given a relative interpretation.

6.4 Lawfulness

6.4.1 Introduction

Previous sections have utilized the concept of meaningfulness for describing the kinds of lawfulness that appeared in the possible psychophysical laws (Sections 2.6, 2.7, 5.7), Weber's Law (Section 5.9) and dimensional analysis (Section 5.10). This section investigates the possibility and utility of employing the more flexible and richer concept relative intrinsicness for these and related purposes. It is argued that relative intrinsicness is the driving force behind the kinds of lawfulness found in the possible psychophysical laws, Weber's Law, Stevens' Power Law, dimensional analysis, and many other laws in science.

6.4.2 Possible Psychophysical Laws

Consider the case of a possible psychophysical law F. In the formulation of Section 5.7 of Chapter 5, this "law" was conceptualized qualitatively as F being an order preserving function from a continuous structure $\mathfrak{X}_1 = \langle X, \succeq_1, R_j \rangle_{j \in J}$ onto a continuous structure $\mathfrak{X}_2 = \langle Y, \succeq_2, S_k \rangle_{k \in K}$ such that $X \cap Y = \varnothing$ and $X \cup Y = A$. Let G be the automorphism group of

$$\mathfrak{Z} = \langle A, F, X, \succeq_1, R_j, Y, \succeq_2, S_k \rangle_{j \in J,\ k \in K} .$$

Lemma 6.4.1 Let γ be an arbitrary element of G and α and β be respectively the restrictions of γ to X and Y. Then α is an automorphism of $\mathfrak{X}_1$, β is an automorphism of $\mathfrak{X}_2$ and $\gamma = \alpha \cup \beta$.

Proof. Immediate from the definitions of α and β and the fact that γ is an automorphism of $\mathfrak{Z}$.

Let

G_2={$\alpha|\alpha$ is an automorphism of $\mathfrak{X}_1$ and there exists an automorphism β of G_2 such that $\alpha \cup \beta$ is in G}

and

G_2={$\beta|\beta$ is an automorphism of $\mathfrak{X}_2$ and there exists an automorphism α of $\mathfrak{X}_1$ such that $\alpha \cup \beta$ is in G}.

Assume axiom system TM with G being the transformation group for M. Then F is meaningful. Thus according the analysis of Section 5.7 of Chapter 5, the "lawfulness" of F consists of it satisfying the following condition:

For each α in G_1 there exists β in G_2 such that for all x in X,

$$F[\alpha(x)] = \beta[F(x)].$$

Note that this condition is intrinsic relative to M, F, G_1, G_2. This statement of relative intrinsicness can be reduced to "intrinsic relative to M, F" by the following theorem.

Theorem 6.4.1 Let F and G be as above and assume axiom system TM with G as the transformation group for M. Then the following two statements are equivalent:

1. Let G_1, G_2, $\mathfrak{X}_1$ and $\mathfrak{X}_2$ be as above. For each α in G_1 there exists β in G_2 such that for all x in X,

$$F[\alpha(x)] = \beta[F(x)].$$

2. Let

   H={$\alpha|\alpha$ is a function from the domain of F onto the domain of F and there exists a function β from the range of F onto the range of F such that $\alpha \cup \beta$ is in G}.

   Then for each α in H, there exists a function β from the range of F onto the range of F such that for all x in X,

$$F[\alpha(x)] = \beta[F(x)].$$

Proof. Suppose Statement 1. By hypothesis X is the domain of F and Y is the range of F. Suppose α is an arbitrary element of H. By the definition of H, let δ be a function from the range of F onto the range of F such that $\alpha \cup \delta$ is in G. Then by Lemma 6.4.1, $\alpha \in G_1$. Thus by Statement 1, there exists β in G_2 such that for each x in X,

$$F[\alpha(x)] = \beta[F(x)].$$

Because β is a function from the range of F onto the range of F and α is an arbitrary element of G_1, Statement *2* has been shown.

Suppose Statement 2. Let α be an arbitrary element of G_1. Then it follows from the definitions of G_1 and H that $\alpha \in H$. Thus by Statement 2, let β be a function from the range of F onto the range of F such that $\alpha \cup \beta$ is in G and for all x in X,

$$F[\alpha(x)] = \beta[F(x)] .$$

Then by Lemma 6.4.1, $\beta \in G_2$, and Statement 1 follows.

Note that in Theorem 6.4.1, Statement 1, which is the "lawful part" of a possible psychophysical law, is logically equivalent to Statement 2, which is formulated in terms of L ($\in$, A, $\emptyset$), F, and G, and thus, because G is intrinsic relative to M, is logically equivalent to a statement that is intrinsic relative to M, F. Also note that Theorem 6.4.1 assumes neither the homogeneity of G_1 nor the homogeneity of G_2.

Consider the special case of Theorem 6.4.1 where

$$\mathfrak{X}_1 = \langle X, \succsim_1, \oplus_1 \rangle \text{ and } \mathfrak{X}_2 = \langle Y, \succsim_2, \oplus_2 \rangle$$

are continuous extensive structures. Let S and T be respectively the sets of isomorphisms of $\mathfrak{X}_1$ and $\mathfrak{X}_2$ onto $\mathfrak{R} = \langle \mathbb{R}^+, \geq, + \rangle$, and let φ be an element of S and Ψ be an element of T. Then by Theorem 6.4.6 below, there exists r and s in $\mathbb{R}^+$ such that for all x in X,

$$\psi[F(x)] = r \cdot \varphi(x)^s . \tag{6.1}$$

In Equation 6.1, r, which depends of the choices of the representations φ and ψ from S and T respectively, is not qualitatively S,T-meaningful; that is, r has no qualitative interpretation in the structure $\mathfrak{Z}$. Therefore, if a substantive conclusion depends on r being a particular real number, then $\mathfrak{Z}$ is not rich enough structure of primitives for that conclusion to have a qualitative interpretation in terms of the primitives of $\mathfrak{Z}$. s is qualitatively S,T-meaningful; it is qualitatively interpretable within $\mathfrak{Z}$ as a certain kind of trade-off between the n-copy operators of $\langle X, \succsim_1, \oplus_1 \rangle$ and the m-copy operators of $\langle Y, \succsim_2, \oplus_2 \rangle$. s, however, is not intrinsic relative to M,F, because it depends on the selections of the extensive structures $\mathfrak{X}_1$ and $\mathfrak{X}_2$ used to measure the domain and range of F.

Recall that Luce's 1959 paper on the possible psychophysical laws was founded on the following two theoretical principles:

> A substantive theory relating two or more variables and the measurement theories for these variables should be that:
>
> 1. *(Consistency of substantive and measurement theories)* Admissible transformations of one or more of the independent variables shall lead, via the substantive theory, only to admissible transformations of the dependent variables.

2. *(Invariance of the substantive theory)* Except for the numerical values of parameters that reflect the effect on the dependent variables of admissible transformations of the independent variables, the mathematical structure of the substantive theory shall be independent of admissible transformations of the independent variables. *(Luce, 1959, Pg. 85)*

Principle 1 of the substantive theory is captured qualitatively by Statement 2 of Theorem 6.4.1, and Principle 2 is captured qualitatively by *(i)* requiring F to be a primitive of the underlying structure, and *(ii)* having Principle 1 to be intrinsic relative to M,F. Thus the above qualitative version of possible psychophysical laws is in close agreement with Luce's original ideas on the subject. In this context, the ramifications of Rozeboom's criticisms of Luce's theory (Subsection 2.5.2) is investigated next.

Recall that Rozeboom argued that Luce's Principles should not be used as a general criterion for lawfulness because they were inconsistent with certain types of physical laws, for example, the law of radioactive decay given by the formula

$$q = ae^{-bt}, \tag{6.2}$$

where q is a positive real number representing a measured quantity of mass, and t is a positive real number representing a measured quantity of time occurring after a specific time, and a and b are empirically determined quantities. Rozeboom's observation is correct about Luce's formulation. To better understand the differences between Luce's possible psychophysical laws and the law of radioactive decay, consider the following qualitative formulation.

Let X be the set of times and $\mathfrak{X}_1 = \langle X, \succsim_1, \oplus_1 \rangle$ be a continuous extensive structure employed to measure time, Y be the set of masses and $\mathfrak{X}_2 = \langle Y, \succsim_2, \oplus_2 \rangle$ be a continuous extensive structure employed to measure mass.

The law of radioactive decay sates a relationship between an initial amount of mass and *the percent of it* that remains after a time interval t. The percentage depends only on t. Because such a percentage is dimensionless, being in this case the ratio of the measurements of two masses, its proper qualitative interpretation is as an automorphism of the structure $\mathfrak{X}_2$ used to measure mass.

Let $\mathcal{A}$ be the set of automorphisms of $\mathfrak{X}_2$ and let $\succsim'$ be the binary relation on $\mathcal{A}$ such that for all for α and β in $\mathcal{A}$,

$$\alpha \succsim' \beta \text{ iff } \alpha(x) \succsim \beta(x) \text{ for all } x \text{ in } X.$$

Then it is not difficult to show that $\succsim'$ is a total ordering on $\mathcal{A}$. (For example, see Lemma 2.2 of Narens, 1991a, or Lemma 4.3 and Definition 4.3 of Narens, 1985.) The set of percentages for mass measurement corresponds to

$$\mathcal{P} = \{\alpha \,|\, \iota \succ' \alpha\},$$

where ι is the identity automorphism of $\mathfrak{X}_2$.

Theorem 6.4.2 Let $\mathfrak{X}_1$, $\mathfrak{X}_2$ and P be as above. Let D be a function from X onto P such that for all t and u in X,

$$t \succeq_1 u \text{ iff } D(t) \succeq' D(u)\,. \tag{6.3}$$

Let φ and Ψ be respectively isomorphisms of $\mathfrak{X}_1$ and $\mathfrak{X}_2$ onto $\langle \mathbb{R}^+, \geq, + \rangle$. And For each t in X, let $p(t)$ be the positive real such that $\psi(D(t))$ is multiplication by $p(t)$. Then the following two statements are equivalent:

1. For each t in X there exists a positive real number r_φ such that

$$p(t) = e^{-r_\varphi \cdot \varphi(t)}\,.$$

2. For all t and u in X,

$$D(t \oplus_1 u) = D(t) * D(u)\,,$$

where, as usual, * is the operation of function composition.

Proof. Suppose Statement 1. Let t and u be arbitrary elements of X. Then, because by hypothesis,

$$\varphi(t \oplus_1 u) = \varphi(t) + \varphi(u)\,,$$

it follows that

$$\begin{aligned} p(t \oplus_1 u) &= e^{-r_\varphi \cdot \varphi(t \oplus_1 u)} = e^{[-r_\varphi \cdot \varphi(t)] + [-r_\varphi \cdot \varphi(u)]} \\ &= e^{-r_\varphi \cdot \varphi(t)} \cdot e^{-r_\varphi \cdot \varphi(u)} = p(t) \cdot p(u)\,. \end{aligned}$$

Because the group of automorphisms of $\langle \mathbb{R}^+, \geq, + \rangle$ is the multiplicative group on $\mathbb{R}^+$, it follows from the above equation, the definition of p, and φ being an isomorphism that

$$\psi(D(t \oplus_1 u) = p(t \oplus_1 u) = p(t) \cdot p(u) = \psi(D(t)) \cdot \psi(D(u)) = \psi(D(t) * D(u)),$$

and thus that

$$D(t \oplus_1 u) = D(t) * D(u).$$

Therefore Statement 2 is true.

Suppose Statement 2. Because $\psi(D)$ is the set of multiplications by positive reals <1, it follows from Equation 6.3 that

$$v \succeq_1 w \text{ iff } p(v) \geq p(w),$$

and it follows from Statement 2 that

$$p(v \oplus_1 w) = p(v) \cdot p(w) \tag{6.4}$$

for all positive v and w<1. It is well known that the only strictly increasing solution to Equation 6.4 for this domain and range of values is

$$p(v) = e^{cv}\,,$$

where c is a negative real constant. Statement 1 easily follows from this.

In Statement 1 of Theorem 6.4.2, the law of radioactive decay is stated in terms of all possible isomorphisms of $\mathfrak{X}_1$ and $\mathfrak{X}_2$ onto $\langle \mathbb{R}^+, \geq, + \rangle$. An equivalent of this is customary in physics. This equivalent form of representation, which amounts to *first* representing

$$\mathfrak{W} = \langle X \cup Y, X, \succeq_1, \oplus_1, Y, \succeq_2, \oplus_2 \rangle$$

through isomorphisms onto a numerical structure and *then* representing D in terms of the representation of $\mathfrak{W}$., is in conflict with the representational theory, which represents

$$\langle X \cup Y, D, X, \succeq_1, \oplus_1, Y, \succeq_2, \oplus_2 \rangle$$

through isomorphisms onto a numerical structure. The approach of physics and its equivalent allows for a unified set of conventions that make dimensional analysis and other forms of "physical inference" user friendly. The cost of this is a sloppy foundation that often requires physical insight to understand what a quantitative statement is asserting.

Luce's possible psychophysical laws are about functions from the domain of a structure onto the domain of another structure. Note that the law of radioactive decay has a quite different form: it is about a function from a domain of a structure into the set of automorphisms of another structure.

Luce's 1959 theory of possible psychophysical laws has a number of natural generalizations in terms of families of functions. The remainder of this subsection investigates a few of these for families of functions of a single variable.

Convention 6.4.1 Let

$$\mathfrak{X} = \langle X, R_j \rangle_{j \in J}, \quad \mathfrak{Y} = \langle Y, S_k \rangle_{k \in K}, \quad \text{and} \quad \mathfrak{Z} = \langle A, T_h \rangle_{h \in H}$$

be structures such that $A = X \cup Y$ and $X \cap Y = \varnothing$. The primitives of $\mathfrak{Z}$ may have primitives in common with $\mathfrak{X}$ or $\mathfrak{Y}$—including the case where each of its primitives is a primitive of either of $\mathfrak{X}$ or $\mathfrak{Y}$—or $\mathfrak{Z}$ may have no primitives in common with $\mathfrak{X}$ and $\mathfrak{Y}$). Let G_X, G_Y, and G be respectively the automorphism groups of $\mathfrak{X}$, $\mathfrak{Y}$ and $\mathfrak{Z}$. Let M be the meaningfulness predicate that results from axiom system TM with A and the set of atoms and G as the transformation group for M.

Definition 6.4.1 Assume Convention 6.4.1. Then the following three definitions hold:

(1) G is said to be *component consistent* if and only if $M(X)$, $M(Y)$, G_X is the restriction of G to X, and G_Y is the restriction of G to Y.

(2) $\mathcal{F}$ is said to be *relatively invariant* if and only if $\mathcal{F}$ be a nonempty set of one-to-one functions from X onto Y, $M(\mathcal{F})$, and

$$\forall \gamma [\gamma \in G \rightarrow \exists F [F \in \mathcal{F} \wedge \gamma(F) = F]] .$$

(3) $\mathcal{F}$ is said to be *universally invariant* if and only if $\mathcal{F}$ be a nonempty set of one-to-one functions from X onto Y and

$$\forall\gamma\forall F[\gamma \in G \wedge F \in \mathcal{F}] \to \gamma(F) = F\,.$$

(Note that if $\mathcal{F}$ is universally invariant, then $M(\mathcal{F})$ by axiom system TM.)

Component consistency may be viewed as a weakened form of Luce's principle of "consistency of substantive and measurement theories." In Definition 6.4.1, "G is component consistent" is intrinsic relative to M, X, Y. Relative and universal invariance are generalized forms of Luce's principle of "invariance of the substantive theory," with universal invariance being a special form of relative invariance. In Definition 6.4.1, relative and universal invariance are intrinsic relative to M, $\mathcal{F}$.

Theorem 6.4.3 Assume Convention 6.4.1. Suppose G is component consistent. Then the following two statements are true:

1. Suppose $\mathcal{F}$ is relatively invariant. Then for each automorphism α of $\mathfrak{X}$, there exist an automorphism β of $\mathfrak{Y}$ and an element F in $\mathcal{F}$ such that for all x in X,

$$F[\alpha(x)] = \beta[F(x)]\,.$$

2. Suppose F is universally invariant. Then for each automorphism α of $\mathfrak{X}$, there exists an automorphism β of $\mathfrak{Y}$ such that for all F in $\mathcal{F}$ and x in X,

$$F[\alpha(x)] = \beta[F(x)]\,.$$

Proof. Immediate from Definition 6.4.1.

Assume F is universally invariant. Then Luce's 1959 theory of the Possible Psychophysical Laws can be viewed as the special case of Theorem 6.4.1 where F consists of a single function.

Intrinsicness also plays an important role in dimensionally invariant laws of dimensional analysis (Section 5.10), and a qualitative treatment of dimensionally invariant laws similar to the above one for possible psychophysical laws can be carried out.

6.4.3 Weber's Law

In Section 5.9, Weber's Law was axiomatized in terms of the structure $\mathfrak{W} = \langle A, \succsim_\star, T, \oplus\rangle$, where $\mathfrak{T} = \langle A, \succsim_\star, T\rangle$ was a continuous threshold structure and $\mathfrak{E} = \langle A, \succsim_\star, \oplus\rangle$ was a continuous extensive structure. Under these assumptions, Weber's Law was expressed qualitatively as follows: For all x and y in A,

$$T(x \oplus y) = T(x) \oplus T(y)\,. \tag{6.5}$$

While Equation 6.5 is an elegant and testable description of the compatibility that needs to hold between a physical structure $\mathfrak{E}$ and the psychological structure

$\mathfrak{T}$ for Weber's Law to hold, it is not very revealing about the "lawful nature" of the compatibility. To achieve the latter, a more intrinsic characterization of Weber's Law is needed, and this is provided by the next theorem.

Theorem 6.4.4 Let $\mathfrak{W} = \langle A, \succeq_\star, T, \oplus\rangle$, G be the automorphism group of $\mathfrak{W}$, $\mathfrak{E} = \langle A, \succeq_\star, \oplus\rangle$, and G' be the automorphism group of $\mathfrak{E}$. Assume $\mathfrak{E}$ is a continuous extensive structure and $\mathfrak{T} = \langle A, \succeq_\star, T\rangle$ is a continuous threshold structure. Then the following two statements are logically equivalent:

1. For all x and y in A,

$$T(x \oplus y) = T(x) \oplus T(y) .$$

2. $G=G'$.

Proof. Theorem 6.9.2.

Assume axiom system TM with G as the transformation group of M. Then Statement 2 of Theorem 6.4.4 is intrinsic relative to M, G'. Because of the equivalence of Statement 2 with Statement 1 in Theorem 6.4.4, it follows that under the hypotheses of the theorem, Weber's Law is logically equivalent to the more generally stated and intrinsic version of it given by Statement 2. Let M' be the meaningfulness predicate that result from assuming axiom system TM with the transformation group G'. Then Statement 2 says that $M=M'$, that is, that $\mathfrak{W}$ and $\mathfrak{E}$ determine the same topic. Thus, from this point of view, the "lawfulness" of Weber's Law may be stated as follows: All the psychology in the topic determined by $\mathfrak{W}$ is characterizable purely in terms of the physical primitives A, $\succeq^*$, and $\oplus$ of $\mathfrak{W}$. It is not difficult to show (through use of Theorems 5.4.3 and 5.4.5) that the physical operation $\oplus$ is not in the topic determined by axiom system TM with transformation group the automorphism group of $\mathfrak{T}$.

6.4.4 Stevens' Psychophysical Power Law

A quantitative characterization of Stevens' psychophysical power law using magnitude estimation and the preservation of equal ratios was discussed in Section 2.8. A qualitative, behavioral axiomatization of magnitude estimation was presented in Section 5.8. The latter is extended in this subsection to include as primitives an additional physical operation and an additional, observable, qualitative axiom. Stevens' psychophysical power law then results when the extended system is represented by isomorphisms onto an appropriate numerical structure. The extended axiomatic characterization is also useful in revealing the intrinsic nature of Stevens' power law.

Theorem 6.4.5 Let

$$\mathfrak{Z} = \langle A, \succeq, \oplus, \overline{1}, \overline{2}, \ldots, \overline{p}, \ldots\rangle_{p \in \mathbb{I}^+}$$

be such that

$$\mathfrak{X} = \langle A, \succsim, \oplus \rangle$$

is a continuous extensive structure of physical stimuli and

$$\mathfrak{B} = \langle A, \succsim, \overline{1}, \overline{2}, \ldots, \overline{p}, \ldots \rangle_{p \in \mathbb{I}^+}$$

is a behavioral structure associated with E (Definition 5.8.3) that satisfies the axioms for behavioral magnitude estimation (Axioms 5.8.1 to 5.8.4). Then the following three statements are equivalent.

1. *(Stevens' Psychophysical Power Law)* Let

$$\mathcal{S} = \{\varphi \mid \varphi \text{ is an isomorphism of } \mathfrak{X} \text{ onto } \langle \mathbb{R}^+, \geq, + \rangle\}$$

and

$$\mathcal{T} = \{\psi \mid \psi \text{ is a multiplicative representing function for } \mathfrak{B} \text{ (Definition 5.8.4).}\}$$

Then for each φ in S and each ψ in T there exist r and s in $\mathbb{R}^+$ such that for all x in A,

$$\psi(x) = r\varphi(x)^s .$$

2. For each p in $\mathbb{I}^+$ and each x and y in A,

$$\overline{p}(x \oplus y) = \overline{p}(x) \oplus \overline{p}(y) .$$

3. Let G be the set of automorphisms of $\mathfrak{Z}$, H be the set of automorphisms of $\mathfrak{X}$, and K be the set of automorphisms of $\mathfrak{B}$. Then G=H=K.

Proof. Theorem 6.9.3.

Using the notation of Theorem 6.4.5, let M, M_H, and M_k be the meaningfulness predicates that result from assuming axioms system TM with respectively G, H, and K as transformation groups. Then by Statement 3, $M = M_H = M_K$.

Statement 1 is Stevens' power law for psychophysical magnitude estimation. Like in the case of possible psychophysical laws above (Subsection 6.4.2), the constant r is not qualitatively S,T-meaningful and therefore has no qualitative interpretation in terms of $\mathfrak{Z}$; whereas the constant s is qualitatively meaningful with respect to $\mathfrak{Z}$ and has an interpretation within $\mathfrak{Z}$. The constant s is neither intrinsic relative to M,$\{\succsim, \overline{1}, \ldots, \overline{p} \ldots\}_{p \in \mathbb{I}^+}$ nor intrinsic relative to M,$\succsim, \oplus$.

Statement 2 corresponds to a simple qualitative observation. It is a qualitative way of testing Stevens' power law for magnitude estimation.

Statement 3 shows that given the hypotheses of the theorem, Stevens' power law for magnitude estimation is intrinsic relative to M. This last fact is given a substantive interpretation in Section 6.5.

6.4.5 Luce's Possible Psychophysical Laws, 1990

Luce (1990) reconsidered his and others work on the possible psychophysical laws. He reformulated qualitatively his earlier theory and applied it to a range of psychological issues. This subsection presents a summary and discussion of the part of Luce (1990) that is concerned with the reformulation of the earlier theory. As is shown in the following quotations, Luce drifts away from his previous position that the possible psychophysical laws were a specialization of dimensional analysis and focuses much more on their qualitative underpinnings, which he views as closely related to empirical testing.

> Historically, the first was my 1959 article "On the Possible Psychophysical Laws," which attempted to account for why two ratio-scaled variables, such as those encountered in the simplest version of cross-modal matching, should be related by power functions. In it, I postulated that if x and y are two ratio-scaled variables that are related by some law $y=f(x)$, where f is a strictly increasing function, and if the units of x are changed by a ratio transformation r, then there is a corresponding ratio change, s(r) of Y such that for all positive x and r,
>
> $$s(r)f(x) = s(r)y = f(rx)\,. \tag{6.6}$$
>
> As is easily demonstrated, this functional equation for f implies that f is a power function.
>
> The key issue surrounding the article, which was first critically discussed by Rozeboom (1962a, 1962b), is, Why should one assume Equation 6.6? I had spoken of it as a "principle of theory construction," thinking it to be on a par with the dimensional invariance of physical laws postulated in the method of dimensional analysis...In the face of Rozeboom's criticism, I (1962) retreated from that position. Later, I (1964); Aczél, Roberts, and Rosenbaum (1986); and Osborne (1970) studied generalizations of Equation 6.6, and Falmagne and Narens (1983) gave a detailed analysis of a collection of closely related principles, showing how they interrelate and what they imply about the forms of the laws. Despite these interesting developments, I have never felt that we have gained a full understanding of what is really involved empirically in assuming Equation 6.6, except I have come to recognize that it is not really the specialization of dimensional invariance that I had thought it was in 1959. *(Luce, 1990, pg. 66)*

Luce formulated his new notion as follows: He first supposed that $\mathfrak{X}_1 = \langle X, \succsim_1, R_j\rangle_{j\in J}$ and $\mathfrak{X}_2 = \langle Y, \succsim_2, S_k\rangle_{K\in K}$ were structures to measure X and Y. He then assumed that these structures were homogeneous and finitely unique, so that Theorem 5.4.16 applied, and thus each of these structures had a homogeneous, 1-point unique group of translations (Definitions 5.1.5,

5.1.6 and 5.1.7). He considered a *matching relation* to be a strictly increasing function from X onto Y. Luce had particular psychological applications in mind, generally where the domain and range of F were physical variables that initiated sensations in different psychological modalities, and F corresponded to the matching of a subject's psychological intensity of a item from X with an item in Y—"cross modal matching." A matching relation F was said to be *translation consistent* if and only if for each translation (Definition 5.1.7) τ of $\mathfrak{X}_1$ there existed a translation σ_τ of $\mathfrak{X}_2$ such that for all x in X,

$$F(x) = y \ \text{ iff } \ F[\tau(x)] = \sigma_\tau[F(x)]\,.$$

Luce then showed the equivalent of the following theorem.

Theorem 6.4.6 Let $\mathfrak{X}_1 = \langle X, \succsim_1, R_j\rangle_{j\in J}$ and $\mathfrak{X}_2 = \langle X, \succsim_2, S_k\rangle_{K\in K}$ be continuous, 1-point homogeneous, finitely point unique structures. Then the following two statements are equivalent:

1. F is a translation consistent matching relation.
2. There exists scales $\mathcal{S}$ and $\mathcal{T}$ of isomorphisms of respectively $\mathfrak{X}_1$ and $\mathfrak{X}_2$ onto numerical structures based on $\mathbb{R}^+$ such that
   (i) the group of translations of $\mathfrak{X}_1$ is mapped by each element of $\mathcal{S}$ onto the multiplicative group of $\mathbb{R}^+$ and the group of translations of $\mathfrak{X}_2$ is mapped by each element of $\mathcal{T}$ onto the multiplicative group of $\mathbb{R}^+$, and
   (ii) for all φ in $\mathcal{S}$ and ψ in $\mathcal{T}$ there exist positive reals r and s such that for all x in X and y in Y,

$$y = F(x) \ \text{ iff } \ \psi(y) = r\varphi(x)^s\,.$$

Proof. Theorem 1 in Appendix B of Luce (1990).

Luce makes the following comment about some of the differences between his previous 1959 formulation and Theorem 6.4.6.

> Note that the concept of translation consistency is closely related to the concept embodied in Equation 6.6, but it is different. It says nothing about representations of the structures or about how changes in the representations relate to one another. Rather, it asserts a way that transformations with the two structures describing the two modalities might systematically relate to one another—if one pair of stimuli match, then to each translation of one structure there is coordinate translation of the second structure that preserves the match. The concept says in the case of cross-modal matches of physical continua that the psychological substance embodied in the matching relation F is consistent with (or invariant under) certain pairs of translations from the two modalities that it relates. Assuming that the modalities are physical, as is often the case, the hypotheses that the

> matching relation is translation consistent amounts to the postulate: The psychology is conceptually consistent with the physics.
>
> Unlike Equation 6.6, for which it is difficult to know what constitutes a suitable test, the form of consistency embodied in [translation consistency] can be tested empirically provided that the translations are known. Of course, the test may be empirically falsified. By contrast, statements about representations and the numerical law relating them are not so clearly empirical. *(Luce, 1990, pg. 70)*

There are similarities and differences between Luce's approach to possible psychophysical laws and the approach given in Subsection 6.4.2, and more generally the approach to laws presented throughout this section. The similarities consist of qualitatively describing the laws as specific kinds of compatibilities between structures of primitives. In Luce (1990) the compatibility is described by translation consistency, and in Subsection 6.4.2 by a slightly more general related principle. Both are the natural qualitative interpretations of Equation 6.6 to their respective underlying qualitative situations. There are several differences. The first is that the ideas and methods of Subsection 6.4.2 apply to many kinds of structures—not to only those with matching relations or to their natural generalizations to functions of several variables. Second, a much more general idea of compatibility is presented in Subsection 6.4.2 than in Luce—a general idea of compatibility that is linked to the concept of intrinsicness, and one that is much more flexible for formulating and *justifying* concepts of lawfulness. These first two differences have their greatest impact in qualitative situations different from those considered by Luce. The next difference applies to matching function situations considered by Luce.

Let $\mathfrak{X}_1 = \langle X, \succsim_1, R_j \rangle_{j \in J}$ and $\mathfrak{X}_2 = \langle Y, \succsim_2, S_k \rangle_{K \in K}$ be continuous structures with disjoint domains and F be a matching function from X onto Y. The third difference concerns the meaningfulness of F. In the treatment of possible psychophysical laws in Subsection 6.4.2, it was required that F be meaningful. The meaningfulness of F versus the non-meaningfulness of F has an impact on the possible numerical forms F can take under measurement (see Section 5.7). In Subsection 6.4.2 the meaningfulness of F was achieved by having it be a primitive of a structure

$$\mathfrak{Z} = \langle X \cup Y, F, X, \succsim_1, R_j, Y, \succsim_2, S_k \rangle_{j \in J,\ k \in K}\,,$$

which contains $\mathfrak{X}_1$ and $\mathfrak{X}_2$ as substructures. In Luce (1990) it is not clear whether the qualitative situation theorized about is the single structure $\mathfrak{Z}$ or the three structures, $\mathfrak{X}_1, \mathfrak{X}_2$, and $\langle X \cup Y, F \rangle$. It makes a difference for measurement and meaningfulness considerations which situation is being theorized about.

The fourth difference is about the desirability of having matching functions be part of the qualitative formulations of psychophysical situations. Luce considers it highly desirable, and in some cases, such as having a subject match sensations from one modality with sensations from another so that the matches represent equal subjective intensity, I agree with him. However, in cases where the matching functions are the natural qualitative analogs of the psychophysical functions-that is, functions that map the physical measurements of stimuli onto derived measurements of psychological intensity—I prefer not to have matching functions be part of the qualitative formulation. My reasons are that a good deal of psychophysics can be formulated qualitatively without them, for example, Weber's and Stevens' Laws of Subsections 6.4.3 and 6.4.4, and that, as is shown in Section 6.5 below, it is often highly desirable to characterize the qualitative psychophysical situation in terms of a structure that cleanly splits into two substructures, one purely physical and one purely psychological, which would be impossible for qualitative structures with a primitive function F with a physical domain and a psychological range.

6.4.6 Meaningfulness Versus Intrinsicness

Meaningfulness and intrinsicness are related concepts. In some situations they coincide and in others they differ. When they differ, intrinsicness exhibits a greater degree of invariance than meaningfulness, and because of the extra invariance, intrinsic relations often appear more "lawful" than meaningful ones.

Meaningfulness is useful for drawing inferences about how a *particular* relation is related to a structure of relations. Intrinsicness, is useful for drawing inferences about the *(lawful) compatibility* of two or more structures. At a more concrete level, these kinds of intrinsic compatibilities are often exhibited by relationships between the automorphism groups of the structures involved, and when this is the case, they often correspond through measurements of the qualitative structures involved to formulae that have a particularly "lawful look" to them. Many revered, quantitative laws in varied parts of science have such a look, and if appropriately axiomatized will reflect the kinds of compatibilities described above.

Intrinsicness and related concepts have only recently entered the measurement literature as formal concepts. The next three sections present some applications of intrinsicness and concepts similar in spirit to it.

6.5 A Theory of the Psychological-Physical Relationship

This section describes a theory due to Narens and Mausfeld (1992) of the relationship of psychology and physics in the subfield of psychology known as "classical psychophysics." Narens and Mausfeld employed relative intrinsicness to formulate what they considered to be an important asymmetry in the psychological-physical relationship. They began as follows:

> Fechner was the first to present a comprehensive psychophysical theory and methodology. He carefully formulated how the psychological and physical were to be treated differently, provided a theory of the psychophysical relationship based in part on this difference in treatments, and used the different treatments as part of a methodology to test his theory. Since his time, the differences in treatment of the psychological and the physical have become less sharp, and psychophysical methodology much more eclectic and not particularly focussed on unique properties of the psychophysical relationship.
>
> In this article, a sharp theoretical distinction between the psychological and physical is reestablished. The distinction is used to formulate a new theoretical principle that restricts the kinds of concepts and analyses that can be applied to the psychological component of the psychophysical situation. The principle is based on concepts of modern logic and measurement theory, and asserts that the manner in which the physical is formulated should not influence conclusions drawn about the psychological. *(Narens and Mausfeld, 1992, pg. 461)*

As will be seen, their new principle asserts that for a wide variety of psychophysical situations (i.e., those in which the primitives can be cleanly divided into "psychological" and "physical"), intrinsicness relative to *M* and the psychological primitives is a necessary condition of psychological significance.

6.5.1 Separable Psychophysical Situations

Definition 6.5.1 Assume ZFA. $\mathfrak{X}_{\Phi\Psi}$ is said to be a *separable psychophysical situation* if and only if $\mathfrak{X}_{\Phi\Psi}$ is a structure with finitely many primitives that has the form

$$\mathfrak{X}_{\Phi\Psi} = \langle A, P_1, \ldots, P_m, R_1, \ldots, R_n \rangle \, ,$$

where $P_1, \ldots, P_m$ are qualitative physical relations and $R_1, \ldots, R_n$ are qualitative psychological relations.

Assume $\mathfrak{X}_{\Phi\Psi} = \langle A, P_1, \ldots, P_m, R_1, \ldots, R_n \rangle$ is a separable psychophysical situation. Then by definition, the substructure $\mathfrak{X}_{\Phi} = \langle A, P_1, \ldots, P_m \rangle$ is called the *physical part* (or *induced physical structure*) of $\mathfrak{X}_{\Phi\Psi}$, and the substructure $\mathfrak{X}_{\Psi} = \langle A, R_1, \ldots, R_n \rangle$ is called the *psychological part* (or *induced psychological structure*) of $\mathfrak{X}_{\Phi\Psi}$.

The restriction of separable psychophysical structures to finitely many primitives is used here to simplify notation and exposition and is not essential for the ideas developed in this section.

Note that the psychological and physical parts of a separable psychophysical situation have the same set of objects *A* as their domains, and thus the domain *A* can be looked at as either a set of physical objects or a set of psychological stimuli. Similarly, some of the physical relations may be looked at as psychological relations, for example, it may be the case that $R_1=P_1$. Although this formulation is natural for many psychophysical situations, it, for example, rules out non-separable situations where domain of the psychological structure is a set of sensations and the domain of physical structure is a set of physical stimuli. It also rules out some important situations where there is a primitive matching function from the domain of the induced physical structure onto the domain of the induced psychological structure that belongs to neither the induced physical structure nor the induced psychological structure. (Instances of the former are considered in Subsection 5.8.4, and instances of the latter in Subsections 6.4.2 and 6.4.3.)

Example 6.5.1 Weber's law (Section 5.9, Subsection 6.4.3) provides a good example of a separable physical situation. Here the primitives consist of a nonempty set of objects A that is to be understood both as a set of physical objects and as a set of psychological stimuli. For example, as physical objects the elements of A may be particular energy densities over the visible spectrum, and as psychological objects, lights to be presented to the subject. The other primitives consist of a physical binary relation $\geq^*$ on A used to order physically A, a physical concatenation operation $\oplus$ on A used to "physically add" elements of A, a psychological binary relation $\succsim'_\star$ used to totally order A, and a psychological function T from A onto A used to describe a subject's discrimination threshold.

There are three kinds of qualitative axioms.

The first kind, called the *physical axioms,* are axioms are about the physical structure $\langle A, \geq_\star, \oplus\rangle$. They say that $\langle A, \geq_\star, \oplus\rangle$ is a continuous extensive structure. (Definition 5.4.5)

The second kind, called the *psychological axioms,* are about the psychological structure $\langle A, \succsim'_\star, T\rangle$. They say that $\langle A, \succsim'_\star, T\rangle$ is a continuous threshold structure (Definition 5.4.2).

The third kind of axioms, called the *psychophysical axioms,* are about the psychophysical situation. They involve both physical and psychological relations. For the present case there are two such axioms:

1. $\geq_\star = \succsim'_\star$.
2. For each x and y in A, $T(x\oplus y)=T(x)\oplus T(y)$.

In the axiomatization of Weber's Law in Example 6.5.1, the primitive A, can be given both psychological and physical interpretations. The primitive $\geq^*$ is to be interpreted as a physical relation; however, because by a psychophysical axiom it is also the relation $\succsim'_\star$, it is also psychologically interpretable. Correspondingly, $\succsim'_\star$ is physically interpretable.

In this situation, $\oplus$ is not interpretable as a psychological concept, because, by the results mentioned at the end of Subsection 6.4.3, $\oplus$ is not set-theoretically definable in terms of the psychological primitives A, $\succsim'_\star$, and T. However, T is set-theoretically definable in terms of the physical primitives A, $\geq^*$, and $\oplus$ by the results at the end of Subsection 6.4.3.

In the axiomatization in Example 6.5.1, the physical primitives and axioms are used to describe the essential properties of the physical stimuli, which in this case are properties that allow it to be measured by a ratio scale of isomorphisms; the psychological primitives and axioms are used to describe the psychological behavior; and the psychophysical axioms are used to describe how the physical and psychological primitives interact.

6.5.2 The Equivalence Principle

Definition 6.5.2 Assume ZFA. Let

$$\mathfrak{X}_{\Phi\Psi} = \langle A, \succsim, P_1, \ldots, P_m, R_1, \ldots, R_n \rangle$$

be a separable psychophysical situation with physical part

$$\mathfrak{X}_{\Phi} = \langle A, \succsim, P_1, \ldots, P_m \rangle$$

and psychological part

$$\mathfrak{X}_{\Psi} = \langle A, R_1, \ldots, R_n \rangle.$$

Let $M_{\Phi,\Psi}$ M_{Φ}, and M_{Ψ} be respectively the meaningfulness predicate that are determined by axiom system TM with the transformation groups being respectively the group of automorphisms of $\mathfrak{X}_{\Phi,\Psi}$, the group of automorphisms of $\mathfrak{X}_{\Phi}$, and the group of automorphisms of $\mathfrak{X}_{\Psi}$. Then the *Equivalence Principle (of Narens and Mausfeld)* states that a necessary condition for an entity a to be psychologically relevant with respect to $\mathfrak{X}_{\Phi\Psi}$ is that it be $M_{\Phi\Psi}$-meaningful and intrinsic relative to M_{Φ}, $\succsim$, and the primitives of X_{Ψ}.

The Equivalence Principle can also be formulated so that it applies to assertions about the psychophysical structure. Suppose $\mathfrak{X} = \langle X, \succsim, P_1, \ldots, P_n \rangle$ is a physical structure, $\mathfrak{Y} = \langle X, R_1, \ldots, R_m \rangle$ is a psychological structure, and $\mathfrak{X}$ and $\mathfrak{Y}$ describe a psychophysical situation. Then the *Equivalence Principle* states that a *necessary* condition for a qualitative or quantitative assertion about the psychophysical structure to have psychological relevance is that the truth value of the assertion does not change when a structure that is L($\in$, *A*, $\emptyset$)-equivalent and isomorphic to $\mathfrak{X}$ is substituted for $\mathfrak{X}$ in the formulation of the assertion.

In psychophysics, one has a very good and clear understanding of primitive physical concepts and axioms involved. These are usually concepts and axioms that allow for the measurement of the physical variables involved. Traditionally, psychophysics has proceeded by formulating the psychological and non-physical psychophysical concepts quantitatively in terms of measurements of relevant physical variables. This practice often makes obscure what is qualitatively being assumed about the psychology and makes it difficult to decide which concepts and assertions are about the psychology inherent in the situation. The

Equivalence Principle, however, is very useful in such contexts in deciding the non-psychological definability of many quantitative concepts and assertions. Narens and Mausfeld make the following comments about this role of the Equivalence Principle:

> The Equivalence Principle is a theory about the relationship of the physical and the psychological in psychophysics. It is interesting to note that this relationship is asymmetric—equivalent structures can be used for describing the underlying physical situation but not the underlying psychological situation. The reason for this is that a primary goal of psychophysics is the characterization of (purely) psychological phenomena in terms of quantitative relationships among the measurements of physical variables. Formally, such psychological phenomena are viewed as *particular* qualitative relationships, which in our formulation are either primitives of the psychological structure or relationships that are definable in terms of the primitives. Because psychophysics is interested in statements about such *particular* psychological relationships, it makes no sense to demand that the truth values of these statements be invariant under substitution of [psychological relationships derived through L(∈, *A*, Ø)-equivalence]. *Particular* physical relationships are not of interest in psychophysics; instead they are used as vehicles for characterizing psychological relationships, and the Equivalence Principle is a description of their role as "vehicles". *(Narens and Mausfeld, 1992, pg. 472)*

6.5.3 Applications of the Equivalence Principle

Sizes of Weber constants

Let $\mathfrak{X}_{\Phi\Psi}$ be the psychophysical Weber's Law structure described in Example 6.5.1. Then by results in Section 5.9, Weber's Law with a Weber constant c and modified Weber constant $1+c$ holds. As discussed in Subsection 5.9.2, the Weber constant c does not correspond to a $\mathfrak{X}_{\Psi}$-meaningful concept, but the modified Weber constant $1+c$ does correspond to the psychological primitive threshold function T, which of course is $\mathfrak{X}_{\Psi}$-meaningful. However, by the Equivalence Principle, both the Weber and modified Weber constants being particular sizes, for example, $c=2$ and $1+c=3$ are not psychologically relevant.

There is no discrepancy here involving the modified weber constant: It is psychologically relevant, because it corresponds qualitatively to the psychological threshold function T. However, it being the "size 3" does not correspond to any Ψ-meaningful concept; instead, it being size 3 corresponds to the Φ-meaningful concept α_3, where α_3 is the automorphism of $\mathfrak{X}_{\Phi}$ defined by

$$\alpha_3(x) = (x \oplus x) \oplus x) \,.$$

It is also worth noting that Weber's Law, formulated as,

There exist an isomorphism φ of $\mathfrak{X}_\Phi$ onto $\langle \mathbb{R}^+, \geq, + \rangle$ and a positive real constant c such that

$$\frac{\Delta_{\varphi(x)}}{\varphi(x)} = c \, ,$$

does satisfy the Equivalence Principle, since its truth value remains unchanged when a structure $\mathfrak{X}'_\Phi$ that is physically equivalent to $\mathfrak{X}_\Phi$ is substituted for $\mathfrak{X}_\Phi$. However, it is not psychologically relevant, because concepts of the physical structure are used in an essential way in its formulation (e.g., see the end of Section 5.9).

Comparison of Weber constants from the same dimension

Example 6.5.2 Let

$$\mathfrak{X}_{\Phi\Psi} = \langle A, \geq_\star, \oplus, \succeq_1, T_1, \succeq_2, T_2 \rangle$$

be a psychophysical structure,

$$\mathfrak{X}_\Phi = \langle A, \geq_\star, \oplus \rangle$$

be its induced physical structure, and

$$\mathfrak{X}_\Psi = \langle A, \succeq_1, T_1, \succeq_2, T_2 \rangle$$

be its induced psychological structure. Let $\mathcal{S}$ be the scale of isomorphisms of $\mathfrak{X}_\Phi$ onto $\langle \mathbb{R}^+, \geq, + \rangle$. Assume

$$\mathfrak{W}_1 = \langle A, \geq_\star, \oplus, \succeq_1, T_1 \rangle \text{ and } \mathfrak{W}_2 = \langle A, \geq_\star, \oplus, \succeq_2, T_2 \rangle$$

are Weber law structures that correspond the discrimination behaviors of Subject 1 and Subject 2 respectively. Let c_1 and c_2 respectively be the Weber constants that result from $\mathfrak{W}_1$ and $\mathfrak{W}_2$ when $\mathfrak{X}_\Phi$ is measured by $\mathcal{S}$.

In psychology the Weber constant has often been employed as an index of sensitivity. For example, for the above situation, in psychology it has often been said,

> Subject 1 discriminates the stimuli in A more finely than Subject 2 if and only if $c_1 < c_2$.

Even though the Weber constants c_1 and c_2 being particular sizes are not psychologically significant by the Equivalence Principle, the ordinal comparison of their magnitudes passes the Equivalence Principle test. This does not mean that such comparisons are psychologically significant, for the Equivalence Principle is only a necessary condition for psychological significance. To show psychological significance, one needs to show that there exists a psychological interpretation. For the statement "$c_1 < c_2$," this is easy: Noting that by psychophysical axioms

$$\geq_\star = \succeq_1 = \succeq_2 ,$$

it follows that

$$c_1 < c_2 \text{ iff } 1 + c_1 < 1 + c_2 \text{ iff } \forall x[T_2(x) \succ_2 T_1(x)] ,$$

and "$\forall x[T_2(x) \succ_2 T_1(x)]$" is an assertion about $\mathfrak{X}_\Psi$ that says that the threshold function T_1 more finely separates stimuli than the threshold function T_2.

Comparisons of Weber constants from different dimensions

A particularly attractive feature of the Weber constant for psychologists is that it is dimensionless, that is, it makes no reference to the physical dimension from which it was derived. This has led psychologists to consider using ordinal comparisons of Weber constants for intermodal comparison of sensitivities, for example, comparing a subject's ability to discriminate loudness with his or her ability to discriminate brightness. This idea goes back to at least Wundt (1911, pg. 648). Since then it can be found in the majority of textbooks on psychophysics, where sometimes it is suggested implicitly by presenting tables of Weber constants for different modalities according to magnitude, and other times spelled out explicitly. Here are three typical examples of the latter:

(1) Engen (1971, pg. 19) states that "the smaller the Weber fraction the keener the sense."
(2) Baird & Noma (1978, pg. 43) argue that because the Weber ratio is dimensionless "one can compare sensitivities for different continua."
(3) Coren & Ward (1989, pg. 36): "Note that K (the Weber fraction) has no units (such as grams), so it does not depend on the physical units to measure I and ΔI. Thus, we can compare Weber fractions across different stimulus dimensions without having to worry about how the stimulus values were measured."

The next example shows that such comparisons of Weber constants do not make psychological sense *unless additional psychological structure and assumptions are present.*

Example 6.5.3 Consider the case of two Weber's Law structures arising from a single subject on two separate modalities, each involving a different physical dimension. This is formulated as follows: Let $A = A_1 \cup A_2$, where $A_1 \cap A_2 = \varnothing$,

$$\mathfrak{X}_{\Phi\Psi} = \langle A, A_1, \geq_1, \oplus_1, \succeq_1, T_1, A_2, \geq_2, \oplus_2, \succeq_2, T_2 \rangle$$

be the psychophysical structure,

$$\mathfrak{X}_{\Phi,1} = \langle A_1, \geq_1, \oplus_1 \rangle$$

be the physical structure on the first physical dimension,

$$\mathfrak{X}_{\Phi,2} = \langle A_2, \geq_2, \oplus_2 \rangle$$

be the physical structure on the second physical dimension,

$$\mathfrak{X}_{\Psi,1} = \langle A_1, \succeq_1, T_1 \rangle$$

be the psychological structure on A_1,

$$\mathfrak{X}_{\Psi,2} = \langle A_2, \succeq_2, T_2 \rangle$$

be the psychological structure on A_2 and $\mathcal{S}_1$ and $\mathcal{S}_2$ be scales of isomorphisms of respectively $\mathfrak{X}_{\Phi}$ and $\mathfrak{X}_{\Phi,2}$ onto $\langle \mathbb{R}^+, \geq, + \rangle$. Assume

$$\mathfrak{W}_1 = \langle A_1, \geq_1, \oplus_1, \succeq_1, T_1 \rangle$$

and

$$\mathfrak{W}_2 = \langle A_2, \geq_2, \oplus_2, \succeq_2, T_2 \rangle$$

are Weber's Law structures that result from the subject's the discrimination behavior and let c_1 and c_2 respectively be the Weber constants that result when $\mathfrak{W}_1$ is measured by $\mathcal{S}_1$ and $\mathfrak{W}_2$ by $\mathcal{S}_2$.

Then $c_1 > c_2$ is an assertion that does not depend on the representations from $\mathcal{S}_1$ and $\mathcal{S}_2$ use to measure, respectively, $\mathfrak{W}_1$ and $\mathfrak{W}_2$. However, unlike the single modality case of Example 6.5.2, it does not satisfy the Equivalence Principle, because the value of the Weber constant $\mathfrak{W}_1$ of changes with respect to structures that are physically equivalent to $\langle X_1, \geq_{\star 1}, \oplus_1 \rangle$, and all positive values are realizable by appropriate choices of physically equivalent structures. Narens and Mausfeld make the following comment about this:

> The upshot of this is that for psychological purposes, one should not compare the order of Weber fractions across modalities on the basis of different physical dimensions *unless additional psychological primitives and axioms are assumed.* A little psychological theorizing leads to the same conclusion: One begins by asking what is needed to appropriately compare psychological sensitivity measures across modalities. In our view there are two obvious, closely interrelated answers. *(Narens and Mausfeld, 1992, pg. 474)*

We only give their "second answer":

It is assumed that an additional primitive m has been added to the psychophysical structure $\mathfrak{X}_{\Phi\Psi}$. This primitive is a binary psychological relation. Its intended interpretation is given by, "$y=m(x)$ if and only if the subjective intensity of y is the same as the subjective intensity of x." And it is assumed that additional psychological axioms are given that say m is a strictly increasing function from A_1 onto A_2. Then the following is an obvious psychological way of comparing the subject's discrimination sensitivities across modalities:

> $\succ_1$ is at least as sensitive as $\succ_2$ if and only if for all x and x' in A_1 and all y and y' in A_2, if $y=m(x)$ and $y'=m(x')$ and $y \succ_2 y'$ then $x \succ_1 x'$.

Assume m is a qualitative "matching law" in the sense of Luce (1990) (and described in Subsection 6.4.5); that is, assume for each automorphism τ of $\mathfrak{X}_{\Phi,1}$ there exists an automorphism σ of $\mathfrak{X}_{\Phi,2}$ such that for each x in A_1,

$$m(x) = y \text{ iff } m[\tau(x)] = \sigma_\tau[m(x)] .$$

Let $\varphi_1 \in \mathcal{S}_1$ and $\varphi_2 \in \mathcal{S}_2$. Then by Theorem 6.4.6, there exist r and s in $\mathbb{R}^+$ such that for all x in A_1 and y in A_2,

$$m(x) = y \text{ iff } \varphi_1(x) = s\varphi_2(y)^r . \tag{6.7}$$

From this equation it easily follows that $\succ_1$ is as least as sensitive as $\succ_2$ if and only if

$$1 + c_1 \leq (1 + c_2)^r , \tag{6.8}$$

where r is as given in Equation 6.7.

Equation 6.8 satisfies the Equivalence Principle. It is also psychological significant, because the following statement is easily derivable from the previous definitions and assumptions:

$1 + c_1 < (1 + c_2)^r$ if and only if there exist x and x' in A_1 and y and y' in A_2 such that $y=m(x)$ and $y'=m(x')$, $x \succ_1 x'$, and $y \not\succ_2 y'$.

Note that the situation in Example 6.5.3 with the additional primitive m and additional axioms represent a possible psychological state of affairs. In this possible state of affairs the ordinal comparison of Weber fractions clearly yields, from a psychological perspective, a wrong measure of sensitivity. The correct measure being the modification of the ordinal measure given in Equation 6.8. If the ordinal comparisons of Weber fractions were a good, general method of comparing sensitivities across modalities, then it should also apply in this possible state of affairs. But Example 6.5.3 show that it does not yield a good comparison of sensitivities in this case.

Comparisons of magnitude estimation exponents

The next example concerns a popular method for making intermodal comparisons of sensitivity based upon magnitude estimation data.

Example 6.5.4 S.S. Stevens and others have carried out hundreds of magnitude estimation experiments on a wide variety of physical continua. The results of these experiments that have generally produced psychophysical functions of the form

$$\psi(x) = a\varphi(x)^r , \tag{6.9}$$

where φ is a representation of some standard scale $\mathcal{S}$ used to measure a continuum B, ψ is a subject's magnitude estimation function of B, a is a positive real number that depends on φ, r is a positive real number, called the *exponent.* In analyzing such experiments (e.g., see Stevens, 1971, 1974), the following index of sensitivity is often used:

> A subject is more sensitive to continuum 1 than continuum 2 if and only if the exponent of his or her psychophysical power function associated with continuum 1 is less than the exponent of his or her psychophysical function associated with continuum 2.

Since the choice of the physically equivalent structure used for measuring a continuum does not influence the subjects' magnitude estimations of that continuum but does influence the size of the exponent of the resulting psychophysical power function, it follows by the Equivalence Principle that the above index of sensitivity comparing exponents of psychophysical functions on different continua does not have psychological significance.

As an independent check of the validity of the exponents, Stevens and others employed a procedure called "the method of transitivity of scales":

Definition 6.5.3 As in Example 6.5.3, let m be a psychological matching relation between the physical continua $\langle A_1, \succsim_1\rangle$ and $\langle A_2, \succsim_2\rangle$, where $A_1 \cup A_2 = A$ and $A_1 \cap A_2 = \varnothing$. Assume a psychophysical structure that has A_1, A_2, $\succsim_1$, $\succsim_2$, and m among its primitives and axioms such that the following three statements are true:

1. Through the physical axioms standard scales of physical measurements $\mathcal{S}$ and $\mathcal{T}$ are established on A_1 and A_2 respectively.
2. For all φ in $\mathcal{S}$ and all γ in $\mathcal{T}$ there exist positive real numbers t and p such that for all x in A_1 and y in A_2,
$$m(x) = y \text{ iff } \varphi(x) = p\gamma(y)^t .$$
3. Magnitude estimations have been made on both continua and these result in psychophysical power functions associated with A_1 and A_2 that have respectively exponents r and s.

Then *transitivity of scales* is said to hold if and only if
$$t = \frac{s}{r} .$$

It is not difficult to show that transitivity of scales is consistent with the Equivalence Principle. However, even if it holds, the ordinal comparison of exponents of psychophysical power functions associated with different continua is still by the Equivalence Principle not a psychologically significant index of sensitivity.

6.6 Structural Archimedeanness

This section extends the concept of an "Archimedean axiom" to continuous structures that may or may not have operations among their primitives. Although the approach of this section shares ideas of a very similar program of Luce and Narens (1992) for capturing a general concept of Archimedeanness, it has important differences with that program.

Luce and Narens (1992) employed a concept closely related to intrinsicness to explore the nature of Archimedeanness. They called their concept "intrinsic Archimedeanness," a term that is usually avoided here because of possible conflicts of the technical concept of "intrinsicness" developed in this chapter. Instead, the term "structural Archimedeanness" is used to refer to a concept of Archimedeanness that applies to general structures.

Luce and Narens presented the following intuitive formulation of their program:

> The Archimedean axiom has its roots in ancient mathematics, where it was used to banish form consideration both infinitely large and infinitesimally small quantities. This was essentially the only rigorous means available to eliminate such quantities until very late in the Nineteenth Century when G.Cantor gave a fully rigorous description of the continuum in terms of an ordering relation. In a great many contexts, Cantor's method provides a different means for eliminating the infinitely large and small. Although the two approaches are quite different, they are interrelated in subtle ways....
>
> In practice, the Cantor axioms have not been widely used in measurement theory because of the nonconstructive nature of the axiom postulating the existence of a denumerable, order-dense subset [Definition 5.4.1]. Preference has been accorded to the more constructive Archimedean approach, when it is available.
>
> Historically, to assure that all magnitudes and differences of magnitudes are commensurable, the concept of Archimedeanness has been defined in terms of an operation, usually assumed to be associative. Its justification in these contexts has consisted in trying to make intuitively clear that, in terms of recursively generated applications of the operation as a method of determining size, no element is infinitely large with respect to another and that no two elements are infinitesimally close together.... We seek to extend the concept of Archimedeanness—of commensurability—to general structures that may have no operation among its defining relations *(primitives)*. In such situations, we see no way to keep Archimedeanness from becoming a much more abstract notion and correspondingly a much more difficult one to justify as correct.
>
> Our approach is to formulate, in a very general fashion, what Archimedeanness should accomplish and then show that this imposes severe restrictions that are satisfied by only one concept (up to logical equivalence). In this approach, the resulting concept of Archimedeanness will be justified by theorems; intuition will play a role only at the beginning stages in stating what should be accomplished....

> Historically one of the principal uses made of Archimedean axioms was to establish the isomorphic imbedding of structures into ones based on a continuum. We take this to be a principal characteristic of Archimedeanness. The basic idea is as follows.
>
> There are certain structures that shall be taken to be intrinsically Archimedean…. Such structures will take the following general form: $\langle X, \succsim, R_1, \ldots, R_i, \ldots \rangle$, where the R_i may be elements of X, relations on X, relations of relations on X, and so on, and $\langle X, \succsim \rangle$ is a continuum, and certain other conditions are satisfied…. None of the R_i need be an operation or partial operation. A structure $\mathfrak{S} = \langle S, \succsim', S_1, \ldots, S_i, \ldots \rangle$ is said to be *Archimedean* if and only if there is an intrinsically Archimedean structure $\mathfrak{X}$ and an isomorphism $ø$ from $\mathfrak{S}$ into $\mathfrak{X}$ such that $S\ ø(S)$ is a dense subset of an open interval of [the domain of $\mathfrak{X}$]. This definition consciously omits cases where the ordering on S may be discrete or have gaps in it. There are obviously discrete structures that are Archimedean (e.g., $\langle \mathbb{I}^+, \geq, + \rangle$,, where $\mathbb{I}^+$ is the positive integers), and the approach presented [in this paper] can be extended to such cases. *(Luce and Narens (1992), pp. 16–17)*

The following definition reviews the basic concepts about concatenation structures that will be used throughout this section.

Definition 6.6.1 Let $\mathfrak{X} = \langle X, \succsim, \oplus \rangle$, where $\succsim$ is a total ordering on X and $\oplus$ is a binary operation on X. Then:

1. $\mathfrak{X}$ is said to be *continuous* if and only if $\langle X, \succsim \rangle$ is a continuum (Definition 5.4.1). Note that this definition does not require the operation $\oplus$ to be continuous.
2. $\mathfrak{X}$ is said to be a *concatenation structure* if and only if $\oplus$ is a $\succsim$-strictly increasing function in each variable.
3. $\oplus$ is said to be *positive* if and only if for all x and y in X, $x \oplus y \succ x$ and $x \oplus y \succ y$.
4. $\oplus$ is said to be *intensive* if and only if for all x and y in X, if $x \succ y$, then $x \succ x \oplus y \succ y$.
5. $\oplus$ is said to be *idempotent* if and only if for all x in X, $x \oplus x = x$.
6. $\oplus$ is said to be *associative* if and only if for all x, y, and z in X, $x \oplus (y \oplus z) = (x \oplus y) \oplus z$.
7. $\oplus$ is said to be *commutative* if and only if for all x and y in X, $x \oplus y = y \oplus x$.
8. $\mathfrak{X}$ is said to be right restrictedly solvable if and only if for all x and y in X, if $x \succ y$ then for some z in X, $x \succ y \oplus z$.
9. $\mathfrak{X}$ is said to be *right solvable* if and only if for all x and y in X, if $x \succ y$ then for some z in X, $x = y \oplus z$.
10. $\mathfrak{X}$ is said to be a *bisection structure* if and only if $\oplus$ is intensive and

commutative and the following two conditions hold for all *x, y, z,* and *w* in *X:*

(i) $(x\oplus y)\oplus(z\oplus w)=(x\oplus z)\oplus(y\oplus w)$, and

(ii) if $x \succ y$, then there exit *u* and *v* in *X* such that $x=u\oplus y$ and $y=x\oplus v$.

11. For each positive integer *n* and each *x* in *X*, *nx* is defined inductively as follows: $1x=x$, and $(n+1)x=(nx)\oplus x$.
12. $\mathfrak{X}$ is said to be *ss-Archimedean* (*Archimedean in standard sequences,* 1*z*, 2*z*, 3*z*,…) if and only if for each *x* and *y* in *X* there exists a positive integer *n* such that $nx \succ y$.

An archetypical example of any intuitive idea of Archimedeanness is the structure $\mathfrak{N} = \langle \mathbb{R}^+, \geq, + \rangle$. By Theorem 6.6.1 a structure is isomorphic $\mathfrak{N}$ to if and only if it is a continuous extensive structure. Therefore, continuous extensive structures should also satisfy any intuitive idea of Archimedeanness. Luce and Narens considered it very reasonable to call *Archimedean* those structures of the from $\mathfrak{Y} = \langle Y, \succeq, \oplus \rangle$, where $\oplus$ was an operation on *Y* and $\mathfrak{Y}$ was isomorphically imbeddable in $\mathfrak{N}$ in a way such that *Y* was densely imbedded $\langle \mathbb{R}^+, \geq \rangle$. Luce and Narens gave an example of a continuous concatenation structure $\mathfrak{X}$ with an operation that is positive, associative, commutative, and such that $\mathfrak{X}$ is ss-Archimedean and is not isomorphically imbeddable into $\langle \mathbb{R}^+, \geq, + \rangle$. They showed that in their example, ss-Archimedeanness was successful in eliminating situations where one element was infinitely large with respect to another, but was not successful in eliminating situations where two elements were infinitesimally close to one another. They also presented the following instructive theorem of what was needed in their example to achieve Archimedeanness.

Theorem 6.6.1 Let $\mathfrak{X} = \langle X, \succeq, \oplus \rangle$ be a continuous concatenation structure and $\oplus$ be a positive and associative operation. Then the following four statements are equivalent:

1. $\mathfrak{X}$ is isomorphic to $\langle \mathbb{R}^+, \geq, + \rangle$.
2. $\mathfrak{X}$ is ss-Archimedean and right restrictedly solvable.
3. $\mathfrak{X}$ is right solvable.
4. $\mathfrak{X}$ is homogeneous.

In Theorem 6.6.1, ss-Archimedeanness and right restricted solvability work together to eliminate the possibility of infinitesimally close elements. Luce and Narens make the following comment about this:

> [Let $x \succ y$] Suppose $x \succ y \oplus z$. Then it is reasonable to say that the "difference between *x* and *y* is greater than *z*." Using this concept of difference, we can formulate the idea of *x* and *y* not being infinitely close by requiring that for each element *w* a positive integer *n* can always be found for which *n* copies *z* exceeds *w*. This approach, which relies entirely

on being able to find a z such that $x \succ y \oplus z$, fails [to have intuitive appeal] when no such z exists, [as is in the case of the example of Luce and Narens described above].

In order to appreciate some of the difficulties inherent in arriving a general definition of Archimedeanness, consider the following three examples. (The three examples are from Luce and Narens, 1992. The first due to M.Cozzens.)

Example 6.6.1 Let $\mathfrak{X}_1 = \langle \mathbb{R}^+, \geq, \oplus_1 \rangle$, where for all x, y, and z in $\mathbb{R}^+$,

$$x \oplus_1 y = z \text{ iff } x + \frac{y}{2} = z\,.$$

It is easy to show by induction that

$$nx = \frac{x(n+1)}{2}\,,$$

and thus that nw is an unbounded sequence for each w in X; that is, $\mathfrak{X}$ is "Archimedean" with respect to the sequences nw as w varies in X. However, it is also easy to show for all w in $\mathbb{R}^+$ that the sequences w_n defined inductively by

$$w_n = w \oplus_1 w_{n-1}$$

have the property that

$$w_n = \frac{w(2^n - 1)}{2^{n-1}} \leq 2w\,,$$

and thus are bounded—that is, $\mathfrak{X}$ is "non-Archimedean" with respect to the sequences w_n. The questions are: Should one of these sequences be given preference over the other for defining "Archimedean?" If not, then how should "Archimedeanness" for the structure $\mathfrak{X}_1$ be defined? Note that these questions do not arise for the case of an associative and commutative operation, because for that case these two and other ways of defining the sequences coincide.

Example 6.6.2 Let $\mathfrak{X}_2 = \langle \mathbb{R}^+, \geq, \oplus_2 \rangle$, where for all x and y in $\mathbb{R}^+$,

$$x \oplus_2 y = \frac{x+y}{2}\,.$$

Then $nx=x$ for all x in $\mathbb{R}^+$, and thus standard sequences are useless for determining what infinitely large and infinitesimally close elements are. How is one going to define Archimedeanness for this case? And more generally, how is one going to define Archimedeanness for structures that do not have operations among its primitives?

Example 6.6.3 Let $\mathfrak{X}_3 = \langle \mathbb{R}^+, \geq, +, \oplus_3 \rangle$, where for all x and y in $\mathbb{R}^+$,

$$x \oplus_3 y = \frac{x+y}{1+xy}\,, \text{ if } x < 1 \text{ and } y < 1;$$

$$x \oplus_3 y = x + y\,, \text{ otherwise.}$$

$\mathfrak{X}_3$ has two operations among its primitives. The $\langle \mathbb{R}^+, \geq, + \rangle$ is clearly Archimedean. It is easy to verify that the structure $\langle \mathbb{R}^+, \geq, \oplus_3 \rangle$ is a concatenation structure with a commutative operation that is positive and right restrictedly solvable. $\langle \mathbb{R}^+, \succeq, \oplus_3 \rangle$ is clearly not "Archimedean" because *n* copies of $\frac{1}{2}$ is <1 for all positive integers *n*, that is, 1 is infinitely large with respect to $\frac{1}{2}$. (Note that $\oplus_3$ when restricted to positive reals <1 is the relativistic "addition" formula for velocities *less than* the velocity of light.) Should we call $\mathfrak{X}_3$ "Archimedean" because $\langle \mathbb{R}^+, \succeq, + \rangle$ is Archimedean, or should we call it "non-Archimedean" because $\langle \mathbb{R}^+, \succeq, \oplus_3 \rangle$ is non-Archimedean?

As a first step to answering these and related questions, we will investigate the possibility of determining the Archimedeanness of a structure based on *A* in terms of the Archimedeanness of another structure that is L(∈, *A*, Ø)-equivalent to it. The intuition for this approach is that although the Archimedeanness/non-Archimedeanness of a structure may be opaque in terms of its primitives, it may nevertheless be very clear in terms of the primitives of another structure that is L(∈, *A*, Ø)-equivalent to it. Because Archimedeanness should be preserved by isomorphisms, in determining the Archimedeanness of the structures $\mathfrak{X}_i$, *i*=1,..., 3 above, we may, without loss of generality, assume that $\mathfrak{X}_i$ are structures based on *A*. (As a technical matter, we need to consider isomorphic copies based on *A* in order to apply the concept of L(∈, *A*, Ø)-equivalence, which by Definition 6.2.2 require structures based on *A*. We can do this for the structures $\mathfrak{X}_i$ because (1) isomorphic copies based on *A* of $\mathfrak{X}_i$ exist. and (2) the relevant definitions given below are in terms of the primitives of the relevant $\mathfrak{X}_i$.) Then (assuming $\mathbb{R}^+$=*A*) in Examples 6.6.1 and 6.6.2 the structures

$$\mathfrak{X}_1 = \langle \mathbb{R}^+, \succeq, \oplus_1 \rangle \text{ and } \mathfrak{X}_2 = \langle \mathbb{R}^+, \succeq, \oplus_2 \rangle$$

are L(∈, *A*, Ø)-equivalent to obviously Archimedean structure $\langle \mathbb{R}^+, \succeq, + \rangle$ by the definitions

$$\begin{aligned} x + y &= x \oplus_1 [(y \oplus_1 y) \oplus_1 y]\,; \\ x \oplus_1 y &= x + \alpha^{-1}(y)\,, \text{ where } \alpha(y) = y + y\,; \\ x + y &= \text{l.u.b.}\{z \,|\, \exists w (z \oplus_2 w < x \oplus_2 y\}\,; \\ x \oplus_2 y &= \alpha^{-1}(x + y)\,, \text{ where } \alpha(y) = y + y\,. \end{aligned}$$

Although the Archimedean structure $\langle \mathbb{R}^+, \succeq, + \rangle$ is L(∈, *A*, Ø)-definable in terms of the primitives of the non-Archimedean structure $\mathfrak{X}_3 = \langle \mathbb{R}^+, \succeq, \oplus_3 \rangle$ by the definition

$$\begin{aligned} x + y = z \text{ iff } \exists w \forall u \forall v [u < w \text{ and } v < w \rightarrow u \oplus_3 v < w) \\ \text{and } (z \oplus_3 w) \oplus_3 w = (x \oplus_3 w) \oplus_3 (y \oplus_3 w)]\,, \end{aligned}$$

it is not, by Theorem 6.6.2 below, L(∈, *A*, Ø)-equivalent to $\langle \mathbb{R}^+, \succeq, + \rangle$.

The previous examples indicate that L(∈, *A*, Ø)-equivalence to structures isomorphic to $\langle \mathbb{R}^+, \succeq, + \rangle$—that is, L(∈, *A*, Ø)-equivalence to a continuous extensive structure—is a promising approach to at least one general type of

Archimedeanness. Of course, additional arguments in the form of theorems are needed to demonstrate convincingly that a general type of Archimedeanness is captured.

To simplify notation and concepts, this chapter only investigates Archimedeanness for general structures with an externally finite (Definition 4.3.7) number of primitives. There are corresponding developments for structures with infinitely many primitives.

Theorem 6.6.2 Suppose $\mathfrak{X} = \langle A, \succsim, R_1, \ldots, R_k \rangle$ is a continuous structure. Then the following two statements are equivalent:

1. $\mathfrak{X}$ is L($\in$, *A*, Ø)-equivalent to a continuous extensive structure of the form $\mathfrak{E} = \langle A, \succsim, \oplus \rangle$.
2. $\mathfrak{X}$ is homogeneous and 1-point unique.

Proof. Theorem 6.9.4.

Note that in Statement 1 of Theorem 6.6.2, the continuous extensive structure $\mathfrak{E}$ and the structure $\mathfrak{X}$ have the same ordering, $\succsim$.

Assume the hypotheses and notation of Theorem 6.6.2. Suppose $\mathfrak{X}$ is homogeneous and 1-point unique. Then the following is an intuitive argument for the "Archimedeanness" of $\mathfrak{X}$: Because $\mathfrak{X}$ is L($\in$, *A*, Ø)-equivalent to $\mathfrak{E}$, they are informationally equivalent; that is axiom systems D″($\succsim$, $R_1, \ldots, R_i$) and D″ ($\succsim, \oplus$) determine the same topic, $\mathcal{T}$. $\mathfrak{E}$ is clearly Archimedean; in fact continuous extensive structures like $\mathfrak{E}$ provide the clearest examples of "Archimedeanness" that are known. Thus viewing the topic $\mathcal{T}$ from the perspective of $\mathfrak{E}$ results in a *positive* view of the commensurability of elements of *A*. This gives rise to the following idea of "Archimedeanness" for $\mathcal{T}$: The topic $\mathcal{T}$ is *Archimedean* if and only there exists a continuous extensive structure that determines $\mathcal{T}$. By this definition, the *non-Archimedeanness* of $\mathcal{T}$ is defined *negatively*, as the failure of finding a way to $\mathcal{T}$ view as Archimedean. With these ideas in mind, we then define a continuous structure $\mathfrak{Y}$ to be *Archimedean* if and only if the topic determined by $\mathfrak{Y}$ is Archimedean. Thus by this latter definition of "Archimedeanness," $\mathfrak{X}$ is Archimedean.

In the previous intuitive argument, one can view $\mathfrak{X}$ as being L($\in$, *A*, Ø)-equivalent to $\mathfrak{E}$ as $\mathfrak{X}$ passing one test of "Archimedeanness", that is, showing that there is a positive way of looking at $\mathfrak{X}$ as being Archimedean. However, unlike in the previous intuitive argument, one might also want to require as a necessary condition for $\mathfrak{X}$ to be Archimedean that it fail *all* tests of "nonArchimedeanness," that is, showing that there does not exist a structure $\mathfrak{S}$ that is L($\in$, *A*, Ø)-equivalent to $\mathfrak{X}$ that is clearly "non-Archimedean" according to some different, intuitive *positive* criterion of "non-Archimedeanness." The problem I find with this latter approach is that I do not think it is likely that one will be able to find a justifiable, general, positive criterion of "non-Archimedeanness." For example, one might try to formulate a positive, general concept of "non-Archimedean concatenation structure" and use these to test for Archimedeanness of $\mathfrak{X}$. But even if one could formulate and justify such a criterion of "non-Archimedeanness" and $\mathfrak{X}$ failed to be L(∈, *A*, Ø)-equivalent to structures that met this criterion, one would still not know if there existed just as good criteria for "non-Archimedeanness" for structures of the form $\mathfrak{Z} = \langle A, \succsim, F \rangle$ where F is a function from $A \times A \times A$ onto A that is strictly $\succsim$-increasing in each variable, and so forth.

This raises the following question about the previous intuitive argument: What if one found a clear example of a structure $\mathfrak{W}$ that was clearly "non-Archimedean" in the positive sense and was L(∈, *A*, Ø)-equivalent to $\mathfrak{X}$. Would this undermine the intuitive argument for the Archimedeanness of $\mathfrak{X}$? First, I do not believe that such a $\mathfrak{W}$ could be found. Second, under the hypothesis that such a $\mathfrak{W}$ exists, one has three options: (i) consider, like in the previous intuitive argument, $\mathfrak{X}$ to be "Archimedean" and define "non-Archimedeanness" negatively; (ii) consider $\mathfrak{X}$ to be "non-Archimedean" and define "Archimedeanness" negatively (in terms of "non-Archimedeanness"); *(iii)* consider $\mathfrak{X}$ to be neither "Archimedean" nor "non-Archimedean." All three options have merit, and to convincingly argue for one of them would require the development of additional concepts and theorems. An argument for *(i)* is given later in this section.

The following theorem shows that all tests of "Archimedeanness" with appropriate associative operations will succeed.

Theorem 6.6.3 Suppose $\mathfrak{X} = \langle A, \succsim, R_1, \ldots, R_k \rangle$ and $\mathfrak{X}$ is L(∈, *A*, Ø)-equivalent to the continuous extensive structure $\langle A, \succsim, \oplus^* \rangle$. Let $\oplus$ be an arbitrary binary operation on A such that $\oplus$ is set-theoretically definable in terms of the primitives of $\mathfrak{X}$. Suppose $\oplus$ is associative and positive and $\mathfrak{C} = \langle A, \succsim, \oplus \rangle$ is a concatenation structure. Then $\mathfrak{C}$ is a continuous extensive structure.

Proof. Because $\oplus$ is set-theoretically definable in terms of the primitives of $\mathfrak{X}$, it follows from Theorem 5.1.2 that $\oplus$ is invariant under the automorphisms of $\mathfrak{X}$. Since $\mathfrak{X}$ is homogeneous by Theorem 6.6.2, it then follows that $\mathfrak{C}$ is homogeneous. Then by Theorem 6.6.1, $\mathfrak{C}$ is a continuous extensive structure.

We now investigate Archimedeanness for 2-point homogeneous, 2-point unique structures. The structure $\mathfrak{N} = \langle \mathbb{R}, \geq, \oplus \rangle$, where $\oplus$ is such that for all r and s in $\mathbb{R}$,

$$r \oplus s = \frac{r+s}{2},$$

will be taken to be an archetypical example of an Archimedean 2-point homogeneous, 2-point unique structure. Theorem 5.4.20 gives the following qualitative characterization of structures isomorphic to $\mathfrak{N}$: A structure is isomorphic to $\mathfrak{N}$ if and only if it is a continuous bisection structure (Definition 5.4.19, Definition 6.6.1).

Theorem 6.6.4 Suppose $\mathfrak{X} = \langle A, \succsim, R_1, \ldots, R_k \rangle$ is a continuous structure. Then the following two statements are equivalent:

1. $\mathfrak{X}$ is L($\in$, A, $\emptyset$)-equivalent to a continuous bisection structure of the form $\mathfrak{C} = \langle A, \succeq, \oplus \rangle$.
2. $\mathfrak{X}$ is 2-point homogeneous and 2-point unique.

Proof. Theorem 6.9.5.

Theorem 6.6.5 Suppose $\mathfrak{X} = \langle A, \succeq, R_1, \ldots, R_k \rangle$, is a 2-point homogeneous, 2-point unique continuous structure Then each concatenation structure of the form $\langle A, \succeq, \oplus \rangle$ that is L($\in$, A, $\emptyset$)-equivalent to $\mathfrak{X}$ is "Archimedean" in the sense that for each pair of distinct elements a and b of A, *(i)* a is not infinitely large with respect to b in terms of $\oplus$, and *(ii)* a is not infinitesimally close to b with respect to $\oplus$ and other elements of A.

Proof. Let $\mathfrak{Y} = \langle A, \succeq, \oplus \rangle$ be a concatenation structure that is L($\in$, A, $\emptyset$)-equivalent to $\mathfrak{X}$. Then by Statement 1 of Lemma 6.9.1 and Theorem 5.4.19, $\mathfrak{Y}$ is isomorphic to

$$\mathfrak{Z} = \langle \mathbb{R}^+, \geq, \oplus' \rangle ,$$

where $\oplus'$ is such that there exist r and s in $\mathbb{R}^+$ such that for all x and y in $\mathbb{R}$,

$$x \oplus' y = rx + (1 - r)y + s|x - y| . \tag{6.10}$$

By inspection of Equation 6.10, it is seen that for all distinct x and y in $\mathbb{R}$, *(i)* x is not infinitely large with respect to y in terms of $+$, and *(ii)* x is not infinitesimally close to y with respect to $+$ and other elements of $\mathbb{R}$; that is, it is seen by inspection that $\mathfrak{Z}$ is "Archimedean." By isomorphism $\mathfrak{X}$ is "Archimedean."

Theorem 6.6.5 is partly about an intuitive concept about 2-point homogeneous, 2-point unique structures—namely, "Archimedeanness"—that has not been properly defined. Because of this, it is perhaps better to call Theorem 6.6.5 and "Intuitive Proposition" instead of a "Theorem." Another strategy is to define all 2-point homogeneous, 2-point unique structures to be "Archimedean," and use Theorems 6.6.4, Theorem 6.6.6 below, and the proof of Theorem 6.6.5 as justification for this definition.

Theorem 6.6.6 Suppose $\mathfrak{X} = \langle A, \succeq, R_1, \ldots, R_k \rangle$ is a continuous, 2-point homogeneous, 2-point unique structure. Suppose $\oplus$ is set-theoretically defined from the primitives of $\mathfrak{X}$ and $\mathfrak{C} = \langle A, \succeq, \oplus \rangle$ is a concatenation structure. Then $\mathfrak{X}$ is L($\in$, A, $\emptyset$)-equivalent $\mathfrak{C}$ to and $\mathfrak{C}$ is a 2-point homogeneous, 2-point unique structure.

Proof. Theorem 6.9.6.

Theorems 6.6.4–6.6.6 establish that 2-point homogeneity, 2-point uniqueness is a workable sufficient condition for structural homogeneity.

Definition 6.6.2 A structure is said to be a *continuous 1, 2-structure* if and only if it is a continuous structure that is 1-point homogenous and 2-point unique but neither 2-point homogeneous nor 1-point unique.

Let $\mathfrak{X} = \langle A, \succsim, R_1, \ldots, R_k \rangle$ be a continuous 1,2-structure. Then by Theorem 5.4.16, the automorphism group of $\mathfrak{X}$ is isomorphic to a group of functions f from $\mathbb{R}^+$ onto $\mathbb{R}^+$ of the form

$$f(x) = rx^s ,$$

where r is an element of $\mathbb{R}^+$ and s is an element of S, where $\langle S, \cdot \rangle$ is some nontrivial, proper, multiplicative subgroup of $\langle \mathbb{R}^+, \cdot \rangle$. For the purposes of this section, the automorphism group of $\mathfrak{X}$ naturally falls into two types: *Type 1,* where $\mathfrak{X}$ has a least element, and Type 2 where $\{s \mid s \in S \text{ and } s > 1\}$ has no least element. Luce and Narens (1985) gave an example of a continuous concatenation structure with a Type 1 automorphism group and showed that no continuous concatenation structure can have a Type 2 automorphism group. The latter implies that if $\mathfrak{X}$ has a Type 2 automorphism group, then $\mathfrak{X}$ cannot be L($\in$,*A*, Ø)-equivalent to a continuous concatenation structure of the form $\langle A, \succsim, \oplus \rangle$; that is, that one cannot test the "Archimedeanness" of $\mathfrak{X}$ through the use of L($\in$,*A*, Ø)-equivalent concatenation structures.

Continuous 1,2-structures have only recently appeared in the measurement literature, and no specific examples of continuous 1,2-structures have been qualitatively axiomatized. A general characterization of "Archimedeanness" for them is not attempted here.

In all, Theorems 6.6.2–6.6.6 establish for continuous structures

$$\langle A, \succsim, R_1, \ldots, R_k \rangle$$

that *(i)* L($\in$,*A*, Ø)-equivalence to a continuous extensive structure with ordering $\succsim$ and *(ii)* L($\in$,*A*, Ø)-equivalence to a continuous bisection structure with ordering $\succsim$ are sound and reasonable criteria for structural Archimedeanness. These two criteria coincide with homogeneity and finite uniqueness (Definition 5.1.5) in all cases except for continuous 1,2-structures. This is in rough agreement with Luce's and Narens' development for "intrinsic Archimedeanness," except they consider all homogeneous, finitely unique continuous structures to be intrinsically Archimedean. Their reason for this was that for such structures, the automorphism groups of the translations were isomorphic to $\langle \mathbb{R}, \geq, + \rangle$ (see Theorem 6.6.7 below) and therefore were Archimedean structures in their own right, and this together with homogeneity suggested the Archimedeanness of the underlying continuous structure.

Theorem 6.6.7 Suppose $\mathfrak{X} = \langle A, \succsim, R_1, \ldots, R_k \rangle$ is a continuous, homogeneous, finitely point unique structure, $\mathcal{T}$ is the set of translations of $\mathfrak{X}$ (Definition 5.1.7), and * is the operation of functional composition restricted to T. Then $\langle T, \succsim', * \rangle$ is isomorphic to $\langle \mathbb{R}^+, \geq, + \rangle$ where $\succsim'$ is the binary relation on T such that for all a and β in T,

$$\alpha \succsim' \beta \text{ iff } \forall x[\text{if } x \in A \text{ then } \alpha(x) \geq \beta(x)] .$$

Proof. By Theorem 2.5 of Narens (1981a) (or Theorem 4.1 of Narens, 1985), $\langle \mathbb{N}, \succsim', * \rangle$ is a totally ordered group and $\langle \mathbb{N}, \succsim' \rangle$ is a continuum. The conclusion of the theorem is then a simple consequence of the well-known representation theorems for such groups ("Hölder's Theorem").

Let $\mathfrak{X} = \langle A, \succsim, R_1, \ldots, R_k \rangle$ be a continuous structure and T be the set of translations of $\mathfrak{X}$. Suppose $\mathfrak{X}$ is homogeneous and finitely point unique. Then using Theorem 6.6.7 it is not difficult to see that *with respect to* $\langle T, \succsim', * \rangle$ $\mathfrak{X}$ has no pair of distinct elements that are infinitely large or infinitesimally close to one another. However, are these ideas of "infinitely large" and "infinitesimally close" proper generalizations of the intuitive ideas of "infinitely large" and "infinitesimally close" that have historically driven the discussion of Archimedeanness in the literature? For the case where $\mathfrak{X}$ is also 1-point unique, I believe they are. This is because for 1-point unique case, *(i)* T is the set of automorphisms of $\mathfrak{X}$, and *(ii)* through using Theorems 6.6.2, 6.6.3, and 5.10.11, it can be shown that

> T=$\{\alpha|$ there exists an associative operation $\oplus$ that is settheoretically defined in terms of the primitives of $\mathfrak{X}$ and for some n *in* $\mathbb{I}^+$ α is the n-copy operator determined by $\oplus\}$.

Through *(ii),* Theorems 6.6.7, 6.6.2 and 6.6.3, there is a direct and strong relationship between the concepts of Archimedeanness of $\langle T, \succsim', * \rangle$ and the Archimedeanness of L($\in$,*A*, Ø)-equivalent concatenation structures of the form $\langle A, \succsim, \oplus \rangle$ where $\oplus$ is an associative operation. For the case where $\mathfrak{X}$ is not 1-point unique, there cannot be a corresponding result, because in that case each L($\in$,*A*, Ø)-equivalent operation $\oplus$ such that $\langle X, \succsim, \oplus \rangle$ is a concatenation structure is idempotent (Theorem 5.4.17), and thus only the identity automorphism is produced through n-copy operators. There is, however, Theorems 6.6.5 and 6.6.6, which establishes for 2-point homogeneous, 2-point unique structures a logical equivalence between the Archimedeanness of $\langle T, \succsim', * \rangle$ and the Archimedeanness of concatenation structures with ordering $\succsim$ that are L($\in$,*A*, Ø)-equivalent to $\mathfrak{X}$. I believe that a similar theorem probably holds for continuous 1,2-structures with Type 1 automorphism groups. However, such an equivalence is not possible for continuous 1,2-structures with Type 2 automorphism groups, because there are no appropriate L($\in$,*A*, Ø)-equivalent concatenation structures.

Luce and Narens did not employ L($\in$,*A*, Ø)-equivalent extensive or bisection structures for testing Archimedeanness of the underlying continuous structure, but instead used a different, weaker notion: testing Archimedeanness in terms of concatenation structures that are invariant with respect to the automorphisms of the underlying continuous structure. In particular, they concluded that a continuous 1,2-structure $\mathfrak{X} = \langle A, \succsim, R_1, \ldots, R_j, \ldots \rangle_{j \in J}$ was Archimedean, because *(i)* there existed a concatenation structure $\langle A, \succsim, \oplus \rangle$ that they considered to be appropriately "Archimedean" and was such that $\oplus$ was invariant under the group of automorphisms of $\mathfrak{X}$, and

(ii) for each concatenation structure $\mathcal{C} = \langle A, \succsim, \oplus' \rangle$ such that $\oplus'$ was invariant under the group of automorphisms of $\mathfrak{X}$ and such that the Archimedeanness/non-Archimedeanness of $\mathcal{C}$ could be appropriately determined, it turned out that $\mathcal{C}$ was Archimedean.

However, when $\mathfrak{X}$ has a Type 2 automorphism group, there are no concatenation structure of the form $\langle A, \succsim, \oplus \rangle$ that is L($\in$,*A*, Ø)-equivalent to $\mathfrak{X}$, and thus the only concatenation structures that met either Conditions *(i)* and *(ii)* were 2-point homogeneous and 2-point unique, because such structures had to be invariant under the automorphisms of $\mathfrak{X}$. Intuitively, because such 2-point homogeneous, 2-point unique concatenation structures have richer automorphism groups than a structure $\mathfrak{X}$ with a Type 2 automorphism group, they contain less information than is present in $\mathfrak{X}$ and therefore the fact that they are "Archimedean" is not sufficient reason to conclude that $\mathfrak{X}$ is "Archimedean." In contrast, the development of this section draws no conclusion about the Archimedeanness of structures with Type 2 automorphism groups.

Luce and Narens also employed conditions that are close in form and spirit to the ss-Archimedean axiom to define Archimedeanness for idempotent concatenation structures. However, to be effective, these conditions required the idempotent concatenation structures to satisfy certain solvability conditions that intuitively are not necessary for Archimedeanness.

Although Luce and Narens considered imbeddability of a structure $\mathfrak{Y}$ into a continuous, "intrinsic Archimedean" structure $\mathfrak{X}$ as an important condition for the Archimedeanness of $\mathfrak{Y}$, they did not give arguments for the sufficiency of this condition. In their paper, imbeddability into an "intrinsic Archimedean" structure is treated more like an intriguing hypothesis to be further investigated than a theory of Archimedeanness for non-continuous structures.

General approaches to Archimedeanness for non-homogenous structures and ∞-unique structures (Definition 5.1.5) have not been considered in the literature.

The approach of this section to Archimedeanness, which is based upon L($\in$,*A*, Ø)-equivalence to archetypical" Archimedean" continuous concatenation structures, can be modified to include other ideas of "equivalence" that are not based on set-theoretic definability. A particularly interesting possibility is to require the test concatenation structures to be both informationally equivalent to the underlying continuous structure $\mathfrak{X}$ and definable from the primitives $\mathfrak{X}$ of through some constructive procedure.

6.7 Dichotomous Data Analysis

Batchelder and Narens (1977) critically examined the interpretations commonly given to the analyses of 0–1 (Yes-No) data sets in terms of a concept that is in many respects similar to the intrinsicness concept considered in this chapter. This section very briefly presents the core of Batchelder's and Narens' theory and illustrates it with an example.

6.7.1 Boolean Equivalent Sets of Properties

Convention 6.7.1 Throughout this section, m and n are assumed to be integers ≥ 2. Y represents the domain of interest, which for the purposes of this section is taken to be a set of n people. Subsets of this domain are characterized qualitatively by a set $\mathcal{Q}$ of m 1-ary predicates $Q_1,\ldots, Q_m$, which throughout this section are generally called "properties." It is assumed that the properties in $\mathcal{Q}$ also apply to a superset X of Y for example, the set of possible people X, and for this superset X, it is assumed for each distinct Q and Q' in $\mathcal{Q}$ there exists an element x in X such that either *(i)* $Q(x)$ is true and $Q'(x)$ is false or *(ii)* Q(x) is false and $Q'(x)$ is true. Furthermore, it is assumed that the structure $\langle X, Y, \mathcal{Q}_1, \ldots, \mathcal{Q}_m \rangle$ describes appropriately some qualitative situation that is under consideration. (Although it is the structure $\langle Y, \mathcal{Q}_1, \ldots, \mathcal{Q}_m \rangle$ that is of interest, the superset X of Y also needs to be considered to ensure that no element of $\mathcal{Q}$ can be interpreted as a purely logical combination of other elements of $\mathcal{Q}$.)

Throughout this section, $\mathcal{R}$ will stand for a set of m properties about elements of Y (and therefore elements of X). $\mathcal{R}$ may be different from $\mathcal{Q}$ or the same as $\mathcal{Q}$.

Also throughout this section, the truth assignment True is denoted by "1"and False by "0".

Definition 6.7.1 Properties about the people in X (i.e., 1-ary relations on X that are definable from $R_1,\ldots, R_m$ through the use of propositional calculus (i.e., are formed from $R_1,\ldots, R_m$ using the connectives $\neg$ and $\wedge$), are called *derived properties* of $R_1,\ldots, R_m$.

Note through use of the propositional calculus that for each derived property R of $\mathcal{R}$, each truth assignment $(a_1,\ldots, a_m)$ to respectively the ordered set of properties $(R_1,\ldots, R_m)$ of $\mathcal{R}$ produces a unique truth value for R, for example, for m=3 and the derived property $R_1 \wedge \neg R_2 \wedge R_3$, the truth assignment (0,1,0) produces the truth value 0.

Convention 6.7.2 Throughout this section, let $\mathbb{T}$ be the set of truth assignments to m properties, that is,

$$\mathbb{T} = \{(a_1, \ldots, a_m) |\ \text{for } i = 1, \ldots, m, \text{either } a_i = 0 \text{ or } a_i = 1\} .$$

(It is implicitly assumed here that the m properties are indexed by the integers i=1,..., m and that a_i is assigned to the ith property.)

Definition 6.7.2 The sets of properties $\mathcal{Q}$ (=(Q_1,...,Q_m)) and $\mathcal{R}$ (= (R_1,..., R_m)) are said to be *Boolean equivalent* if and only if each property in $\mathcal{R}$ is a derived property of $\mathcal{Q}$ and each property of $\mathcal{Q}$ is a derived property of $\mathcal{R}$.

The following two theorems are due to Batchelder and Narens (1977).

Theorem 6.7.1 Suppose $\mathcal{Q}$ and $\mathcal{R}$ are Boolean equivalent. Let $\mathbb{T}$ be the set of truth assignments to m properties (Convention 6.7.2). For each truth assignment a=(a_1,...,a_m) to $\mathcal{Q}$ let $F(a)$=(b_1,..., b_m), where for i= 1,..., m, b_i is the truth value of the derived property R_i of $\mathcal{R}$ produced by the truth assignment a. Then F is a permutation of $\mathbb{T}$.

Proof. Theorem 6.9.7.

Theorem 6.7.2 Suppose F is a permutation of $\mathbb{T}$. Then a set of m properties S about the set of people Y (and the set of possible people X) can be found such that the following two statements are true:

1. S=(S_1,..., S_m) is Boolean equivalent to $\mathcal{Q}$..
2. For each a in $\mathbb{T}$ and each i, i=1,..., m, the ith component of $F(a)$ is the truth value of $S_i$produced by the truth assignment a.

Proof. Theorem 6.9.8.

Theorem 6.7.1 says the Boolean Equivalent sets of properties $\mathcal{Q}$ and $\mathcal{R}$ code the same information: That is, for each person p, the properties in $\mathcal{Q}$ true about p completely determine those in $\mathcal{R}$ that are true about p, and vice versa. (The "vice versa" follows by considering F^{-1}.) And Theorem 6.7.2 says that any recoding of the information contained in $\mathcal{Q}$ can be accomplished through a Boolean equivalent set of properties to $\mathcal{Q}$.

6.7.2 An Illustrative Example

The nonmetric multidimensional scaling of the properties $\mathcal{Q}$ proceeds by first generating a similarity measure σ between pairs of properties in $\mathcal{Q}$ as follows:

$\sigma(Q,Q')$=the proportion of people in Y who have both property Q and Q'.

Then a "best fitting" representation φ into a Euclidean space of some specific dimension is found, that is, a function φ into a Euclidean space E of dimension $k\geq 1$ is found such that for all Q, Q', T, and T' in Q,

$$\sigma(Q,Q') > \sigma(T,T') \text{ iff } d(\varphi(Q),\varphi(Q')) < d(\varphi(T),\varphi(T')) \ ,$$

where d is the distance function for E. Thus by this representation, the more similar two properties in $\mathcal{Q}$ are with respect to the population Y the closer their representations are in the Euclidean space.

The intuitive idea behind this form of modeling is that there is a true similarity between the properties in $\mathcal{Q}$ with respect to the some intended population of people of which Y is a sample, and that the above form of nonmetric multidimensional scaling is an appropriate procedure for modeling this true similarity. Whether or not it is reasonable to believe that this is the case is not be discussed here. What is important for the considerations of this section is whether transforming $\mathcal{Q}$ into a Boolean equivalent set of properties $\mathcal{R}$ invalidates the intuition underlying this kind of modeling procedure. It does not, because the transformed properties in $\mathcal{R}$ also have similarity measures between them, and in general have representations as *new* points in an Euclidean space E', which in general are different from the Euclidean space E into which $\mathcal{Q}$ is imbedded. The similarity measures of $\mathcal{Q}$ and $\mathcal{R}$ can be jointly represented in a Euclidean space E'' (in general, of higher dimension than either of the spaces E or E' and in E'' the orderings of distances between elements of $\mathcal{Q}$ are the same as in E, and similarly for $\mathcal{R}$ and E'. Thus in particular, the similarity measures between properties of $\mathcal{Q}$ about people in Y are unchanged by transforming $\mathcal{Q}$ into $\mathcal{R}$ or by augmenting $\mathcal{Q}$ with $\mathcal{R}$. A rather different pattern of results occurs for similarity measures *between people* in Y.

The similarity measure $\mathcal{T}$ between persons p and p' in Y is computed as follows:

> $\tau(p,p')$=the proportion of properties Q in $\mathcal{Q}$ such that either *(i)* Q is true about both p and p' or *(ii)* Q is false about p and p'.

Then the people (and the similarity measure $\mathcal{T}$) is represented in Euclidean space in the manner like before.

However this time, when the set $\mathcal{Q}$ is transformed in an equivalent set $\mathcal{R}$, *the similarity measures on pairs of people in Y generally change, usually inverting the relative sizes of some pairs of similarity measures.* This problem cannot be solved by taking average similarity measures of pairs of people over all Boolean equivalent sets: Batchelder and Narens (1977) show that for this case the average similarity of persons p and p' is 1 if exactly the same properties in $\mathcal{Q}$ hold for p and p', and the average similarity of p and p' is

$$\frac{m(2^{m-1}-1)}{2^m-1}$$

if some property holds for one of persons p, p' but fails for the other person. Thus such averages are clearly useless similarity measures.

6.7.3 Conclusions

Batchelder and Narens make a distinction between a domain $\mathcal{B}$ of information about a set Y of people and a set of properties $\mathcal{Q}$ that generates the information *in* the domain $\mathcal{B}$. $\mathcal{B}$ is thought of as a set of properties that forms a Boolean algebra under $\neg$ ("not") and $\wedge$ ("and"). (I.e., the properties T and F that are respectively true and false about all people in Y are in $\mathcal{B}$, and for all E and F in $\mathcal{B}$, $\neg E$ and $E \wedge F$ are in $\mathcal{B}$.) $\mathcal{Q}$ is thought of as a subset of $\mathcal{B}$ that generates with respect to $\neg$ and $\wedge$ the properties in $\mathcal{B}$ and is such that no element of $\mathcal{Q}$ is logically equivalent to a logical combination of other elements of $\mathcal{Q}$. To emphasize the different roles that $\mathcal{B}$ and $\mathcal{Q}$ play in this discussion, call $\mathcal{B}$ the *space of properties* and $\mathcal{Q}$ *a basis for* $\mathcal{B}$.

The space of properties $\mathcal{B}$ is interpreted as the information contained in the basis set of properties $\mathcal{Q}$. Sets of properties $\mathcal{Q}'$ that are Boolean equivalent to $\mathcal{Q}$ also generate $\mathcal{B}$ and therefore contain the same information as $\mathcal{Q}$; the main difference between $\mathcal{Q}$ and $\mathcal{Q}'$ being that the information contained in $\mathcal{Q}'$ is organized differently than $\mathcal{Q}$. Thus the data D consisting of which properties in $\mathcal{Q}$ hold for which people in Y can be used either *(i)* for an analysis *of* the information contained in $\mathcal{Q}$, that is, as an analysis of $\mathcal{B}$, or *(ii)* for the analysis of a particular way of organizing the information $\mathcal{B}$ that is, as an analysis of $\mathcal{Q}$. For *(i),* the results of data processing routines must show invariance across Boolean equivalent sets of properties—a condition that often fails for many data processing routines found in the literature. (See Batchelder and Narens, 1977, and Kaiwi, 1978, for many examples; and see Watanabe, 1969, circa pg. 376 for a related perspective.)

When the goal of the research is to use D to find how particular properties in $\mathcal{B}$ are related, there is usually no problem in using Boolean equivalent sets of properties for interpreting the results. However, when the goal is to use D to structure Y, difficulties arise for almost all the relevant data processing routines in the literature, because the results of such processing routines generally vary with Boolean equivalent sets of properties. The difficulties do not concern the correctness of the methods as parts of statistical methodology, but rather concern the uses to which the results of such methods are put. Because the results of such an analysis generally vary with sets of properties equivalent to $\mathcal{Q}$, the resultant structuring of Y depends on the particular organization of information $\mathcal{Q}$. Thus a claim that the resulting structuring of Y is a relevant and substantive structuring rests on justifying the selection of $\mathcal{Q}$ and the rejection of most other sets of properties that are Boolean equivalent to $\mathcal{Q}$. In the behavioral and social sciences literatures such justifications are almost never provided.

Biology provides an informative example of the depth of this problem. Consider a population Y of organisms and a set of observable characteristics $\mathcal{Q}$ of these organisms. Biology often provides a theory about how the population Y may be structured in terms of genotypes so that the observable phenotypical characteristics in $\mathcal{Q}$ would produce the data set D of which elements of Y have which properties in $\mathcal{Q}$. It is the lack of analogous substantive theories that in Batchelder's and Narens' view make suspect many methods used in the social and behavioral sciences for structuring populations.

6.8 Conclusions

Intrinsicness and meaningfulness are closely related concepts, with intrinsicness implying meaningfulness. It is often a subtle matter to decide which to use in an application, because such decisions are not based on the correctness of a derivation or of a method of data analysis, but on theoretical and philosophical perspectives about the correctness of an interpretation of a derivation or the interpretation of the results of a method of analysis. Thus, for example, in stating that intrinsicness, rather than meaningfulness, is appropriate for a given situation, one is proposing a theory about the nature of the phenomena of the situation, and formation and justification of such theories are beyond empirical considerations and the representational theory of measurement. Such theories are generally formed through insights into the nature of phenomena, often tempered by philosophical considerations.

Weber's law in Section 6.4 presents an example of a situation where such considerations are relevant. Using the concepts, notation, and conventions of Section 6.4, consider the formulation of Weber's law,

$$(1)\ \forall x \forall y[T(x \oplus y) = T(x) \oplus T(y)],$$

where $\oplus$ and T are meaningful. (1) yields a meaningful version of Weber's law, with a specific Weber constant that is meaningful in terms of the primitives, which include the operation $\oplus$. Next consider the formulation,

$$(2)\ G = G'.$$

(2) yields a version of Weber's law that is intrinsic relative to M, G'. (2) does not specify a particular Weber constant. For some purposes, for example, using the Weber constant as a parameter in a model that predicts other psychophysical behavior, (1) is preferable, and for other purposes, for example, understanding what aspects of Weber's law can be interpreted purely in terms purely psychological primitives, (2) is preferable. Thus the decision of whether the meaningful or the intrinsic formulation of Weber's law is appropriate in an application depends critically on the nature of the application.

The capability of intrinsicness to be a unifying concept for several ideas of "lawfulness" has been demonstrated in Section 6.4. Because intrinsicness is distinguishable from meaningfulness and is formulable in terms of definability concepts, it appears to me to provide a better grounding for theories of laws than the ones previously proposed, which relied on invariance concepts, usually formulated quantitatively. The value of intrinsicness as a theoretical principle has been demonstrated in Section 6.5, where it played a critical role in the formulation of a theory about how psychology and physics were linked in psychophysics.

A key idea in the definition of intrinsicness is that of "L($\in$, A, Ø)-equivalent structure" (Definition 6.2.3). Because L($\in$, A, Ø)-equivalent structures are inter-definable, they contain exactly the same information about the underlying qualitative situation. There are related concepts similar to intrinsicness, and applications of two of these—to Archimedeanness and to the structuring of populations in terms of characteristics of their members—were discussed respectively in Sections 6.6 and 6.7. The applications of the related concepts were based on the idea that there are several "equivalent" ways of describing a qualitative situation, and that certain kinds of concepts and analyses do not depend on which of the "equivalent" ways is used for describing the situation.

One way of characterizing "equivalent descriptions" and "invariance under equivalent descriptions" is through the concept of $\mathcal{E}$-intrinsicness. Section 6.2 shows that $\mathcal{E}$-intrinsicness is a viable concept that generalizes both relative intrinsicness and meaningfulness. As a generalization of meaningfulness, it allows for the flexible formulation of topics in situations where the usual meaningfulness concepts produce unwanted and trivial topics.

Deciding when to employ intrinsicness or a related concept is often a deep, important, and difficult matter—and curiously, one that has received almost no attention in the philosophy of science literature or in the foundational literatures of the various sciences. I believe that these concepts have enormous theoretical potential—equaling at least that of meaningfulness. I expect that in the future there will be many and varied applications of them in the literature.

6.9 Additional Proofs and Theorems

Theorem 6.9.1 *(Theorem 6.2.2)* Let $\mathcal{M}$ be the collection of ε-intrinsic entities. Then $\mathcal{M} \subseteq M$ and $\langle V, \in, A, \varnothing, \mathcal{M}\rangle$ satisfies axiom system D'.

Proof. By Definition 4.3.1, we need to only show Axioms *MP, AL* and *MC'*. Let p be an arbitrary pure set. Let $\eta(x)$ be the formula $x=p$. Then $p \subseteq \mathcal{M}$ by Definition 6.2.5. Because p is an arbitrary pure set, it follows that axiom *MP* is true.

Let c be an arbitrary entity such that $\{c\}$ is in $\mathcal{M}$. Let $\langle X, R_j\rangle_{j\in J}$ be an element of $\mathcal{E}$. By Definition 6.2.5, let $\psi(x, x_1, \ldots, x_n, y_1, \ldots, y_m)$ be a formula of L($\in$, A, $\varnothing$), $.j_1, \ldots, j_n$ elements of J, and $b_1, \ldots, b_m$ pure sets such that

$$\psi(\{c\}, R_{j(1)}, \ldots, R_{j(n)}, b_1, \ldots, b_m) \wedge$$
$$\forall x[\psi(x, R_{j(1)}, \ldots, R_{j(n)}, b_1, \ldots, b_m) \rightarrow x = \{c\}].$$

Let z be a variable not in the formula $\psi(x, x_1, \ldots, x_n, y_1, \ldots, y_m)$ and let

$$\theta(z, x_1, \ldots, x_n, y_1, \ldots, y_m)$$

be the formula

$$\exists x \psi(x, x_1, \ldots, x_m, \ldots, y_1, \ldots, y_m) \wedge z \in x\,.$$

Then

$$\theta(c, R_{j(1)}, \dots, R_{j(n)}, b_1, \dots, b_m) \land$$
$$\forall x[\theta(x, R_{j(1)}, \dots, R_{j(n)}, b_1, \dots, b_m) \to x = c].$$

Let $\langle X, S_j \rangle_{j \in J}$ be an arbitrary element of $\mathcal{E}$. Then since $\{c\} \in \mathcal{M}$, it follows from Definition 6.2.5 and the choice of ψ that

$$\psi(\{c\}, S_{j(1)}, \dots, S_{j(n)}, b_1, \dots, b_m) \land$$
$$\forall x[\psi(x, S_{j(1)}, \dots, S_{j(n)}, b_1, \dots, b_m) \to x = \{c\}].$$

Thus

$$\theta(c, S_{j(1)}, \dots, S_{j(n)}, b_1, \dots, b_m) \land$$
$$\forall x[\theta(x, S_{j(1)}, \dots, S_{j(n)}, b_1, \dots, b_m) \to x = c].$$

Because $\langle X, S_j \rangle_{j \in J}$ be an arbitrary element of $\mathcal{E}$, it then follows from Definition 6.2.5 that $c \in \mathcal{M}$. Thus axiom *AL* is true.

Suppose $\varphi(x, x_1, \dots, x_n, y_1, \dots, y_m)$ is a formula of L($\in$,*A*, Ø), $a_1,\dots,a_n$ are elements of *M*, $b_1,\dots,b_m$ are pure sets, and

$$a = \{x | \varphi(x, a_1, \dots, a_n, b_1, \dots, b_m)\}.$$

To simplify notation, we will assume n=2 and m=1. The case for general m and n follow by an almost identical argument. Thus

$$a = \{x | \varphi(x, a_1, a_2, b_1)\}. \tag{6.11}$$

Let $\langle X, R_j \rangle_{j \in J}$ be an element of $\mathcal{E}$. Because a_1 and a_2 are intrinsic, let, by Definition 6.2.5,

$$\varphi_1(u, v_1, \dots, v_q, w_1, \dots, w_r) \text{ and } \varphi_2(x, y_1, \dots, y_s, z_1, \dots, z_t)$$

be formulas of L($\in$,A,$\varnothing$), $j(1), \dots, j(q), k(1), \dots, k(s)$ be in *J,* and

$$c_1,\dots,c_r,\ d_1,\dots,d_t$$

be pure sets such that

$$\varphi_1(a_1, R_{j(1)}, \dots, R_{j(q)}, c_1, \dots, c_r) \land$$
$$\forall x[\varphi_1(x, R_{j(1)}, \dots, R_{j(q)}, c_1, \dots, c_r) \to x = a_1]$$

and

$$\varphi_2(a_2, R_{k(1)}, \dots, R_{j(s)}, d_1, \dots, d_t) \land$$
$$\forall x[\varphi_2(x, R_{k(1)}, \dots, R_{j(s)}, d_1, \dots, d_t) \to x = a_2].$$

Let

$$\gamma(x, R_{j(1)}, \dots, R_{j(q)}, R_{k(1)}, \dots, R_{j(s)}, c_1, \dots, c_r, d_1, \dots, d_t, b_1)$$

be the formula

$$\forall y[y \in x \leftrightarrow \exists z_1 \exists z_2[\varphi[y, \varphi_1(z_1, R_{j(1)}, \dots, R_{j(q)}, c_1, \dots, c_r),$$
$$\varphi_2(z_2, R_{k(1)}, \dots, R_{j(s)}, d_1, \dots, d_t), b_1]].$$

Then by Equation 6.11,

$$\gamma(a, R_{j(1)}, \ldots, R_{j(q)}, R_{k(1)}, \ldots, R_{j(s)}, c_1, \ldots, c_r, d_1, \ldots, d_t, b_1) \wedge$$
$$\forall x[\gamma(x, R_{j(1)}, \ldots, R_{j(q)}, R_{k(1)}, \ldots, R_{j(s)}, c_1, \ldots, c_r, d_1, \ldots, d_t, b_1) \rightarrow x = a]$$

is true. Let $\langle X, S_j \rangle_{j \in J}$ be an arbitrary element of $\mathcal{E}$. Then by the choices of φ_1 and φ_2, the intrinsicness of a_1 and a_2, and Definition 6.2.5,

$$\varphi_1(a_1, S_{j(1)}, \ldots, S_{j(q)}, c_1, \ldots, c_r) \wedge$$
$$\forall x[\varphi_1(x, S_{j(1)}, \ldots, S_{j(q)}, c_1, \ldots, c_r) \rightarrow x = a_1]$$

and

$$\varphi_2(a_2, S_{k(1)}, \ldots, S_{j(s)}, d_1, \ldots, d_t) \wedge$$
$$\forall x[\varphi_2(x, S_{k(1)}, \ldots, S_{j(s)}, d_1, \ldots, d_t) \rightarrow x = a_2].$$

Thus by Equation 6.11, the definition of the formula γ, and Definition 6.2.5,

$$\gamma(a, S_{j(1)}, \ldots, S_{j(q)}, S_{k(1)}, \ldots, S_{j(s)}, c_1, \ldots, c_r, d_1, \ldots, d_t, b_1) \wedge$$
$$[\forall x[\gamma(x, S_{j(1)}, \ldots, S_{j(q)}, S_{k(1)}, \ldots, S_{j(s)}, c_1, \ldots, c_r, d_1, \ldots, d_t, b_1) \rightarrow x = a]]$$

is true. Since $\langle X, S_j \rangle_{j \in J}$ is an arbitrary element of $\mathcal{E}$, it follows from Definition 6.2.5 that $a \in \mathcal{M}$. Thus axiom D′ is true.

Theorem 6.9.2 *(Theorem 6.4.4)* Let $\mathfrak{W} = \langle A, \succsim_\star, T, \oplus \rangle$, G be the automorphism group of $\mathfrak{W}$, $\mathfrak{E} = \langle A, \succsim_\star, \oplus \rangle$ and G' be the automorphism group of $\mathfrak{E}$. Assume $\mathfrak{E}$ is a continuous extensive structure and $\mathfrak{T} = \langle A, \succsim_\star, T \rangle$ is a continuous threshold structure. Then the following two statements are logically equivalent:

1. For all x and y in A,

$$T(x \oplus y) = T(x) \oplus T(y).$$

2. $G = G'$.

Proof. Assume Statement 1. Clearly $G \subseteq G'$. Thus to show Statement 2, it needs to be only shown that $G' \subseteq G$. Let α be an arbitrary element of G'. Since by hypothesis it follows that $\mathfrak{T}$ is a continuous threshold structure, it follows that for all x and y in A,

$$x \succsim_\star y \text{ iff } T(x) \succsim_\star T(y)\,. \tag{6.12}$$

By Statement 1, for all x and y in A,

$$T(x \oplus y) = T(x) \oplus T(y)\,. \tag{6.13}$$

Thus by Equations 6.12 and 6.13, T is an automorphism of $\mathfrak{E}$, that is, $T \subseteq G'$. It is an easy consequence of Theorem 5.4.7 that the automorphisms of a continuous extensive structure commute, and therefore that the elements of G' commute. Let α be an arbitrary element of G'. Then $\alpha * T = T * \alpha$., that is, for all x and y in A,

$$\alpha[T(x)] = T[\alpha(x)]\,. \tag{6.14}$$

Because a is an automorphism of $\mathfrak{C} = \langle A, \succsim_\star, \oplus\rangle$ it follows from Equation 6.14 that α is an automorphism of $\mathfrak{W} = \langle A, \succsim_\star, T, \oplus\rangle$, that is, $\alpha \in G$.

Assume Statement 2. Then each α in G' is an automorphism of $\mathfrak{W}$, in particular, for each x in A,

$$\alpha[T(x)] = T[\alpha(x)]\,. \tag{6.15}$$

Since $\mathfrak{T}$ is a continuous threshold structure, for all x in A,

$$x \succsim_\star y \text{ iff } T(x) \succsim_\star T(y)\,. \tag{6.16}$$

By Theorem 5.4.7, let φ be an isomorphism of $\mathfrak{C}$ onto $\langle \mathbb{R}^+, \geq, +\rangle$ and let $T^* = \varphi(T)$. By Theorem 5.4.7 and the choice of $\varphi, \varphi(G)$ is the set of multiplications by positive reals. Thus by the choice of φ and Equations 6.16 and 6.15, T^* is strictly increasing and for all r and s in $\mathbb{R}^+$,

$$r \cdot T^\star(s) = T^\star(r \cdot s)\,. \tag{6.17}$$

It is well-known that any strictly increasing T^* satisfying Equation 6.17 for all r and s in $\mathbb{R}^+$ is of the form

$$T^*(s) = a \cdot s$$

for some fixed positive real a; that is, T^* is a multiplication by a positive real. Therefore T^* is in $\varphi(G)$, and thus $T \in G$; that is, T is an automorphism of $\mathfrak{W}$. Therefore, for each x and y in A,

$$T(x \subseteq y) = T(x) \subseteq T(y).$$

Theorem 6.9.3 (Theorem 6.4.5) Let

$$\mathfrak{Z} = \langle A, \succsim, \oplus, \overline{1}, \overline{2}, \ldots, \overline{p}, \ldots\rangle_{p \in \mathbb{I}^+}$$

be such that

$$\mathfrak{X} = \langle A, \succsim, \oplus\rangle$$

is a continuous extensive structure of physical stimuli and

$$\mathfrak{B} = \langle A, \succsim, \overline{1}, \overline{2}, \ldots, \overline{p}, \ldots\rangle_{p \in \mathbb{I}^+}$$

is a behavioral structure associated with E (Definition 5.8.3) that satisfies the axioms for behavioral magnitude estimation (Axioms 5.8.1 to 5.8.4). Then the following three statements are equivalent.

1. *(Stevens' Psychophysical Power Law)* Let

$$\mathcal{S} = \{\varphi \mid \varphi \text{ is an isomorphism of } \mathfrak{X} \text{ onto } \langle \mathbb{R}^+, \geq, +\rangle\}$$

and

$T=\{\psi|\psi$ is a multiplicative representing function for $\mathfrak{B}$ (Definition 5.8.4)$\}$.

Then for each φ in S and each ψ in T there exist r and s in $\mathbb{R}^+$ such that for all x in A,

$$\psi(x)=r\varphi(x)^s.$$

2. For each p in $\mathbb{I}^+$ and each x and y in A,

$$\bar{p}(x \oplus y) = \bar{p}(x) \oplus \bar{p}(y)\,.$$

3. Let G be the set of automorphisms of $\mathfrak{Z}$, H be the set of automorphisms of $\mathfrak{X}$, and K be the set of automorphisms of $\mathfrak{B}$. Then $G=H=K$.

Proof. We first show Statement 1 implies Statement 2. Suppose Statement 1. Let p be an arbitrary element of $\mathbb{I}^+$ Let ψ be an arbitrary element of T. By Theorem 5.4.7, S is a ratio scale. Thus let φ in S be such that for all x in A,

$$\psi(x)=\varphi(z)^s.$$

Since ψ is a multiplicative representation for $\mathfrak{B}$, let a in $\mathbb{R}^+$ be such that for all x in A,

$$\psi[\bar{p}(x)] = a \cdot \psi(x)\,.$$

Then for all x in A,

$$\begin{aligned} y = \bar{p}(x) \quad &\text{iff} \quad \psi(y) = \psi(\bar{p}(x)) \\ &\text{iff} \quad \psi(y) = a \cdot \psi(x) \\ &\text{iff} \quad \varphi(y)^s = a \cdot \varphi(x)^s \\ &\text{iff} \quad \varphi(y) = a^{\frac{1}{s}} \cdot \varphi(x)\,; \end{aligned}$$

that is, for all x in A,

$$y = \bar{p}(x) \ \text{ iff } \ \varphi(y) = a^{\frac{1}{s}} \cdot \varphi(x)\,. \tag{6.18}$$

Let $m = \varphi(\bar{p})$ Then by Equation 6.18, m is the function that is multiplication by $a^{\frac{1}{s}}$ Thus for all x and y in A,

$$m(\varphi(x) + \varphi(y)) = m(\varphi(x)) + m(\varphi(y))\,. \tag{6.19}$$

Then, by taking j^{-1} of Equation 6.19, for all x and y in A,

$$\bar{p}(x \oplus y) = \bar{p}(x) \oplus \bar{p}(y)\,.$$

We next show that Statement 2 implies Statement 3. Assume Statement 2. Let G be the set of automorphisms of $\mathfrak{Z}$, H be the set of automorphisms of $\mathfrak{X}$, and K be the set of automorphisms of $\mathfrak{B}$ For each p in $\mathbb{I}^+$ if $p>1$ then $\bar{p}(x) \succ x$

for each x in A. Thus $\langle A, \succsim, \bar{p}\rangle$ is a continuous threshold structure for each p in $\mathbb{I}^+$ such that $p>1$. By an analog of the proof of Statement 3 of Theorem 6.9.2, it then follows that for each p in $\mathbb{I}^+$ such that $p>1$, H is the automorphism group of $\langle A, \succsim, \oplus, \bar{p}\rangle$ and from the latter it easily follows that H is the automorphism of $\mathfrak{Z}$. Thus $H=G$. Clearly, $G\subseteq K$. Thus to complete the proof of Statement 2 implying Statement 3, we need only show that $K\subseteq G$. Let $\alpha\in K$. Let x be an element of A. Then, because $H=G$ is 1-point homogeneous, let β in G be such $\beta(x)=\alpha(x)$. Because $G\subseteq K$, $\beta\in K$. But K is 1-point unique by Theorem 5.8.2. Therefore $\beta=\alpha$, that is, $\alpha\in G$.

We will finally show that Statement 3 implies Statement 1. Assume Statement 3. Let S and T be as in the hypothesis of Statement 1. Then by hypothesis S and T are ratio scales. Let $\psi\in T$ and $+'=\psi(\oplus)$. Let t, u, and v be arbitrary elements of $\mathbb{R}^+$. it will be shown a little later that

$$t(u +' v) = (tu) +' (tv)\,. \tag{6.20}$$

Because in Equation 6.20 t, u, and v are arbitrary elements of $\mathbb{R}^+$, it follows by Theorem 5.10.11 (applied to the continuous extensive structure $\langle\mathbb{R}^+, \geq, +'\rangle$ that there exists s in $\mathbb{R}^+$ such that for all a and b in $\mathbb{R}^+$,

$$a +' b = [a^{\frac{1}{s}} + b^{\frac{1}{s}}]^s$$

Thus ψ^s is an isomorphism of $\mathfrak{X}$ onto $\langle\mathbb{R}^+, \geq, +\rangle$. Therefore, $\psi^s\in S$. Since S is a ratio scale, $\varphi=r\cdot\psi^s$ for some $r\in\mathbb{R}^+$ Therefore, Statement 1 will follow once Equation 6.20 is shown:

To show Equation 6.20, let x and y be elements of A such that

$$u=\psi(x) \text{ and } v=\psi(y),$$

Because T is a ratio scale, let α in K be such that for all z in A,

$$\psi * \alpha(z)=t\cdot\psi(z).$$

Because $\oplus$ is invariant under H, and by hypothesis $H=K$, it follows that $\oplus$ is invariant under K. Thus

$$\begin{aligned} t(u +' v) &= t(\psi(x) +' \psi(y)) = t\psi(x \oplus y) = \psi * \alpha(x \oplus y) \\ &= \psi(\alpha(x) \oplus \alpha(y)) = \psi[\alpha(x)] +' \psi[\alpha(y)] \\ &= \psi * \alpha(x) +' \psi * \alpha(y) = (t\psi(x)) +' (t\psi(y)) \\ &= (tu) +' (tv)\,, \end{aligned}$$

that is, Equation 6.20 holds.

Lemma 6.9.1 Suppose $\mathfrak{X} = \langle A, \succsim, R_1, \ldots, R_k\rangle$ is a continuous structure with automorphism group G, $\mathfrak{C} = \langle A, \succsim, \oplus\rangle$ is a concatenation structure (Definition 6.6.1) with automorphisms group H, $\mathfrak{X}$ and is L($\in$,A, Ø)-equivalent to $\mathfrak{C}$. Then the following four statements are true:

1. $G{=}H$.
2. $\mathfrak{X}$ is homogeneous if and only if $\mathfrak{C}$ is homogeneous.
3. If $\mathfrak{X}$ is homogeneous, then $\mathfrak{X}$ is 2-point unique.
4. If $\oplus$ is non-idempotent and either $\mathfrak{X}$ or $\mathfrak{C}$ is homogeneous, then $\mathfrak{X}$ is 1-point unique.

Proof. 1. By hypothesis, $\mathfrak{X}$ is L($\in$,A, Ø)-equivalent to $\mathfrak{C}$. Thus each primitive of $\mathfrak{X}$ is set-theoretically definable in terms of the primitives of $\mathfrak{C}$. Therefore, by Theorem 5.1.2, each primitive of $\mathfrak{X}$ is invariant under the automorphisms of $\mathfrak{C}$. Thus $H \subseteq G$. Since each primitive of $\mathfrak{C}$ is also set-theoretically definable in terms of the primitives of $\mathfrak{X}$, it follows from a similar argument that $G \subseteq H$. Thus $G{=}H$.

2. Statement 2 is an immediate consequence of Statement 1.
3. Statement 3 is an immediate consequence of Statement 1 and Theorem 5.4.17.
4. Statement 4 is an immediate consequence of Statement 1 and Theorem 5.4.17.

Theorem 6.9.4 (Theorem 6.6.2) Suppose $\mathfrak{X} = \langle A, \succeq, R_1, \ldots, R_k \rangle$ is a continuous structure. Then the following two statements are equivalent:

1. $\mathfrak{X}$ is L($\in$,A, Ø)-equivalent to a continuous extensive structure of the form $\mathfrak{E} = \langle A, \succeq, \oplus \rangle$.
2. $\mathfrak{X}$ is homogeneous and 1-point unique.

Proof. uppose Statement 1. It follows from Theorem 5.4.7 the $\mathfrak{E}$ is homogeneous. Thus by Statements 2 and 4 of Lemma 6.9.1, $\mathfrak{X}$ is homogeneous and 1-point unique. Thus Statement 2 has been shown.

Suppose Statement 2. Then by Theorem 5.4.10 let $\oplus$ be an entity that is set-theoretically defined from the primitives of $\mathfrak{X}$ such that $\mathfrak{E} = \langle A, \succeq, \oplus \rangle$ is a continuous extensive structure. Thus to show Statement 1, one needs to only show that each primitive of $\mathfrak{X}$ is set theoretically definable in terms of the primitives of $\mathfrak{E}$. Let G be the automorphism group of $\mathfrak{E}$ and K be the automorphism group $\mathfrak{X}$. Since $\oplus$ is set-theoretically defined from $\mathfrak{X}$ it follows by Theorem 5.1.2 that $\oplus$ is invariant under K. Thus $K \subseteq G$. Because, by hypothesis, K is homogeneous, and it follows from Theorem 5.4.7 that G is 1-point unique, it easily follows that $K{=}G$. Thus each primitive of $\mathfrak{X}$ is invariant under G. By Theorem 5.1.2, each primitive of $\mathfrak{X}$ is set-theoretically definable in terms of the primitives of $\mathfrak{E}$. Thus $\mathfrak{X}$ is L($\in$,A, Ø)-equivalent to $\mathfrak{E}$.

Theorem 6.9.5 *(Theorem 6.6.4)* Suppose $\mathfrak{X} = \langle A, \succeq, R_1, \ldots, R_k \rangle$ is a continuous structure. Then the following two statements are equivalent:

1. $\mathfrak{X}$ is L($\in$,A, Ø)-equivalent to a continuous bisection structure of the form $\mathfrak{C} = \langle A, \succeq, \oplus \rangle$.
2. $\mathfrak{X}$ is 2-point homogeneous and 2-point unique.

Proof. Suppose Statement 1. It easily follows from Theorem 5.4.20 that $\mathfrak{C}$ is 2-point homogeneous, 2-point unique. Thus by Statement 1 of Lemma 6.9.1, $\mathfrak{X}$ is 2-point homogeneous, 2-point unique.

Suppose Statement 2. By Theorem 5.4.18, let $\mathfrak{N} = \langle \mathbb{R}, \geq, T_1, \ldots, T_k \rangle$ be a numerical structure such that the set S of isomorphisms of $\mathfrak{X}$ onto $\mathfrak{N}$ form an interval scale. (Theorem 5.4.18 chooses the numerical structure so that the scale of isomorphisms onto the numerical structure is a log-interval scale. From this, it easily follows that the structure $\mathfrak{N}$ above exists.) Let $\odot$ be the following operation on $\mathbb{R}$: For all r and s in $\mathbb{R}$,

$$r \odot s = \frac{r+s}{2}\,.$$

Then $\mathfrak{B} = \langle \mathbb{R}, \geq, \odot \rangle$ is a continuous bisection structure. Since S is an interval scale of isomorphisms, it is easy to show that $\mathfrak{B}$ is invariant under the automorphisms of $\mathfrak{N}$.

Let $\varphi \in S$ and

$$\oplus = \varphi^{-1}(\odot)\,.$$

Let ψ be an arbitrary element of S. Then for some automorphism η of $\mathfrak{N}$, $\psi = \eta * \varphi$. Thus

$$\psi(\oplus) = \eta * \varphi(\oplus) = \eta(\odot) = \odot.$$

The above shows that $\oplus$ has the following definition: $\oplus$ is the operation on A such that

$$\oplus = \gamma^{-1}(\odot)$$

for each γ in S. Because S is set-theoretically defined in terms the primitives of $\mathfrak{X}$, $\oplus$ is set-theoretically defined in terms of the primitives of $\mathfrak{X}$ by the just above definition. Furthermore, because S consists of isomorphisms and $\odot$ is invariant under the automorphisms of $\mathfrak{N}$, it follows that $\oplus$ is invariant under the automorphisms of $\mathfrak{X}$. Similarly, because the elements of S are isomorphisms and $\mathfrak{B}$ is a continuous bisection structure, it follows that $\mathfrak{C} = \langle A, \succsim, \oplus \rangle$ is a continuous bisection structure. By Theorem 5.1.2, each relation that is invariant under the automorphisms of $\mathfrak{X}$ is set theoretically defined in terms of the primitives of $\mathfrak{X}$. Thus the primitives of $\mathfrak{C}$ are set-theoretically defined in terms of the primitives of $\mathfrak{X}$.. Therefore, to show Statement 1, we need only show that the primitives of $\mathfrak{X}$ are set-theoretically defined in terms of the primitives of $\mathfrak{C}$. It is easy to verify that the automorphism group of $\mathfrak{B}$ is the automorphism group of $\mathfrak{N}$. By isomorphism, the automorphism group G of $\mathfrak{C}$ is the automorphism group of $\mathfrak{X}$ Therefore each primitive of $\mathfrak{X}$ is invariant under G. Thus by Theorem 5.1.2, each primitive of $\mathfrak{X}$ is settheoretically definable in terms of the primitives of $\mathfrak{C}$.

Theorem 6.9.6 *(Theorem 6.6.6)* Suppose $\mathfrak{X} = \langle A, \succsim, R_1, \ldots, R_k \rangle$ is a continuous, 2-point homogeneous, 2-point unique structure. Suppose $\oplus$ is settheoretically defined from the primitives of $\mathfrak{X}$ and $\mathfrak{C} = \langle A, \succsim, \oplus \rangle$ is a concatenation structure.

Then $\mathfrak{X}$ is L($\in$,*A*, Ø)-equivalent to $\mathfrak{C}$ and $\mathfrak{C}$ is a 2-point homogeneous, 2-point unique structure.

Proof. Because each primitive of $\mathfrak{C}$ is set-theoretically definable in terms of the primitives of $\mathfrak{X}$, by Theorem 5.1.2, each primitive of $\mathfrak{C}$ is invariant under the set K of automorphisms of $\mathfrak{X}$. Let G be the automorphism group of $\mathfrak{C}$. Then $K \subseteq G$, and thus $\mathfrak{C}$ is 2-point homogeneous. By Theorem 5.4.17, $\mathfrak{C}$ is 2-point unique. It easily follows from the facts that K and G are 2point homogeneous, 2-point unique and $K \subseteq G$ that $K=G$. Thus $\mathfrak{C}$ is a 2-point homogeneous, 2-point unique structure. Because $K=G$, it follows from Theorem 5.1.2 that each primitive of $\mathfrak{X}$ is set-theoretically defined in terms of the primitives of $\mathfrak{C}$. By hypothesis, each primitive of $\mathfrak{C}$ is defined in terms of the primitives of $\mathfrak{X}$. Therefore, $\mathfrak{X}$ is L($\in$,*A*, Ø)-equivalent to $\mathfrak{C}$.

Theorem 6.9.7 *(Theorem 6.7.1)* Suppose $\mathfrak{Q}$ and R are Boolean equivalent. Let $\mathbb{T}$ be the set of truth assignments to m properties (Convention 6.7.2). For each truth assignment $a=(a_1,\ldots,a_m)$ to $\mathfrak{Q}$ let $F(a)=(b_1,\ldots,b_m)$, where for $i=1,\ldots, m$, b_i is the truth value of the derived property R_i of R produced by the truth assignment a. Then F is a permutation of $\mathbb{T}$.

Proof. Clearly F is a function from $\mathbb{T}$ into $\mathbb{T}$ Thus, because $\mathbb{T}$ is a finite set, it is only necessary to show that F is one-to-one. Suppose F were not oneto-one. A contradiction will be shown. Let a and c in $\mathbb{T}$ be such that $a \neq c$ and $F(a)=F(c)$. Since $a \neq c$, let $1 \leq i \leq m$ be such that a_i and c_i are different truth values. Consider Q_i as a derived proposition of R, that is, consider Q_i as a propositional formula θ in terms of the variables $R_1,\ldots,R_m$. Similarly, each of the R_j, $1 \leq j \leq m$, is a propositional formula Δ_j in terms of the variables $Q_1,\ldots, Q_m$. Substituting Δ_j for R_j into θ for each $j=1,\ldots, m$ then yields a propositional formula Σ in terms of the variables $Q_1,\ldots,Q_m$ such that Q_i and Σ are logically equivalent, and therefore Q_i and Σ have the same truth values for a and the same truth values for c. Therefore, since a and c yield different truth values for Q_i, they yield different truth values for Σ But this is impossible, because a and c yield the same truth values for each R_j, $1 \leq j \leq m$, and therefore the same truth values for Δj, and therefore the same truth values for Σ.

Theorem 6.9.8 *(Theorem 6.7.2)* Suppose F is a permutation of $\mathbb{T}$. Then a set of m properties S about the set of people Y (and the set of possible people X) can be found such that the following two statements are true:

1. $S=(S_1,\ldots, S_m)$ is Boolean equivalent to $\mathfrak{Q}$.
2. For each a in $\mathbb{T}$ and each $i=1,\ldots, m$, the ith component of $F(a)$ is the truth value of S_i produced by the truth assignment a.

Proof. Let $1 \leq i \leq m$. Let

$$\mathbb{F}_i = \{a \,|\, a \in \mathbb{T} \text{ and the } i^{th} \text{ component of } F(a) = 1\}.$$

For each a $\mathbb{F}$ in let C_a be the property that is the *conjunction corresponding to a of elements* of $\mathfrak{Q}$; for example, for $m=3$ and $a=(1,0,0)$,

$$C_a = Q_1 \wedge \neg Q_2 \wedge \neg Q_3 .$$

Let S_i be the disjunctionof all the C_a such that a is in $\mathbb{F}_i$, for example, for $\mathbb{F}_i = \{C_e, C_f, C_g, C_h\}$,

$$S_i = C_e \vee C_f \vee C_g \vee C_h .$$

Then, by construction, Statement 2 of the theorem has been shown.

To show Statement 1, suppose $1 \leq k \leq m$. Let

$$\mathbb{T}_k = \{a \,|\, a \in \mathbb{T} \text{ and } a_k = 1\} .$$

For each a in $\mathbb{T}_i$ let $D_{F(a)}$ be the property that is the conjunctive normal form corresponding to $F(a)$ of elements of S; for example, for m=3 and $F(a)$=(0,1,0),

$$D_a = \neg S_1 \wedge S_2 \wedge \neg S_3 .$$

Let Q'_k be the disjunction of all the D_a such that a is in $\mathbb{T}_k$; for example, for $\mathbb{T}_k = \{e, f, g, h\}$,

$$Q'_i = D_e \vee D_f \vee D_g \vee D_h .$$

We will show that Q'_k is logically equivalent to Q_k, thus establishing Statement 1.

Let a be an arbitrary element of $\mathbb{T}$ such that a makes Q_k true. Then $a \in \mathbb{T}_k$. By Statement 2 (which already has been shown), a assigns to $(S_1,\ldots, S_m)$ the truth assignment $F(a)$. But, because a $a \in \mathbb{T}_k$ it follows from the definition of Q'_k that Q'_k is true under the assignment of $F(a)$ to S and therefore is true under the assignment of a to $\mathcal{Q}$.

Let a be an arbitrary element of $\mathbb{T}$ such that a makes Q_k false. Then $a \notin \mathbb{T}_k$. By Statement 2 (which already has been shown), a assigns to $(S_1,\ldots, S_m)$ the truth assignment $F(a)$. But, because $a \notin \mathbb{T}_k$, it follows from the definition of Q'_k that Q'_k is false under the assignment of $F(a)$ to S and therefore is false under the assignment of a to $\mathcal{Q}$.

7. Qualitativeness

7.1 Introduction

In Chapters 4, 5, and 6, a topic *T* under consideration was captured by a structure $\mathfrak{X} = \langle A, R_j \rangle_{j \in J}$ based on *A*, and relations and concepts that belonged to T were called "meaningful." Chapter 6 investigated subtopics of *T* that historically and intuitively corresponded to lawful subparts of *T*, and the relations and concepts of such subtopics were called "(relatively) intrinsic." In this section a different kind of subtopic of *T* is considered. The relations and concepts of this new form of subtopic are called "qualitative." *The reader should note that the concept of "qualitativeness" considered in this chapter is a formal one that makes specific some of the informal uses of "qualitative" of the previous chapters.*

Like meaningfulness and intrinsicness, qualitativeness satisfies strong definability principles, and like intrinsicness it may be viewed formally as a meaningfulness concept. The key difference between it and the meaningfulness concepts of Chapter 4 is that meaningfulness allows pure mathematics—or a large portion of it—to belong to the (meaningfully generated) topic, whereas qualitativeness allows only an extremely restricted portion of pure mathematics to belong to the (qualitatively generated) topic. Because historically topics within science have permitted the free use of pure mathematics in their generations and specifications, meaningfulness appears to me to be a correct and natural concept to identify with "scientific topic." Philosophers and others engaged in the foundations of science have considered other notions of "scientific topic," usually ones that purportedly had metaphysically sounder bases than those that have evolved in science. These purportedly "metaphysically sounder" kinds of scientific concepts are often based on ideas of qualitativeness, and they sometimes require the mathematics used by the topic to be developed within it.

Another concept frequently invoked in the foundation of science is empiricalness. Empiricalness is not be systematically developed in this book. However, in order to show how it differs with meaningfulness and qualitativeness, a very rudimentary version of it is developed in Section 7.11. There, empiricalness is viewed as a methodology formulated in terms of the concepts of verifiability, refutability, testability, observability, and so forth, instead of a process that together with "empirical primitives" determines a topic.

Previous chapters have treated numbers as platonic objects whose existences were independent of the qualitative or empirical situations under considerations. For example, in the axiomatic generalizations of the Erlanger Program of Chapter 4, pure mathematics was identified with structures based on pure sets, for example, the identification of a structure based on platonic real numbers with a structure based on the set-theoretic reals; and in Chapter 5, measurement was conceptualized as specific kinds of mappings from a qualitative domain into a system of platonic numbers, usually a system based on a subset of the set-theoretic reals. Because platonic numbers are not dependent on qualitative and empirical situations, they have obvious appeal in the conduction of science: they provide a common element that allows results from various scientific domains to be compared or combined; and they, as well as relations and concepts based on them, are invariant under automorphisms of structures based on qualitative or empirical domains.

Some researchers perceive platonic numbers as bringing into science a metaphysical component that is irrelevant and/or undesirable. They would like to replace platonic numbers with numbers that are qualitatively or empirically based, thus reducing the metaphysical content inherent in the conduction of science. It is not the intention of this chapter to give a systematic account of the literature about this interesting subject. Instead, the concept of "qualitative number" is only developed from perspectives similar to those presented in Chapters 4, 5, and 6, and only a brief discussion of the possible involvement of qualitative numbers in obtaining metaphysical reductions in science is proffered. For additional perspectives see Niederée (1987, 1992a, 1992b,) Michell (1990), and Field (1980).

7.2 Axiom System $Q_\omega(a)$

From many points of view, the concept of "qualitativeness" should not depend in essential ways on "pure mathematics," as, for example, embodied in structures based on the set-theoretic reals. In Definition 7.2.1 below, an axiomatic theory of qualitativeness, $Q_\omega(a)$, is given that *(i)* excludes most of "pure mathematics" from being qualitative, and *(ii)* prevents purely mathematical concepts from being employed in essential ways for establishing the qualitativeness of entities. The axiomatic theory $Q_\omega(a)$ assumes a set a of qualitative primitives is given, and it considers an entity to be "qualitative" if and only if the entity is "definable" in terms of a and its elements. The concept of qualitativeness provided by axiom system $Q_\omega(a)$ is in many ways similar to the concept of meaningfulness given by axiom system D″*(a)* (Definition 4.3.6) with the following exceptions: (1) in axiom system D″*(a),* the set a of primitives need not be meaningful, whereas in axiom system $Q_\omega(a)$ the set a of primitives is qualitative; and (2) in axiom system D″*(a)* all pure sets are meaningful, whereas in axiom system $Q_\omega(a)$ the qualitative pure sets are severely limited.

Definition 7.2.1 *Axiom system $Q_\omega(a)$* is said to hold if and only if Q is a 1-place predicate and the following three conditions hold:

(i) Axiom system ZFA.
(ii) $a \in V_\omega$ (Definition 3.5.1).
(iii) For all entities b, $Q(b)$ if and only if there exist entities $b_1,\ldots,b_n$ and a formula $\varphi(x,v_1,\ldots,v_n)$ of L($\in$, A, Ø) such that $b_1 \in a \cup \{a\},\ldots, b_n \in a \cup \{a\}$ and

$$\varphi(b, b_1, \ldots, b_n) \wedge \forall x[x = b \leftrightarrow \varphi(x, b_1, \ldots, b_n) \text{ and } x \in V_\omega].$$

Lemma 7.2.1 Assume axiom system $Q_\omega(a)$. Then the following five statements are true:

1. $Q(A)$.
2. $Q(\varnothing)$.
3. $Q(a)$.
4. For all entities c, if $c \in a$ then $Q(c)$.
5. For all entities c, if $Q(c)$ then $c \in V_\omega$.

Proof. Statement 1 follows from Condition *(i)* of Definition 7.2.1 and the statement

$$\forall x[x = A \leftrightarrow x = \boldsymbol{A}],$$

and Statements 2, 3, and 4 follow by similar arguments. Statement 5 is an immediate consequence of Condition *(iii)* of Definition 7.2.1.

Convention 7.2.1 Assume axiom system $Q_\omega(a)$. The phraseologies "b is an element of Q," "b is in Q," "$b \in Q$," and so forth, is often used to stand for "b is an entity (within ZFA) and $Q(b)$." Also, by convention, Q will denote the collection of entities b such that $Q(b)$.

Definition 7.2.2 Assume axiom system $Q_\omega(a)$. Then Q is called the *collection of qualitative entities* and elements of Q are said to be *qualitative*. By Statement 5 of Lemma 7.2.1, Q is a subcollection of the set V_ω.

Lemma 7.2.2 Q is not qualitative.

Proof. Suppose Q were qualitative. Then Q would be an entity and $Q \in Q$, the latter being impossible by Theorem 3.3.6.

Assume axiom system $Q_\omega(a)$. By Statement 5 of Lemma 7.2.1 all qualitative pure sets are in V_ω. The pure sets in V_ω consists of the elements of P_ω (Definition 3.5.1), and it follows from results of Chapter 2 that each element of P_ω is a finite set inductively built out of Ø, the singleton operation {}, and the union operation ∪. In particular, each finite ordinal and each finite fragment of arithmetic is in V_ω. However, not every finite ordinal need be qualitative. This is because it is a well-known result from mathematical logic that models of ZFA exist in which some finite ordinals—that is, some elements of ω—are infinite collections when they are viewed from outside of the model. It is not difficult to show from this result that there are models of ZFA such that for many choices of the qualitative primitives (e.g., $a=\{\emptyset\}$), ω has non-qualitative elements. Also, it is not difficult to show that each externally finite element (Definition 4.3.7) of P_ω is qualitative.

When a is externally finite (Definition 4.3.7), there is no problem in associating Q in a qualitative way with a structure that has domain A and the set of primitive relations $\{A\}\cup a$. For example, if $a=\{R,S,T\}$, then $\langle A, R, S, T\rangle$ is an appropriate qualitative structure, where $\langle A, R, S, T\rangle$ is interpreted as the ordered 4-tuple (A,R,S,T). When a is infinite difficulties are encountered. These arise because indexings of a, $a=\{a_j|j\in J\}$, may not be qualitative; that is, either (1) J or some of its elements may not be qualitative or (2) the function F from J onto a such that for each $j\in J$, $F(j)=a_j$ may not be qualitative. It is interesting to note that the corresponding problem for meaningfulness does not arise with axiom system TM: With TM one could always be assured of a meaningful indexing of a by, for example, taking J to be an appropriate pure set, for example, an appropriate set of ordinals. Then each element of F is an ordered pair (c, d), where c is meaningful by the axiom of Meaningful Pure Sets and d is meaningful by assumption because it is an element of a. Thus (c, d) is meaningful by the axiom of Meaningful Comprehension', and F meaningful by the axiom of Meaningful Inheritability.

Many concepts that use a portion of pure mathematics in their formulation can often be given a qualitative reformulation. The following two examples illustrate this for continua.

Example 7.2.1 Let $\succeq$ be a binary relation on A. Let $a=\{\succeq\}$ and assume $Q_\omega(a)$. Then using only qualitative concepts it is easy to say that $\succeq$ is a total ordering on A that is unbounded and dense (Definition 5.1.2). Suppose this is done. Then by Definition 5.1.2, $\langle A, \succeq\rangle$ is a continuum if and only if it satisfies Dedekind completeness and denumerable density. The following is a qualitative formulation of Dedekind completeness:

Let B be the set of all nonempty subsets b of A such that there exists an element y of A such that

$$\forall x(\text{if } x \in b \text{ then } y \succeq x)$$

and

$$\forall x \forall z(\text{if } x \in b \text{ and } x \succeq z \text{ then } z \in b).$$

Then B is qualitative, and $\langle A, \succeq\rangle$ is Dedekind complete if and only if for each b in B, there exists an element u in A such that *(i)* for all $x\in b$, $u \succeq x$, and *(ii)* for all y in A, if $y \succeq x$ for all x in b, then $y \succeq u$.

Because $\omega \notin V_\omega$, the customary definitions "denumerable density" (e.g., Definition 5.10.1) do not provide for a *qualitative* formulation of the denumerable density of $\langle A, \succeq \rangle$. However, the denumerable density of $\langle A, \succeq \rangle$ has the following qualitative formulation:

> Let E be the set of all associative operations $\oplus$ on A that are monotonic, positive, and solvable with respect to $\succeq$ (Definition 5.4.5). Then E is qualitative. By Theorem 5.4.7 and the remarks following it, it then follows (assuming total ordering, unboundedness, density, and Dedekind completeness) that $E \neq \emptyset$ if and only if $\langle A, \succeq \rangle$ satisfies denumerable density (Definition 5.10.1). Thus *in this context,* "$E \neq \emptyset$" is a qualitative formulation of "denumerable density."

In the next example, a continuum results from qualitative assumptions about the relations in a:

Example 7.2.2 Let $a = \{\succeq\} \cup \{R_j \mid j \in J\} \cup \{F\}$, where F is the function on J such that $F(j)=R_j$ for each j in J, and suppose $Q_\omega(a)$. Then saying $\mathfrak{X} = \langle A, \succeq, R_j \rangle_{j \in J}$ is a continuous scalar structure (Definition 5.4.6) can be done in a straightforward qualitative way through the obvious qualitative formulations of the concepts presented in Definition 5.4.6. Suppose that this is done. Then by Theorem 5.4.8, $\langle A, \succeq \rangle$ is a continuum. Thus in this case, the higher-order qualitative properties of the homogeneity of $\mathfrak{X}$ the 1-point uniqueness of $\mathfrak{X}$, and the Dedekind completeness of $\langle A, \succeq \rangle$ are used to show that $\langle A, \succeq \rangle$ satisfies denumerable density—a property whose customary definition (Definition 5.10.1) involves pure mathematics.

Like in the definitional theories of meaningfulness discussed in Chapter 4, invariance under transformation groups plays an important role in the theory of qualitativeness. The following definition and two theorems characterize some of the fundamental properties of this kind of invariance for qualitativeness.

Theorem 7.2.1 Assume axiom system $Q_\omega(a)$. Let

$$G_a = \{f \mid f \text{ is a permutation on } A \text{ and } f(x) = x \text{ for all } x \text{ in } a\}.$$

Then G_a is qualitative and $\langle G_a, *' \rangle$ is a group, where $*'$ is the restriction of the operation of functional composition $*$ to G_a. Furthermore, $*'$ is qualitative.

Proof. The set of permutations Π on A is qualitative by Definitions 7.2.1 and 7.2.2. Because

$$G_a = \{f \mid f \in \Pi \text{ and } f(x) = x \text{ for all } x \text{ in } a\},$$

G_a is qualitative by Definitions 7.2.1 and 7.2.2. Because G_a is qualitative, it easily follows from Definition 7.2.1 that the restriction of $*$ to G_a, $*'$, is qualitative. The verification that $\langle G_a, *' \rangle$ is a group is straightforward and is left to the reader.

Definition 7.2.3 Assume axiom system Q_ω *(a)*. Let

$$G_a = \{f \mid f \text{ is a permutation on } A \text{ and } f(x) = x \text{ for all } x \text{ in } a\}.$$

Then G_a is called the *transformation group* of Q.

Theorem 7.2.2 Assume axiom system Q_ω *(a)*. Let G_a be the transformation group of Q (Definition 7.2.3), and define the meaningfulness predicate M as follows: For all entities b,

$$M(b) \text{ iff } f(b) = b \text{ for all } f \in G_a.$$

Then the following three statements are true:

1. Axiom system TM holds and G_a is the transformation group for M (Definition 4.2.1).
2. $Q \subset M$.
3. $G_a \in Q$.

Proof. 1. Statement 1 follows immediately by the definition of M in the hypothesis of the theorem.

2. By Statement 1 axiom system TM holds. Thus by Theorem 4.2.10, axioms *MI, MC'*, and *AL* hold. It follows from the definitions of M and G_a that $M(x)$ for each x in a. Thus by axiom *MI*, $M(a)$. $Q \subseteq M$ then follows by axioms MC′ and *AL* and Definition 7.2.1. Because $M(w)$ and $\omega \neq V_\omega$, it follows from Statement 5 of Lemma 7.2.1 that $Q \neq M$. Thus $Q \subset M$.

3. $G_a \in Q$ by Theorem 7.2.1.

Theorem 7.2.3 Assume axiom system $Q_\omega(a)$. Then for each qualitative R and for each β in the transformation group G_a of Q,

$$\beta(R) = R.$$

Proof. Define the meaningfulness predicate M as follows: For all entities b,

$$M(b) \text{ iff } f(b) = b \text{ for all } f \in G_a.$$

Then by Statement 1 of Theorem 7.2.2, axiom system TM holds and G_a is the transformation group for M (Definition 4.2.1). By Statement 2 of Theorem 7.2.2, $M(R)$. Thus R is G_a-invariant, that is,

$$\beta(R) = R.$$

for each β in G_a.

Theorem 7.2.4 Let $a = \{R_j \mid j \in J\}$ and assume axiom system $Q_\omega(a)$. Then the following two statements are true:

1. The transformation group for Q (Definition 7.2.3) is the automorphism group of the structure

$$\langle A, R_j \rangle_{j \in J}.$$

2. If $b = \{S_k | k \in K\}$, $a \supseteq b$, and $S_k \in Q$ for each k in K, then the transformation group for Q (Definition 7.2.3) is the automorphism group of the structure

$$\langle A, S_k \rangle_{k \in K} \, .$$

Proof. Statement 1 immediately follows from Definition 7.2.3. Statement 2 follows from Statement 1 and Theorem 7.2.3.

Theorem 7.2.5 Assume axiom system $Q_\omega(a)$. Suppose α is a one-to-one function from A onto A and $\alpha \in Q$. Then the for each β in the transformation group G_a of Q,

$$\beta * \alpha = \alpha * \beta \, .$$

Proof. Let β be an arbitrary element of G_a. Then by Theorem 7.2.3, $\beta(\alpha)=\alpha$, and from this it easily follows that

$$\beta * \alpha = \alpha * \beta \, .$$

7.3 Integral Domains and Fields

The numerical structures $\mathfrak{I} = \langle \mathbb{I}, \geq, +, \cdot, 0, 1 \rangle$ and $\mathfrak{R} = \langle \mathbb{R}, \geq, +, \cdot, 0, 1 \rangle$ are of enormous importance to mathematics and science. Undoubtedly a major reason for this is that they permit easy calculations and serve as the basis for the formulation of important and applicable algebraic concepts. The major thrust of this section is to axiomatize qualitatively structures that are algebraically identical to $\mathfrak{I}$ and $\mathfrak{R}$.

Definition 7.3.1 $\mathfrak{X} = \langle X, \succsim, \oplus, \odot, i_0, i_1 \rangle$ is said to be a *totally ordered integral domain (with additive operation* $\oplus$*, multiplicative operation* $\odot$*, additive identity* i_0*, and multiplicative identity* i_1*)* if and only if the following eight conditions hold:

(i) $\succsim$ is a total ordering.

(ii) $\oplus$ and $\odot$ are commutative and associative operations on X, that is, $\oplus$ and $\odot$ are operations on X and for all x, y, and z in X,

$$x \oplus y = y \oplus x \text{ and } (x \oplus y) \oplus z = x \oplus (y \oplus z)$$

and

$$x \odot y = y \odot x \text{ and } (x \odot y) \odot z = x \odot (y \odot z) \, .$$

(iii) $\odot$ distributes over $\oplus$; that is, for all x, y, and z in X,

$$x \odot (y \oplus z) = (x \odot y) \oplus (x \odot z) \, .$$

(iv) i_0 and i_1 are elements of X,

$$i_0 \neq i_1 ,$$

and for all x in X,

$$i_0 \oplus x = x \ , \ i_1 \odot x = x \ , \text{ and } i_0 \odot x = i_0 \, .$$

(v) For each x in X there exists y in X such that $x \oplus y = i_0$.
(vi) For all x and y in X, *if* $x \odot y = i_0$, then either $x = i_0$ or $y = i_0$.
(vii) For all x, y, and z in X,

$$x \succeq y \text{ iff } x \oplus z \succeq y \oplus z \, .$$

(viii) For all x, y, and z in X, if $x \succ i_0$, and, $z \succ i_0$ then

$$x \odot z \succ i_0$$

and

$$x \succeq y \text{ iff } x \odot z \succeq y \odot z \, .$$

The integers, rational numbers, and real numbers form integral domains with their usual orderings, addition and multiplication operations, and their additive and multiplicative identities, 0 and 1. The following definitions and theorem characterize the totally ordered integral domain of integers.

Definition 7.3.2 Let $\mathfrak{X} = \langle X, \succeq, \oplus, \odot, i_0, i_1 \rangle$ be a totally ordered integral domain. Then, by definition,

$$X^+ = \{x | x \in X \text{ and } x \succ i_0\} \, .$$

Definition 7.3.3 $\mathfrak{X} = \langle X, \succeq, \oplus, \odot, i_0, i_1 \rangle$ is said to be a *well-ordered* integral domain if and only if $\mathfrak{X}$ is a totally ordered integral domain and $\langle X^+, \succeq \rangle$ is a well-ordered set, that is, each nonempty subset of X^+ has a $\succeq$-least element.

The following theore is a well-known result of algebra.

Theorem 7.3.1 Let $\mathfrak{X} = \langle X, \succeq, \oplus, \odot, i_0, i_1 \rangle$ be a well-ordered integral domain. Then there exists an isomorphism of $\mathfrak{X}$ onto $\langle \mathbb{I}, \geq, +, \cdot, 0, 1 \rangle$.

Integral domains may lack solvability with respect to multiplication. For example, in the totally ordered integral domain of integers, the multiplicative equation $3 \cdot x = 2$ has no solution for x. By assuming solvability with respect to multiplication a richer structure called a "field" results.

Definition 7.3.4 $\mathfrak{X} = \langle X, \succeq, \oplus, \odot, i_0, i_1 \rangle$ is said to be a *totally ordered field* if and only if $\mathfrak{X}$ is a totally ordered integral domain and for all x in X if $x \neq i_0$, then there exists y in X such that $x \odot y = i_1$.

The real numbers form a totally ordered field with its usual ordering, addition and multiplication operations, and its additive and multiplicative identities, 0 and 1. The following definition and theorem characterize the totally ordered field of real numbers.

Definition 7.3.5 $\mathfrak{X} = \langle X, \succeq, \oplus, \odot, i_0, i_1 \rangle$ is said to be a *continuum field* if and only if $\mathfrak{X}$ is a totally ordered field and $\langle X, \succeq \rangle$ is a continuum (Definition 5.4.1).

The following theorem is a well-known result of algebra.

Theorem 7.3.2 Let $\mathfrak{X} = \langle X, \succeq, \oplus, \odot, i_0, i_1 \rangle$ be a continuum field. Then there exists an isomorphism of $\mathfrak{X}$ onto $\langle \mathbb{R}, \geq, +, \cdot, 0, 1 \rangle$.

The totally ordered field of rational numbers lack solutions to many polynomial equations with rational coefficients, for example, the equation $x^2=2$ has no rational solution although it has a real solution. Subfields $\mathfrak{F} = \langle F, \geq, +, \cdot, 0, 1 \rangle$ of the totally ordered field of real numbers $\mathfrak{R}$ for which each polynomial equation with coefficients in F has all of its real roots in F are called “real closed subfields of $\mathfrak{R}$.” It is well-known in algebra that $\mathfrak{F}$ is a real closed subfield of $\mathfrak{R}$ if and only if for each polynomial $p(x)$ of odd degree with coefficients in F, the equation $p(x)=0$ has a solution in F. This fact gives rise to the following definition:

Definition 7.3.6 $\mathfrak{X} = \langle X, \succeq, \oplus, \odot, i_0, i_1 \rangle$ is said to be a *real closed field* if and only if $\mathfrak{X}$ is a totally ordered field and for each polynomial $p(x)$ of odd degree with coefficients in X, the equation $p(x)=0$ has a solution in X.

Because the field of real numbers is a real closed field, all continuum fields are real closed fields by Theorem 7.3.1. It is a well-known theorem of logic that real closed fields are first-order complete, that is, any two real closed fields have the same set of true first-order statements. The ordered field of real algebraic numbers is a well-known example of a denumerable real closed field.

Integral domains and fields are algebraically rich systems. Part of their richness consists in their having domains that contain “negative” as well as “positive” elements. As discussed in Section 1.7 of Chapter 1, the metaphysics of negative integers and negative real numbers has been a basis of controversy. In physics and other parts of science, qualities are usually measured in terms of positive numbers. Because of this, the positive subpart integral domains and fields often appear in science as the natural structures on which to base a qualitative theory of numbers.

Definition 7.3.7 $\langle X, \succeq, \oplus, \odot, i_1 \rangle$ is said to be a *positive totally ordered field* if and only if there exists a totally ordered field $\langle Y, \succeq^\star, \oplus^\star, \odot^\star, i_0, i_1 \rangle$ such that

$$X = \{x \mid x \in Y \text{ and } x \succ^\star i_0\}$$

and $\succsim$, $\oplus$, and $\odot$ are respectively the restrictions of $\succsim^\star$, $\oplus^\star$, and $\odot^*$ to X.

Similar definitions hold *positive continuum field, positive real closed field,* and so forth.

Convention 7.3.1 As the reader can verify, all of the "positive" structures referred to in Definition 7.3.7 have simple algebraic characterizations in terms of their primitives, and Theorems 7.3.1 and 7.3.2 have obvious reformulations for positive structures.

The following two theorems are useful in proofs.

Theorem 7.3.3 Suppose $\langle X, \succsim, \oplus, \odot, i_0, i_1\rangle$ and $\langle X, \succsim, \oplus, \odot', i_0, i_1\rangle$ are well-ordered integral domains. Then $\odot = \odot'$.

Proof. It will first be shown by induction that for positive x and y,

$$x \odot y = x \odot' y .$$

Let x be an arbitrary positive element. Because i_1 is the multiplicative identity of both $\odot$ and $\odot'$,

$$x \odot i_1 = x \odot' i_1 .$$

Suppose $y \succ i_1$ and $x \odot y = x \odot' y$. Then

$$(x \odot y) \oplus x = (x \odot' y) \oplus x ,$$

that is,

$$x \odot (y \oplus i_1) = x \odot' (y \oplus i_1) .$$

The general case follows from the positive case above, the commutativity of $\odot$ and $\odot'$, and the following three observations: For all x and y in X,

(1) $x \odot i_0 = i_0 = x \odot' i_0$;
(1) if $x \succ i_0 \succ y$, then

$$x \odot y = -[x \odot (-y)] = -[x \odot' (-y)] = x \odot' y; \text{ and}$$

(3) if $i_0 \succ x$ and $i_0 \succ y$, then

$$x \odot y = (-x) \odot (-y) = (-x) \odot' (-y) = x \odot' y .$$

Theorem 7.3.4 Suppose $\langle X, \succsim, \oplus, \odot, i_0, i_1\rangle$ and $\langle X, \succsim, \oplus, \odot', i_0, i_1\rangle$ are continuum fields. Then $\odot = \odot'$.

Proof. By Theorem 7.3.2, it is sufficient to show the theorem for the case where $X = \mathbb{R}$, $\succsim = \geq$, $\oplus = +$, $\odot' = \cdot$, $\iota_0 = 0$, and $\iota_1 = 1$. It will first be shown that $\odot = \cdot$ on $\mathbb{R}^+$. Let r be an arbitrary element of $\mathbb{R}^+$. Define f_r on $\mathbb{R}^+$ as follows: for all x in $\mathbb{R}^+$,

$$f_r(x) = r \odot x .$$

f_r is strictly increasing and for all x and y in $\mathbb{R}^+$,

$$f_r(x+y) = r \odot (x+y) = (r \odot x) + (r \odot y)\,,$$

that is,

$$f_r(x+y) = f_r(x) + f_r(y)\,. \tag{7.1}$$

It is well-known that the only strictly increasing solution to Equation 7.1 is f_r=multiplication by a positive constant. Because $f_r(1)=r \odot 1=r$, f_r= multiplication by r. Thus $\odot$=$\cdot$on $\mathbb{R}^+$.

The general case follows from the positive case above, the commutativity of $\odot$, and the following three observations: For all x and y in $\mathbb{R}^+$

(1) $x \odot 0=0=x\cdot 0$;

(1) if x is positive and y is negative, then

$$x \odot y = -[x \odot (-y)] = -[x \cdot (-y)] = x \cdot y; \text{ and}$$

(3) if x and y are negative, then

$$x \odot y = (-x) \odot (-y) = (-x) \cdot (-y) = x \cdot y\,.$$

7.4 Qualitative Systems of Magnitude Numbers

Throughout this chapter, a form of qualitative "numbers" called "magnitude numbers" is developed. The theoretical roots of magnitude numbers reach back to Eudoxus' theory of proportions in Book *V* of Euclid's *Geometry.* Eudoxus' theory has inspired several theories of qualitative numbers, including those developed for the foundations of measurement by Niederée (1987, 1992a, 1992b) and Michell (1990).

For the considerations of this chapter, the most important systems of qualitative numbers are those whose domains consist of functions f from the set of atoms A onto itself such that f is strictly increasing with respect to some qualitative total ordering. The following definition formalizes this concept of "number."

Definition 7.4.1 Assume axiom system $Q_\omega(a)$. Then α is said to be a $\succeq$-*magnitude number* if and only if $\succeq$ is a qualitative total ordering on A and α is a strictly $\succeq$ -increasing function from A onto A.

Definition 7.4.2 Assume axiom system $Q_\omega(a)$. Suppose $\succeq$ is a qualitative total ordering on A. Then, by definition, $N_\succeq$ is the set of $\succeq$-magnitude numbers. Also by definition, $\mathbb{Q}_\succeq$ is the collection of qualitative $\succeq$-magnitude numbers.

It is immediate from Definition 7.4.2 that each element of $\mathbb{Q}_\succeq$ is an element of $N_\succeq$. Note that because the collection of qualitative entities Q need not be a set in ZFA and $\mathbb{Q}_\succeq$ is defined partially in terms of Q, $\mathbb{Q}_\succeq$ need not be a set in ZFA. In particular, $\mathbb{Q}_\succeq$ need not be qualitative.

Assume axiom system $Q_\omega(a)$ and suppose $\succeq$ is a qualitative total ordering on A. Then the identity function $\imath$ on A is always a qualitative $\succeq$-magnitude number because,

$$\imath = \{x | x=(y, y) \text{ for some } y \in A\}.$$

Convention 7.4.1 Assume axiom system $Q_\omega(a)$ and suppose $\succeq$ is a qualitative total ordering on A. The identity function on A is sometimes called the *identity $\succeq$-magnitude number*. More usually, it is called the *identity* and denoted by $\imath$. In fields of $\succeq$-magnitude numbers, it is usually either the additive identity and denoted by i_0 or the multiplicative identity and denoted by i_1.

The term "$\succeq$-magnitude number" was selected in part because of its use in the theory of magnitude estimation presented in Section 7.7. Why "$\succeq$-magnitude numbers" should be considered as "numbers" will become apparent later in the chapter.

Definition 7.4.3 Assume axiom system $Q_\omega(a)$ and suppose $\succeq$ is a qualitative total ordering on A. Then $\succeq'$ is said to be the *$\succeq$-induced magnitude ordering* on $N_\succeq$ if and only if $\succeq'$ is the binary relation on $N_\succeq$ such that for all α and β in $N_\succeq$,

$$\alpha \succeq' \beta \text{ iff } \alpha(x) \succeq \beta(x) \text{ for all } x \text{ in } A.$$

Assume axiom system $Q_\omega(a)$. It is easy to show for each qualitative total ordering $\succeq$ on A that the $\succeq$-induced magnitude ordering $\succeq'$ is a transitive and antisymmetric relation. However, $\succeq'$ may not be a total ordering because it may not be connected. Even on the restricted set $\mathfrak{Q}_\succeq$, $\succeq'$ may not be connected.

$\succeq$-induced magnitude orderings are used throughout this section to totally order domains of algebraic systems of $\succeq$-magnitude numbers. Because of the possible lack of connectivity of induced magnitude orderings, this can only be effectively achieved by either *(i)* imposing additional assumptions about $Q_\omega(a)$, or *(ii)* appropriately limiting the choice of the domain of qualitative algebraic system.

Definition 7.4.4 $\mathfrak{N} = \langle \mathbb{N}, \succeq', \oplus, \odot, i_0, i_1 \rangle$ is said to be a *qualitative integral domain of $\succeq$-magnitude numbers* if and only if the following four statements are true:

1. $\mathfrak{N}$ is a totally ordered integral domain.
2. $\mathbb{N}$ is a set of $\succeq$-magnitude numbers.
3. $\succeq'$ is the restriction to $\mathbb{N}$ of the $\succeq$-induced magnitude ordering of some qualitative total ordering $\succeq$ on A.
4. $\mathbb{N}$, $\succeq'$, $\oplus$, $\odot$, i_0, and i_1 are qualitative.

Similar definitions hold for *qualitative totally ordered field of* $\succeq$*-magnitude numbers, qualitative continuum field of* $\succeq$*-magnitude numbers, qualitative positive continuum field of* $\succeq$*-magnitude numbers,* and so forth.

Convention 7.4.2 To simplify notation, the convention has been adopted throughout this book of using the same symbol of an operation or relation to stand for its restriction, particularly when the restriction is a primitive of a structure, for example, the symbol "$\geq$" stands for an ordering on $\mathbb{R}$ as well as its restriction to $\mathbb{R}^+$. With qualitativeness another ambiguity is added to this convention: a relation may be qualitative and its restriction not. Sometimes this latter ambiguity is directly avoided by employing a different symbol for the restriction, but more usually it is avoided indirectly by having context indicate which operation or relation is intended.

The qualitative algebraic systems of $\succeq$-magnitude numbers can be employed to achieve a qualitative form of measurement via the following idea: A totally ordered integral domain $\langle \mathbb{N}, \succeq', \oplus, \odot, i_0, i_1 \rangle$ is selected for which $\mathbb{N}$ is a qualitative set of $\succeq$-magnitude numbers, $\succeq'$ is the restriction to $\mathbb{N}$of the $\succeq$ -induced magnitude ordering of some qualitative total ordering $\succeq$ on *A*, $\mathbb{Q}$ is dense in $\langle \mathbb{N}, \succeq' \rangle$ and $\oplus$, $\odot$, i_0, and i_0 are qualitative entities. An element *u* of *A* is selected and is assigned the $\succeq$-magnitude number $\imath_1$. $\succeq$-magnitude numbers are then assigned to other elements of *A* through the algebra of the integral domain, for example, an element *x* of *A* for which

$$i_2(x) = (i_1 \oplus i_1)(u)$$

is assigned the $\succeq$-magnitude number $i_2 = i_1 \oplus i_1$. This form measurement is developed in the next four sections.

7.5 Qualitative Homogeneity

Much of this chapter is concerned with "qualitative systems of numbers" that are totally ordered integral domains or fields with qualitative domains, qualitative ordering relations, qualitative operations, and qualitative identity elements. Most of the investigations concerning these qualitative systems of numbers can easily be accomplished within axiom system $Q_\omega(a)$. However, there are a few concepts involving qualitativeness that are difficult to formulate in natural ways using only the language $\mathsf{L}(\in, A, \varnothing)$ and axiom system ZFA. These bear some resemblance to "Archimedean axioms." They arise because ZFA has an enormous variety of models. For example, as mentioned earlier, there are models of ZFA in which each of the "standard" ordinals, 0, 1, 2, …, is definable in terms of $\mathsf{L}(\in, A, \varnothing)$ and w consists only of such "standard" elements; and there are models of ZFA in which there are elements of *w* that are "nonstandard", that is, in which there are elements of *w* that are not definable in terms of $\mathsf{L}(\in, A, \varnothing)$. When it is necessary to eliminate models like the latter, an additional assumption of a "standard model of ZFA" is made.

Convention 7.5.1 Most mathematicians believe that the axioms of ZFA are true statements about a particular mathematical reality $\langle V, \in, A, \varnothing \rangle$. In this mathematical reality, $\in$ is more than a binary relation with certain properties: it *is* the membership relation among elements of V, all of which in reality are either sets or elements of A. This particular mathematical reality, $\langle V, \in, A, \varnothing \rangle$, is called a *standard model of* ZFA. It is considered to be unique up to the choice of A.

There are various ways to develop the theory of qualitative systems of numbers presented in this chapter that employ far weaker assumptions than that of a standard model of ZFA. However, for the purposes of brevity and exposition, I have chosen the more convenient assumption of a standard model of ZFA. Places where this assumption is used are explicitly noted in the text. In this section standard models are used to guarantee that $\mathbb{Q}_{\succeq}$ is dense in the relevant domain of magnitude numbers, that is, that there are qualitative magnitude numbers close to each magnitude number in the domain.

Definition 7.5.1 Assume axiom system $Q_\omega(a)$. Then A is said to be *qualitatively homogeneous* if and only if for each x and y in A there exists α in the transformation group of Q (Definition 7.2.3) such that $\alpha(x) = y$. And A is said to be *qualitatively 1-point unique* if and only if for all β and γ in the transformation group of Q, if $\beta(x)=\gamma(x)$ for some x in A, the $\beta=\gamma$.

Theorem 7.5.1 Assume axiom system $Q_\omega(a)$. Suppose A is qualitatively homogeneous, $\succeq$ is qualitative total ordering on A, $\succeq'$ be the $\succeq$-induced magnitude ordering on the set of $\succeq$-magnitude numbers, $N_{\succeq}$, (Definition 7.4.3). Then the restriction of $\succeq'$ to $\mathbb{Q}_{\succeq}$ is a total ordering on $\mathbb{Q}_{\succeq}$.

Proof. Let a and β be in $\mathbb{Q}_{\succeq}$ It easily follows from Definition 7.4.3 that $\succeq'$ is a transitive and antisymmetric relation. Thus it needs to only be shown that $\alpha \succeq' \beta$ or $\beta \succeq' \alpha$. Let x and y be arbitrary elements of A. Without loss of generality, suppose $\alpha(x) \succeq \beta(x)$. By the qualitative homogeneity of A, let γ in the transformation group of Q be such that $\gamma(x)=y$. By Theorem 7.2.3, γ preserves $\succeq$. Thus by Theorem 7.2.5,

$$\begin{aligned} \gamma[\alpha(x)] \succeq \gamma[(\beta(x)] \text{ iff } & \alpha[\gamma(x)] \succeq \beta[(\gamma(x)] \\ \text{iff } & \alpha(y) \succeq \beta(y). \end{aligned}$$

Therefore, by Definition 7.4.3, $\alpha \succeq' \beta$.

Theorem 7.5.2 Assume axiom system $Q_\omega(a)$. Suppose $\mathbb{N}$ is the transformation group of Q, A is qualitatively homogeneous and qualitatively 1-point unique, and $\succeq$ is qualitative total ordering on A. Then the following four statements are true:

1. Let $\succeq'$ be the $\succeq$-induced magnitude ordering on the set of $\succeq$-magnitude numbers. Then the restriction of $\succeq'$ to $\mathbb{N}$ is a total ordering of $\mathbb{N}$, and for all α and β in $\mathbb{N}$,

$$\alpha \succeq' \beta \text{ iff } \alpha(x) \succeq \beta(x) \text{ for all } x \text{ in } A$$
$$\text{iff } \alpha(x) \succeq \beta(x) \text{ for some } x \text{ in } A.$$

2. Suppose $\langle A, \succeq \rangle$ is a continuum. Then $\{\iota\} \subset \mathbb{N}$ Let β in $\mathbb{N}$ be such that $\beta \succ' \iota$. Then $\langle \mathbb{N}, \succeq', *, \iota, \beta \rangle$ is isomorphic to $\langle \mathbb{R}, \geq, +, 0, 1 \rangle$ and to $\langle \mathbb{R}^+, \geq, \cdot, 1, 2 \rangle$.
3. Suppose $\langle A, \succeq \rangle$ is a continuum. Then for all α and β $\mathbb{N}$ in $\mathbb{N}$, $\alpha * \beta = \beta * \alpha$.
4. Suppose $\langle A, \succeq \rangle$ is a continuum. Then $\mathbb{Q}_{\succeq} \subseteq \mathbb{N}$.

Proof. Theorem 7.13.1.

The following four theorems characterizes algebraic systems of qualitative magnitude numbers under the assumptions of $Q_\omega(a)$, $\langle A, \succeq \rangle$ is a continuum, and A is qualitatively homogenous. They show that the resulting systems of magnitude numbers are either well-ordered integral domains or positive continuum fields.

In each of the four theorems, a restriction of the operation of functional composition * is used either as a addition operation or as a multiplication operation. Note that in each of these theorems, the restriction of * is qualitative if and only if the domain on which it is defined is qualitative.

The first two theorems characterize the well-ordered integral domain case.

Theorem 7.5.3 Assume axiom system $Q_\omega(a)$. Suppose A is qualitatively homogeneous, $\succeq$ is qualitative total ordering on A, $\langle A, \succeq \rangle$ is a continuum, and $\succeq'$ is the $\succeq$-induced magnitude ordering. Let $\mathbb{Q}^+_{\succeq}$ be the collection of all elements a of $\mathbb{Q}_{\succeq}$ such that $\alpha \succ' \iota$. Suppose $\langle \mathbb{Q}^+_{\succeq}, \succeq' \rangle$ has a least element, i_1. Let $i_0 = \iota$. For each n in w, let ni_1 denote n *-compositions of i_1 with itself, for example, $2i_1 = i_1 * i_1$, and let $(-n)i_1$ be the inverse of the function ni_1, and let $0i_1 = \iota$. Let

$$\mathbb{N} = \{mi_1 \mid m \in \omega \text{ or } -m \in \omega \text{ or } m = 0\}.$$

Define $\odot$ on $\mathbb{N}$ as follows: For all pi_1 and qi_1 that are in $\mathbb{N}$, let

$$pi_1 \odot qi_1 = (p \cdot q)i_1 .$$

Then the following two statements are true:

1. $\mathfrak{N} = \langle \mathbb{N}, \succeq'', *, \odot, i_0, i_1 \rangle$ is a qualitative well-ordered integral domain of magnitude numbers, where $\succeq''$ is the restriction of $\succeq'$ to $\mathbb{N}$..
2. Assume a standard model for ZFA (Convention 7.5.1). Then $\mathbb{Q}_{\succeq} = \mathbb{N}$.

Proof. Theorem 7.13.2.

Note that both the definitions of $\mathbb{N}$ and $\odot$ given in the hypothesis of Theorem 7.5.3 use the set of finite ordinals, ω. Thus to establish the qualitativeness of $\mathbb{N}$ and $\odot$, other kinds of definitions for them must be given, and this is done in the proof of Theorem 7.5.3.

The qualitativeness of an entity is established through the use of a formula of the language $\mathsf{L}(\in, A, \varnothing)$. Inherent in the description of syntax of this language is a concept of a "natural number." A standard model of ZFA guarantees that this metalinguistic use of "natural number" matches in a precise manner (i.e., is isomorphic to) the concept of natural number as an element of w within the standard model.

Note that under the assumption of a standard model for ZFA, Statements 1 and 2 of Theorem 7.5.3 imply $\mathbb{Q}_{\succeq}$ is qualitative.

Theorem 7.5.3 can be interpreted as an "existence theorem" that gives conditions for the existence of a qualitative, well-ordered integral domain of magnitude numbers. The next theorem can similarly be interpreted as the corresponding "uniqueness theorem" for Statement 2 of Theorem 7.5.3 and a standard model of ZFA.

Theorem 7.5.4 Assume axiom system $Q_{\omega}(a)$. Suppose A is qualitatively homogeneous, $\succeq$ is qualitative, $\langle A, \succeq \rangle$ is a continuum,

$$\langle \mathbb{Q}_{\succeq}, \succeq', \oplus, \odot', i_0', i_1' \rangle$$

is a qualitative, well-ordered integral domain of magnitudes numbers, $\succeq'$ is the restriction of the $\succeq$-induced magnitude ordering to $\mathbb{Q}_{\succeq}$, and $i_0' = \iota$. Let $*$, $\odot$, ι_0, and ι_1 be as in the statement of Theorem 7.5.3. Then

$$\oplus = *,\ \odot' = \odot,\ \iota_0' = \iota_0, \text{ and } \iota_1' = \iota_1.$$

Proof. Theorem 7.13.3.

The following two theorems provide similar "existence" theorems for positive continuum fields of magnitude numbers. The statements and proofs of their corresponding "uniqueness" theorems are left to the reader The two theorems are similar, except in the first the operation of function composition, *, is taken to be the qualitative addition operation, whereas in the second theorem * is taken to be the qualitative multiplication operation.

Theorem 7.5.5 Assume axiom system $Q_{\omega}(a)$. Suppose A is qualitatively homogeneous and qualitatively 1-point unique, $\succeq$ is qualitative total ordering on A, $\langle A, \succeq \rangle$ is a continuum. Let $\langle \mathbb{N}, * \rangle$ be the transformation group of Q and $\succeq'$ be the restriction of $\succeq$-induced magnitude ordering to $\mathbb{N}$. Let i_1 in $\mathbb{Q}_{\succeq}$ be such that $i_1 \succ' \iota$. Let $i_0 = \iota$. Then the following three statements are true:

1. $\mathfrak{N} = \langle \mathbb{N}, \succeq', *, \odot, i_0, i_1 \rangle$ is a qualitative, continuum field of magnitude numbers, where $\odot$ is the unique operation on $\mathbb{N}$ such that

$$\mathfrak{N} = \langle \mathbb{N}, \succeq', *, \odot, i_0, i_1 \rangle$$

is a continuum field.

2. $\langle \mathbb{Q}_{\succeq}, \succeq', *, \odot, i_0, i_1 \rangle$, where $\odot$ is as in Statement 1, is a real closed subfield of $\mathfrak{N}$.

3. Assume a standard model of ZFA. Then $\mathbb{Q}_{\succeq}$ is order dense in $\langle \mathbb{N}, \succeq' \rangle$, that is, for all α and γ in $\mathbb{N}$, if $\alpha \succ' \gamma$, then there exists β in $\mathbb{Q}_{\succeq}$ such that $\alpha \succ' \beta \succ' \gamma$.

Proof. Theorem 7.13.4.

Theorem 7.5.6 Assume axiom system $Q_\omega(a)$. Suppose A is qualitatively homogeneous and qualitatively 1-point unique, $\succeq$ is qualitative total ordering on A, $\langle A, \succeq \rangle$ is a continuum. Let $\langle \mathbb{N}, * \rangle$ be the transformation group of Q and $\succeq'$ be the restriction of $\succeq$-induced magnitude ordering to $\mathbb{N}$. Let $i_1 = \iota$. Let α_2 in $\mathbb{Q}_{\succeq}$ be such that ak $\alpha_2 \succ' \iota$. Then the following three statements are true:

1. $\mathfrak{N} = \langle \mathbb{N}, \succeq', \oplus, *, i_1 \rangle$ is a qualitative, positive continuum field of magnitude numbers, where $\oplus$ is the unique operation on $\mathbb{N}$ such that

(i) $\langle \mathbb{N}, \succeq', \oplus \rangle$ is a continuous extensive structure;
(ii) $\alpha_2 = i_1 \odot i_1$; and
(iii) for all x, y, and z in $\mathbb{N}$,

$$z * (x \oplus y) = (z * x) \oplus (z * y) .$$

2. $\langle \mathbb{Q}_{\succeq}, \succeq', \oplus, *, i_1 \rangle$ is a positive real closed subfield of $\mathfrak{N}$, where $\oplus$ is as in Statement 1.
3. Assume a standard model of ZFA. Then $\mathbb{Q}_{\succeq}$ is order dense in $\langle \mathbb{N}, \succeq' \rangle$, that is, for all α and γ in $\mathbb{N}$, if $\alpha \succ' \gamma$, then there exists β in $\mathbb{Q}_{\succeq}$ such that $\alpha \succ' \beta \succ' \gamma$.

Proof. Theorem 7.13.5.

7.6 Qualitative Canonical Measurement

The previous section showed that there are algebraically rich qualitative systems of magnitude numbers, and thus it is natural to inquire into qualitative forms of measurement into these systems and ask if anything of value may be gained through their employment. In this section, a qualitative method is given for representing the elements of the domain of a 1-point homogenous, 1-point unique qualitative structure into a positive field of magnitude numbers. The usefulness of this form of measurement is discussed and illustrated in later sections.

Definition 7.6.1 Assume axiom system $Q_\omega(a)$. Suppose $\mathbb{N}$ is the transformation group of Q, A is qualitatively homogeneous and qualitatively 1-point unique, and $\succeq$ is qualitative total ordering on A. For each x in A and each α in $\mathbb{N}$, let α_x be the function from A into $\mathbb{N}$, such that

$$\alpha_x(x) = \iota \text{ and } \forall y[y \in A \rightarrow \alpha_x(y) = \beta] ,$$

where, by the qualitative homogeneity and qualitative 1-point uniqueness of A, β is the unique element of $\mathbb{N}$ such that

$$\beta(x) = y .$$

Then

$$\{\alpha_x \,|\, \alpha \in \mathbb{N} \text{ and } x \in A\}$$

is called the *(qualitative) canonical scale* for A.

The following theorem shows that when A is qualitatively homogeneous and qualitatively 1-point unique and $\langle A, \succeq \rangle$ is a continuum, the canonical scale behaves qualitatively like a ratio scale and the qualitative relations on A behave like they are qualitatively invariant (Definition 5.5.3) with respect to this scale.

Theorem 7.6.1 Assume axiom system $Q_\omega(a)$. Suppose $\mathbb{N}$ is the transformation group of Q, A is qualitatively homogeneous and qualitatively 1-point unique, $\langle A, \succeq \rangle$ is a continuum, and $\succeq$ is qualitative. Let S be the qualitative canonical scale for A (Definition 7.6.1). Then the following eight statements are true:

1. S is qualitative.
2. For each φ in S, φ is a one-to-one function from A onto $\mathbb{N}$.
3. For all φ and ψ in S, there exists α in $\mathbb{N}$ such that

$$\varphi = \psi * \alpha .$$

4. For all φ in S and all β in $\mathbb{N}$, $\varphi * \beta$ is in S.
5. For all φ and ψ in S, if $\varphi(x)=\psi(x)$ for some x in A, then $\varphi=\psi$.
6. For all x in A and β in $\mathbb{N}$ there exists φ in S such that $\varphi(x)=\beta$.
7. For all qualitative R and all φ and ψ in S,

$$\varphi(R) = \psi(R) .$$

8. For all φ in S,

$$\varphi(\succeq) = \succeq' ,$$

where $\succeq'$ is the restriction to $\mathbb{N}$ of the $\succeq$-induced magnitude ordering.

Proof. 1. Statement 1 is an immediate consequence of Theorem 7.2.1 and Definitions 7.2.1, 7.2.2, and Definition 7.6.1.

2. Let $\varphi \in S$. Then it is an immediate from Definition 7.6.1 that φ is a function from A into $\mathbb{N}$. By Definition 7.6.1, let α be the element of $\mathbb{N}$ and x be an element of A such that $\varphi=a_x$. Suppose y and z are elements of A such that $\varphi[y]=\varphi[z]$. But by Definition 7.6.1,

$$\varphi[y](x) = y \text{ and } \varphi[z](x) = z ,$$

and thus $y=z$. Therefore, φ is one-to-one. To show that φ is onto $\mathbb{N}$, let β be an arbitrary element $\mathbb{N}$. of Let $\omega=\beta(x)$. Then by Definition 7.6.1, $\varphi(\omega)=\beta$.

3. Let φ and ψ be arbitrary elements of S. By Definition 7.6.1, let x and y be elements of A such that $\varphi(x)=\iota$ and $\psi(y)=\iota$. By the qualitative homogeneity of A, let α in $\mathbb{N}$ be such that $\alpha(x)=y$. Statement 3 will be shown by showing $\psi^{-1} * \varphi=\alpha$. It follows from Statement 2 that $\psi^{-1} * \varphi$ is a one-to-one function from A onto A. Let z be an arbitrary element of A. By the qualitative homogeneity of A, let θ in $\mathbb{N}$ be such that $\theta(x)=z$. Then by Definition 7.6.1,

$$\varphi(z) = \theta \text{ and } \psi[\theta(y)] = \theta\,.$$

By Statement 3 of Theorem 7.5.2,

$$\theta * \alpha = \alpha * \theta\,.$$

Thus,

$$\psi^{-1}\varphi(z) = \psi^{-1}\theta = \theta(y) = \theta * \alpha(x) = \alpha * \theta(x) = \alpha(z)\,.$$

4. Let $\varphi\in S$ and $\beta \in \mathbb{N}$. By Definition 7.6.1, let y in A be such that $\varphi(y)=\iota$. Let $x=\beta^{-1}(y)$. By Definition 7.6.1, β_x is in S. By Statement 3, let γ in $\mathbb{N}$ be such that $\beta_x=\varphi*\gamma$. Since β_x is in S, it needs to only be shown that $\beta=\gamma$. Because

$$\beta(x) = y \text{ and } \gamma(x) = \varphi^{-1} * \beta_x(x) = \varphi^{-1}(\iota) = y\,,$$

it follows from the qualitative 1-point uniqueness of A that $\beta=\gamma$.

5. Suppose φ and ψ are in S, $x\in A$, and $\varphi(x)=\psi(x)$. By Statement 3, let α in N be such that $\varphi=\psi * \alpha$. Then

$$\alpha(x) = \psi^{-1} * \varphi(x) = x = \iota(x)\,,$$

which by the 1-point uniqueness of S yields $\alpha=\iota$, and therefore, $\varphi=\psi$.

6. Let $x \in A$, $\beta \in \mathbb{N}$ and $\psi\in S$. Because by Statement 2 ψ is onto $\mathbb{N}$., let y in A be such that $\psi(y)=\beta$. By the qualitative homogeneity of A, let α in $\mathbb{N}$ be such that $\alpha(x)=y$. Let $\varphi=\psi * \alpha$. Then by Statement 4, $\varphi \in S$. Because

$$\varphi(x) = \psi * \alpha(x) = \psi(y) = \beta\,,$$

Statement b has been shown.

7. Let φ and ψ be in S and R be qualitative. By Statement 3, let α in $\mathbb{N}$ be such that $\beta=\psi * \alpha$. By Theorem 7.2.3, $\alpha(R)=R$. Therefore,

$$\psi(R) = \psi * \alpha(R) = \varphi(R)\,.$$

8. Let φ be an element of S. By Definition 7.6.1, let α in $\mathbb{N}$ and x in X be such that $\varphi=a_x$. Let y and z be arbitrary elements of A. By the qualitative homogeneity of A, let β and γ in $\mathbb{N}$ be such that

$$\beta(x) = y \text{ and } \gamma(x) = z\,.$$

Then

$$\begin{aligned}(y,z) \in \succsim \text{ iff } & y \succsim z \\ \text{iff } & \beta(x) \succsim \gamma(x) \\ \text{iff } & \beta \succsim' \gamma \\ \text{iff } & \alpha_x(y) \succsim' \alpha_x(z) \\ \text{iff } & (\alpha_x(y), \alpha_x(z)) \in \succsim' .\end{aligned}$$

Because by Statement 2 α_x is onto $\mathbb{N}$, the above series of logical equivalences shows that $\alpha_x(\succsim) = \succsim'$.

Note that the measurement through the canonical scale S in Theorem 7.6.1 is *not* a qualitative form of the representational theory. In fact no "representing structure" is given. However, using Theorem 7.6.1 it is easy to construct *(i)* a qualitative structure $\mathfrak{X}$ whose primitives consist of A and the elements of a, and *(ii)* a structure $\mathfrak{M}$ with domain $\mathbb{N}$ such that S is the set of isomorphisms of $\mathfrak{X}$ onto $\mathfrak{M}$:

Theorem 7.6.2 Assume the hypothesis of Theorem 7.6.1 and let $\{R_j | j \in J\}$ be an indexing of the elements of a. Let φ be an element of S,

$$\mathfrak{X} = \langle A, \succsim, R_j \rangle_{j \in J}$$

and

$$\mathfrak{M} = \langle \mathbb{N}, \succsim', \varphi(R_j) \rangle_{j \in J} .$$

Then the primitives of $\mathfrak{X}$ and the primitives of $\mathfrak{M}$ are qualitative, and S is the set of isomorphisms of $\mathfrak{X}$ onto $\mathfrak{M}$.

Proof. By hypothesis, $\succsim$ is qualitative. By Lemma 7.2.1, A and each element of a are qualitative. Thus the primitives of $\mathfrak{X}$ are qualitative. By Definition 7.2.3 and Theorem 7.2.1, $\mathbb{N}$ is qualitative. From the qualitativeness of $\mathbb{N}$ and the definition of "$\succsim$-induced magnitude ordering" (Definition 7.4.3), it easily follows from Definitions 7.2.1 and 7.2.2 that $\succsim'$ is qualitative. By Statement 7 of Theorem 7.6.1, $\varphi(R)=\psi(R)$ for each R in a and each ψ in S. Since S is qualitative by Statement 1 of Theorem 7.6.1, it then follows by Definitions 7.2.1 and 7.2.2 that $\varphi(R)$ is qualitative for each R in a. Thus the primitives of $\mathfrak{M}$ are qualitative.

By Statement 2 of Theorem 7.6.1, φ is a one-to-one function from A onto $\mathbb{N}$. Thus by Statements b and 7 of Theorem 7.6.1 that φ is an isomorphism of $\mathfrak{X}$ onto $\mathfrak{M}$. Since φ is an arbitrary element of S, each element of S is an isomorphism of $\mathfrak{X}$ onto $\mathfrak{M}$. Let θ be an arbitrary isomorphism of $\mathfrak{X}$ onto $\mathfrak{M}$. To show that S is the set of isomorphisms of $\mathfrak{X}$ onto $\mathfrak{M}$, it is sufficient to show that $\theta \in S$. $\varphi^{-1} * \theta$ is an automorphism of $\mathfrak{X}$ and therefore by Theorem 7.2.3 is an element α. of $\mathbb{N}$. From $\varphi^{-1} * \theta = \alpha$, it follows that $\theta = \varphi * \alpha$, and thus by Statement 4 of Theorem 7.6.1, $\theta \in S$.

For purposes of exposition, call a relation R based on A "S-domain invariant" if and only if for all ψ and θ in S, $\psi(R)=\theta(R)$. Then conceptually and structurally the concept of "S-domain invariant" is very similar to the representational concept of "qualitatively T-invariant" (Definition 5.5.3). Note that Statement 6 implies that all qualitative relations based on A are S-domain invariant. In general the converse does not hold. For example, if a is a finite or denumerable set, then there are at most denumerably many qualitative entities and therefore at most denumerably many qualitative relations based on A, but there are at least $2^{\aleph_0}$ many "S-domain invariant" first-order relations on A. (The latter is shown by the following calculation: Let ψ be an element of S. Using Theorem 7.6.1 it can be shown that *(i)* $\mathbb{N}$ has the same cardinality as A, which is $2^{\aleph_0}$, *(ii)* $\psi^{-1}(\alpha)$ is "S-domain invariant" for each α in $\mathbb{N}$, and *(iii)* for each a and β in $\mathbb{N}$, if $\alpha \neq \beta$, then $\psi^{-1}(a) \neq \psi^{-1}(\beta)$. *(i)*, *(ii)*, and *(iii)* establish that there are at least $2^{\aleph_0}$ many "S-domain invariant" first-order relations on A.)

Statement 2 of Theorem 7.5.2 and Statements 3 and 4 of Theorem 7.6.1 suggest viewing S as a qualitative version of a ratio scale. However, to argue for this rigorously, it is necessary to establish that * is interpretable as a multiplication operation on a qualitative positive field with domain $\mathbb{N}$, total ordering $\succeq'$, and multiplicative identity ι. The following theorem establishes this when $\mathbb{Q}_{\succ} \neq \{\iota\}$.

Theorem 7.6.3 Assume axiom system $Q_{\omega}(a)$. Suppose $\mathbb{N}$ is the transformation group of Q, A is qualitatively homogeneous and qualitatively 1-point unique, $\langle A, \succeq\rangle$ is a continuum, and $\succeq$ is qualitative total ordering on A. Let $\succeq'$ be the restriction to $\mathbb{N}$ of the $\succeq$-induced magnitude ordering, β be a qualitative element of $\mathbb{N}$ such that $\beta \succ' \iota$, and S be the qualitative canonical scale for A. Then the following three statements are true:

1. There is exactly one operation $\oplus$ on $\mathbb{N}$ such that
$$\mathfrak{F} = \langle \mathbb{N}, \succeq', \oplus, *, \iota\rangle$$
is a qualitative positive continuum field and $\beta = \iota \oplus \iota$.
2. There is exactly one qualitative operation $\bigcirc$ on A such that $\langle A, \succeq, \bigcirc\rangle$ is a continuous extensive structure and for each x in X,
$$\beta(x) = x \bigcirc x\,.$$
3. Let $\oplus$ be as in Statement 1 and $\bigcirc$ as in Statement 2. Then for all φ in S, $\varphi(\bigcirc)=\oplus$.

Proof. Theorem 7.13.6

Measurement through qualitative canonical scales is applied in the next two sections to provide new insights into two classical issues in the theory of measurement.

7.7 Qualitative Magnitude Estimation

In Section 5.8, representational measurement theories were given for certain magnitude estimation paradigms. This section presents a completely qualitative version of magnitude estimation where stimuli are assigned magnitude numbers from a qualitative, positive continuum field. The assignment process is done in a way such that each numeral is assigned the natural magnitude number with respect to the continuum field; for example, the numeral **2** is assigned the magnitude number $i_1 \oplus i_1$, where $\oplus$ is the continuum field's additive operation and i_1 its multiplicative identity.

Let $\langle X, \succeq \rangle$ be a continuum. The subject is given instructions to produce for each stimulus x a numeral $\boldsymbol{p}$ that represents his or her subjective evaluation of the intensity of x compared to u. Let $\varphi u(x)$ be the "number" p that is the "number" that is scientifically named by the numeral $\boldsymbol{p}$. By letting the unit u vary over the stimuli, a scale $\mathcal{S} = \{\varphi_u \,|\, u \in A\}$ on X results.

Of course there are several possibilities for the "scientific numbers p" named by $\boldsymbol{p}$. Most researchers who used magnitude estimation would probably want p to be the same number that pure mathematicians use as the referent of $\boldsymbol{p}$, which for the purposes of this section is taken to be the appropriate set-theoretic real number. I am not sure if Stevens would have endorsed a choice which took a platonic number as a referent of $\boldsymbol{p}$. Because his writings about the nature of number and measurement are, by the standards of contemporary mathematical logic, vague and somewhat confused, it is difficult to judge which kind of formal theory best fits his methods and definitions. In this section, consideration is given to assigning a magnitude number to $\boldsymbol{p}$. This latter choice of "number" appears to me to capture a spirit of Stevens' approach to magnitude estimation and measurement that has been generally ignored by measurement theorists.

Definition 7.7.1 Suppose a magnitude estimation situation in which X is the set of stimuli, $\succeq$ is a total ordering on X, and $\langle X, \succeq \rangle$ is a continuum. For each numeral $\boldsymbol{p}$ used in the magnitude estimation situation, define the function $\alpha\boldsymbol{p}$ on the set of stimuli X as follows: for all x and t in X, $\alpha p(t)=x$ if and only if the subject magnitude estimates x as being $\boldsymbol{p}$ times t.

Assume $\succeq$ is qualitative and the behavioral data is such that for each $\boldsymbol{p}$ used in the magnitude estimation situation, $\alpha\boldsymbol{p}$ is a strictly increasing function from X onto X. Then for each $\boldsymbol{p}$ used in the magnitude situation, $\alpha\boldsymbol{p}$ is a magnitude number. Qualitative measurement is accomplished through a qualitative scale S that has several properties, including the following: For each element φ in S there exists u in X such that

(i) $\varphi(u)=\iota$, and
(ii) for all numerals $\boldsymbol{p}$ used in the magnitude estimation situation and all x in X,

$$\varphi(x) = \alpha_{\boldsymbol{p}} \text{ iff } \alpha_{\boldsymbol{p}}(u) = x$$
$$\text{iff the subject estimates } x \text{ as } \boldsymbol{p} \text{ times } u.$$

Properties *(i)* and *(ii)* are especially desirable for measuring stimuli through magnitude estimation, because they make the natural assignment of (magnitude) numbers to those numerals expressed by a subject in a magnitude estimation paradigm. They are also characteristic of the qualitative canonical scale for X (Definition 7.6.1), when *(a)* X is the set of atoms, *(b)* X is qualitatively homogeneous, and *(c)* X is qualitatively 1-point unique. This suggests that we should look for conditions in terms of the magnitude estimation data that imply the qualitative homogeneity of X and the qualitative 1-point uniqueness of X, and that we should explore more deeply the implications of measurement through the qualitative canonical scale for X for magnitude estimation data. This is done in this section. However, the current situation is complicated by the fact that there are numerals as well as stimuli present and both need to be taken as atoms.

Axiom 7.7.1 $A=X\cup \boldsymbol{N}$, where X and $\boldsymbol{N}$ are nonempty sets.

Axiom 7.7.2 $\langle X, \succeq\rangle$ is a continuum (Definition 5.4.1).

Axiom 7.7.3 $\geq^*$ is a well-ordering on $\boldsymbol{N}$ and is such that each element of $\boldsymbol{N}$ that is not the $\geq^*$-first element of $\boldsymbol{N}$ has an immediate $\geq^*$-predecessor in $\boldsymbol{N}$.

Definition 7.7.2 Assume Axiom 7.7.3. Let 1 be the first element $\langle \boldsymbol{N}, \geq_\star\rangle$ Define the function $\boldsymbol{S}$ on $\boldsymbol{N}$ as follows: For each $\boldsymbol{p}$ in $\boldsymbol{N}$, $\boldsymbol{S}(\boldsymbol{p})$ is the immediate $\geq^*$-successor of $\boldsymbol{p}$. Let **2**=$\boldsymbol{S}$(**1**).

Definition 7.7.3 Assume Axiom 7.7.3. Then, because $\geq^*$ is a well-ordering on $\boldsymbol{N}$, the structure $\mathfrak{W} = \langle \boldsymbol{N}, \boldsymbol{S}\rangle$ satisfies Peano's well-known axioms for arithmetic. By using Peano's method of defining addition and multiplication for $\mathfrak{W}$, let+and • be respectively the resulting addition and multiplication operations defined on $\boldsymbol{N}$. (Although Peano's method uses mathematical induction to define+and •, it can easily be modified so that+and • can be defined using the well-ordering $\geq^*$ in place of mathematical induction. The latter approach uses only $\boldsymbol{N}$, $\boldsymbol{S}$, $\geq^*$, and concepts in $V_\omega(a)$ for defining+ and •.)

It follows by Peano's construction and Definition 7.7.3 that

$$\langle \boldsymbol{N}, \geq_\star, +, \bullet, \mathbf{1}\rangle$$

is a positive, well-ordered integral domain, and therefore by Theorem 7.3.1 it is isomorphic to $\langle \mathbb{I}^+, \geq, +, \cdot, 1\rangle$ Peano's method of defining+and • in terms of $\boldsymbol{N}$ and $\boldsymbol{S}$ yields that+and • are qualitative whenever $\boldsymbol{N}$ and $\geq^*$ are qualitative.

The following is a well-known theorem of mathematics:

Theorem 7.7.1 Suppose $\mathfrak{X} = \langle X, \succeq \rangle$ and $\mathfrak{Y} = \langle Y, \succeq' \rangle$ are structures such that *(i)* $\succeq$ and $\succeq'$ are well-orderings, *(ii)* X and Y are infinite, and *(iii)* each element of X (respectively Y) that is not the first element of $\mathfrak{X}$ (respectively $\mathfrak{Y}$ has an immediate predecessor. Then $\mathfrak{X}$ and $\mathfrak{Y}$ are isomorphic.

The following theorem is an immediate consequence of Theorem 7.7.1.

Theorem 7.7.2 Assume Axioms 7.7.1 and 7.7.3. Then there is a unique isomorphism δ of $\langle \boldsymbol{N}, \geq_{\star} \rangle$ onto $\langle \omega - \{0\}, \geq \rangle$.

Definition 7.7.4 By Theorem 7.7.2, let δ be the unique isomorphism of $\langle \boldsymbol{N}, \geq_{\star} \rangle$ onto $\langle \omega - \{0\}, \geq \rangle$ Because of the existence and uniqueness of δ, there is an unambiguous assignment numerals to numbers. By definition, for each element p in w—$\{0\}$ let $\boldsymbol{p}$ be the unique numeral in $\boldsymbol{N}$ such that

$$\delta(p) = p$$

Then p is said to be the *name* of p.

Definition 7.7.4 gives a natural association between numerals, which here are taken to be atoms, and set-theoretic numbers. An important technical difference between numerals and numbers is that the set of numerals $\boldsymbol{N}$ is an element of V_ω, and therefore is potentially a qualitative entity, whereas the set of positive (set-theoretic) integers, ω—$\{0\}$, is not an element of V_ω, and therefore is not potentially a qualitative entity.

Axiom 7.7.4 $\{\alpha_{\boldsymbol{p}} \mid \boldsymbol{p} \in \boldsymbol{N}\}$ is a set such that the following six statements hold for each $\boldsymbol{p}$ and $\boldsymbol{q}$ in $\boldsymbol{N}$:

1. $\boldsymbol{ap}$ is a $\succeq$-strictly increasing function from X onto X.
2. $\boldsymbol{ap(x)} \succeq x$.
3. α_1 is the identity function ι on X.
4. For all t in $\boldsymbol{X}$, $\boldsymbol{ap(t)} \succ \boldsymbol{aq(t)}$ iff $\boldsymbol{p} >^* \boldsymbol{q}$.
5. For all x and t in X, if $x \succ t$, then there exist $\boldsymbol{m}$ in $\boldsymbol{N}$ and z in X such that

$$x \succ z \succ t \text{ and } \alpha_{\boldsymbol{m}} + {\bf 1}(t) = \alpha_{\boldsymbol{m}}(z).$$

6. If $\boldsymbol{r} = \boldsymbol{q} \bullet \boldsymbol{p}$, then $\alpha_r = \boldsymbol{aq} * \boldsymbol{ap} * \boldsymbol{ap}$. (Axiom 5.8.9)

Note that Statements 1 to 6 of Axiom 7.7.4 are restatements of Axioms 5.8.1, 5.8.2, 5.8.3, and 5.8.9 of Section 5.8, that is, Axiom 7.7.4 is a restatement of behavioral axioms for magnitude estimation as given in Section 5.8.

Definition 7.7.5 Throughout the rest of this section let

$$a = \{X, \succeq, \boldsymbol{N}, \geq_{\star}\} \cup \{\alpha_{\boldsymbol{p}} \mid \boldsymbol{p} \in \boldsymbol{N}\}$$

and

$$b = \{X, \succeq\} \cup \{\alpha p \mid p \in N\}$$

and

$$G_b = \{f \mid f \text{ is a one-to-one function from } X$$
$$\text{onto } X \text{ and for all } x \text{ in b}, f(x) = x\}.$$

(Thus G_b is the set of permutations on X that leave invariant X, $\succeq$, and αp for each p in N.)

Lemma 7.7.1 $\langle G_b, *\rangle$ is a 1-point homogeneous, 1-point unique group of transformations on X (Definition 5.1.5).

Proof. Immediate consequence of Theorem 5.8.2.

Convention 7.7.1 Consider for the moment the situation where X is the set of atoms and assume axiom system $Q_\omega(b)$. Let Q' be the qualitativeness predicate associated with $Q_\omega(b)$. Then G_b is the transformation group of Q'. Let $\succeq'$ be the $\succeq$-induced magnitude ordering restricted to G_b. By Lemma 7.7.1, X is Q'-qualitatively homogenous, and by Lemma 7.7.1, X is Q'-qualitatively 1-point unique. Thus Definition 7.4.3 of "canonical scale" applies to this situation. Therefore, for the rest of this section, let S be the canonical scale for X that results from assuming X is the set of atoms and axiom system $Q_\omega(b)$.

The Theorem 7.7.3 below provides a method of measuring of X through a qualitative canonical scale. Note that in the theorem, A and *not* X is taken to be the set of atoms, that axiom system $Q_\omega(a)$ and *not* axiom system $Q_\omega(b)$ is assumed, and that qualitativeness refers to Q and *not* Q'.

Theorem 7.7.3 Assume Axioms 7.7.1 to 7.7.4 and axiom system $Q_\omega(a)$. Let S and $\succeq'$ be as in Convention 7.7.1, and let $\mathbb{N} = G_b$. Then the following ten statements are true:

1. S is qualitative.
2. For each φ in S, φ is a one-to-one function from A onto $\mathbb{N}$.
3. For all φ and ψ in S, there exists α in $\mathbb{N}$ such that

$$\varphi = \psi * \alpha .$$

4. For all φ in S and all γ in $\mathbb{N}$, $\varphi * \gamma$ is in S.
5. For all φ and ψ in S, if $\varphi(x) = \psi(x)$ for some x in A, then $\varphi = \psi$.
6. For all x in A and α in $\mathbb{N}$ there exists φ in S such that $\varphi(x) = \alpha$.
7. For all Q'-qualitative R and all φ and ψ in S,

$$\varphi(R) = \psi(R).$$

8. Let $\varphi \in S$. Then S is the set of isomorphisms of $\langle X, \succeq, \alpha p\rangle_{p \in N}$ onto $\langle \mathbb{N}, \succeq', \varphi(\alpha p)\rangle_{p \in N}$.

9. There is exactly one qualitative operation $\oplus$ on $\mathbb{N}$ such that

$$\mathfrak{F} = \langle \mathbb{N}, \succeq', \oplus, *, \iota \rangle$$

is a qualitative positive continuum field and $\alpha_2 = \iota \oplus \iota$.

10. Assume a standard model of ZFA. Let $\oplus$ be as in Statement 9. Let φ be an arbitrary element of *S*, and by Statement 2 let *u* in *X* be such that

$$\varphi(u) = \iota\,.$$

Then for each $\boldsymbol{p}$ in $\boldsymbol{N}$ and each *x* in $\boldsymbol{X}$, $\boldsymbol{\alpha p}$ is qualitative and

$$x = \alpha \boldsymbol{p}(u) \text{ iff } \varphi(x) = \alpha \boldsymbol{p}$$

$$\text{iff } \varphi(x) = \overbrace{\iota \oplus \iota \oplus \cdots \oplus \iota}^{p \text{ concatenations}}\,.$$

Proof. Because $\boldsymbol{N}$ and $\geq^*$ are not mentioned in Statements 1 to 7 and $Q' \subseteq Q$, where Q' is the qualitativeness predicate of $Q_\omega(b)$ and Q is the qualitativeness predicate of $Q_\omega(a)$, it is sufficient to show Statements 1 to 7 under the hypothesis of $Q_\omega(b)$. Then Statements 1 to 7 are restatements of Statements 1 to 7 of Theorem 7.6.1 applied to the situation described by Theorem 7.7.3.

Statement 8 is shown as follows: Because $\boldsymbol{\alpha p}$ is in *b* for each $\boldsymbol{p}$ in $\boldsymbol{N}$, it is both Q'- and Q-qualitative. Statement 8 then follows from Theorem 7.6.2 with the assumption of axiom system $Q_\omega(b)$.

To show Statement 9, we first show that $\alpha_2 \in \mathbb{N}$. Let *c* be a fixed element of *X*. By the Q'-qualitative homogeneity of *X*, let γ in $\mathbb{N}$ be such that $\gamma(c) = \alpha_2(c)$. Let *x* be an arbitrary element of *X*, and by the Q'-qualitative homogeneity of *X*, let θ in $\mathbb{N}$ be such that $\theta(c) = x$. Then, because $\langle \mathbb{N}, * \rangle$ is 1-point homogeneous and 1-point unique (Lemma 7.7.1), it follows from Statement 1 of Theorem 7.5.5 that * is a commutative operation on $\mathbb{N}$. Thus by the commutativity of * and the qualitative homogeneity of α_2,

$$\begin{aligned}
\gamma(x) &= \gamma[\theta(c)] \\
&= \theta[\gamma(c)] \\
&= \theta[\alpha_2(c)] \\
&= \alpha_2[\theta(c)] \\
&= \alpha_2(x).
\end{aligned}$$

Thus, because *x* is an arbitrary element of *X*, $\gamma = \alpha_2$, and therefore ak α_2, is in $\mathbb{N}$. Then Statement 9 follows by the previously established fact in the proof of Statement 8 that $\alpha_2 \in \boldsymbol{N}$, Lemma 7.7.1, and Statement 1 of Theorem 7.6.3.

Statement 10 is shown in Theorem 7.13.7.

Statement 10 of Theorem 7.7.3 says that if the subject evaluates "x as $\boldsymbol{p}$ times u," then in the measurement of X by the qualitative canonical scale S, the representation φ in $\boldsymbol{S}$ for which u is the unit ι, that is, for which $\varphi(u)=\iota$, measures x as p qualitative additions of the qualitative multiplicative unit, that is,

$$\varphi(x) = \overbrace{\iota \oplus \iota \oplus \cdots \oplus \iota}^{p \text{ concatenations}} .$$

This appears to me to capture in a precise way a coherent version of Stevens' method of magnitude estimation.

The assumption of a standard model of ZFA guarantees that for $p \in w$, "p concatenations of ι" can be expressed in terms of a formula of $\mathsf{L}(\in, A, \varnothing)$ and elements of $a \cup \{a\}$.

Stevens used magnitude estimation to conclude that the psychophysical function linking physical intensity to subjective intensity was a power function. (See Subsections 2.8.1 and 6.5.3.) The following examines this in terms of qualitative canonical measurement. We start first with the canonical measurement of the physical stimuli.

Convention 7.7.2 Throughout the rest of this section, assume that the stimuli in X are also physical, $\mathfrak{X} = \langle X, \succsim_1, \bigcirc \rangle$ is a continuous extensive structure, and

$$c = \{\succsim_1, \bigcirc\} .$$

c is to be interpreted as the set of physical primitives used to physically measure X. Taking X to be the set of atoms, assume axiom system $Q_\omega(c)$. Let Q_1 be the qualitative predicate associated with axiom system $Q_\omega(c)$. Let $\mathbb{N}_1$ be the transformation group of Q_1 and $\succsim_1'$ be the $\succsim_1$-induced magnitude ordering restricted to $\mathbb{N}_1$. By Theorem 5.4.7, X is Q_1-qualitatively homogeneous and Q_1-qualitatively 1-point unique. Therefore, by Theorem 7.6.1 and Definition 7.6.1, for the rest of this section let $\boldsymbol{S}_1$ be the canonical scale for X determined by Q_1.

Note that in the following theorem A and *not* X is taken to be the set of atoms, that axiom system $Q_\omega(d)$ and *not* $Q\omega(c)$ is assumed, and thus the qualitativeness predicate Q_2 associated with $Q_\omega(d)$ and *not* Q_1 is used to determine qualitativeness.

Theorem 7.7.4 Let X, $\succsim_1$, $\bigcirc$, $\mathbb{N}_1$, $\succsim_1'$, and $\boldsymbol{S}_1$ be as in Convention 7.7.2. Let

$$d = \{\succsim_1, \bigcirc\} \cup \boldsymbol{N} \cup \{\geq_\star\} .$$

Assume axiom system $Q\omega(d)$. Let β be the function on X such that for all x in X,

$$\beta(x) = x \bigcirc x .$$

Then the following nine statements are true:

1. *Si* is qualitative.
2. For each φ in S_1, φ is a one-to-one function from X onto $\mathbb{N}_1$.
3. For all φ and ψ in S_1, there exists α in $\mathbb{N}_1$ such that

$$\varphi = \psi * \alpha .$$

4. For all φ in S_1 and all γ in $\mathbb{N}_1$, $\varphi * \gamma$ is in S_1.
5. For all φ and ψ in S_1, if $\varphi(x)=\psi(x)$ for some x in X, then $\varphi=\psi$.
6. For all x in X and all α in $\mathbb{N}_1$ there exists φ in S_1 such that $\varphi(x)=\alpha$.
7. For all Q_1-qualitative relations R based on X and all φ and ψ in S_1,

$$\varphi(R)=\psi(R).$$

8. There is exactly one qualitative operation $\oplus_1$ on $\mathbb{N}$ such that $\beta=\iota \oplus_1 \iota$ and

$$\mathfrak{F} = \langle \mathbb{N}_1, \succeq_1', \oplus_1, *, \iota \rangle$$

is a qualitative positive continuum field.
9. Let $\oplus_1$ be as in Statement 7. Then for all φ in S_1, φ is an isomorphism of the extensive structure $\langle X, \succeq_1, \bigcirc \rangle$ onto $\langle \mathbb{N}_1, \succeq_1', \oplus_1 \rangle$.

Proof. Because N and $\geq^*$ are not mentioned in the theorem and $Q_1 \subseteq Q_2$, where Q_1 is the qualitativeness predicate of $Q_\omega(c)$ and Q_2 is the qualitativeness predicate of $Q_\omega(d)$, it is sufficient to show the theorem under the hypothesis of $Q_\omega(c)$.

1–7. Statements 1 to 6 are restatements of Statements 1 to 7 Theorem 7.6.1 applied to the situation described by Theorem 7.7.4.

8. Because $\bigcirc$ is qualitative by hypothesis, it immediately follows by Definitions 7.2.1 and 7.2.2 that β is qualitative. Because $\langle X, \succeq_1, \bigcirc \rangle$ is a continuous extensive structure, $\beta(x) = x \bigcirc x \succ_1 x$ for each x in X. Thus $\beta \succ_1' \iota$. Statement 8 then follows from Statement 1 of Theorem 7.6.3.

9. Define $\circledcirc$ on $\mathbb{N}$ as follows: For all α, β, and γ in $\mathbb{N}$,

$$\alpha \circledcirc \beta = \gamma \text{ iff for all } x \text{ in } X \ \alpha(x) \bigcirc \beta(x) = \gamma(x) .$$

Let φ be an arbitrary element of S_1. Then Definition 2.9 and Lemmas 2.6 and 2.7 of Narens (1981a) show that *(i)* φ is an isomorphism of $\langle X, \succeq_1, \bigcirc \rangle$ onto $\langle \mathbb{N}, \succeq_1', \circledcirc \rangle$, and *(ii)* for all α, β, and γ in $\mathbb{N}$,

$$\alpha * (\beta \circledcirc \gamma) = (\alpha * \beta) \circledcirc (\alpha * \gamma) .$$

Thus $\langle \mathbb{N}, \succeq_1', \circledcirc, *, \iota \rangle$ is a positive continuum field. Because for each x in X,

$$\beta(x) = x \bigcirc x = \iota(x) \bigcirc \iota(x),$$

it follows by the definition of $\circledcirc$ that

$$\beta = \iota \circledcirc \iota .$$

Thus by Theorem 7.6.3, $\circledcirc = \oplus_1$. Therefore, φ is an isomorphism of $\langle \mathbb{N}, \succeq_1', \oplus_1 \rangle$ onto $\langle X, \succeq_1, \bigcirc \rangle$, and Statement 9 has been shown.

The power law is a consequence of assumptions relating the set of psychological primitives a to the set of physical primitives c. A number of different sets of psychophysical assumptions produce the power law. The ones employed in the following theorem are discussed in detail in Section 6.4.4.

Theorem 7.7.5 Let X, $\{\alpha_{\boldsymbol{p}} \mid \boldsymbol{p} \in N\}$, $\succsim$, $\mathbb{N}$, $\succsim'$, $\oplus$ and S be as in Theorem 7.7.3, and let $\succsim_1$, $\bigcirc$, $\mathbb{N}_1$, $\succsim_1'$, $\oplus_1$, and S_1 be as in Theorem 7.7.4. Suppose the following two psychophysical axioms:

(i) $\succsim = \succsim_1$.
(ii) For all $\boldsymbol{p}$ in N and all x and y in X,

$$\alpha_{\boldsymbol{p}}(x \bigcirc y) = \alpha_{\boldsymbol{p}}(x) \bigcirc \alpha_{\boldsymbol{p}}(y)\,.$$

Then the following two statements are true:

1. $\mathbb{N} = \mathbb{N}_1$, $\succsim' = \succsim_1'$, and $S=S_i$.
2. There is an automorphism θ of the structure $\langle \mathbb{N}, \succsim', * \rangle$ such that θ $(\oplus)= \oplus_1$.

Proof. Theorem 7.13.8

In Theorem 7.7.5 qualitative canonical measurement is used to qualitatively measure both the extensive structure $\mathfrak{X} = \langle X, \succsim_1, \bigcirc \rangle$ and the magnitude estimation structure $\mathfrak{M} = \langle X, \succsim, \alpha_{\boldsymbol{p}} \rangle_{\boldsymbol{p} \in N}$. By Statement 1 of Theorem 7.7.5, both $\mathfrak{X}$ and $\mathfrak{M}$ are measured canonically by the same scale S. This is in sharp contrast to representational case, where $\mathfrak{X}$ is measured by a scale $\mathcal{U}$ of isomorphisms onto $\langle \mathbb{R}^+, \geq, + \rangle$ and $\mathfrak{M}$ is measured by a scale $\mathcal{U}'$ of isomorphisms that represent $\alpha_{\boldsymbol{p}}$ as multiplication by p, that is, by a numerical multiplicative scale (Definition 5.8.8). In the representational case, a power law relates elements of $\mathcal{U}$ to $\mathcal{U}'$. Because the same scale S qualitatively measures both $\mathfrak{X}$ and $\mathfrak{M}$, the qualitative version of this power law needs to be expressed by an entirely different means. This is done in Statement 2 of Theorem 7.7.5, where, as in the numerical case, the "power" θ is an automorphism of an ordered multiplicative group from a positive continuum field. For the qualitative case, this automorphism is determined by an explicit relationship between the addition operation $\oplus$ associated with the canonical measurement of $\mathfrak{X}$ and the addition operation $\oplus_1$ associated with the canonical measurement of $\mathfrak{M}$.

7.8 Method of Standard Sequences

The method of standard sequences has played a prominent role in the theory of measurement as a means for constructing representations. It is based on the following idea:

For each object x from a domain X, a sequence,

$$x_1, \ldots, x_n, \ldots$$

is constructed such that $x{=}x_1$ and the elements of the sequence are "equally spaced" for elements of a continuum. In the literature, equally spaced sequences are often defined in terms of the primitives of an underlying qualitative structure, and the "equally spaced" nature of a sequence is justified on intuitive grounds, based on considerations about the construction of the sequence. Further justifications are then given by theorems that show the existence of a function φ from the domain X into the positive reals such that for all "equally spaced" sequences $x_1, \ldots, x_i, \ldots$ and all positive integers n,

$$\varphi(x_{n+1}) - \varphi(x_n) = \varphi(x_1).$$

This section presents a qualitative theory of standard sequences. Essentially, the theory states that the method of standard sequences can be viewed formally as a form of magnitude estimation.

Definition 7.8.1 Let $\mathfrak{X} = \langle X, \succsim \rangle$ be a continuum. Then $\mathfrak{S}$ is said to be a *system of standard sequences of* $\mathfrak{X}$ if and only if the following three conditions hold:

1. Each element of $\mathfrak{S}$ is a sequence of elements of X.
2. For each element of x of X there exists an element σ of $\mathfrak{S}$ such that x is the first element of σ.
3. There exists a function φ from X onto $\mathbb{R}^+$ such that for all y and z in X,

$$y \succsim z \text{ iff } \varphi(y) \geq \varphi(z),$$

and for each sequence $x_1, \ldots, x_n, \ldots$ in $\mathfrak{S}$ and each n in $\mathbb{I}^+$,

$$\varphi(x_{n+1}) - \varphi(x_n) = \varphi(x_1).$$

Let $\mathfrak{S}$ be a system of standard sequences for the continuum $\mathfrak{X}$.. Then, by definition, functions φ satisfying Condition 3 above are called $\mathfrak{S}$-*representing functions*.

The following theorem characterizes systems of standard sequences.

Theorem 7.8.1 Let $\mathfrak{X} = \langle X, \succsim \rangle$ be a continuum and $\mathfrak{S}$ be a nonempty set of sequences of elements of X. For each p in $\mathbb{I}^+$ and each t and y in X, let

$\alpha_p(t){=}y$ if and only if there exists σ in $\mathfrak{S}$ such that t is the first element of σ and y is the pth element of σ.

Then the following two statements are equivalent:

1. $\mathfrak{S}$ *is* a system of standard sequences for $\mathfrak{X}$.

2. Interpreting "$\alpha_p(t)=y$" as "the subject estimates that y is p times as intense as t," the following behavioral axioms for magnitude estimation hold: Axioms 5.8.1, 5.8.2, 5.8.3 and 5.8.9 of Section 5.8. (Note that these axioms are the same as Axiom 7.7.4 of Section 7.7 when α_p is substituted for α_p and $\boldsymbol{N}$ is substituted for $\mathbb{I}^+$.)

Proof. Assume Statement 1. Let φ be a $\mathfrak{S}$-representing function (Definition 7.8.1). Let t be an arbitrary element of X. Then it easily follows that

$$\varphi[\alpha_1(t)] = 1 \cdot \varphi(t) .$$

Suppose n is in $\mathbb{I}^+$ and $\varphi[\alpha_n(t)]=n\ .\ \varphi(t)$. Then by Definition 7.8.1,

$$\varphi[\alpha_{n+1}(t)] = \varphi[\alpha_n(t)] + \varphi(t) = n \cdot \varphi(t) + \varphi(t) = (n+1) \cdot \varphi(t) .$$

Thus by induction, $\varphi[\alpha_m(t)]=m \cdot \varphi(t)$, for each m in $\mathbb{I}^+$ and each t in X. From this and the hypotheses of the Theorem, Statement 2 easily follows. (The details are left to the reader.)

Assume Statement 2. By Theorem 5.8.8, let φ be a numerical multiplicative representation (Definition 5.8.8) of

$$\mathfrak{X} = \langle X, \succsim, \alpha_n \rangle_{n \in \mathbb{I}^+} .$$

It only needs to be shown that φ is a $\mathfrak{S}$-representing function. To do this, it is sufficient to show that for each t in X and each n in $\mathbb{I}^+$,

$$\varphi[\alpha_{n+1}(t)] = \varphi[\alpha_n(t)] + \varphi(t) .$$

Because φ is a multiplicative representation for $\mathfrak{X}$,

$$\varphi[\alpha_{n+1}(t)] = (n+1) \cdot \varphi(t) = n \cdot \varphi(t) + \varphi(t) = \varphi[\alpha_n(t)] + \varphi(t) .$$

Theorem 7.8.2 Suppose $\mathfrak{X} = \langle X, \succsim \rangle$ is a continuum and $\mathfrak{S}$ is a system of standard sequences for $\mathfrak{X}$. Let S be the set of $\mathfrak{S}$-representing functions (Definition 7.8.1). Then S is a ratio scale.

Proof. Theorems 7.8.1 and 5.8.2

There are a variety of ways standard sequences can be constructed. The following theorem gives the classical method for continuous extensive structures.

Theorem 7.8.3 Let $\mathfrak{X} = \langle X, \succsim, \oplus \rangle$ be a continuous extensive structure. Then for each x in X, the sequence,

$$\beta_1(x), \beta_2(x), \ldots, \beta_i(x), \ldots ,$$

is a standard sequence, where for each positive integer n, β_n is the n-copy operator of $\oplus$ (Definition 5.4.8).

Proof. The proof follows by showing that $\oplus$ is commutative and then using the commutativity and associativity of $\oplus$ to show Axioms 5.8.1, 5.8.2, 5.8.3 and 5.8.9 of Section 5.8. The details are left to the reader. (Theorem 5.4.7 is useful for the proof.)

The following theorem gives another method for establishing standard sequences.

Theorem 7.8.4 Let $\mathfrak{X} = \langle X, \succsim, R_j \rangle_{j \in J}$ be a homogenous, 1-point unique continuous structure. Suppose β is a strictly increasing function from X onto X that is invariant under the automorphism group G of $\mathfrak{X}$, and suppose $\beta(y) > y$ for some y in X. Then

$$\beta(x), \beta^2(x), \ldots, \beta^i(x), \ldots$$

is a standard sequence for each x in X, where for each positive integer n, $\beta^{n+1} = \beta^n * \beta$.

Proof. It easily follow from previous theorems that $\langle G^+, \succsim', * \rangle$ is a continuous extensive structure, where $\succsim'$ is the $\succsim$-induced magnitude ordering restricted to G and

$$G^+ = \{\alpha \,|\, \alpha \in G \text{ and } \alpha \succ' \iota\}.$$

By appropriately adapting (and substituting β for ***ap*** in) the proof of Theorem 7.7.3, it follows that $\beta \in G$. Since $\beta(y) \succ y$ for some y in X, it follows from Statement 1 of Theorem 7.5.2 that $\beta \in G^+$. The theorem then follows from Theorem 7.8.3, since by hypothesis $\langle G^+, \succsim', * \rangle$ is a continuous extensive structure and β^n is the n-copy operator of $*$.

The tradition of measurement derived from Helmholtz (1887) and Hölder (1901) and popularized by the physicist N.R.Campbell (1920, 1928) was based on standard sequences. The standard sequences, in turn, were constructed from continuous extensive structures $\langle X, \succsim, \oplus \rangle$ (or variants that used Archimedean axioms in place of $\langle X, \succsim \rangle$ being a continuum). Because of the critical role the associative and commutative operation played in the construction of systems of standard sequences, Campbell believed having an observable associative and commutative operation was critical for measurement, and he strongly criticized other forms of "measurement" that were not based on such operations. As discussed in Section 2.2, this view of Campbell led the psychologist S.S.Stevens (1946, 1951) to formulate his theory of measurement. As an example of a fundamentally different kind of measurement, he created forms of magnitude estimation. In my view and others, Stevens' theory of magnitude estimation—at least as he presented it—was non-rigorous and somewhat confused. The behavioral and cognitive theories of (ratio) magnitude estimation of Section 5.8 and the theory of qualitative (ratio) magnitude estimation of Section 7.7 are alternative approaches to Stevens' theory that appear, to me, to capture in a coherent and rigorous manner the main ideas inherent in Stevens' approach. Ironically, by Theorem 7.8.1, these versions of magnitude estimation produce systems of standard sequences that lead naturally to ratio scale representations (Theorem 7.8.2). Even more ironically, the proofs of the relevant theorems reveal that these systems of standard sequences result from applying the classical method of measurement to an associative and commutative operation, that is, these theories of magnitude estimation are variants of the form of measurement espoused by Campbell! The main difference between these theories and the form of measurement used by Campbell is that the associative and commutative operation used in magnitude estimation is the operation * of functional composition on the behavioral magnitude estimation functions, whereas in the cases of physical measurement considered by Campbell, the associative and commutative operation is on the domain. The operation of functional composition * is automatically associative. In the (ratio) magnitude estimation situation, the commutativity of * is an immediate consequence of the axiom

$$\forall p \forall q [\text{if } p \in \mathbb{I}^+ \text{ and } q \in \mathbb{I}^+ \text{ then } \alpha_p * \alpha_q = \alpha_{pq}] \,.$$

It is also worthwhile to note that using Theorem 7.8.1 and the methods of Section 7.7, it is not difficult to show that slightly modified systems of standard sequences can be used to establish measurement onto positive continuum fields of magnitude numbers through canonical scales.

7.9 Qualitative Homogeneity: Other Cases

Assume $Q_\omega(a)$ and suppose $\succeq$ is qualitative, $\langle A, \succeq \rangle$ is a continuum, and the transformation group G for Q is homogeneous and finite point unique (in terms of the ordering $\succeq$). Then by Theorem 5.4.16, either G is 1-point unique or G is 2-point unique but not 1-point unique. The algebraic systems of magnitude numbers for the 1-point unique case was discussed in Section 7.5. Assume G is 2-point unique but not 1-point unique. Then by Statement 3 of Theorem 5.4.16 it follows that G has an element ν that is a nontranslation and an element τ that is a translation different from the identity ι. Using that G is isomorphic to a subscale of an interval scale (Theorem 5.4.16), it is easy to establish that for each translation $\alpha \neq \iota$, $\nu * \alpha \neq \alpha * \nu$, and for each nontranslation β, $\tau * \beta \neq \beta * \tau$. From these results it then follows from Theorem 7.2.5 that $\mathbb{Q}_{\succeq}=\{\iota\}$. A similar argument shows that if G is isomorphic to an ordinal scale, then ι is the only element of $\mathbb{Q}_{\succeq}$

The following argument shows that for situations where $\mathbb{Q}_{\succeq}=\{\iota\}$, there can be neither a qualitative integral domain of magnitude numbers nor a qualitative positive integral domain of magnitude numbers: Such qualitative number systems must have a qualitative multiplicative identity ι_1 and a qualitative addition operation $\oplus$, and therefore by Definition 7.2.1 must have at least two qualitative elements—ι_1 and $\iota_1 \oplus \iota_1$. But this contradicts $\mathbb{Q}_{\succeq}$ having only one element.

One conclusion to draw from these observations is that qualitative systems of magnitude numbers are improvised algebraically in interval and ordinal scalable situations.

7.10 Qualitative Numbers and Metaphysical Reduction

As discussed in Chapter 1, views about the nature of number and its role in science have undergone a number of transformations. Since the nineteenth century, a commonly held view in the philosophy of science has been that a better and more rigorous fragment of science results when one reduces the fragment's metaphysical content. This section briefly investigates, through consideration of a few simple illustrative examples, methods of obtaining metaphysical reductions by replacing platonic number systems with qualitatively based ones.

To simplify the discussion, the roles of qualitatively and platonically based numerical systems are illustrated for situations based on extensive measurement. Accordingly, the following three conventions are observed throughout the section:

(1) $\mathfrak{X} = \langle A, \succsim, \oplus \rangle$ is a continuous extensive structure (Definition 5.4.5);
(2) $u \in A$, and $\mathfrak{X}_u = \langle A, \succsim, \oplus, u \rangle$; and
(3) $\mathfrak{R}^+ = \langle \mathbb{R}^+, \geq, +, \cdot, 1 \rangle$.

Assume axiom system $\mathrm{Q}_\omega(\{\succsim, \oplus, u\})$. It follows from Theorem 7.5.2 that $\mathfrak{X}_u$ is isomorphic to $\mathfrak{S} = \langle \mathbb{R}^+, \geq, +, 1 \rangle$. Then $\mathfrak{S}$ is a subpart of $\mathfrak{R}^+$—a subpart that has important and applicable algebraic structure. This suggests extending $\mathfrak{X}_u$ in a qualitative manner to a structure $\mathfrak{F}^+$ that is isomorphic to $\mathfrak{R}^+$ and employ elements of A (=the domain of $\mathfrak{F}^+$) in the place of numbers to achieve a partial "metaphysical reduction."

Let f be an isomorphism from $\mathfrak{X}_u$ onto $\mathfrak{S} = \langle \mathbb{R}^+, \geq, +, 1 \rangle$, $\otimes_u = f^{-1}(\cdot)$, and

$$\mathfrak{F}^+ = \langle A, \succsim, \oplus, \otimes_u, u \rangle .$$

Then by isomorphism $\mathfrak{F}^+$ is a positive continuum field with multiplicative unit u. By Theorem 7.3.4 $\otimes_u$ has the following alternative definition:

$\otimes_u$ is the unique operation on A such that

$$\langle A, \succsim, \oplus, \otimes_u, u \rangle$$

is a positive continuum field.

It then follows from Definitions 7.2.1 and 7.2.2 that $\otimes_u$ is qualitative.

Although isomorphic, $\mathfrak{R}^+$ and $\mathfrak{F}^+$ are founded on different kinds of entities. This metaphysical difference between $\mathfrak{R}^+$ and $\mathfrak{F}^+$ may affect the kind of inference rules used in scientific theorizing and deduction. Obvious examples are rules using invariants: The platonic numbers are invariant under the automorphisms of $\mathfrak{X}$, while the elements of A—considered as metaphysically reduced "numbers"—are not. Thus quantitative theories and methods of inference formulated in terms of $\mathfrak{R}^+$ that attribute special status to invariants of $\mathfrak{X}$ do not directly translate via the isomorphism of $\mathfrak{R}^+$ and $\mathfrak{F}^+$ into similar theories and methods formulated in terms of $\mathfrak{F}^+$. In other words, there are important scientific uses of the platonic numbers $\mathbb{R}^+$, for example, their invariance under automorphisms of the underlying qualitative structure, that are not captured by formal algebraic properties of structures such as $\mathfrak{F}^+$.

Of course, the $\succsim$-magnitude numbers associated with $\mathfrak{X}$ are individually invariant and therefore are, from many perspectives, more like platonic numbers than elements of A. This potential usefulness of $\succsim$-magnitude numbers as scientific numbers is investigated next.

Throughout the remainder of this section the following conventions are observed:

(1) $\mathbb{N}$ is the set of automorphisms of $\mathfrak{X}$.
(2) $\succsim'$ is the $\succsim$-induced magnitude ordering restricted to $\mathbb{N}$.
(3) By Statement 1 of Theorem 7.5.6, let $\oplus'$ be the unique operation on $\mathbb{N}$ such that

$$\mathfrak{N} = \langle \mathbb{N}, \succsim', \oplus', *, \iota \rangle$$

is the positive continuum field such that for all x in A,

$$[\iota \oplus' \iota](x) = x \oplus x .$$

(4) When relevant, for i=1,..., n, $\mathfrak{X}_i$ are extensive structures with disjoint domains A_i such that

$$A = \bigcup_{i=1,\ldots,n} A_i$$

with qualitative positive continuum fields $\mathfrak{N}_i$ that are defined in manners similar to (1), (2), and (3) above.

Then $\mathfrak{N}$ has many desirable attributes as a number system for representing scientific ideas that are lacking in the system $\mathfrak{F}^+$ previously described. Many of these are entailed by the following two statements:

(i) The numbers—that is, the elements of $\mathbb{N}$—are invariant under the automorphisms of $\langle A, \succsim, \oplus \rangle$.
(ii) Measurement from A onto $\mathfrak{N}$ can be accomplished through the canonical scale (Definition 7.6.1) and the rich sets of measurement properties of Theorems 7.6.1 and 7.6.3 apply.

The representational theory, however, utilizes the platonic system $\mathfrak{R}^+$ in two important ways that are not captured by Statements *(i)* and *(ii)*: *(a)* The addition operation+and the elements of $\mathbb{R}^+$ are intrinsic, whereas the in $\mathfrak{N}$ the operation $\oplus'$ and the elements of $\mathbb{N}$ are not intrinsic (relative M under axiom system D″($\succsim$, $\oplus$)); and *(b)* when there are several extensive structures $\mathfrak{X}_i$ to be measured, each structure having its own distinct domain, then the platonic system $\mathfrak{R}^+$ can be used to simultaneously measure all these structures, whereas qualitative measurement would use different systems of numbers $\mathfrak{N}_i$ to measure different $\mathfrak{X}_i$. Because of *(i)* and *(ii)*, the platonic number system $\mathfrak{R}^+$ allows for more convenient calculation and simpler notation than the corresponding qualitative number systems $\mathfrak{N}$ and $\mathfrak{N}_i$ (But it should be noted, as discussed in Chapter 6 and Section 5.10, the nonintrinsic nature of the addition operation of $\mathfrak{N}$ and the fact that different $\mathfrak{N}_i$ have different sets of automorphisms as domains are often useful in understanding subtle theoretical interpretations of platonically based quantitative concepts. Thus the more cumbersome multiple systems of qualitative numbers may in some situations provide better insights than the single platonic system.)

If one is willing to have a qualitative system of numbers outside the topic of consideration, then *(i)* and *(ii)* can also be achieved using elements of A as numbers. For example, suppose the topic of interest is captured by the structure

$$\mathfrak{Y} = \langle \bigcup_{i=2,\ldots,n} A_i, \succsim_i, \oplus_i, R_j \rangle_{i=2,\ldots,n,\ j\in J}\,.$$

Then for the qualitative number system $\mathfrak{N}$, $\oplus_1$, the elements of the domain of $\mathfrak{N}_1$, all relations on the elements of the domain of $\mathfrak{N}_1$, and so forth, are invariant under the automorphisms of $\mathfrak{Y}$. Also these concepts and relations based on the domain of $\mathfrak{N}_1$ can be made to assume the role that the related concepts and relations base on $\mathbb{R}^+$ have in applications that use intrinsicness, because they are not affected by changing $\mathfrak{Y}$ to another set of equivalent primitives. ("Equivalent set of primitives" here is of course with respect to the meaningfulness concept determined by $\mathfrak{Y}$ Because $\mathfrak{N}_1$ is outside the topic determined by $\mathfrak{Y}$, the use of $\mathfrak{N}_1$ instead of $\mathfrak{R}^+$ in a fragment of science should, in my view, be construed as a *change* in metaphysics rather than a *reduction*, because the platonic numbers in $\mathfrak{R}^+$ are not being reduced to concepts based on $\mathfrak{Y}$.

It should be noted that many of the deficiencies with using $\mathfrak{F}^+$ as a qualitative number system also disappear for similarly constructed number systems when the latter are used as systems outside the topic under consideration. This is because the latter number systems would exhibit the same kinds of invariance and intrinsicness properties with respect to the topic as $\mathfrak{N}_1$ did in the previous example.

The above examples illustrate some of the gains and losses encountered when the conduction or formulation of a fragment of science attempts metaphysical reductions by replacing platonic numbers with structured systems based on scientific objects like lengths, times, physical intensities. It appears to me that these replacements are unworkable for the efficient and effective conduction of science, and from a foundational point of view, I am doubtful that they are likely to provide insights that lead to improved scientific methods. However, it also appears to me that from a philosophical perspective such replacements are important, because they may be used to argue that (platonic) numbers, although convenient for the conduction and formulation of science, are theoretically not necessary for these enterprises. While this would be an interesting philosophical result, it does not appear to me to attack the more fundamental and deeper issue of what scientific numbers are.

I believe that with respect to the latter we are currently in a position similar to Gauss when he gave his geometric interpretation to complex numbers (Section 1.7): Gauss realized that his interpretation, while providing for the geometric consistency of the complex numbers, did not yield the "true metaphysics of $\sqrt{-1}$." Similarly, although qualitative number systems like those above may be helpful in establishing the "scientific consistency" of various mathematical techniques, they do not yield the "true metaphysics of numbers as they are used in science."

7.11 Meaningfulness Versus Qualitativeness Versus Empiricalness

7.11.1 Meaningfulness Versus Qualitativeness

Qualitativeness and meaningfulness are related concepts. For the purposes of this subsection, it is convenient to treat qualitativeness as an extreme form of meaningfulness in order to contrast it with the other forms of meaningfulness discussed in the book.

When compared with the other forms of meaningfulness, qualitativeness (as a meaningfulness concept) is severely impoverished in its means of producing meaningful (=qualitative) entities from primitives. The major reason for this is that for axiom system $Q_\omega(a)$, only pure sets in V_ω, that is, pure sets of finite rank, can be employed to produce meaningful relations via definitions from primitives, whereas the other systems of meaningfulness allow much freer, and usually unlimited, use of pure sets for this purpose.

Mathematics and science are founded on different kinds of metaphysical principles. One intended role of meaningfulness systems developed in this book is to make the meaningful relations of a mathematical science more coordinate with the metaphysical principles inherent in the underlying science. For example, axiom system TM allows through the use of the axiom of Meaningful Inheritability the formation of meaningful concepts based in part on the set-theoretic axiom of Choice. Axiom system D′ does not permit such formations. Thus for sciences that eschew principles like the axiom of Choice because of metaphysical concerns, the employment of D′ for meaningfulness considerations instead of TM may be appropriate, because D′ makes the metaphysics in the mathematical parts of a science more coordinate with the metaphysics in other parts. Qualitativeness goes further in this direction than D′ and the other axiomatic systems of meaningfulness of this book by eliminating many additional highly platonic concepts and entities from being meaningful.

However, it should be emphasized that the past advancements made by science through the elimination of unnecessary or unacceptable metaphysics concerned kinds of metaphysics that were of an entirely different character than the metaphysics inherent in platonic mathematics. As discussed in Section 7.10, I find it doubtful that the elimination from science of the platonic metaphysics inherent in mathematics would help in advancing science, although it is of course a subject of great philosophical interest whether or not such eliminations can take place without altering scientific contents. This, however, does not mean that certain byproducts of efforts to eliminate such metaphysics are not important for science. For example, qualitative axiomatizations of fragments of science initially designed to exorcise metaphysics often produce byproducts that *(i)* provided insights into how various concepts are related in ways that were not obvious in mathematical, non-qualitative formulations, *(ii)* suggested new experiments, and *(iii)* helped in the formulation of alternative theories. *(i)* to *(iii)* generally result from constraints imposed by qualitative axiomatization processes, which by its nature often forces sharper and more insightful theories about the interactions of primitive concepts.

On the other hand, the richness of platonic mathematics makes for ease of formulation and inference of scientific ideas that might be otherwise constrained in qualitative presentations. Having access to this kind of richness is particularly important at the beginnings of a fragment of science, where not only non-qualitative but also non-meaningful formulations of models and ideas often abound.

Thus in the end what appears to matter is the stage the science is in and nature and kind of problem being worked on. Qualitative, meaningful, and non-meaningful formulations of a fragment of mathematical science all have important roles in the conduction of science on the fragment. As the fragment matures, the tendency of progression usually moves from non-meaningful to meaningful to qualitative formulations. The primary role of meaningfulness in this progression is to guarantee that a correct description of the underlying situation is being employed in a mathematical model or application and to occasionally exploit this fact. The primary role of qualitativeness is to provide a precise and insightful description of the underlying qualitative situation—a description that often generates new ideas and clarifies controversies.

7.11.2 Empiricalness

In the measurement literature, the property "empirical" and the concept of "empirical structure" have not been given consistent meanings. Properly speaking, empiricalness refers to the observation of real-world phenomena and the verification or refutability of propositions about the real-world, and because of this, it is often concerned with concepts of error. This concept of "real-world empiricalness" has consistently eluded adequate formal treatments by philosophers and scientists.

This subsection briefly discusses an idealized version of "empiricalness" in which error plays no part and in which infinite structures may appear as idealizations of finite situations. Idealized versions of empiricalness have often been informative in scientific-philosophical discussions, and it is these forms of "empiricalness" that have been generally invoked in measurement-theoretic discussions.

Convention 7.11.1 Throughout the rest of this chapter, unless otherwise indicated explicitly or by context, the terms "empirical" and "empiricalness" will refer to idealized forms of "empirical" and "empiricalness" in which error plays no part and in which infinite structures may appear as idealizations of finite situations.

The primary concern of this subsection is to distinguish empiricalness from meaningfulness and qualitativeness. Because of this concern's limited nature, a detailed development of "empiricalness" is not needed, and accordingly, the concept of "empiricalness" developed here is not intended to apply to a wide range of foundational and philosophical issues.

Sometimes in the literature the empiricalness of concepts and relations has been identified with first-order definability in terms of the primitives of a structure consisting of first-order empirical relations. I see no rationale for the automatic elimination of higher-order concepts from empirical considerations, nor do I see any justification for the automatic inclusion of all first-order definable concepts and relations for empirical considerations. I have never found in the literature anything approaching a cogent argument for the identification or limitation of empiricalness of relations and concepts with first-order definability.

One reason general concepts of qualitativeness or meaningfulness are easier to formulate than concepts of empiricalness is that they are preserved by isomorphism. For example suppose $\mathfrak{Y} = \langle Y, R_1, \ldots, R_n \rangle$ and $\mathfrak{Z} = \langle X, S_1, \ldots, S_n \rangle$ are structures with primitives satisfying the same general concept of qualitativeness or meaningfulness and φ is an isomorphism of $\mathfrak{Y}$ onto $\mathfrak{Z}$. Then it is reasonable to require that a relation R based on Y be qualitative (respectively, meaningful) with respect to $\mathfrak{Y}$ if and only if $\varphi(R)$ is qualitative (respectively, meaningful) with respect to $\mathfrak{Z}$. This property of being preserved under isomorphism allows for the easy formulation of general concepts of qualitativeness or meaningfulness in terms of purely structural properties of the structure of primitives.

Empiricalness is not similarly preserved under isomorphism. This is because the methods for establishing the empiricalness of, for example, a relation T on the domain of an empirical structure $\mathfrak{E} = \langle E, E_1, \ldots, E_n \rangle$ usually depends in part on relations and methods not contained in the topic determined by $\mathfrak{E}$ For example, consider the case where E is a set of physical stimuli, for example, lights of different frequencies and intensities, $E_1, \ldots, E_n$ are behavioral relations based on E, involving a subject's behavior in a psychological experiment, and $\mathfrak{E} = \langle E, E_1, \ldots, E_n \rangle$. Then empirical relations and processes *from physics* may be used freely in establishing the empiricalness of a relation H on E. These auxiliary empirical physical relations need not be based on E, but can come from parts of physics that are not exclusively concerned with the characterization of physical lights. Suppose $\mathfrak{E}' = \langle E', E'_1, \ldots, E'_n \rangle$ is another empirical structure, and suppose *f* is an isomorphism of $\mathfrak{E}$ onto $\mathfrak{E}'$. Note that the existence of the isomorphism *f* between $\mathfrak{E}$ and $\mathfrak{E}'$ does not imply that the *empirical* physical environment in which *E* is imbedded has an isomorphic *empirical* counterpart in which $\mathfrak{E}'$ is imbedded. Thus *f(H)* may not be empirical.

The main point of the previous example is that for defining or constructing a particular empirical relation on the domain *E, any other* empirical relation on *E, or on other domains,* may be used in the defining or the constructing. This is a key characteristic of the concept "empirical" that is in wide variance with properties of the concept "qualitative".

Metaphysically, the concept of "qualitative" is founded on the classical concept of truth. Through this concept of truth, the familiar notion of logical consequence can be explicitly formulated, and when the concept of logical consequence is restricted to qualitative axioms about a qualitative structure, it yields a precise description of logical consequences of those axioms, and a framework—a fragment of higher-order logic—for calculating some of those consequences, and a formal language for describing which statements about the structure and which relations on the domain, including higher-order ones, should also be called "qualitative"; that is, it yields procedures of deduction and formulation that preserve qualitativeness.

Metaphysically, the concept of "empirical" is founded on the concepts of observability, verifiability, and refutability—concepts very different from the classical, platonic concept of truth. No one has yet produced a workable concept of "empirical consequence" anywhere near approaching the clarity, sophistication, and power of "logical consequence," nor has any one produced an adequate formal language for the general construction of empirical statements from other empirical statements. These remain central problems in the philosophy of science—problems that are likely not to have elegant solutions like higher-order logic, and perhaps have no satisfactory solutions at all.

7.11.3 Conclusions

The following is a summary of the key differences presented here between empiricalness and the concepts "qualitativeness" and "meaningfulness": Qualitativeness and meaningfulness are connected with a specific structure, whose primitives form the basis for forming other qualitative and meaningful concepts. The methods of formation are forms of higher-order definability. The relations that can enter into a formula through these definability concepts are very restricted, and for qualitativeness can be restricted to the primitives of the underlying qualitative structure. However, the means for constructing the "defining formulas"—a fragment of higher-order logic—are very powerful, and intuitively are not very restricted. Empiricalness, on the other hand, is connected with methods of observation, verification, and refutability. It can be connected with a specific *empirical* structure—a structure in which all the primitives are empirical. The methods of producing additional empirical relations on the domain of the structure may, like in the qualitative case, use primitives of the underlying empirical structure, but unlike the qualitative case, may also use additional empirical relations, including ones based on other domains. Thus intuitively, the empirical case is much less restrictive than the qualitative or meaningful cases in what relations can be used as the basis for producing additional relations. However, intuitively, the form of production for the qualitative and meaningful cases—definability through higher-order languages—appears to be much less restricted than the forms of production in the empirical case—definability-like concepts based on observability, verifiability, and refutability, instead of truth.

A structure of primitives together with a qualitativeness or meaningfulness concept create a topic, for example, Euclidean plane geometry is created by taking the usual two dimensional Euclidean structure of primitives of plane, point, line, angle, incidence, and congruence, and the meaningfulness concept of invariance under transformations generated by rotations, translations, and reflections. The same domain can be used as a basis for other topics, for example, hyperbolic geometry of two dimensions, through specification of a different structure of primitives and meaningfulness (or qualitativeness) concepts. The topic is created by the restrictions imposed by meaningfulness or qualitativeness applied to the structure of primitives. A structure of empirical primitives together with an empiricalness concept does not create any thing like a "topic" in the above sense, because the empiricalness concept is not relativizable to a structure. However, a structure of empirical primitives together with empiricalness *and* qualitativeness concepts (respectively, empiricalness *and* meaningfulness concepts) creates something much more topic-like in this respect. The addition of qualitativeness or meaningfulness can disallow relations from being considered relevant to the topic just because they are empirical; they must be both empirical and qualitative or meaningful to be relevant. It appears to me that when authors want to talk about an empirical situation, they usually have in mind those relations and statements that are empirical *and* are formulated in terms of the primitives of an appropriate higher-order language.

7.12 Summary of Main Points

In the theories of meaningfulness of Chapter 4, pure sets could be freely employed in establishing the meaningfulness of concepts and relations. This use was intended to reflect the science's free employment of mathematics in formulating concepts and relations. The qualitativeness concept developed in this chapter enormously curtails the employment of pure sets for formulating qualitative concepts and relations, and correspondingly topics generated through qualitativeness are much more restricted than those generated through the meaningfulness concepts of Chapter 4.

Sections 7.4 to 7.6 demonstrated that rich qualitative algebraic systems of "numbers" can be established for certain general classes of homogeneous continuous structures. These systems of "numbers" had the algebraic properties of well-ordered integral domains, continuum fields, or the positive parts of such integral domains and fields. The "numbers" within them were $\succsim$-strictly increasing functions, where $\succsim$ was a specific qualitative total ordering, $\succsim$. Such "numbers" were called "$\succsim$-magnitude numbers." Theorems were presented that showed $\succsim$-magnitude numbers were qualitative correlates of one of two important systems of (platonic) numbers that occur repeatedly in the theory of measurement: *(i)* integral multiples of a modified Weber constant, and *(ii)* (dimensionless) ratios of dimensional quantities. Because of the importance of *(i)* and *(ii)* in measurement, it was natural to inquire if measurement through assigning $\succsim$-magnitude numbers to objects would produce theoretical perspectives that might be informative about foundational issues.

Let $\mathfrak{X} = \langle A, \succsim, R_j \rangle_{j \in J}$ be a homogeneous, 1-point unique continuous structure and $\mathfrak{z} = \{\succsim\} \cup \{R_j | j \in J\}$ Assume axiom system $Q_\omega(a)$. Let $\mathbb{N}$ be the set of automorphisms of $\mathfrak{X}$ and $\alpha \in \mathbb{N}$. Then α is a $\succsim$-magnitude number. Discussions in Chapters 5 and 6 established that under measurement by a ratio scale S for $\mathfrak{X}$, α was the qualitative interpretation of a ratio $\varphi(x)/\varphi(y)$, where φ is in S and x and y are elements of A; that is, α is the qualitative correlate of a "dimensionless number." Qualitative canonical measurement (Definition 7.6.1) is a method of assigning elements of $\mathbb{N}$ to elements of A. Theorems of Section 7.6 established that this method of measuring $\mathfrak{X}$ was productive and shared a number of analogies with the representational measurement of $\mathfrak{X}$ through a ratio scale. Section 7.7 exploited a formal similarity between canonical measurement and Stevens' method of (ratio) magnitude estimation to give an entirely qualitative foundation to magnitude estimation, and Section 7.8 employed this qualitative foundation to draw a tight analogy between (ratio) magnitude estimation and the measurement-theoretic method of standard sequences. The latter analogy was surprising, because it has been traditional to consider magnitude estimation and the method of standard sequences as rather different kinds of measurement processes.

The above investigations involving qualitative canonical measurement make clear that measurement into qualitative number systems sometimes reveal important structure and relationships that are not readily apparent in quantitative formulations. They also provide for an alternative foundation for the theory of measurement for situations captured by a homogenous, 1-point unique continuous structure.

As discussed in Section 7.9, the other homogeneous, finitely unique cases cannot have similarly rich qualitative algebraic systems based on $\succsim$-magnitude numbers, when $\succsim$ is qualitative and $\langle A, \succsim \rangle$ is a continuum. Of course, for these cases other kinds of qualitative numbers could be sought, for example, ones based on some special variety of 3-ary relations on A. To my knowledge, this has not been pursued in any systematic way. The challenge is to find for these other homogeneous cases algebraically rich qualitative systems whose domain and relations have *useful and natural* interpretations as numerical domains and numerical relations, particularly from the points of view of measurement and science. The goal is not to just produce an ad hoc, qualitative, algebraically rich structure that happens to be isomorphic to some well-known numerical structure.

Providing qualitative axiomatizations of quantitative models is an important step in the maturing of a fragment of science, often yielding new methods and insights that advance the fragment. Reducing the metaphysics in a fragment of science by replacing platonic numbers with qualitative numbers does not appear to yield useful new methods or insights that are likely to advance the fragment. In fact, the adoption of this strategy would likely have the opposite effect. However, in philosophy the option for such replacements may be of importance, particularly in discussions concerning the metaphysics inherent in a given fragment of science.

As developed in this chapter, the concept of empiricalness, which is based on the ideas of observability, verifiability, and refutability, is quite distinct from the concept of qualitativeness, which is based on the idea of truth. Also, qualitativeness is connected to a particular subject matter, that is, a topic, whereas empiricalness is not. This formulation of "empirical" is at odds with many of the formulations and uses of "empirical" in the literature. The concept "empirical and qualitative" appears to capture the intent behind many of formulations and uses of "empirical" in the literature.

7.13 Additional Proofs and Theorems

Theorem 7.13.1 *(Theorem 7.5.2)* Assume axiom system $Q_\omega(a)$. Suppose $\mathbb{N}$ is the transformation group of Q, A is qualitatively homogeneous and qualitatively 1-point unique, and $\succsim$ is qualitative total ordering on A. Then the following four statements are true:

1. Let $\succsim'$ be the $\succsim$-induced magnitude ordering on the set of $\succsim$-magnitude numbers. Then the restriction of $\succsim'$ to $\mathbb{N}$ is a total ordering of $\mathbb{N}$, and for all α and β in $\mathbb{N}$,

$$\begin{aligned} \alpha \succeq' \beta \ \text{ iff } \ & \alpha(x) \succeq \beta(x) \text{ for all } x \text{ in } A \\ \text{iff } \ & \alpha(x) \succeq \beta(x) \text{ for some } x \text{ in } A. \end{aligned}$$

2. Suppose $\langle A, \succeq \rangle$ is a continuum. Then $\{\iota\} \subset \mathbb{N}$ Let β in $\mathbb{N}$ be such that $\beta \succeq' \iota$. Then $\langle \mathbb{N}, \succeq', *, \iota, \beta \rangle$ is isomorphic to $\langle \mathbb{R}, \geq, +, 0, 1 \rangle$ and to $\langle \mathbb{R}^+, \geq, \cdot, 1, 2 \rangle$.

3. Suppose $\langle A, \succsim\rangle$ is a continuum. Then for all α and β in $\mathbb{N}$ $\alpha * \beta = \beta * \alpha$.
4. Suppose $\langle A, \succsim\rangle$ is a continuum. Then $\mathbb{Q}_{\succsim} \subseteq \mathbb{N}$.

Proof. 1. Statement 1 follows from Lemma 2.2 of Narens (1981a) (or from Lemma 4.3 and Definition 4.3 of Narens, 1985).

2. By Theorem 2.5 of Narens (1981a) (or Theorem 4.1 of Narens, 1985), $\langle \mathbb{N}, \succsim', *\rangle$ is a totally ordered group and $\langle \mathbb{N}, \succsim'\rangle$ is a continuum. Statement 2 is then a simple consequence of the well-known representation theorems for such groups ("Hölder's Theorem").

3. Statement 3 is an immediate consequence of Statement 2 and the fact that+is commutative.

4. Let $\gamma \in \mathbb{Q}_{\succsim}$. Then to show Statement 4 it is sufficient to show $\gamma \in \mathbb{N}$. By Statement 2, let f be an isomorphism of $\langle \mathbb{N}, \succsim', *\rangle$ onto $\langle \mathbb{R}^+, \geq, \cdot\rangle$ and let $\gamma' = f(\gamma)$. By Theorem 7.2.5,

$$a * \gamma = \gamma * a.$$

for each α in $\mathbb{N}$. Therefore, by the isomorphism f,

$$r.\gamma'(s) = \gamma'(r \cdot s) \tag{7.2}$$

for all r and s in $\mathbb{R}^+$ It is well-known that the only order preserving functional solution to Equation 7.2 is

$$\gamma'(s) = t \cdot s,$$

where t is a particular element of $\mathbb{R}^+$ But because multiplication by t is the image under f of some element of $\mathbb{N}$ it follows that $\gamma \in \mathbb{N}$.

Theorem 7.13.2 *(Theorem 7.5.3)* Assume axiom system $Q_\omega(a)$. Suppose A is qualitatively homogeneous, $\succsim$ is qualitative total ordering on A, $\langle A, \succsim\rangle$ is a continuum, and $\succsim'$ is the $\succsim$-induced magnitude ordering. Let $\mathbb{Q}^+_{\succsim}$ be the collection of all elements α of $\mathbb{Q}_{\succsim}$ such that $\alpha \succsim' \iota$ Suppose $\langle \mathbb{Q}^+_{\succsim}, \succsim'\rangle$ has a least element, i_1. Let $i_0 = \iota$. For each n in ω, let ni_1 denote n compositions of i_1 with itself, for example, $2i_1 = i_1 * i_1$, and let $(-n)i_1$ be the inverse of the function ni_1, and let $0i_1 = \iota$. Let

$$\mathbb{N} = \{mi_1 \mid m \in \omega \text{ or } -m \in \omega \text{ or } m = 0\}.$$

Define $\odot$ on $\mathbb{N}$ as follows: For all pi_1 and qi_1 that are in $\mathbb{N}$, let

$$pi_1 \odot qi_1 = (p \cdot q)i_1 .$$

Then the following two statements are true:

1. $\mathfrak{N} = \langle \mathbb{N}, \succsim'', *, \odot, i_0, i_1\rangle$ is a qualitative well-ordered integral domain of magnitude numbers, where $\succsim''$ is the restriction of $\succsim'$ to $\mathbb{N}$.
2. Assume a standard model for ZFA (Convention 7.5.1). Then $\mathbb{Q}_{\succsim} = \mathbb{N}$.

Proof. *Statement 1.* It is a straightforward verification that $\mathfrak{N}$ is a wellordered integral domain. ι_0 and ι_1 are qualitative, because they are in $\mathbb{Q}_{\succeq}$.

It is easy to verify that $\mathfrak{X} = \langle A, \succeq, i_1 \rangle$ is a continuous threshold structure (Definition 5.4.2). Let H be the automorphism group of $\mathfrak{X}$. Then H is qualitative by Definitions 7.2.1 and 7.2.2. Let

$$B = \{\alpha \mid \alpha \in H \text{ and } \forall\beta[\text{if } \beta \in H \text{ then } \alpha * \beta = \beta * \alpha]\}\,.$$

Then B is qualitative. By using the characterization of the automorphism group of the canonical, numerical, continuous threshold structure given in Convention 5.4.1 and Theorem 5.4.5, it is not difficult to show that B=$\mathbb{N}$, and therefore $\mathbb{N}$ is qualitative. Because $\mathbb{N}$ is qualitative, it then follows that $\succeq''$, the restriction of $\succeq'$ to $\mathbb{N}$ is qualitative.

By Theorem 7.3.3 $\odot$ has the following alternative definition:

$\odot$ is the unique operation on B such that

$$\mathfrak{B} = \langle B, \succeq', *, \odot, i_0, i_1 \rangle$$

is a well-ordered integral domain.

Because $\mathbb{N}$, $\succeq''$, $*$, ι_0, and ι_1 are qualitative, it follows that $\odot$ is qualitative by its alternative definition and Definitions 7.2.1 and 7.2.2. Thus $\mathfrak{N}$ is a qualitative well-ordered integral domain.

Statement 2. i_0 and i_1 are qualitative by Statement 1. Since $2i_1$=i_1 $*$ i_1 and i_1 and $*$ are qualitative, $2i_1$ is qualitative by Definitions 7.2.1 and 7.2.2. Since $(-2)i_1$=$(2i_1)^{-1}$, $(-2)i_1$ is qualitative by Definitions 7.2.1 and 7.2.2. Similarly, ni_1 and $(-n)i_1$ are qualitative for each *externally* finite (Definition 4.3.7) n in ω. Because ZFA is a standard model, all elements of ω are externally finite. Thus $\mathbb{N} \subseteq \mathbb{Q}_{\succeq}$. It will be shown by contradiction that $\mathbb{N} = \mathbb{Q}_{\succeq}$.

Suppose α is in $\mathbb{Q}_{\succeq} - \mathbb{N}$ Then $\alpha \neq \iota$ and α^{-1} is in $\mathbb{Q}_{\succeq} - \mathbb{N}$. Therefore, without loss of generality suppose $\alpha \succeq' \iota$. Because $\succeq'$ is a total ordering on $\mathbb{Q}_{\succeq}$ (Theorem 7.5.1), there are two cases to consider. *Case 1.* There exists m in ω—$\{0\}$ such that

$$mi_1 \succ' \alpha \succ' (m-1)i_1\,.$$

Then $i_1 \succeq' (1—m)i_1 *\alpha \succeq' 0i$=$\iota$, contradicting the choice of i_1 as the least element of $\mathbb{Q}_{\succeq}$ that is $\succeq'$-greater than ι. *Case 2.* $\alpha \succeq' mi_1$ for each $m \in \omega$. Let x be an element of A. Then by Definition 7.4.3,

$$\alpha(x) > mi_1(x) \text{ for all } m \in \omega. \tag{7.3}$$

However, as shown in Statement 1, $\mathfrak{X} = \langle A, \succeq, i_1 \rangle$ is a continuous threshold structure, and by Theorem 5.4.3 it easily follows that there exists k in ω such that $ki_1(x) \succ' \alpha(x)$, contradicting Equation 7.3.

Theorem 7.13.3 *(Theorem 7.5.4)* Assume axiom system $Q_\omega(a)$. Suppose A is qualitatively homogeneous, $\succeq$ is qualitative, $\langle A, \succeq \rangle$ is a continuum,

$$\langle \mathbb{Q}_{\succeq}, \succeq', \oplus, \odot', i_0', i_1' \rangle$$

is a qualitative, well-ordered integral domain of magnitudes numbers, $\succeq'$ is the restriction of the $\succeq$-induced magnitude ordering to $\mathbb{Q}_{\succeq}$ and $i_0' = \iota$. Let $*$, $\odot$, ι_0, and ι_1 be as in the statement of Theorem 7.5.3. Then

$$\oplus = *, \ \odot' = \odot, \ \iota_0' = \iota_0, \text{ and } \iota_1' = \iota_1.$$

Proof. $i_1'=i_0=\iota$ by hypothesis. In all well-ordered integral domains the multiplicative identity is the immediate successor with respect to the ordering of the additive identity. Thus i_1' and i_1 are the immediate $\succeq'$-successor of ι, that is, $i_1'=i_1$.

Let β be an arbitrary element of $\mathbb{Q}_{\succeq}$. It will be shown by induction that $\beta*\alpha=\beta\oplus\alpha$ for all α in $\mathbb{Q}_{\succeq}^{+}$. Because $i_1'=i_1$ and in all well-ordered integral domains the immediate successor of an element is that element added with the multiplicative identity, it follows that for all δ in $\mathbb{Q}_{\succeq}$,

$$\delta * i_1 = \delta \oplus i_1, \tag{7.4}$$

and in particular,

$$\beta * i_1 = \beta \oplus i_1.$$

Suppose it has been shown that

$$\beta_* \gamma = \beta \odot \gamma.$$

Then by Equation 7.4,

$$(\beta*\gamma) * i_1 = (\beta\oplus\gamma)\oplus i_1,$$

which by the associativity of * and $\oplus$ yields

$$\beta*(\gamma*i_1)=\beta\oplus(\gamma\oplus i_1).$$

Thus by induction, $\beta * \alpha=\beta\oplus\alpha$ for all β in $\mathbb{Q}_{\succeq}$ and all α in $\mathbb{Q}_{\succeq}^{+}$. By using elementary properties of well-ordered integral domains, it easily follows that $*=\oplus$.

By the "alternative definition of $\odot$" given in the proof of Statement 1 of Theorem 7.13.2, $\odot'=\odot$.

Theorem 7.13.4 *(Theorem 7.5.5)* Assume axiom system $\mathcal{Q}_{\omega}(a)$. Suppose A is qualitatively homogeneous and qualitatively 1-point unique, $\succeq$ is qualitative total ordering on A, $\langle A, \succeq \rangle$ is a continuum. Let $\langle \mathbb{N}, * \rangle$ be the transformation group of Q and $\succeq'$ be the restriction of $\succeq$-induced magnitude ordering to $\mathbb{N}$. Let i_1 in $\mathbb{Q}_{\succeq}$ be such that $i_1 \succeq' \iota$. Let $i_0=\iota$. Then the following three statements are true:

1. $\mathfrak{N} = \langle \mathbb{N}, \succeq', *, \odot, i_0, i_1 \rangle$ is a qualitative, continuum field of magnitude numbers, where $\odot$ is the unique operation on $\mathbb{N}$ such that

$$\mathfrak{N} = \langle \mathbb{N}, \succeq', *, \odot, i_0, i_1 \rangle$$

is a continuum field.
2. $\langle \mathbb{Q}_{\succeq}, \succeq', *, \odot, i_0, i_1 \rangle$ where $\odot$ is as in Statement 1, is a real closed subfield of $\mathfrak{N}$.
3. Assume a standard model of ZFA. Then $\mathbb{Q}_{\succeq}$ is order dense in $\langle \mathbb{N}, \succeq' \rangle$ that is, for all α and γ in $\mathbb{N}$ if $\alpha \succ' \gamma$, then there exists β in $\mathbb{Q}_{\succeq}$ such that $\alpha \succeq' \beta \succeq' \gamma$.

Proof. 1. Since $\iota_1 \in \mathbb{Q}_{\succeq}$, it follows from Statement 4 of Theorem 7.5.2 that $\iota_1 \in \mathbb{N}$. By Statement 2 of Theorem 7.5.2, let g be an isomorphism of $\langle \mathbb{N}, \succeq', *, i_0, i_1 \rangle$ onto $\mathfrak{R} = \langle \mathbb{R}^+, \geq, +, 0, 1 \rangle$. Let $\odot = g^{-1}(\cdot)$. Then g is an isomorphism of $\mathfrak{N} = \langle \mathbb{N}, \succeq', *, \odot, i_0, i_1 \rangle$ onto $\langle \mathbb{R}, \geq, +, \cdot, 0, 1 \rangle$ and therefore $\mathfrak{N}$ is a continuum field. Thus by Theorem 7.3.4, $\odot$ is the unique operation such that $\mathfrak{N} = \langle \mathbb{N}, \succeq', *, \odot, i_0, i_1 \rangle$ is a continuum field. Therefore, since by hypothesis $\mathbb{N}$, $\succeq'$, $*$, i_0, and i_1 are qualitative, $\odot$ qualitative by Definitions 7.2.1 and 7.2.2. Thus, $\mathfrak{N}$ is a qualitative continuum field of magnitude numbers.

2. Let $\mathfrak{N}$ be as in Statement 1. $\mathbb{Q}_{\succeq} \subseteq \mathfrak{N}$ by Statement 4 of Theorem 7.5.2. It is easy to verify that $\mathfrak{Q} = \langle \mathbb{Q}, \succeq', \oplus, \odot, i_0, i_1 \rangle$ is a field. Let p be an arbitrary polynomial of odd degree with coefficients in $\mathbb{Q}$. Then, because $\mathfrak{N}$ is a real closed field, p has a root in $\mathbb{N}$. Because each element of $\mathbb{Q}$ has an explicit definition in terms of L($\in$, A, Ø) and elements of $a \cup \{a\}$, the polynomial $p(x)$ is formulable in terms of L($\in$, A, Ø) and elements of $a \cup \{a\}$. Since $p(x)$ has finitely many roots in $\mathbb{N}$, let b be the unique element of $\mathbb{N}$ such that

$$p(b) \wedge \forall y (y \in \mathbb{N} \text{ and } p(y) \rightarrow y \succeq' b)\,.$$

By Definitions 7.2.1 and 7.2.2, $b \in \mathbb{Q}$. Thus $\mathfrak{Q}$ is a real closed field.

3. Let $\mathfrak{N}$ be as in Statement 1. Let f be an isomorphism of $\mathfrak{R} = \langle \mathbb{R}, \geq, +, \cdot, 0, 1 \rangle$ onto $\mathfrak{N}$. Call an element α of $\mathbb{N}$ "rational" if and only if $f^{-1}(\alpha)$ is rational. Because a standard model of ZFA is being assumed, each rational number in $\mathbb{R}$ has a first-order definition in terms of $\geq$, $+$, $\cdot$, 0, and 1. Therefore the image under f of each rational number in $\mathbb{R}$ has a first-order definition in terms of $\succeq$, $\oplus$, $\odot$, i_0, and i_1, e.g., $f(\frac{2}{3})$ has the definition

$$(i_1 \oplus i_1) \odot (i_1 \oplus i_1 \oplus i_1)^{-1}.$$

Thus each "rational" in $\mathbb{N}$ is qualitative. Because the rational numbers in $\mathbb{R}^+$ are order dense $\langle \mathbb{R}^+, \geq \rangle$ it follows by the isomorphism f that the "rationals" in $\mathbb{N}$ are order dense in $\langle \mathbb{N}, \succeq \rangle$.

Theorem 7.13.5 *(Theorem 7.5.6)* Assume axiom system $Q_\omega(a)$. Suppose A is qualitatively homogeneous and qualitatively 1-point unique, $\succeq$ is qualitative total ordering on A, $\langle A, \succeq \rangle$ is a continuum. Let $\langle \mathbb{N}, * \rangle$ be the transformation group of Q and $\succeq'$ be the restriction of $\succeq$-induced magnitude ordering to $\mathbb{N}$. Let $i_1 = \iota$. Let α_2 in $\mathbb{Q}_{\succeq}$ be such that $\alpha_2 \succeq' \iota$ Then the following three statements are true:

1. $\mathfrak{N} = \langle \mathbb{N}, \succeq', \oplus, *, i_1 \rangle$ is a qualitative, positive continuum field of magnitude numbers, where $\oplus$ is the unique operation on $\mathbb{N}$ such that
   (i) $\langle \mathbb{N}, \succeq', \oplus \rangle$ is a continuous extensive structure;
   (ii) $\alpha_2 = i_1 \oplus i_1$; and
   (iii) for all x, y, and z in $\mathbb{N}$,
   $$z * (x \oplus y) = (z * x) \oplus (z * y).$$
2. $\langle \mathbb{Q}_{\succeq}, \succeq', \oplus, *, i_1 \rangle$ is a positive real closed subfield of $\mathfrak{N}$, where $\oplus$ is as in Statement 1.
3. Assume a standard model of ZFA. Then $\mathbb{Q}_{\succeq}$ is order dense in $\langle \mathbb{N}, \succeq' \rangle$, that is, for all α and γ in $\mathbb{N}$ if $\alpha \succeq' \gamma$, then there exists β in $\mathbb{Q}_{\succeq}$ such that $\alpha \succeq' \beta \succeq' \gamma$.

Proof. 1. Since $\iota_1 \in \mathbb{Q}_{\succeq}$, it follows from Statement 4 of Theorem 7.5.2 that $\iota_1 \in \mathbb{N}$ By Convention 7.3.1 and Theorem 7.3.2, let g be an isomorphism of $\langle \mathbb{N}, \succeq', *, i_1 \rangle$ onto $\mathfrak{R} = \langle \mathbb{R}^+, \geq, \cdot, 1 \rangle$ By Theorem 5.10.11, all operations $+'$ on $\mathbb{R}^+$ such that $\langle \mathbb{R}^+, \geq, +' \rangle$ is a continuous extensive structure and for all r, x, and y in $\mathbb{R}^+$,

$$r\text{-}(x +' y) = (r \cdot x) +' (r \cdot y),$$

are of the form

$$x +' y = (x^s + y^s)^{\frac{1}{s}},$$

where s is a positive real. Thus let t be the unique positive real such that

$$g(\alpha_2) = (1^t + 1^t)^{\frac{1}{t}} = 2^{\frac{1}{t}},$$

and let $+''$ be the operation on $\mathbb{R}^+$ such that $\langle \mathbb{R}^+, \geq, +'' \rangle$ is a continuous extensive structure and for all x and y in $\mathbb{R}^+$,

$$x +'' y = (x^t + y^t)^{\frac{1}{t}}.$$

Let $\oplus = g^{-1}(+'')$. Then Statement 1 is true by the isomorphism g and Theorem 5.10.11.

Statements 2 and 3 follow by arguments similar to the ones for Statements 2 and 3 of Theorem 7.13.4.

Theorem 7.13.6 *(Theorem 7.6.3)* Assume axiom system $Q_\omega(a)$. Suppose $\mathbb{N}$ is the transformation group of Q, A is qualitatively homogeneous and qualitatively 1-point unique, $\langle A, \succeq \rangle$ is a continuum, and $\succeq$ is qualitative total ordering on A. Let $\succeq'$ be the restriction to $\mathbb{N}$ of the $\succeq$-induced magnitude ordering, β be a qualitative element of $\mathbb{N}$ such that $\beta \succ' i$, and S be the qualitative canonical scale for A. Then the following three statements are true:

1. There is exactly one operation $\oplus$ on $\mathbb{N}$ such that

$$\mathfrak{F} = \langle \mathbb{N}, \succsim', \oplus, *, \iota \rangle$$

is a qualitative positive continuum field and $\beta=\iota\oplus\iota$.

2. There is exactly one qualitative operation $\bigcirc$ on A such that $\langle A, \succsim, \bigcirc \rangle$ is a continuous extensive structure and for each x in X,

$$\beta(x)=x \bigcirc x.$$

3. Let $\oplus$ be as in Statement 1 and $\bigcirc$ as in Statement 2. Then for all φ in S, $\varphi(\bigcirc)=\oplus$.

Proof. 1. Statement 1 is an immediate consequence of Theorem 7.5.6.

2. Let $\{R_j | j \in J\}$ be an indexing of the elements of a. Then by hypothesis and Theorem 7.2.2 $\mathbb{N}$ is the set of automorphisms of $\langle A, R_j \rangle_{j \in J}$. Because $\succsim$ is qualitative, it follows by Theorem 7.2.2 that $\succsim$ is $\mathbb{N}$-invariant. Thus $\mathbb{N}$ is the set of automorphisms of

$$\mathfrak{X} = \langle A, \succsim, R_j \rangle_{j \in J},$$

and $\mathfrak{X}$ is a homogeneous and 1-point unique structure. Thus by Theorem 5.4.9, let $\mathfrak{N} = \langle \mathbb{R}^+, \geq, R'_j \rangle_{j \in J}$ be a structure such that the set of isomorphisms of $\mathfrak{X}$ onto $\mathfrak{N}$ is a ratio scale. Then the set of automorphisms H of $\mathfrak{N}$ is the set of multiplications by positive reals. It is well-known that H is also the automorphism group of $\langle \mathbb{R}^+, \geq, + \rangle$. Thus by Theorem 2.7 of Narens (1981a), each H-invariant operation $+'$ such that $\langle \mathbb{R}^+, \geq, +' \rangle$ is an extensive structure has the form

$$x +' y = (x^r + y^r)^{\frac{1}{r}},$$

where $r \in \mathbb{R}^+$.

Let f be an isomorphism of $\mathfrak{X}$ onto $\mathfrak{N}$ and let $\beta'=f(\beta)$. Then $\beta' \in H$. Let $+''$ be the operation on $\mathbb{R}^+$ such that for all x and y in $\mathbb{R}^+$,

$$x +'' y = (x^r + y^r)^{\frac{1}{r}},$$

where r is such that

$$\beta'(1) = (1 +'' 1) = (1^r + 1^r)^{\frac{1}{r}} = 2^{\frac{1}{r}}.$$

Then, because $\beta' \in H$,

$$\beta'(x) = 2^{\frac{1}{r}} x = x +'' x,$$

for all x in $\mathbb{R}^+$ From these facts it follows that $+''$ has the following definition:

> $+''$ is the unique operation on $\mathbb{R}^+$ such that $\langle \mathbb{R}^+, \geq, +'' \rangle$ is a continuous extensive structure, $+''$ is H-invariant, and for all x in $\mathbb{R}^+$, $\beta'(x)=x+''x$.

Let $\bigcirc = f^{-1}(+'')$. Then by isomorphism, $\bigcirc$ has the following definition:

> $\bigcirc$ is the unique operation on A such that $\langle A, \succeq, \bigcirc \rangle$ is a continuous extensive structure, $\bigcirc$ is $\mathbb{N}$ -invariant, and for all x in A, $\beta(x) = x \bigcirc x$

Then by Definitions 7.2.1 and 7.2.2, $\bigcirc$ is qualitative. Thus Statement 2 has been shown.

3. Let $\bigcirc$ be as in Statement 2. Let φ be an element of S and $\oplus' = \varphi(\bigcirc)$. Then $\oplus'$ is qualitative, since *(i)* $\bigcirc$ is qualitative by Statement 2, *(ii)* S is qualitative by Statement 1 of Theorem 7.6.1, and *(iii)* by *(i)* above and Statement 7 of Theorem 7.6.1 $\oplus'$ is the unique operation on $\mathbb{N}$ such that $\oplus' = \psi(\bigcirc)$ for each ψ in S. By Statement 2 of Theorem 7.6.1, φ is a one-to-one function from A onto $\mathbb{N}$ and by Statement 8 of Theorem 7.6.1, $\varphi(\succeq) = \succeq'$. Thus φ is an isomorphism of $\langle A, \succeq, \bigcirc \rangle$ onto $\langle \mathbb{N}, \succeq', \oplus' \rangle$. Therefore, $\langle \mathbb{N}, \succeq', \oplus' \rangle$ is a continuous extensive structure, since by hypothesis $\langle A, \succeq, \bigcirc \rangle$ is a continuous extensive structure. Thus, since by Statement 2 of Theorem 7.5.2, $\langle \mathbb{N}, \succeq', * \rangle$ is isomorphic to $\langle \mathbb{R}^+, \geq, \cdot \rangle$ to show that $\langle \mathbb{N}, \succeq', \oplus', *, \iota \rangle$ is a positive continuum field, it is sufficient to show that for all θ, γ, and δ in $\mathbb{N}$,

$$\theta * (\gamma \oplus' \delta) = (\theta * \gamma) \oplus' (\theta * \delta).$$

This will be done next.

By Definition 7.6.1, let a in A be such that

$$\varphi(a) = \iota.$$

By the homogeneity and 1-point uniqueness of A, for each x and b in A, let $\alpha_{b,x}$ be the unique element of $\mathbb{N}$ such that

$$\alpha_{b,x}(a) = x.$$

Then by Definition 7.6.1, for each x in A,

$$\varphi(x) = a_{\alpha,x.}$$

Thus $\alpha_{a,\theta(x)}$ and $\theta * \alpha_{a,x}$ are elements of $\mathbb{N}$ for each x in A and each θ in $\mathbb{N}$. Therefore, because $\mathbb{N}$ is 1-point unique and for each x in A and θ in $\mathbb{N}$,

$$\alpha_{a,\theta(x)}(a) = \theta(x) = \theta * \alpha_{a,x(a)},$$

it follows that

$$\alpha_{a,\theta(x)} = \theta * \alpha_{a,x} \tag{7.5}$$

for each x in A and each θ in $\mathbb{N}$.

Let x, y, and z be arbitrary elements of A. Because φ is onto $\mathbb{N}$ (Statement 2 of Theorem 7.6.1) and x, y, and z, are arbitrary elements of A, we may without loss of generality assume that $\gamma = \alpha_{a,x,} \delta = \alpha_{a,y,}$ and $\eta = \alpha_{a,z}$ are arbitrary elements of $\mathbb{N}$. Let θ be an arbitrary element of $\mathbb{N}$. Since by Statement 2 $\bigcirc$ is qualitative, it follows from Theorem 7.2.3 that $\bigcirc$ is $\mathbb{N}$-invariant.

Thus from the definition of $\oplus'$, Equation 7.5, and the already established fact that φ is an isomorphism of $\langle A, \succsim, \bigcirc\rangle$ onto $\langle \mathbb{N}, \succsim', \oplus'\rangle$

$$\begin{aligned}
\gamma \oplus' \delta = \eta \ \text{ iff } & \alpha_{a,x} \oplus' \alpha_{a,y} = \alpha_{a,z} \\
\text{iff } & \varphi^{-1}(\alpha_{a,x}) \bigcirc \varphi^{-1}(\alpha_{a,y}) = \varphi^{-1}(\alpha_{a,z}) \\
\text{iff } & x \bigcirc y = z \\
\text{iff } & \theta(x) \bigcirc \theta(y) = \theta(z) \\
\text{iff } & \varphi(\theta(x)) \oplus' \varphi(\theta(y)) = \varphi(\theta(z)) \\
\text{iff } & \alpha_{a,\theta(x)} \oplus' \alpha_{a,\theta(y)} = \alpha_{a,\theta(z)} \\
\text{iff } & (\theta * \alpha_{a,x}) \oplus' (\theta * \alpha_{a,y}) = \theta * \alpha_{a,z} \\
\text{iff } & (\theta * \gamma) \oplus' (\theta * \delta) = \theta * \eta\,.
\end{aligned}$$

In summary, the above shows that for all γ, δ, and θ in $\mathbb{N}$

$$\theta(\gamma \oplus' \delta) = (\theta * \gamma) \oplus' (\theta * \delta), \tag{7.6}$$

establishing that $\langle \mathbb{N}, \succsim', \oplus', \iota\rangle$ is a positive continuum field.

Let $\oplus$ be as in Statement 1. Then by Statement 1, $\mathfrak{F} = \langle \mathbb{N}, \succsim', \oplus, *, \iota\rangle$ is a continuum field. Therefore by Theorem 7.3.2, let f be an isomorphism of $\mathfrak{F}$ onto $\mathfrak{R} = \langle \mathbb{R}^+, \geq, +, \cdot, 1\rangle$. Let $+_1 = f(\oplus')$. Then by isomorphism, for each u, v, and t in $\mathbb{R}^+$,

$$t.(u +_1 v) = (t \cdot u) +1\ (t \cdot v)$$

if and only if

$$f^{-1}(t) * (f^{-1}(u) \oplus' f^{-1}(v)) = (f^{-1}(t) * f^{-1}(u)) \oplus' (f^{-1}(t) * f^{-1}(v)))\,,$$

and the latter equation follows by Equation 7.6. Therefore by Theorem 2.7 of Narens (1981a), $+_1$ is of the form

$$x +_1 y = (x^r + y^r)^{\frac{1}{r}}\,, \tag{7.7}$$

where $r \in \mathbb{R}^+$. By hypothesis,

$$\beta = \iota \oplus \iota \text{ and } \beta(x) = x \bigcirc x \text{ for all } x \text{ in } A.$$

Thus by the definition of $\alpha_{a,x}$ and Equation 7.5

$$\begin{aligned}
\iota \oplus' \iota &= \varphi(a) \bigcirc \varphi(a) = \varphi(a \bigcirc a) = \varphi(\beta(a)) \\
&= \alpha_{a,\beta(a)} = \beta * \alpha_{a,a} = \beta * \iota = \beta \\
&= \iota \oplus \iota\,.
\end{aligned}$$

Therefore by isomorphism and Equation 7.7,

$$1 + 1 = 1 +_1 1 = (1^r + 1^r)^{\frac{1}{r}}\,,$$

and thus $r=1$. Therefore by Equation 7.7, $+ = +_1$, which by f^{-1} yields $\oplus = \oplus'$.

Lemma 7.13.1 For all positive reals $r>s>2$, there exist positive integers k and m such that $r^k>2^m>s^k$.

Proof. Let r and s be positive reals such that $r>s>2$. Then $\log r> \log s>\log 2$. Choose $k \in \mathbb{I}^+$ such that

$$k(\log r-\log s)>\log 2 .$$

Then a positive integer m can be found such that

$$k \log r>m \log 2>k \log s,$$

that is,

$$\log r^k >\log 2_m>\log s^k$$

that is,

$$r^k>2^m>s^k .$$

Theorem 7.13.7 *(Statement 10 of Theorem 7.7.3)* Assume Axioms 7.7.1 to 7.7.4 and axiom system $Q_\omega(a)$. Let S and $\succeq'$ be as in Convention 7.7.1, and let $\mathbb{N}=G_b$. Assume a standard model of ZFA. Let φ be an arbitrary element of S, and by Statement 1 of Theorem 7.7.3, let u in X be such that

$$\varphi(u)=\iota .$$

Then for each p in N and each x in X, αp is qualitative and

$$x = \alpha p(u) \text{ iff } \varphi(x) = \alpha p$$
$$\text{iff } \varphi(x) = \overbrace{\iota \oplus \iota \oplus \cdots \oplus \iota}^{p \text{ concatenations}} .$$

Proof. For each p in N and each x in X, αp is qualitative by Definition 7.7.5 and the hypothesis $Q_\omega(a)$, and it follows from Definition 7.6.1 that

$$x=\alpha p(u) \text{ iff } \varphi(x)=\alpha p .$$

Thus it needs to be only shown that

$$\alpha p = \overbrace{\iota \oplus \iota \oplus \cdots \oplus \iota}^{p \text{ concatenations}} .$$

By Statement 9 of Theorem 7.7.3,

$$\mathfrak{F} = \langle \mathbb{N}, \succeq', \oplus, *, \iota \rangle$$

is a positive continuum field such that

$$\alpha 2=\iota\oplus\iota. \tag{7.8}$$

For each positive integer m, let $m\iota$ be m $\oplus$-concatenations of ι, that is, $1\iota=\iota$, $2\iota=\iota\oplus\iota$, and so forth, and let

$$J=\{m\iota | m \text{ is a positive integer}\}.$$

Because $\mathfrak{F}$ is a positive concatenation field, it follows that for all positive integers m and n,

$$m\iota \succeq' n\iota \text{ iff } m \geq n.$$

Therefore, $\succeq'$ is a well-ordering on J.

It will next be shown that for each positive integer $\boldsymbol{p}$, $\boldsymbol{ap=p\iota}$ Suppose p is a positive integer such that $\boldsymbol{ap} \neq p\iota$. *A* contradiction will be shown. Without loss of generality, we may assume that

$$\boldsymbol{ap} \succeq' p\iota. \tag{7.9}$$

(The case where $p\iota \succeq' \alpha p$ follows by a similar argument.) $\alpha_1=1\iota$ by Statement 3 of Axiom 7.7.4, and $\alpha_2=2\iota$ by Equation 7.8. Thus $p>2$. It then follows from Axiom 7.7.3, Definitions 7.7.2 and 7.7.4, and Statement 4 of Axiom 7.7.4 and the fact that $\mathfrak{F}$ is a positive continuum field that

$$\boldsymbol{ap} \succeq' 2\iota \text{ and } p\iota \succeq' 2\iota. \tag{7.10}$$

By Theorem 7.3.2 and Convention 7.3.1, let f be an isomorphism of $\mathfrak{F}$ onto $\mathfrak{R} = \langle \mathbb{R}^+, \geq, +, \cdot, 1\rangle$. Then by $f(\iota)=1$, $f(\oplus)=+$, and Equations 7.9 and 7.10,

$$f(\alpha p)>f(p\iota)=P>f(2\iota)=2.$$

Thus by Lemma 7.13.1, let k and m be positive integers such that

$$f(\alpha p)^k>f(2\iota)^m=2^m>f(P\iota)^k=P^k\,. \tag{7.11}$$

For each α in $\mathbb{N}$ and each positive integer n, define α^n inductively as follows: $\alpha^1=\alpha$, and $\alpha^{n+1}=\alpha^n*\alpha$. It follows from Statement 6 of Axiom 7.7.4 and the fact that $\alpha_2=2\iota$ that

$$\alpha_{\boldsymbol{p}^k} = \alpha_{\boldsymbol{p}}^k \text{ and } \alpha_{\boldsymbol{2}^m} = \alpha_{\boldsymbol{2}}^m = (2\iota)^m\,.$$

Thus by Equation 7.11 and the isomorphism f^{-1},

$$\alpha_{\boldsymbol{p}^k} = \alpha_{\boldsymbol{p}}^k \succ' (2\iota)^m = \alpha_{\boldsymbol{2}}^m = \alpha_{\boldsymbol{2}^m}\,,$$

that is, $\alpha_p k \succeq' \alpha_2 m$, which by Statement 4 of Axiom 7.7.4 yields,

$$p^k>2^m. \tag{7.12}$$

However, it follows from Equation 7.11 that

$$2^m>p^k,$$

which contradicts Equation 7.12.

Theorem 7.13.8 (*Theorem* 7.7.5) Let X, $\{\alpha p \mid p \in N\}$, $\succsim$, $\mathbb{N}$, $\succsim'$, $\oplus$, and S be as in Theorem 7.7.3, and let $\succsim_1$, $\bigcirc$, $\mathbb{N}_1$, $\succsim_1'$, $\oplus_1$ and S_1 be as in Theorem 7.7.4. Suppose the following two psychophysical axioms:

(i) $=\succsim\succsim 1$.
(ii) For all p in N and all x and y in X,

$$\alpha p(x \bigcirc y) = \alpha p(x) \bigcirc \alpha p(y)\,.$$

Then the following two statements are true:

1. $\mathbb{N} = \mathbb{N}_1$, $\succsim\, = \,\succsim_1'$ and $S{=}S_1$.
2. There is an automorphism θ of the structure $\langle \mathbb{N}, \succsim', * \rangle$ such that θ $(\oplus){=}\oplus 1$.

Proof. 1. $\mathbb{N} = \mathbb{N}_1$ follows from Statement 3 of Theorem 6.4.5. Because $\mathbb{N} = \mathbb{N}_1$, it follows from psychophysical axiom *(i)* that $\succsim'{=}\succsim'$. Because $\mathbb{N} = \mathbb{N}_1$, it follows from the definition of "canonical scale" (Definition 7.6.1 that $S{=}S_1$. Thus Statement 1 has been shown.

Let f be an isomorphism of $\langle \mathbb{N}, \succsim', * \rangle$ onto $\langle \mathbb{R}^+, \geq, \cdot \rangle$. Let $+'{=}f(\oplus)$ and $+'1{=}f(\oplus_1)$. By Statement 9 of Theorem 7.7.3,

$$\alpha * (\gamma \oplus \delta) = (\alpha * \gamma) \oplus (\alpha * \delta)$$

for each α in $\mathbb{N}$. Similarly, by $\mathbb{N} = \mathbb{N}_1$ and Statement 8 Theorem 7.7.4,

$$\alpha * (\gamma \oplus_1 \delta) = (\alpha * \gamma) \oplus_1 (\alpha * \delta),$$

for each α in $\mathbb{N}$. Then by the isomorphism f, $\langle \mathbb{R}^+, \geq, +' \rangle$ and $\langle \mathbb{R}^+, \geq, +_1' \rangle$ are continuous extensive structures, and

$$r \cdot (u +' v) = (r \cdot u) +' (r \cdot v)$$

for all r, u, and v in $\mathbb{R}^+$ and

$$r \cdot (u +_1' v) = (r \cdot u) +_1' (r \cdot v)$$

for all r, u, and v in $\mathbb{R}^+$. It then follows from Theorem 5.10.11 that there exist s and t in $\mathbb{R}^+$ such that for all v and v in $\mathbb{R}^+$,

$$u +' v = (u^s + v^s)^{\frac{1}{s}} \text{ and } u +_1' v = (u^t + v^t)^{\frac{1}{t}}\,.$$

From this it follows that ω in $\mathbb{R}^+$ can be found such that

$$(u^w +_1' v^w)^{\frac{1}{w}} = u +' v\,.$$

Let T be the function from $\mathbb{R}^+$ to $\mathbb{R}^+$ such that for all u in $\mathbb{R}^+$, $T(u){=}u^{\omega}$. Then T is a one-to-one function $\mathbb{R}^+$ from onto itself. Consider $+'$ and $+_1'$ as sets of

3-ary relations, that is, $(u, v, z) \in +'$ if and only if $u +' v = z$, and similarly for $+'_1$. Then for all u, v and z in $\mathbb{R}^+$,

$$(u, v, z) \in +' \text{ iff } (T(u), T(v), T(z)) \in +'_1$$

that is, $T(+') = +'_1$. T is also an automorphism of $\langle \mathbb{R}^+, \geq, \cdot \rangle$. Thus by the isomorphism f^{-1}, $\theta = f^{-1}(T)$ is an automorphism of $\langle \mathbb{N}, \succeq', * \rangle$ and $\theta(\oplus) = \oplus_1$.

8. Meanin gfulness and the Axiom of Choice

8.1 Introduction

The axiom of Choice has generated more controversy than any axiom in the history of mathematics. It was formally introduced and used by Zermelo (1904) to prove that every set could be well-ordered—that is, to prove that for each set a there exists a one-to-one function from a onto an ordinal. It met stiff and great resistance by many—if not most—of the outstanding mathematicians and philosophers of mathematics of the time, and debates about its acceptability as a valid mathematical principle raged widely throughout the mathematical journals. The resulting controversies generated some long lasting issues about the nature of mathematical entities that to this day have not been satisfactorily resolved. As the axiom became increasingly crucial for the types of arguments needed in the development of twentieth century topology and analysis, it became much more accepted among mathematicians, especially in light of its producing no outright logical contradictions despite its widespread and varied use. Along with its new acceptability came quiescence about the more serious philosophical reservations and objections that were raised earlier about its validity. The axiom finally became so commonplace that it achieved the status of an unquestioned, valid mathematical principle, and today is treated as such by almost all mathematical textbooks.

The debates about the foundations of mathematics that took place at the beginning of the twentieth century were about nothing less than the nature and form that modern mathematics should take. The individuals involved were of the highest caliber and prestige. Curiously enough, by contemporary standards many of their arguments appear somewhat weak and out of place. In my opinion, a good part of this is due to the fact that these mathematicians and philosophers did not have an appropriate language for formulating the kinds of distinctions they needed to rigorously present their views.

In rereading the debates, I was struck by the close similarity of some of the more prominent positions and the formal theories of meaningfulness presented in Chapter 4. In this chapter, some of these similarities are explored.

Much of the historical material presented here comes from Moore (1982), an excellent source of information about the axiom of Choice.

8.2 The Axiom of Choice

The axiom of Choice says that given a set *a* of nonempty sets another set *b* can be formed consisting of an element from each set in *a*. This is essentially Zermelo's 1904 formulation of the axiom. As a mathematical principle, it—or various forms of it—was used without comment in a wide and varied number of mathematical arguments prior to Zermelo's explicit formulation. Zermelo realized that an assumption like it was needed to show the principle that every set could be well-ordered—a principle whose validity was doubted by a great number of mathematicians of the time. Because of its use in obtaining the well-ordering result, the axiom of Choice naturally drew a great deal of attention and scrutiny. Zermelo, Cantor, and several others interested in the development of infinitistic mathematics saw the axiom as an intuitively clear, valid mathematical principle. Others were much more doubtful. Their objections were generally similar to the following concerns raised earlier by Peano and Bettazzi.

Before Zermelo, the Italian mathematician G.Peano recognized the utility of principles like the axiom of Choice, but rejected them as valid: Moore (1982) writes,

> In 1886 Peano published a new demonstration of the theorem, due to Cauchy, that the differential equation $y'=f(x, y)$ has a unique solution. Here Peano weakened Cauchy's hypotheses to require only that $f(x, y)$ be continuous. Four years later Peano returned to this theorem and generalized his proof to finite systems of first-order equations. When he arrived at a step that required a single element to be chosen from each set in a certain sequence $A_1, \ldots, A_i, \ldots$ of subsets of $\mathbb{R}$, he remarked carefully: "But since one cannot apply infinitely many times an *arbitrary* rule by which one assigns to a class A an individual of this class, a *determinate* rule is stated here, by which, under suitable hypotheses, one assigns to each class A a member of the class." To obtain his rule, he employed least upper bounds. Thereby he became the first mathematician who, while accepting infinite classes, categorically rejected the use of infinitely many arbitrary choices. Even though he was familiar with Cantor's researches, apparently it did not occur to him, or to anyone else at the time, that Cantor had often used such arbitrary choices. After Zermelo's proof appeared, Peano's suspicions were aroused, and he vigorously criticized the Axiom of Choice as well as earlier results depending on it implicitly. (Moore, 1982, pg. 76)

The axiom was also rejected by another Italian mathematician prior to Zermelo's formulation: Bettazzi (1892) writes:

> A point can be taken arbitrarily from a given set of points [in $\mathbb{R}$], or from one of its subsets, or from a finite number of its subsets. But when one has

> to consider infinitely many of its subsets and to construct a subset formed by choosing in each of these subsets any point whatever (as will be the case in what follows), it does not suffice to say that one forms this set by taking a point *arbitrarily* in each of these subsets. For one cannot regard as determinate an *infinite* number of objects all chosen arbitrarily in given classes. This follows clearly when one notes that giving them arbitrarily is equivalent to defining them separately one at a time… (Bettazzi, 1892, pg. 176, quoted in Moore 1982, pg. 77.)

What Peano, Bettazzi, and others objected to was the lack of a specific rule that defined the choice set in the axiom of Choice. Of course it was exactly because of such a lack that the axiom of Choice had to be stated as a specific axiom, for otherwise it could be deduced from principles like the axiom of Comprehension.

Peano's and Bettazzi's objections have obvious analogies in the axiomatic systems of meaningfulness of Chapter 4. Peano and Bettazzi are describing what can properly be said to exist. If we were to substitute "meaningfully exists" (i.e., "exists as a meaningful entity") for "exists", then their remarks about the axiom of Choice essentially become this: Given a meaningful set a of nonempty sets, one can form a set b consisting of an element of each set in a, but although b exists it does not necessarily meaningfully exist. I am not sure what Peano and Bettazzi would have made of this analogy. However, other critics of the axiom seem to have been struggling to try to get across an idea like it, but were only marginally successful because they did not have sufficiently precise ways of formulating their ideas.

There was a famous exchange of letters by the French mathematicians J-S. Hadamard (1865–1963), E.Borel (1871–1956), R-L.Baire (1874–1932), and H-L.Lebesgue (1875–1941) about the axiom of Choice and Zermelo's proof of the well-ordering theorem. These letters were published in the *Société Mathématique de France* (1905), and English translations of them are available in Appendix 1 of Moore (1982). In these letters, the writers are highly critical of Zermelo's axiom of Choice. The least vehement are Hadamard and Lebesgue, who adopted views that—as we will see—make rather good sense from a meaningfulness perspective.

This is how the letters were generated:

E.Borel was invited by Hilbert to write a brief article for the journal *Mathematische Annalen* about Zermelo's proof of the well-ordering of the real numbers. Borel accepted and wrote an article that basically raised the same kind of objections that Peano and Bettazzi made earlier. He found particularly objectionable Zermelo's use of uncountably many arbitrary choices, a principle he considered to be "outside mathematics."

Hadamard read Borel's article and sent him a dissenting opinion. Moore (1981) writes the following about Hadamard's letter:

> After he read Borel's article in *Mathematische Annalen,* Hadamard wrote Borel a dissenting letter. First of all, Hadamard distinguished Zermelo's proof sharply from reasoning which required an infinite number of successive choices, each of which depended on those made previously. Zermelo's proof was acceptable, he emphasized, precisely because the choices were *independent* of each other...
>
> Was it possible, Hadamard inquired, to make such independent choices effectively, that is, in a way that someone could actually perform? Certainly Zermelo had given no method for doing so, and it seemed unlikely that anyone could provide one. What Zermelo had done was to state an existence proof. The essential distinction, which Hadamard credited to Jules Tannery, was between establishing:
> *(i)* that a function exists and
> *(ii)* that it can be specified uniquely.
> Hadamard added that many mathematical questions would have a completely different meaning, and different solutions, if *(i)* were replaced by *(ii)*. Furthermore, he continued, even Borel had used functions which he proved to exist but which could not be defined uniquely, especially in certain theorems on the convergence of complex series. As for the notion of unique definability, it was, to borrow Borel's phrase, "outside mathematics" since it belong to the psychology of the human mind.

Borel then sent Hadamard's letter to R.Baire for comment. Baire agreed with Borel, but went further towards constructiveness by regarding any infinity as a potential infinity. He believed that Zermelo's result was consistent but meaningless:

> we do not perceive a contradiction in supposing that, in each set which is defined for us, the elements are positionally related to each other in exactly the same way as the elements of a well-ordered set. In order to say, then, that one has established that every set can be put in the form of a well-ordered set, the meaning of these words must be extended in an extraordinary way and, I would add, a fallacious one. (Baire et al, 1905, pg. 264; also in Moore, 1982, Appendix 1.)

Borel next requested Lebesgue's opinion on these matters. Lebesgue's reply revolved around the answer to the following question: "Can one prove the existence of a mathematical object without defining it." Moore (1982) summarizes Lebesgue's ideas about this question as follows:

> In particular, Lebesgue inquired whether an existence proof is legitimate if it does not specify uniquely an object of the type purported to exist. While he recognized that it was a matter of convention whether one restricted existence proofs in this way, and admitted that he himself had deviated at times from such usage, he remained convinced that one could prove

> the existence of a mathematical object only by defining it uniquely. What Lebesgue rejected, in other words, were proofs that show the existence of a non-empty class of objects of a certain kind rather than a specific object of that kind.

Later in 1905, Lebesgue wrote a monograph (not published until 1971) that made more explicit his position stated above. He divided mathematicians into two camps, the Idealists and the Empiricists: The Empiricists admitted functions that could only be uniquely defined, while the Idealist admitted other kinds of functions as well. This divergence also extended to mathematical proofs: Moore (1982) writes,

> The proof that every continuous function is integrable was, for the Empiricist, only "a form devoid of meaning" but which acquired meaning when one restated the argument for a specific function. On the other hand such a general proof was
>
> > the definitive and complete line of reasoning for the Idealist, since for him a function is [uniquely] determined when he affirms it is... When an Idealist wants to determine a function, he does not seek a characteristic property which would permit him, as well as others, to be sure of always thinking of the same function; [rather] he contents himself with saying that he chooses this function...; he affirms...that he is always thinking of the same function. This affirmation, which the Idealist recognizes and declares unverifiable, appears meaningless to the Empiricist who places these [functions] peculiar to the Idealist outside of mathematics. (Lebesgue, 1971, pg. 39).
>
> Thus Lebesgue made definability the touchstone for his Empiricist philosophy of mathematics.
>
> ... Nevertheless, as in his letter, Lebesgue remained more ambivalent than his fellow Empiricists [Baire and Borel]. He recognized that some eminent mathematicians were Idealists and even that the Idealist position might have practical consequences some day. Furthermore, in the past those who wished to extend the concept of function had always been in the right, and perhaps the same would hold true for the extension proposed by the Idealists. In any case, the matter would probably be resolved not by theoretical arguments but by the degree to which the Idealist's functions proved to be useful. (Moore 1982, pg. 100)

Hadamard and Lebesgue had similar views, but since Lebesgue's was more detailed and better articulated, it will be the primary focus of the discussion that follows.

Lebesgue talks about two approaches to the existence of mathematical objects. I find it preferable to talk instead about one kind of existence and two kinds of objects. The shift from two kinds of existences to a single kind of existence and two kinds of objects avoids certain logical difficulties and makes the analysis of Lebesgue's position easier. In addition, this shift neither constricts the kinds of distinctions that Lebesgue wanted to make nor alters the force of his arguments.

Using the above shift in terminology, Lebesgue's position can be reformulated as follows:

> An object exists Empirically if and only if a mathematical expression can be found that uniquely specifies it. An object exists Ideally if and only if either it exists Empirically or exists by some process that uses non-Empirical principles such as the axiom of Choice.

When stated this way, Lebesgue's view looks very similar to a definitional concept of meaningfulness. The following illustrates its similarity with the axiomatic systems of meaningfulness of Chapter 4:

Let $E(x)$ stand for "x is Empirical." Let us restrict our attention to the mathematical field of analysis, Lebesgue's specialty. From other of his writings, it is reasonable to believe that Lebesgue would consider an entity of analysis to be Empirical if and only if it is definable in terms of the real numbers and its arithmetical operations of addition and multiplication. Because Lebesgue probably would not want to identify real numbers with pure sets, for this example let us let the set of atoms A be the set of real numbers. Lebesgue at times worried about irrational real numbers, because not all of them are explicitly definable in terms of the set of reals and the operations of addition and multiplication, but Lebesque did not wish to eliminate them, because these nonspecifiable real numbers arose naturally in so many questions of analysis. (Lebesgue 1971, pp. 37–39). So let us consider all individual real numbers as meaningful. Then the set of primitives T for axiomatizing E is given by

$$T = A \cup \{\geq, +, \cdot\} \cup \{A\} .$$

Now let us take "definable from T" to be axiom *MC'*. This probably includes more than Lebesgue's constructiveness tendencies would like, but it does not include the axiom of Choice, the principle that Lebesgue most wanted to avoid. Axiom *AL* is immediately fulfilled in this scheme, because each element of A is Empirical. So that the Idealist's position can also be incorporated, axiom system ZFA is assumed. Axiom *MP* presents a problem. One might want to argue that even though the existence of certain pure sets can only be established through applications of the axiom of Choice, they nevertheless can be taken to be Empirical, because they function in the present context more like logical entities than as entities of analysis. However, this cannot be done while maintaining the kind of distinction between Empirical and Ideal (non-Empirical) Lebesgue wanted: If *MP* were assumed, then with the other assumptions it would follow from Theorem 4.3.6 that all entities would be Empirical, since it can be shown that the structure $\langle A, \geq, +, \cdot \rangle$ has the identity as it only automorphism. Thus axiom *MP* has to be either modified or dropped. Let us drop it. Then pure sets definable from the primitives are Empirical. These include the set of finite ordinals as well as each individual finite ordinal. Finally, let's declare a set to be non-Empirical if and only if it is not definable from the primitives. Then the predicate *E(x)* is defined for all entities *x*.

It seems to me that the above concept of Empiricalness is probably very close to what Lebesgue intended. There is however a consequence that Lebesgue and some others would have found objectionable, but one that I believe is inevitable if the Empiricist/Idealist distinction is going to be rigorously developed:

Consider the set of real numbers. It is Empirical. Let τ be the least ordinal that is equipollent with the real numbers if such an ordinal exists; otherwise let $\tau=\emptyset$. τ is Empirical because it is definable in terms of the real numbers. Let *b* be the set of all one-to-one functions from the real numbers onto τ. *b* is Empirical because it is definable in terms of Empirical sets. *b* is nonempty by the axiom of Choice. Its elements are paradigms of what Lebesgue believed to be non-Empirical. He and many other critics of the axiom saw no problem with a single arbitrary choice from a nonempty set. (It was an infinity of such choices that they objected to.) That is, a single arbitrary selection from an Empirical set should not (as I understand the debate) produce a nonEmpirical object. This is clearly inconsistent with *b* being Empirical and its elements being non-Empirical. Thus to maintain the Empiricist/Idealist distinction, the principle of a single arbitrary selection from a nonempty set would have to be modified.

The need for such a modification in arguments involving the axiom of Choice was suggested in 1904 by F.Bernstein: Moore (1982) writes,

> … Bernstein introduced his concept of many-valued equivalence. He termed two sets *A* and *B* "many-valued equivalent" if there exists a non-empty family *C* of bijections from *A* onto *B* such that no member of *C* is "distinguished." Apparently he meant that no member of *C* could be uniquely defined, and hence that *C* was infinite. If a theorem asserted that $\underline{A}$ and *B* were equipollent, then the theorem was said to have "multiplicity $\overline{\overline{C}}$. The logical significance of such a theorem was less, he granted, than that of a theorem where a member of *C* could be uniquely specified. In particular, his theorem that the set of all closed subsets of $\mathbb{R}$ has the power of the continuum had multiplicity $2^{\aleph_0}$. Recognizing that a theorem's multiplicity depended on present knowledge, he concluded with the hope that future research would reduce it as far a possible [Bernstein, 1904, pg. 558]. Except for W.H. and G.C.Young, then at Göttingen, Bernstein's notion of many-valued equivalence found no adherents. *(Moore, 1982, pg. 109.)*

(Note the similarity of Bernstein's ideas to the theory of meaningful cardinals presented in Section 5 of Chapter 4.)

8.3 Lebesgue's Measure Problem

One of the main objections to the axiom of Choice was that it sometimes produced counter-intuitive results. In this section and the next the most famous of these, the existence of non-Lebesgue measurable sets and the paradoxes of Hausdorff and of Banach and Tarski, are discussed.

The following concepts are fundamental for the developments of this section:

Definition 8.3.1 A set of sets Γ is called a *ring* if and only if for all a and b in Γ, $a \cup b$ and $a\text{-}b$ are in Γ.

Suppose Γ is a ring. Then m is said to be a *(finitely additive) measure* on Γ if and only if m is a function from Γ into the nonnegative real numbers and for all a and b in Γ, if $a \cap b \neq \emptyset$, then $m(a \cup b)=m(a)+m(b)$. m is said to be a *σ-measure* on Γ if and only if for all countable sequences of elements of Γ, $a_1,\ldots, a_n,\ldots$, if $a_i \cap a_j = \emptyset$ for all $i \neq j$, then $\bigcup_{i=1}^{\infty} a_i$ is in Γ and

$$m\left(\bigcup_{n=1}^{\infty} a_n\right) = \sum_{n=1}^{\infty} m(a_n) .$$

Lebesgue (1902), in a famous and far reaching paper on the theory of integration, formulated the following problem, which subsequently became known as **Lebesgue's measure problem:** *Does there exists a σ-measure on the subsets of points of Euclidean p-space such that the measure is positive for some bounded subset and such that it assigns the same number to congruent subsets?*

(For a metric space with metric d, two sets a and b are *congruent* if and only if there exists a one-to-one function f from a onto b such that for all x and y in a, $d(x, y)=d[f(x), f(y)]$. For bounded subsets in Euclidean p-space, this formulation of "congruent" is equivalent to the following: a and b are congruent if and only if there is an euclidean motion that takes a into b.)

In attempting to solve his measure problem, Lebesgue defined a σ-measure and a ring of subsets that today has become known as "Lebesgue measure" and "Lebesgue measurable sets":

Definition 8.3.2 Let E^p denote p dimensional Euclidean space. E^p is later considered as a qualitative entity with its points being the set of atoms A. However, for this lengthy definition, it is expedient to disregard qualitativequantitative distinctions so that the important ideas can be stated succinctly. This also allows for easy restatements of many known results. So, for the purposes of this definition, E^p is considered as a real vector space of dimension p with the usual Euclidean metric for describing distances. Thus the "points" of E^p are represented as ordered p-tuples of real numbers.

An *interval* in E^p is either a set of points $x=(x_1,\ldots,x_p)$ of E^p such that there exists real numbers r_i, and s_i, $i=1,\ldots, p$, and

$$r_i \le x_i \le s_i \tag{8.1}$$

or a set of points that is similarly characterized by Equation 8.1 with any or all of the relations $\le$ replaced by $<$. Note that the possibility of $r_i = s_i$ for some i is not ruled out and that the empty set is an interval.

An *elementary set* is a finite union of intervals.

Throughout this definition ε denotes the set of elementary sets, and throughout this definition μ denotes the following function on ε: If a is an interval (as defined through Equation 8.1 with perhaps one or more replacements of $\le$ by $<$), then by definition,

$$\mu(a) = (s_1 - r_1) \cdot (s_2 - r_2) \cdots (s_p - r_p) \ ,$$

and if $b = \bigcup_{i=1}^{k} a_i$, where $a_1, \ldots, a_k$ are finitely many pairwise disjoint inter vals, then by definition

$$\mu(b) = \mu(a_1) + \cdots + \mu(a_k) \ . \tag{8.2}$$

The following are easy to establish: *(i)* ε is a ring; *(ii)* if c is in ε, then c is a union of a finite number of disjoint intervals; *(iii)* if b is in ε, then $\mu(b)$ is well-defined by Equation 8.2 in the sense that if two different decompositions of b into disjoint intervals are used, then both give $\mu(b)$ the same value; and *(iv)* μ is a measure on ε.

Let b be an arbitrary bounded subset of E^p. Then, by definition, a *σ-cover C* of b is a countable set of bounded sets of E^p such that $b \subseteq \bigcup C$. μ^*, called the *outer measure based on* μ, is defined at b as follows:

$$\mu^*(b) = \text{g.l.b.} \sum_i \mu(c_i) \ ,$$

where the c_i are elements of a σ-cover C of b such that each element of C is an open elementary set, and where the g.l.b. is taken over all such C.

It is easy to show that for all a in ε, $\mu^*(a) = \mu(a)$. Thus μ^* extends μ as a function.

For all bounded subsets a and b of E^P, let

$$d(a, b) = \mu^*[(a - b) \cup (b - a)] \ .$$

A bounded subset a of E^p is said to be *finitely μ-measurable* if and only if there exists a denumerable sequence a_k of elementary sets such that

$$\lim_{k \to \infty} d(a, a_k) = 0 \ .$$

A bounded subset b is said to be *Lebesgue measurable* if and only if it is a countable union of finitely μ-measurable bounded subsets of E^p.

It can be shown that *(i)* the Lebesgue measurable, bounded subsets of E^p form a ring; *(ii)* μ^* is a σ-measure on this ring; and *(iii)* for all congruent sets a and b in this ring, $\mu^*(a) = \mu^*(b)$.

μ^* restricted to the ring B of Lebesgue measurable bounded subsets of E^p is called *Lebesgue measure* on B.

Lebesgue did not know whether his measure solved his measure problem; that is, he did not know if every bounded subset of E^p was Lebesgue measurable. In 1905, G.Vitali showed through the use of the axiom of Choice that Lebesgue's measure problem had no solution, and so in particular that there existed bounded subsets of E^p that were not Lebesgue measurable.

8.4 Hausdorff's Measure Problem

Many mathematicians and philosophers during the early 1900's believed that the axiom of Choice would ultimately lead to inconsistencies, and some saw Zermelo's theorem about the existence of well-orderings of the real numbers and Vitali's theorem about the existence of non-lebesgue measurable sets as harbingers of future contradictions. In 1914, F.Hausdorff (1869–1942) utilized the axiom Choice to produce a result that seemed so counter-intuitive that some prominent mathematicians considered it sufficient grounds for the total rejection of the axiom. This theorem, known today as *Hausdorff's Paradox,* was the result of Hausdorff 's investigation of a weakened form of Lebesgue's Measure Problem, which today is often called **Hausdorff's Measure Problem.** It is formulated as follows: *Does there exist a measure on the bounded subsets of E^p such that the measure of the p-dimensional unit cube is 1, and such that bounded congruent subsets are assigned by the measure the same number?* Note that the critical difference between Hausdorff's and Lebesgue's versions of the measure problem is that Hausdorff weakens σ-additivity of the measure to finite additivity.

Hausdorff was not able to solve his problem for Euclidean 1 and 2dimensional spaces. However, for Euclidean p-dimensional space with $p \geq 3$, he showed that no measure existed that satisfied the conditions of his problem. This was done by showing that one half of a sphere was congruent to one third of the same sphere, or more precisely, that the sphere could be decomposed into four (disjoint) sets *A, B, C,* and *D,* where *A, B, C,* and *B»C* were congruent and *D* was denumerable. Moore (1982) writes the following about the reception of this "paradox":

> The first to respond to Hausdorff's paradox—indeed, the first to characterize it as a paradox—was Borel, who felt quite certain that the culprit was the Axiom of Choice. In the second edition of his *Leçons sur la théorie des fonctions* [1914], Borel concluded his exposition of the paradox with a polemic against the Axiom:
>
> If, then, we designate by *a, b, c* the probability that a point in *S* belongs to *A, B,* or *C* respectively and if we grant that the probability of a point belonging to a set *E* is not changed by a rotation around a diameter (this

> is what Lebesgue expresses by saying that two congruent sets have the same measure), one obtains the contradictory equalities: $a+b+c=1$, $a=b$, $a=c$, $a=b+c$.
> The contradiction has its origin in the application…of Zermelo's *Axiom of Choice*. The set *A* is homogeneous on the sphere; but it is at the same time a half and a third of it … The paradox results from the fact that *A is not defined,* in the logical and precise sense of the word *defined.* If one scorns precision and logic, one arrives at contradictions. [Borel 1914, 255–256.]

S.Banach (1892–1945) and A.Tarski (1902–1983) provided in 1924 a deeper analysis of Hausdorff's Measure Problem. Two sets *a* and *b* in a metric space are said to be *equivalent by a finite decomposition* if and only if there exists an integer *k* such that *a* and *b* can be respectively partitioned into subsets $a_1,\ldots, a_k$ and $b_1,\ldots, b_k$, where for $i=1,\ldots, k$, a_i is congruent to b_i. Banach and Tarski showed that any two bounded subsets of E^p, $p \geq 3$, with nonempty interiors are equivalent by a finite decomposition. A consequence of this, frequently cited in popular expositions, is that a sphere of radius *r* can be decomposed into a finite number of pieces and reassembled into two spheres of radius *r*. Banach and Tarski also showed the existence of measures for the Euclidean line and plane that satisfied the conditions of Hausdorff's Measure Problem.

A still deeper analysis of Hausdorff 's Measure Problem was provided by J.von Neumann (1903–1957) in 1929, where he generalized the problem as follows: *m* is called a *(S, H)-measure* if and only if the following four conditions hold: *(i)* *S* is a nonempty set and *H* is a subgroup (under function composition) of one-to-one functions from *S* onto itself; *(ii)* *m* is a (finitely additive) measure on the ring of all subsets of *S; (iii)* $m(a)=1$ for some subset *a* of *S*; and *(iv)* for all *f* in *H* and all subsets *b* of *S*, $m(b)=m[\{f(x)|x\in b\}]$. Von Neumann generalized Hausdorff's problem to the following: (**von Neumann's Measure Problem**): *When does a (S,H)-measure exist?* He showed that if *H* had a free subgroup with two generators—which is always the case for euclidean motions in Euclidean *p*-space with $p\geq 3$—then a *(S, H)*-measure did not exist. He also showed that for the case where *S* is the set of real numbers *H* can be selected so that no *(S, H)*-measure exists.

8.5 Results by Banach and Ulam

Banach (1929) produced a different sort of answer to Lebesgue's measure problem: He showed, assuming the continuum hypothesis, that no σ-measure exists on the subsets of Euclidean p-space ($p\geq 1$) such that the measure of the unit cube is 1. Later in 1939 S.M.Ulam (1909–1984) produced a similar result using a condition that can be viewed as a very weakened form of the continuum hypothesis. These latter results, which assume the stronger condition of σ-additivity, do not assume that congruent sets are assigned the same number by the measure.

8.6 Discussion

σ-additivity is a very desirable and useful condition. However, one can ask—and in fact should ask—is it a reasonable, intuitively plausible condition for the concept of area or volume? It seems to me that it is precisely the kind of principle that one would prefer to derive from more basic and intuitively plausible axioms rather than take as an axiom. Unfortunately, this is not done in mathematics, where σ-additivity is taken as a defining property of two very important concepts: volume and probability. Many mathematicians and philosophers recognized the great difficulty in justifying σ-additivity, and this might be part of the reason why Vitali's result—as striking as it was at the time—did not cause the same kind of consternation as Hausdorff 's.

σ-additivity is, mathematically, a very powerful condition, and the results by Banach and Ulam show how constraining it can be. Finite additivity is a much milder, less constraining concept: for example, by using the axiom of Choice, it can be shown that Lebesgue measure can be extended to a finitely additive measure defined on all bounded subsets of E^p. (Of course, it follows from Hausdorff 's result that for $p \geq 3$ such a measure cannot assign to every pair of congruent, bounded subsets the same number.)

σ-additivity is often invoked as a fundamental characteristic of probability spaces. In 1933 A.N.Kolmogorov (1903–1987) made it an essential ingredient in his influential axiomatization of probability, which has become the standard for mathematics and most of science. Kolmogorov only justified this principle in the weakest—and in my view, vaguest—terms. (More recent books on the subject have not, in my opinion, done much better). The objection to σ-additivity is not its usefulness in many mathematical and scientific applications, but its assumption as a *necessary* condition for the concept of probability. For example, it clearly rules out many probabilistic situations that are natural idealizations of finite processes, for example, those that are describable by algebras ε of subsets of a nonempty denumerable set X such that the singleton set $\{a\}$ is in ε for each a in X, and such that the "probabilities" of such singleton elements are 0.

The paradoxes of Hausdorff and of Banach and Tarski present a different sort of difficulty. Let p be an integer ≥ 3 and let A be the set of points of Euclidean p-space and G be the group of Euclidean p-motions on A. Let us assume axiom system TM. Then it is easy to show that the set B of bounded subsets of A is meaningful. Let m be a finitely additive measure on B such that $m(\{a\})=0$ for each a in A and such that $m(c)=1$ for some nondegenerate cube in B. (It is not difficult to show, using the axiom of Choice, that such a measure exists.) Then the Banach-Tarski paradox, rephrased slightly, says that m is not meaningful. Put this way the result is not so startling. After all, there are many nonmeaningful concepts, and why should not a very general volume concept that applies to all bounded subsets be one of them?

From the point of view of Transformational Meaningfulness, individual subsets of A (other than A and Ø) are not meaningful, but their orbits are. Such orbits are congruence classes, that is, any two elements in the orbit are congruent. One of the main ideas of the measure problems of Lebesgue and Hausdorff was that elements of such an orbit should be assigned the same number by the measure. If we identify "numbers" with particular pure sets, then Transformational Meaningfulness requires the same constraint. Also since by Transformational Meaningfulness all orbits of elements of B are meaningful, those that arise from Lebesgue measurable sets are for meaningfulness purposes indistinguishable from those that arise from non-Lebesgue measurable ones.

Since by the Banach-Tarski paradox no meaningful finitely additive measure exists on B, it is natural to look at those meaningful finitely additive measures that exist on subrings of subsets of B. Lebesgue measure is one such. It is a particularly attractive one, because all the sets of points that are (in a first-order way) defined from finitely many spheres and lines are in the domain of the measure. But from a transformational meaningfulness point of view, this does not make it any better than other meaningful measures that do not include sets definable (in a first-order way) from spheres and lines. This is because although intuitively spheres and lines might appear to be among the most basic of Euclidean sets of points, from the point of view of transformational meaningfulness they can be no better than any set of concepts that meaningfully specifies the group G of euclidean p-motions, and such a set may not include "the most basic Euclidean sets" while at the same time be a ring with a meaningful measure defined on it.

I personally view Lebesgue measurability to be too broad of a concept for sensible meaningfulness considerations. I believe Lebesgue really wanted the measure function to result from a "construction" out of the measures of intervals (which are constructive) and negligible sets (which are possibly "nonconstructive") that will have 0 as their measure. He failed to achieve this in his definition of measure, because he allowed for the measurability of arbitrary countable unions of measurable sets—including those whose existence could only be established through an application of the axiom of Choice.

Even though one may believe that Lebesgue measurability is not the right meaningfulness concept for "volume", one can still ask if it has the formal properties of a reasonable meaningfulness concept. I believe the answer to this depends on whether or not one wants the meaningfulness concept extended to measurable functions. Before going into details, it is useful to very briefly review some of the history of the concept of "function". For simplicity, the discussion is restricted to the case of E_1.

What is an allowable real-valued function has been highly controversial in the history of mathematics and has undergone many transformations:

> Although the notion of a function did not originate with Euler, it was he who first gave it prominence by treating the calculus as a formal theory of functions. In his *Introductio in analysin infinitorum* of 1748 he defined a function of a variable quantity as "an analytical expression" composed in any way of that variable and constants. The key to this definition is the notion of an analytical expression, which Euler evidently understood to be the common characteristic of all known functions. It was also Euler, however, who initiated a viewpoint that eventually led to the introduction of the modern concept of a function. In his pioneering study of partial differential equations of 1734, Euler admitted "arbitrary functions" into the integral solutions. And, in answer to Jean d'Alembert—who maintained that these arbitrary functions must be given by a single algebraic or transcendental equation in order to be the proper object of mathematical analysis—Euler clarified his earlier pronouncement by contending that the curves which the arbitrary functions represent need not be subject to any law but may be "irregular" and "discontinuous," i.e., formed from the parts of many curves or traced freehand in the plane.
>
> It is important to observe that the term "discontinuous" as used by Euler and his contemporaries refers to a discontinuity in the analytical form of expression of the functional relationship: A function can be continuous in the modern sense and "discontinuous" in the sense of Euler. On the other hand, the possibility of arbitrary functions which are discontinuous in the modern sense at more than a finite number of points in a finite interval does not appear to have been seriously considered by anyone at this time. Attention was focused upon the fact that arbitrary functions are not determined by a single equation rather than upon their properties as correspondences $x \to f(x)$ between real numbers. [Burkill, 1951, pg. 3]

In the early nineteenth century, J.B.Fourier (1768–1830) and others studied limit functions of series of functions, where each term of the series was a function on an interval that corresponded to a single equation. It turned out that in general such limit functions could not be identified piece-wise with single equations. A new concept of function was needed, and P.G.Dirichlet (1809–1859) provided, after a flurry of controversy, what has become the modern definition of "function" as a correspondence that associates with each real number x a unique real number y.

One of the fundamental concerns of nineteenth century mathematics was understanding which limits of trigonometric series could be properly integrated. The research on this issue ultimately culminated in the concepts of Lebesgue integral and Lebesgue measurable function.

(A function f from E^1 to E^1 is said to be *Lebesgue measurable* if and only if for each interval a in E^1, $\{x|f(x)\in a\}$ is a Lebesgue measurable subset of E^1. Lebesgue measurable functions have Lebesgue integrals and are, from many perspectives, quite well behaved. They are of fundamental importance to modern analysis.)

The following theorem shows that Lebesgue measurability is consistent with a reasonable meaningfulness concept, and the theorem that follows it shows that this consistency is necessarily shattered by the addition of the meaningfulness of Lebesgue measurable functions:

Theorem 8.6.1 Consider E^1 as a qualitative structure and let A be the set of points of E^1. Then there exists a subcollection M of V such that $\langle V, \in, A, \varnothing, M\rangle$ satisfies axiom system D′ and for each subset a of A, a is Lebesgue measurable if and only if $M(a)$.

Proof. Theorem 8.7.2.

Theorem 8.6.2 Consider E^l as a qualitative structure and let A be the set of points of E^1. Suppose M is a subcollection of V such that for each subset a of A, $M(a)$ if and only if a is Lebesgue measurable, and for each function f from A onto A, $M(f)$ if and only if f is a Lebesgue measurable function. Then axiom *MC'* is false about $\langle V, \in, A, \varnothing, M\rangle$.

Proof. Theorem 8.7.3.

8.7 Lebesgue Measurability and Meaningfulness

Definition 8.7.1 Assume ZFA. Suppose S is an algebra of subsets of A. Then p is said to be a *S-partition* of A if and only if p is a finite subset of S, $a\cap b=\emptyset$ for all a and b in p such that $a\neq b$, and $\bigcup p = A$.

Definition 8.7.2 Assume ZFA. Then $\langle V, \in, A, \varnothing, M\rangle$ is said to be *the model* $\mathfrak{B}$ *(S)* if and only if S is an algebra of subsets of A and M is defined as follows: For each S-partition p of A, let

$$G(p)=\{f|f\in \Pi \text{ and } f(x)=x \text{ for all } x \text{ in } p\}.$$

Then define M on V as follows: for all entities x,

> $M(x)$ if and only if there exists a S-partition p of A such that for all f in $G(p)$, $f(x)=x$.

Theorem 8.7.1 Assume ZFA. Suppose S is an algebra of subsets of A. Let $\langle V, \in, A, \varnothing, M\rangle$ be the model $\mathfrak{B}(S)$.Then the following two statements are true:

1. The model $\langle V, \in, A, \varnothing, M\rangle$ satisfies axiom system D′.
2. For each subset x of A, $M(x)$ if and only if $x \in S$.

Proof. 1. Let

$$H=\{G(p) \mid p \text{ is a } S\text{-partition of } A\} .$$

It will be shown that H is a transformational family for M (Definition 4.4.2). $H \neq \varnothing$, since $t=\{A, \varnothing\}$ is a S-partition of A and $G(t)=\Pi$. Let p and q be arbitrary S-partitions of A and

$$r=\{x \cap y \mid x \in p \text{ and } y \in q\} .$$

Then r is a S-partition of A Suppose f is an arbitrary element of $G(r)$ and x is an arbitrary element of p. It will be shown that $f(x)=x$. Let a be an arbitrary element of x. Since q is a S-partition of A, let y in q be such that $a \in x \cap y$. Then

$$f(a) \in f(x \cap y) \subseteq x,$$

that is, $f(a) \in x$. Since a is an arbitrary element of x, it has been shown that $f(x) \subseteq x$. Since $G(r)$ is a group, $f^{1} \in G(r)$, and by a similar argument, $f^{1}(x) \subseteq x$. Thus for each b *in* x, $f^{1}(b) \in x$, and thus $f\,[f^{1}(b)]=b$, that is, $f(x)=x$. Since f is an arbitrary element of $G(r)$, it has been shown that $I_{G(r)}(x)$'. And since x is an arbitrary element of p, it follows that $I_{G(r)}(x)$ is true for each x in p, that is, that $G(r) \subseteq G(p)$. Similarly, $G(r) \subseteq G(q)$. It then follows from Definitions 4.4.2 and 4.4.1 that H is a transformational family for M. Thus by Theorem 4.4.1, $\mathfrak{B}(S)$ satisfies axiom system D′.

2. Let x be an arbitrary element of 5. Let $t=\{x, A—x\}$. Then t is a S-partition of A. Thus by Definition 8.7.2, $I_{G(t)}(x)$ and therefore $M(x)$. Now suppose that y is an arbitrary subset of A such that $y \notin S$. It will be shown by contradiction that $\neg M(y)$, and this together with just the above will establish Statement 2. For suppose $M(y)$. $y \neq \varnothing$, since $\varnothing \in S$. By Definition 8.7.2, let p be a S-partition of A such that $I_{G(p)}(y)$. Let

$$p'=\{x \mid x \in p \text{ and } x \cap y \neq \varnothing\}.$$

$p' \neq \varnothing$ since p is a S-partition of A and y is a nonempty subset of A. Now for some x in p', $y \cap x \neq x$, since if $\forall x(x \in p' \rightarrow y \cap x=x)$, then because p is a S-partition of A, $y=\bigcup p'$, and since p' is a subset of a finite set (namely p) of elements of S, it follows from S being an algebra of subsets of A that $y \in S$, which is contrary to the choice of y. So let z in p' be such that $y \cap z \neq z$. Then $y \cap z$ is a nonempty, proper subset of z. Thus let a and b be entities such that $a \in y \cap z$ and $b \in z—y$. Let f be the following permutation on A: $f(w)=w$ if $\omega \in A—\{a, b\}$, $f(a)=b$, and $f(b)=a$. Then, because p is a partition and z e p and $\{a, b\} \subseteq z$, it follows that for each x in p, $f(x)=x$. Thus $f \in G(p)$. However, $f(y) \neq y$, since $a \in y$ and $f(a)=b \notin y$. This contradicts $I_{G(p)}(y)$.

Theorem 8.7.2 (Theorem 8.6.1) Consider E^1 as a qualitative structure and let A be the set of points of E^1. Then there exists a subcollection M of V such that $\langle V, \in, A, \varnothing, M \rangle$ satisfies axiom system D′ and for each subset a of A, a is Lebesgue measurable if and only if $M(a)$.

Proof. Let S be the set of all Lebesgue measurable subsets of A. Then S is an algebra of subsets of A. The theorem then follows from Theorem 8.7.1.

Theorem 8.7.3 (Theorem 8.6.2) Consider E^1 as a qualitative structure and let A be the set of points of E^1. Suppose M is a subcollection of V such that for each subset a of A, $M(a)$ if and only if a is Lebesgue measurable, and for each function f from A onto A, $M(f)$ if and only if f is a Lebesgue measurable function. Then axiom MC' is false about $\langle V, \in, A, \varnothing, M \rangle$.

Proof. See Rudin (1976) pp. 312–313, which cites an example from pg. 241 of McShane (1944) of measurable functions f and g such that $f_* g$ is not measurable.

References

Aczél, J., Roberts, F.S. (1989). On the possible merging functions. *Math. Soc. Sci., 17,* 205–243.

Aczél, J., Roberts, F.S., and Rosenbaum, Z. (1986). On scientific laws without dimensional constants. *Journal of Mathematical Analysis and Applications, 119,* 389–416.

Adams, E. (1966). On the nature and purpose of measurement. *Synthese, 16,* 125–169.

Alper, T.M. (1985). A note on real measurement structures of scale type (m, m+1 1). *Journal of Mathematical Psychology, 31,* 135–154.

Alper, T.M. (1987). A classification of all order-preserving homomorphism groups of the reals that satisfy finite uniqueness. *Journal of Mathematical Psychology, 31,* 135–154.

Baird, J.C. & Noma, E. (1978). *Fundamentals of scaling and psychophysics.* New York: Wiley.

Baire, R., Borel, E., Hadamard, J., and Lebesgue, H. (1905). Cinq lettres sur la théorie des ensembles. *Bulletin de la Société Mathématique de France (Comptes Rendus), 33,* 261–273. (English translation in Appendix I of Moore, 1982.)

Batchelder, W. and Narens, L. (1997). A critical examination of the analysis of dichotomous data. *Philosophy of Science, 44,* 113–135.

Bell, E.T. (1935). *Men of Mathematics.* New York: Simon and Schuster.

Bettazzi, R. (1892). Sui punti di discontinuità delle funzioni di variabile reale. Circolo Matematico di Palermo, Rendiconti.

Bridgman, P. (1931) *Dimensional Analysis.* New Haven: Yale University Press.

Bridgman, P. (1968). Dimensional Analysis. *Encyclopedia Britannica, 1968 Edition,* 439–449.

Brunner, N. and Mihara, H.R. (2000). Arrow's Theorem, Weglorz' Models and the Axiom of Choice. *Mathematical Logic Quarterly, 46,* 335–359.

Buckingham, E. (1914). On physically similar systems: illustrations of the use of dimensional equations. *Phys. Rev., 4,* 345–376.

Burkill, J.C. (1951). *The Lebesgue Integral.* Cambridge: Cambridge University Press.

Burtt, E.A. (1954) *The Metaphysical Foundations Of Modern Physical Science.* Garden City, N.Y.: Doubleday

Campbell, N.R. (1920). *Physics: The Elements.* Cambridge: Cambridge University Press. Reprinted as *Foundations of Science: The Philosophy of Theory and Experiment.* New York: Dover, 1957.

Campbell, N.R. (1928). *An account of the principles of measurement and calculation.* London: Longmans.

Cantor, G. (1895). Beiträge zur Begründung der transfiniten Mengenlehre. *Math. Ann., 46,* 481–512.

Causey, R.L. (1967). Derived measurement and the foundations of dimensional analysis. *Technical Report No. 5, University of Oregon, Measurement Theory and Mathematical Models Reports,*

Causey, R.L. (1969). Derived measurement, dimensions, and dimensional analysis. *Phil Sci., 36,* 252–270.

Cohen, M. &; Narens, L. (1979). A theory of ratio scalability: fundamental unit structures. *Journal of Mathematical Psychology, 20,* 193–232.

Coren, S. &; Ward, L.M. (1989). *Sensation and Perception.* Third Edition. San Diego: Harcourt Brace.

Dauben, J.W. (1979). *Georg Cantor.* Cambridge, Mass.: Harvard University Press.

de Jong, F.J. (1967). *Dimensional Analysis for Economists.* Amsterdam: North-Holland.

Descartes, R. (1952). *The Geometry of René Descartes.* Translated from French and Latin by D.E.Smith and M.L.Latham. La Salle, Illinois: The Open Court Publishing Co.

Dzhafarov, E.N. 1995. Empirical Meaningfulness, Measurement-Dependent Constants, and Dimensional Analysis. In Luce, R.D., D'Zmura, M., Hoffman, D., Iverson, G.J., and Romney, A.K. (Eds.), *Geometric Representations of Perceptual Phenomena.*

Ellis, B. (1966). *Basic Concepts of Measurement* London: Cambridge University Press.

Enderton, H.B. (1972). *A Mathematical Introduction to Logic.* Academic Press: New York.

Engen, T. (1971). Psychophysics. I. Discrimination and detection. In J.W.Kling & L.A.Riggs (Eds.), *Woodworth & Schlosberg's, Experimental Psychology,* 11–46. New York: Holt, Rinehard & Winston.

Falmagne, J.C.& Narens, L. (1983). Scales and meaningfulness of quantitative laws. *Synthese, 55,* 287–325.

Field, H.H. (1980). *Science Without Numbers.* Oxford: Basil Blackwell.

Fine, T.L. (1973). *Theories of probability; an examination of foundations.* New York, Academic Press.

Fraenkel, A. (1922a). Uber den Begriff 'definit' und die Unabhängigkeit des Auswahlaxioms. *Sitzungsberichte der Preussichen Akademie der Wissenwschaften, Physik.-math. Klasse,* 253–257.

Fraenkel, A. (1922b). Zu den Grundlagen der Cantor-Zermeloschen Mengenlehre. *Mathematische Annalen, 86,* 230–237.

Fraenkel, A. (1935). Sur l'axiome du choix. *Enseignement Mathématique, 34,* 32–51.

Freudenthal, H. (1965). Lie groups in the foundations of Geometry. *Advances in Mathematics, 1,* 145–190.

Gödel, K. (1931). Uber formal unentscheidbare Sätze per Principia Mathematica und verwandter Systeme I. *Monatshefte für Mathematik und Physik, 38,* 173–98.

Gödel, K. (1983). What is Cantor's continuum problem? In Benacerraf, P. and Putnam, H. (Eds.), *Philosophy of Mathematics, Selected Readings* (Second Edition). Cambridge: Cambridge University Press.

Hilbert, D. (1921). *The Foundations of Geometry, 2nd ed.* Chicago: Open Court.

Helmholtz, H.v. (1887). Zählen and Messen erkenntnistheoretisch betrachtet. *Philosophische Aufsätze Eduard Zeller gewidmet.* Leipzig.

Hölder, O. (1901). Die Axiome der Quantität und die Lehre vom Mass. *Berichte über die Verhandlungen der Königlich Sächsischen Gesellschaft der Wissenschaften zu Leipzig, Mathematisch-Physikaliche Classe, Bd. 53,* 1–64. (Part I translated into English by J.Michell and C.Ernst, "The axioms of quantity and the theory of measurement," *Journal of Mathematical Psychology, 1996, Vol 40,* 235–252.)

Householder, R.S. and Young, G. (1940). Weber laws, the Weber Law, and Psychophysical Analysis. *Psychometrik, 5,* 183–193

Iverson, G.J. and Pavel, M. (1981). Invariant properties of masking phenomena in psychoacoustics and their theoretical consequences. In S.Grossberg, (Ed.), *Symposium in Applied Mathematics of AMS-SIAM, Vol 13,* 17–24. Providence, RI: American Mathematical Society.

Jaspers, K. (1976). *Anselm and Nicholas of Cusa.* New York: Hardcourt Brace Jovanovich. Translated by Ralph Manheim from *The Great Philosophers: The Original Thinkers,* Edited by Hannah Arendt, 1957.

Jeans, J.H. (1948). *The mysterious universe.* Cambridge, Eng. : The University Press.

Jech, T.J. (1973). *The Axiom of Choice.* New York: North-Holland.

Kaiwi, J.L. (1978). *An analysis of the effects of logical transformations on qualitatively based measures of distance and similarity.* Ph.D dissertation. Irvine: University of California.

Klein, F. (1872). Vergleichende Betrachtungen über neuere geometrische Vorschungen, Programm zu Eintritt in die philosophische Facultät und den Senat der Universität zu Erlangen, Erlangen, Deichert.

Koestler, A. (1968). *The Sleepwalkers.* New York: Macmillan.

Krantz, D.H., R.D.Luce, P.Suppes, & A.Tversky (1971). *Foundations of Measurement, Vol. I.* New York: Academic Press.

Lebesgue, H. (1971). A propos de quelques travaux mathématiques récents. *Emseignement mathématique, (2) 17,* 1–48.

Luce, R.D. (1956). Semiorders and a theory of utility discrimination. *Econometrica, 24,* 178–191.

Luce, R.D. (1959). On the possible psychophysical laws. *Psychological Review, 66,* 81–95.

Luce, R.D. (1962). Comments on Rozeboom's criticisms of "On the possible psychophysical laws." *Psychological Review, 69,* 548–551.

Luce, R.D. (1964). A generalization of a theorem of dimensional analysis. *Journal of Mathematical Psychology, 1,* 278–284.

Luce, R.D. (1978). Dimensionally invariant laws correspond to meaningful qualitative relations. *Philosophy of Science, 45,* 1–16.

Luce, R.D. (1987). Measurement structures with Archimedean ordered translation groups. *Order, 4,* 165–189.

Luce, R.D. (1990). "On the possible psychophysical laws" revisited: remarks on cross-modal matching. *Psychological Review, 97,* 66–77.

Luce, R.D., D.H. Krantz, P.Suppes, & A.Tversky (1990). *Foundations of Measurement, Vol. III.* New York: Academic Press.

Luce, R.D. and Narens, L. (1978). Qualitative independence in probability theory. *Theory and Decision, 9,* 225–239.

Luce, R.D. and Narens, L. (1985). Classification of concatenation structures by scale type. *Journal of Mathematical Psychology, 29,* 1–72.

Luce, R.D. and Narens, L. (1992). Intrinsic archimedeanness and the continuum. Wade Savage and Philip Ehrlich (Eds. *The Nature and Purpose of Measurement.* Lawrence Erlbaum and Associates. 15–38.

McShane, E.J. (1944) *Integration.* Princeton, N.J.: Princeton University Press.

Michell, J. (1990). *An introduction to the Logic of Measurement.* Hillsdale: Erlbaum.

Moore, G.H. (1982). *Zermelo's Axiom of Choice.* Heidelberg: Springer-Verlag.

Nagel, E. (1935). Impossible numbers. In Columbia University, Department of Philosophy (Ed.), *Studies in the History of Ideas, Vol. 3.* New York: Columbia University Press. 429–476.

Narens, L. (1976). Utility-uncertainty trade-off structures. *Journal of Mathematical Psychology, 13,* 1976, 296–322.

Narens, L. (1980). A qualitative treatment of Weber's Law. *Journal of Mathematical Psychology, 13,* 88–91.

Narens, L. (1981 a). A general theory of ratio scalability with remarks about the measurement-theoretic concept of meaningfulness. *Theory and Decision, 13,* 1–70.

Narens, L. (1981 b). On the scales of measurement. *Journal of Mathematical Psychology, 24,* 249–275.

Narens, L. (1985). *Abstract Measurement Theory.* Cambridge, Mass.: The MIT Press.

Narens, L. (1988). Meaningfulness and the Erlanger Program of Felix Klein. *Mathématiques Informatique et Sciences Humaines, 101,* 61–72.

Narens, L. (1994). The measurement theory of dense threshold structures. *Journal of Mathematical Psychology, 38,* 301–321.

Narens, L. (1996). A theory of magnitude estimation. *Journal of Mathematical Psychology, 40,* 109–129.

Narens, L. (2001). A meaningful justification for the representational theory of measurement. To appear in the *Journal of Mathematical Psychology.*

Narens, L. and Luce, R.D. (1983). How we may have been misled into believing in the intercomparability of utility. *Theory and Decision, 15,* 247–260.

Narens, L. and Mausfeld, R. (1992). On the relationship of the psychological and the physical in psychophysics. *Psychological Review, 99,* 467–479.

Niederée, R. (1987). On the reference to real numbers in fundamental measurement: A model-theoretic approach. In E.Roskam and R.Suck (Eds.), *Progress in Mathematical Psychology.* Amsterdam: North-Holland. Pp. 3. 23.

Niederée, R. (1992a). What do numbers measure? A new approach to fundamental measurement. *Mathematical Social Sciences, 24,* 237–276.

Niederée, R. (1992b). *Mass und Zahl. Logisch-modelltheoretische Untersuchungen zur Theorie der fundamentalen Messung.* Frankfurt a.M: Lang.

Osborne, D.K. (1970). Further extensions of a theorem of dimensional analysis. *Journal of Mathematical Psychology, 7,* 236–242.

Osborne, D.K. (1976). Unified theory of derived measurement. *Synthese, 33,* 455–81.

Osborne, D.K. (1978). On dimensional invariance. *Quality and Quantity, 12,* 75–89.

Pavel, M. (1980). Homogeneity in complete and partial masking. Unpublished doctoral dissertation, New York University.

Pfanzagl, J. (1959). Die axiomatischen Grundlagen einer allgemeinen Theorie des Messens. *Schriftenreihe Statist. Inst. Univ. Wien, Vol. 1.* Würzburg: Physica-Verlag.

Pfanzagl, J. (1968). *Theory of Measurement.* New York: John Wiley &: Sons.

Pollatsek, A. and Tversky, A. (1970). A theory of risk. *Journal of Mathematical Psychology, 7,* 540–553.

Ramsay, J.O. (1976). Algebraic representation in physical and behavioral sciences. *Synthese, 32,* 419–453.

Lord Rayleigh (1915). The principle of similitude. *Nature, 95,* 66–68, 202, 644.

Roberts, F.S. (1979) *Measurement theory with applications to decision making, utility, and the social sciences.* Reading, Mass. : Addison-Wesley.

Roberts, F.S. (1985). Applications of the theory of meaningfulness to psychology. *Journal of Mathematical Psychology, 29,* 311–332.

Roberts, F.S. (1990). Meaningfulness of conclusions from combinatorial optimization. *Discrete Applied Math., 29,* 221–241.

Roberts F.S, & Rosenbaum, Z. (1986). Scale type, meaningfulness, and the possible psychophysical laws. *Mathematical Social Sciences, 12,* 77–95.

Robinson, A. (1966). *Non-Standard Analysis.* Amsterdam: North-Holland.

Roskam, E.E. (1989). Formal models and axiomatic measurement. In E. E.Roskam (Ed.), *Mathematical Psychology in Progress.* Berlin: SpringerVerlag. 49–67.

Rozeboom, W.W. (1962 a). Comment. *Psychological Review, 69,* 552.

Rozeboom, W.W. (1962 b). The untenability of Luce's principle. *Psychological Review, 69,* 542–547.

Rudin, W. (1967). *Principles of Mathematical Analysis, Third Edition.* New York: McGraw-Hill Book Company.

Scott, D. and Suppes, P. (1958). Foundational aspects of theories of measurement. *J.Symbolic Logic, 23,* 113–128.

Sedov, L.I. (1959) *Similarity and Dimensional Methods in Mechanics.* English translation of the 1956 Russian edition by Holt, M. and Friedman, M., New York: Academic Press.

Sierpiński, W. (1965). *Cardinal and Ordinal Numbers.* Translated from Polish by J.Smólska. Warszawa: PWN-Polish Scientific Publishers.

Stevens, S.S. (1946). On the theory of scales of measurement. Science, 103, 677–680.

Stevens, S.S. (1951). Mathematics, measurement and psychophysics. In S.S.Stevens (Ed.), *Handbook of experimental psychology* (pp. 1–49). New York: Wiley.

Stevens, S.S. (1971). Issues in psychophysical measurement. *Psychological Review, 78,* 426–450

Stevens, S.S. (1974). Perceptual magnitude and its measurement. In E.C.Carterette & M.P.Friedman, *Handbook of perception, Vol. II.* New York: Academic Press.

Suppes, P., D.H.Krantz, R.D.Luce, & A.Tversky (1990). *Foundations of measurement, Vol. II.* New York: Academic Press.

Suppes, P. &; Zinnes, J. (1963). Basic measurement theory. In R.D.Luce, R.R.Bush, &; E.Galanter (Eds.), *Handbook of mathematical psychology, Vol. 1.* New York: Wiley.

Tolman, R.C. (1914). The principle of similitude. *Phys. Rev., 3,* 244–255.

Torgerson, W.S. (1961). Distances and ratios in psychological scaling. *Acta Psychologica, 19,* 201–205.

Veblen, O. and Young, J.W. (1946). *Projective Geometry, Vol. II.* Boston: Ginn and Company.

Ward, L.M. (1990). Cross modality additive conjoint structures and psychophysical scale convergence. *Journal of Experimental Psychology: Gen-eral, 119,* 161–175.

Watanabe, S. (1969). *Knowing and guessing.* New York: Wiley.

Whittaker, E.T. (1958). *From Euclid to Eddington..* New York: Dover Publications.

Wundt, W. (1908). *Grundzüge der physiologischen Psychologie, Bd. I (6.Aufl.).* Leipzig: Engelmann.

Zermelo, E. (1904). Beweis, dass jede Menge wohlgeordnet werden kann (As einem an Herrn Hilbert gerichteten Briefe. *Mathematische Annalen, 59,* 514–516.

Index

For Product Safety Concerns and Information please contact our EU representative GPSR@taylorandfrancis.com
Taylor & Francis Verlag GmbH, Kaufingerstraße 24, 80331 München, Germany

www.ingramcontent.com/pod-product-compliance
Ingram Content Group UK Ltd.
Pitfield, Milton Keynes, MK11 3LW, UK
UKHW021719240926
475702UK00029BA/1424
* 9 7 8 0 4 1 5 6 5 4 5 6 2 *